Color Atlas
of Biochemistry

Jan Koolman
Institute of Physiological Chemistry
Philipps University of Marburg, Germany

Klaus-Heinrich Röhm
Institute of Physiological Chemistry
Philipps University of Marburg, Germany

Translated by Kathryn Schuller
194 color plates by Jürgen Wirth

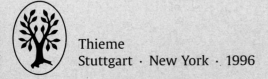

Thieme
Stuttgart · New York · 1996

Jan Koolman, Ph. D.
Professor, Institute of Physiological Chemistry
Philipps University of Marburg
Deutschhausstrasse 1–2
35033 Marburg, Germany

Klaus-Heinrich Röhm, Ph. D.
Professor, Institute of Physiological Chemistry
Philipps University of Marburg
Karl-von-Frisch-Strasse 1
35033 Marburg, Germany

Kathryn Schuller, Ph. D.
Flinders University
Adelaide, Australia

Jürgen Wirth
Professor of Visual Communication
Fachhochschule Darmstadt, Germany

Library of Congress Cataloging-in-Publication Data

Koolman, Jan.
  [Taschenatlas der Biochemie, Englisch]
  Color atlas of biochemistry / Jan Koolman, Klaus-Heinrich Röhm ;
translated by Kathryn Schuller ; color plates by Jürgen Wirth.
    p. cm.
  Includes bibliographical references and index.
  ISBN 3-13-100371-5 (GTV). -- ISBN 0-86577-584-2 (TMP)
  1. Biochemistry--Handbooks, manuals etc, 2. Biochemistry--Atlases. I. Röhm, Klaus-Heinrich. II. Title.
  [DNLM; Biochemistry--atlases. QU 17 K82t 1996a]
QP514.2.K6613 1996
574.19'2--dc20
DNLM/DLC
for Library of Congress   96-16527
                                                         CIP

Important Note: Medicine is an ever-changing science undergoing continual development. Research and clinical experience are continually expanding our knowledge, in particular our knowledge of proper treatment and drug therapy. Insofar as this book mentions any dosage or application, readers may rest assured that the authors, editors and publishers have made every effort to ensure that such references are in accordance with the state of knowledge at the time of production of the book.
Nevertheless this does not involve, imply, or express any guarantee or responsibility on the part of the publishers in respect of any dosage instructions and forms of application stated in the book. Every user is requested to examine carefully the manufacturers' leaflets accompanying each drug and to check, if necessary in consultation with a physician or specialist, whether the dosage schedules mentioned therein or the contraindications stated by the manufacturers differ from the statements made in the present book. Such examination is particularly important with drugs that are either rarely used or have been newly released on the market. Every dosage schedule or every form of application used is entirely at the user's own risk and responsibility. The authors and publishers request every user to report to the publishers any discrepancies or inaccuracies noticed.

This book is an authorized, updated, and expanded translation of the German edition published and copyrighted 1994 by Georg Thieme Verlag, Stuttgart, Germany.
Title of the German edition: Taschenatlas der Biochemie.

© 1996 Georg Thieme Verlag, Rüdigerstrasse 14, 70469 Stuttgart, Germany
Thieme Medical Publishers, Inc., 381 Park Avenue South, New York, NY 10016
Typesetting by primustype Robert Hurler GmbH, 73274 Notzingen, Germany
Printed in Singapore by Imago Productions Pte Ltd.

ISBN 3-13-100371-5 (GTV, Stuttgart)
ISBN 0-86577-584-2 (TMP, New York)

About the Authors

Jan Koolman was born in Lübeck on the German coast, and grew up familiar with the wind from the Baltic Sea. His classical education in this Hanseatic town had a major influence on him. From 1963 to 1969, he studied biochemistry at the University of Tübingen. He completed his diploma studies with a thesis in inorganic biochemistry under the supervision of Ulrich Weser. He then moved to the University of Marburg to study for his Ph. D with the biochemist Peter Karlson. This was where his interest in the biochemistry of insects and other invertebrates began. He completed his Habilitation (the German postdoctoral degree) in 1977 in the Faculty of Human Medicine, and in 1984 he became professor. Nowadays, his research area is biochemical endrocinology. He is also interested in the teaching of biochemistry, and has written a short biochemistry textbook together with Peter Karlson and Detlef Doenecke. Jan Koolman is married to an art teacher.

Klaus-Heinrich Röhm comes from Stuttgart in the south of Germany. After matriculating from the Lutheran Theological Seminary in Urach, which, like Jan Koolman's *alma mater*, provides a classical education, he briefly turned to physics before finally settling to study biochemistry at the University of Tübingen. There the two authors met for the first time. Since 1970, Klaus-Heinrich Röhm has also been a member of the Faculty of Medicine at the University of Marburg. He completed his Ph. D with Friedhelm Schneider, and took his Habilitation degree in the Department of Chemistry. Since 1986, he has been a professor. His research areas are the mechanisms of enzyme catalysis, protein chemistry, and the application of computers in biochemistry. He is married and has two children. Although he claims to have no time for hobbies, his friends say he is keen on photography and astronomy.

Jürgen Wirth was born in Wöllstadt, Hessen. After his matriculation in Friedberg and practical training as typesetter, he commenced his tertiary studies at an art college in Offenbach, majoring in graphic arts, before transferring to the university (Bildende Künste) in Berlin. His diploma thesis at the Offenbach Academy of Art and Design focused on the development and function of the scientific drawing. From 1963, he was responsible for exhibition design during the reorganization of the Senckenberg Museum in Frankfurt (Main) and the development of innovative design concepts and techniques. At the same time, Jürgen Wirth was freelancing for various publishers (book design, graphic art for school textbooks, non-fiction, and scientific publications). In 1978, he became a professor at the academy of Design in Schwäbisch Gmünd, and, since 1986, he has held a professorship in design in Darmstadt (FH). His first computerized graphics in the field of electronic representation and technique were produced in 1987. Jürgen Wirth is married with three children.

Preface

Biochemistry is a dynamic, rapidly growing field, and the goal of this color atlas is to illustrate this fact visually. The precise boundaries between biochemistry and related fields, such as cell biology, anatomy, physiology, genetics, and pharmacology are difficult to define and, in many cases, arbitrary. This overlap is not coincidental. The object being studied is often the same – a nerve cell or a mitochondrion, for example – and only the point of view differs.

For a considerable period of its history, biochemistry was strongly influenced by chemistry, and concentrated on investigating metabolic conversions and energy transfers. Explaining the composition, structure, and metabolism of biologically important molecules has always been in the foreground. However, new aspects inherited from biochemistry's other parent, the biological sciences, are now increasingly being added: the relationship between chemical structure and biological function, the pathways of information transfer, attention to the ways in which biomolecules are spatially and temporal distributed in cells and organisms, and an awareness of evolution as a biochemical process. These new aspects of biochemistry are bound to become more and more important.

Due to space limitations, we have concentrated here on the biochemistry of humans and mammals, although the biochemistry of other animals, plants, and microorganisms is no less interesting. In selecting the material for this book we have put the emphasis on subjects relevant to students of human medicine. The main purpose of the atlas is to serve as an overview and to provide visual information quickly and efficiently. Any gaps can easily be filled by referring to textbooks. For readers encountering biochemistry for the first time, some of the plates may look too complex. It must be emphasized, therefore, that the atlas is not intended as a substitute for a comprehensive textbook of biochemistry.

As the subject matter is often difficult to visualize, symbols, models, and other graphic elements had to be found that make complicated phenomena appear tangible. Some of the forms chosen, therefore, may seem rather subjective and oversimplified. The graphics, many of which became rather complex, were designed conservatively, the aim being to avoid illustrations that might look too spectacular or exaggerated. Our goal was to achieve a visual and aesthetic way of representing scientific facts that would be simple and at the same time effective for teaching purposes. Using graphics software helped to maintain consistency in the use of shapes, colors, dimensions, and labels, in particular. Formulae and other repetitive elements and structures could be handled easily and precisely with the assistance of the computer.

Color coding has been used throughout to aid reader, and the key to this is given in two color plates of inner cover of the atlas. For example, each of the more important atoms has a particular color – gray for carbon, white for hydrogen, blue for nitrogen, red for oxygen, and so on. The different classes of biomolecules are also distinguished by color – proteins are always shown in brown tones, carbohydrates in violet, lipids in yellow, DNA in blue, and RNA in green. In addition, specific symbols are used for the important coenzymes, such as ATP and NAD^+.

The compartments in which biochemical processes take place are color-coded as well. For example, the cytoplasm is shown in yellow, while the extracellular space is blue. Arrows indicating a chemical reaction are always black, and those representing a transport process are gray. Although we have tried to use the code consistently, it certainly has its limitations.

In terms of the visual clarity of its presentation, biochemistry has still to catch up with anatomy and physiology. In this book, we often use simplified ball-and-stick models instead of the classical chemical formulae, as we think these are clearer when one is dealing with small molecules. In addition, many compounds are represented by space-filling models. In these cases, we have tried to be as realistic as possible. The models of small molecules are based on conformations calculated by computer-based "molecular modeling". In illustrating macromolecules, we used structural information obtained by X-ray crystallography that is stored in the Protein Data Bank. In naming enzymes, we have followed the official nomenclature recommended by the IUBMB. For quick identification, EC numbers (in italics) are included with enzyme names.

To help students to assess the relevance of the material (while preparing for an examination, for example), we have included symbols on the text pages next to the section headings to indicate how important each topic is. A filled circle stands for "basic knowledge" a half-filled circle indicates "standard knowledge," and an empty circle stands for "in-depth knowledge." Of course, this classification only reflects our subjective view.

We are indebted to our colleagues, especially Dr. Harold Evans, Portland, and Dr. Mechthild Röhm, for their comments and valuable criticisms during the preparation of this book. Of course, we would also welcome further comments and suggestions from readers.

December 1995

Jan Koolman, Klaus-Heinrich Röhm
Marburg

Jürgen Wirth   Kathryn Schuller
Darmstadt   Adelaide

Contents

Basics

Chemistry
Periodic Table 2
Chemical Bonds 4
Molecular Structure 6
Biomolecules 8

Physical Chemistry
Energetics 10
Free Energy 12
Enthalpy and Entropy 14
Reaction Kinetics 16
Catalysis 18
Water as Solvent 20
Hydrophobic Interactions 22
Acids and Bases 24
Buffer Systems 26
Redox Processes 28

Biomolecules

Carbohydrates
Stereochemistry 30
Sugar Derivatives 32
Monosaccharides 34
Disaccharides and Polysaccharides 36
Plant Polysaccharides 38
Glycoproteins and Glycosaminogly-
cans 40

Lipids
Overview 42
Fatty Acids 44
Fats 46
Phospholipids and Glycolipids 48
Isoprenoids 50
Steroid Structure 52
Classes of Steroids 54

Amino Acids
Properties 56
Proteinogenic Amino Acids 58
Amino Acid Analysis 60

Peptides and Proteins
Functions of Proteins 62
The Peptide Bond 64
Secondary Structures 66
Structural Proteins 68
Globular Proteins 70
Protein Folding 72
Insulin: Structure 74
Molecular Models: Insulin 76

Nucleotides, Nucleic Acids
Nucleobases, Nucleotides, RNA 78
DNA 80
Functions of Nucleic Acids 82
Molecular Models: DNA, tRNA 84

Metabolism

Enzymes
Classification 86
Enzyme Catalysis 88
Enzyme Kinetics 90
Inhibitors 92
Lactate Dehydrogenase: Structure 94
Lactate Dehydrogenase: Mechanism 96
Enzymatic Analysis 98
Redox Coenzymes 100
Group-Transferring Coenzymes 102

Metabolic Regulation
Intermediary Metabolism 104
Regulatory Mechanisms 106
Transcriptional Control 108
Hormonal Control 110
Allosteric Regulation 112

Energy Metabolism
ATP 114
Energy Conservation at Membranes 116
Photosynthesis: Light Reactions 118
Photosynthesis: Dark Reactions 120
Molecular Models: Bacteriorhodopsin,
Reaction Center 122

VIII   Contents

Oxoacid Dehydrogenases 124
Citric Acid Cycle: Reactions 126
Citric Acid Cycle: Metabolic Functions .. 128
Respiratory Chain 130
ATP Synthesis 132
Regulation of Energy Metabolism I 134
Regulation of Energy Metabolism II 136
Fermentations 138

Catabolic Carbohydrate Metabolism
Glycolysis 140
Hexose Monophosphate Pathway 142

Catabolic Lipid Metabolism
Fat Metabolism: Overview 144
Degradation of Fatty Acids: β-Oxidation 146
Minor Pathways of Fatty Acid Degradation 148

Catabolic Protein Metabolism
Nitrogen Balance, Peptidases 150
Transamination and Deamination 152
Amino Acid Degradation 154

Catabolic Nucleic Acid Metabolism
Nucleotide Degradation 156

Anabolic Carbohydrate Metabolism
Gluconeogenesis 158
Glycogen Metabolism 160
Regulation of Carbohydrate Metabolism 162
Diabetes Mellitus 164

Anabolic Lipid Metabolism
Fatty Acid Synthesis 166
Biosynthesis of Complex Lipids 168
Biosynthesis of Cholesterol 170

Anabolic Protein Metabolism
Amino Acid Biosynthesis 172

Anabolic Nucleic Acid Metabolism
Purine and Pyrimidine Synthesis 174
Nucleotide Biosynthesis 176

Porphyrin Metabolism
Heme Biosynthesis 178
Degradation of Porphyrins 180

Organelles

Basics
Structure of Cells 182
Cell Fractionation 184
Centrifugation 186
Cell Components, Cytoplasm 188

Cytoskeleton
Composition 190
Structure and Functions 192

Nucleus 194

Mitochondria
Structure and Functions 196
Transport Systems 198
Mitochondrial DNA, Peroxisomes 200

Membranes
Membrane Components 202
Functions 204
Transport Processes 206
Transport Proteins 208

Endoplasmic Reticulum and Golgi Apparatus 210

Lysosomes 212

Intracellular Traffic
Protein Translocation 214
Protein Sorting 216

Molecular Genetics
The Genome 218
Replication 220
Transcription 222
RNA Maturation 224
Genetic Code, Amino Acid Activation ... 226
Ribosomes, Initiation of Translation 228
Elongation, Termination 230
Antibiotics 232
Mutation and Repair 234

Gene Technology
DNA Cloning 236
DNA Sequencing 238
PCR, RFLP 240

Tissues and Organs

Digestive System
Digestion: Overview 242
Digestive Secretions 244
Digestive Processes 246
Resorption 248

Blood
Composition and Functions 250
Plasma Proteins 252
Lipoproteins 254
Acid-Base Balance 256
Gas Transport 258
Hemoglobin, Erythrocyte Metabolism .. 260
Hemostasis 262
Immune Response 264
Antibodies 266
Molecular Models: Hemoglobin, Immunoglobulin G 268
Antibody Biosynthesis 270
MHC Proteins 272
Monoclonal Antibodies, Immunoassay . 274

Liver
Functions 276
Metabolism in the Well-Fed State 278
Metabolism During Starvation 280
Carbohydrate Metabolism 282
Lipid Metabolism 284
Urea Cycle 286
Bile Acid Metabolism 288
Biotransformations 290
Cytochrome P450 Systems 292
Ethanol Metabolism 294

Kidneys
Functions 296
Urine 298
Proton and Ammonia Excretion 300
Electrolyte and Water Recycling 302
Hormones of the Kidneys 304

Muscle
Contraction 306
Control of Muscle Contraction 308
Metabolism 310

Connective Tissue
Collagens 312
Composition of the Extracellular
Matrix 314

Brain and Sensory Organs
Nerve Tissue 316
Resting Potential, Action Potential 318
Synapses 320
Neurotransmitters 322
Sight 324

Nutrition

Nutrients
Organic Nutrients 326
Minerals and Trace Elements 328

Vitamins
Lipid-Soluble Vitamins 330
Water-Soluble Vitamins I 332
Water-Soluble Vitamins II 334

Hormones

Hormone Systems
Principles 336
Plasma Levels and Hormone Hierarchy . 338

Classification
Liphophilic Hormones 340
Hydrophilic Hormones 342

Lipophilic Hormones
Metabolism 344
Mechanism of Action 346

Hydrophilic Hormones
Metabolism 348
Mechanism of Action 350
Second Messengers 352

Mediators
Eicosanoids 354

Growth and Development
Cell Proliferation
Cell Cycle 356
Oncogenes 358
Tumors 360

Viruses 362

Morphogenesis 364

Metabolic Charts
Calvin Cycle 366
Carbohydrate Metabolism 367
Biosynthesis of Fats and Membrane
Lipids 368

X   Contents

Synthesis of Ketone Bodies and
Steroids 369
Amino Acid Degradation I 370
Amino Acid Degradation II 371
Biosynthesis of the Essential Amino
Acids 372
Biosynthesis of the Non-Essential
Amino Acids 373
Biosynthesis of Purine Nucleotides 374
Biosynthesis of the Pyrimidine Nu-
cleotides, C_1-Metabolism 375

Annotated Enzyme List 376

Abbreviations 387

Quantities and Units 390

Suggested Reading 391

Source Credits 392

Index 393

Introduction

This color atlas is designed for students of medicine and the biological sciences. It provides an introduction to biochemistry but, because of its modular structure, can also be used as a reference book. The 191 color plates contain much of our current knowledge in the area of biochemistry. The illustrations are accompanied by brief explanations on the facing pages. Some general rules regarding the design of the illustrations are summarized on the inside of the cover. Keywords, definitions, explanations of special terminology, and chemical formulae can be found with the help of the index.

The book begins with the **basics** of biochemistry (pp. 2–29). There is a brief explanation of the key concepts in *chemistry* (pp. 2–9). These include the periodic table of the elements, chemical bonds, the general rules governing molecular structure, and the structures of important classes of compounds. Several basic concepts of *physical chemistry* are also essential for an understanding of biochemical processes. Therefore, pp. 10–29 are devoted to the different energy forms and their interconversion, reaction kinetics and catalysis, the properties of water, acids and bases, and redox processes.

The basic concepts are followed by a section on the structure of the important **biomolecules** (pp. 30–85). This part of the book is arranged according to the different classes of metabolites. It discusses carbohydrates, lipids, amino acids, peptides and proteins, nucleotides, and nucleic acids.

In the next part (pp. 86–181), the reactions involved in the interconversion of these metabolites are presented. This part of biochemistry is commonly referred to as **metabolism**. The section starts with a discussion of the enzymes and coenzymes, followed by the mechanisms of metabolic regulation, and the so-called "energy metabolism." After this, the central metabolic pathways are discussed, once again organized according to the class of metabolite. The degradative (*catabolic*) pathways are found on pp. 140–157, followed by the synthetic (*anabolic*) processes.

The second half of the book begins with a discussion the functional compartments within the cell, the **cellular organelles** (pp. 182–217). On pp. 218–241, there follows the up-to-date area of **molecular genetics** (*molecular biology*). A further, quite detailed, section is devoted to the biochemistry of individual **tissues** and **organs** (pp. 242–325). Here, it has only been possible to focus on the most important systems (digestive system, blood, immune system, liver, kidneys, muscles, connective and supportive tissues, and the brain).

Other topics include the biochemistry of **nutrition**, the structure and function of important **hormones,** and **growth and differentiation** (pp. 326–365).

At the end of the pocket atlas, we present a series of highly schematic **metabolic "charts"** (pp. 366–375). These illustrations, which are not accompanied by explanatory text, show simplified versions of the most important synthetic and degradative pathways. The enzymes catalyzing the various reactions are only indicated by their EC numbers. Their names can be found in the systematically organized and annotated enzyme list (pp. 378–388). These metabolic charts are primarily meant to serve as reference material, summarizing the metabolic section of the book.

Periodic Table

A. Biologically important elements ◑

In nature, there are 81 stable elements. Fifteen of these are present in all living things, whereas a further 8–10 are only found in particular organisms. The illustration shows the first half of the **periodic table**, which contains all of the biologically important elements. In addition to physical and chemical data, it also provides information about the distribution of the elements in the living world and their abundance in the human body. The basic rules underlying the periodic table are discussed in chemistry textbooks.

Almost 99% of the atoms in the bodies of animals can be accounted for by just four elements—*hydrogen* (H), *oxygen* (O), *carbon* (C) and *nitrogen* (N). Hydrogen and oxygen are the constituents of water, which alone makes up 60–70% of cell mass (see p. 188). Together with carbon and nitrogen, hydrogen and oxygen are the major constituents of the organic compounds on which most living processes depend. Many biomolecules also contain *sulfur* (S) or *phosphorus* (P). The above **macroelements** are all essential for life.

A second group of biologically important elements, which make up only about 0.5% of the body mass, are present, almost without exception, in the form of inorganic ions. This group includes the alkali metals *sodium* (Na) and *potassium* (K), and the alkaline earth metals *magnesium* (Mg) and *calcium* (Ca). The halogen *chlorine* (Cl) is also always ionized in the cell. All other elements, which are important for life are present in such small quantities that they are referred to as **trace elements**. These include transition metals, such as *iron* (Fe), *zinc* (Zn), *copper* (Cu), *cobalt* (Co) and *manganese* (Mn). A few non-metals, such as *iodine* (I) and *selenium* (Se), can also be classified as essential trace elements (see p. 328).

B. Electronic configurations: examples ○

The chemical properties of the atoms and the types of bonds, which they can form with one another are determined by their electron shells. Simplified representations of the **electronic configurations** of the elements are shown in Fig. **A**. Fig. **B** explains the symbols and abbreviations used. More detailed discussions of the subject are found in chemistry text books.

The possible states of electrons are called **orbitals**. They are denoted by a so-called quantum number and a letter, i.e., s, p, or d. The orbitals are filled one by one as the number of electrons increases. Each orbital can hold a maximum of two electrons, which must have oppositely directed 'spins' (↑ and ↓ , respectively). Fig. **A** shows the distribution of the electrons among the orbitals for each of the elements. For example, the 6 electrons of carbon (**B1**) occupy the 1s orbital, the 2s orbital, and the 2p orbitals. A filled 1s orbital has the same electronic configuration as the noble gas helium (He). This region of the electron shell of carbon is, therefore, abbreviated as "He" in Fig. **A**. Below this, the numbers of electrons in the other filled orbitals are shown (2s and 2p in the case of carbon). The electron shell of chlorine (**B2**) consists of that of neon (Ne) and 7 additional electrons in 3s and 3p orbitals. In iron (Fe), a transition metal of the first series (**B3**), electrons occupy the 4s orbital even though the 3d orbitals are still partly empty. Many reactions of the transition metals involve empty d orbitals, e.g., redox reactions or complex formation with bases.

Meaning of symbols

- ● Basic knowledge
- ◑ Standard knowledge
- ○ In-depth knowledge

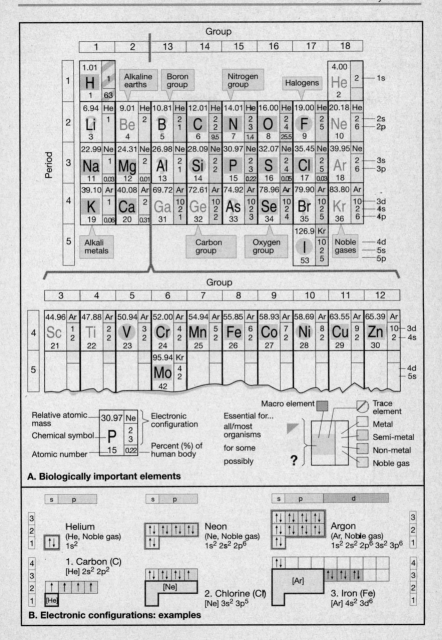

A. Biologically important elements

B. Electronic configurations: examples

Chemical Bonds

A. Orbital hybridization and chemical bonding ○

Most biomolecules are compounds of carbon with hydrogen, oxygen, nitrogen, sulfur, or phosphorus. Stable, covalent bonds between these non-metal atoms arise as a result of the combination of certain of their orbitals (see p. 2) to form **molecular orbitals** that are occupied by one electron from each of the atoms. Thus, the four bonding electrons of the carbon atom occupy 2s and 2p atomic orbitals (**1a**). The 2s orbital is spherical in shape, while the three 2p orbitals are shaped like dumbbells arranged along the x, y, and z axes. It might, therefore, be assumed that carbon atoms should form at least two different types of molecular orbitals. However, this is not normally the case. The reason is an effect known as **orbital hybridization**. Combination of the s orbital and the three p orbitals of carbon gives rise to four equivalent, tetrahedrally arranged **sp³ atomic orbitals**. When these overlap with the 1s orbitals of H atoms, four σ-**molecular orbitals** (**1b**) are formed, i.e., carbon has a **valence** of four. The methane molecule (CH_4) contains four σ or **single bonds**. Single bonds between other atoms arise in a similar way. Thus, the phosphate anion and the ammonium cation are also tetrahedral in structure (**1c**).

A second common type of orbital hybridization involves the 2s orbital and only two of the three 2p orbitals (**2a**). This process is, therefore, referred to as **sp² hybridization**. The result is three equivalent sp² hybrid orbitals lying in one plane at an angle of 120° to one another. The remaining $2p_x$ orbital is oriented perpendicular to this plane. In contrast to their sp³ counterparts, sp²-hybridized atoms form two different types of bonds when they combine into molecular orbitals (**2b**): The three sp² orbitals engage in σ bonds, as described above, while the electrons in the two $2p_x$ orbitals, the so-called π **electrons**, combine to give an additional, elongated π **molecular orbital**, which is located above and below the plane of the σ bonds. Bonds of this type are called **double bonds**. They con-

sist of a σ bond and a π bond, and arise only when both of the atoms involved are capable of sp² hybridization. In contrast to single bonds, double bonds are not freely rotatable, since rotation would distort the π molecular orbital. This is why all of the atoms lie in one plane (**2c**); in addition, *cis-trans* isomers exist in such cases (see p. 44).

B. Resonance ◑

Certain molecules containing several double bonds are much less reactive than might be expected. This is, because the double bonds in these molecules cannot be localized unequivocally. Their π orbitals are not confined to the space between individual atoms, but form a shared, extended π molecular orbital. Structures with these properties are referred to as **resonance hybrids**, because it is impossible to describe their actual bonding structure by standard formulae (see chemistry textbooks for details).

In this book, broken lines are sometimes used to represent the location of the delocalized π orbitals. The resonance-stabilized systems include carboxylate groups as in **formate**, hydrocarbons with conjugated double bonds, such as **1,3-butadiene**, and so-called *aromatic ring systems*. The best-known aromatic compound is **benzene**, which has 6 delocalized π electrons in its ring.

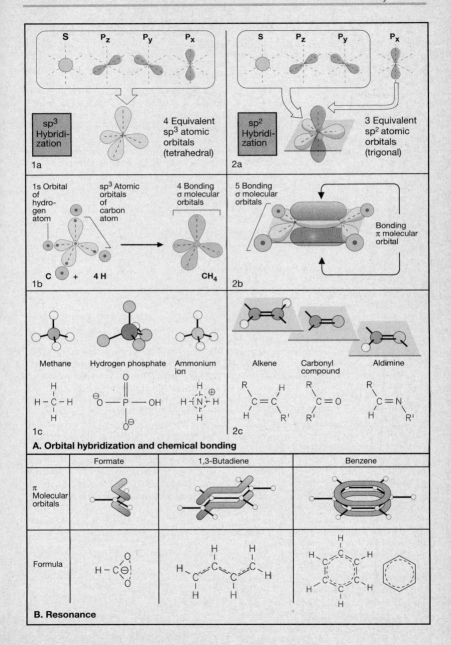

S **P_z** **P_y** **P_x**

sp³ Hybridization 4 Equivalent sp³ atomic orbitals (tetrahedral)

1a

S **P_z** **P_y** **P_x**

sp² Hybridization 3 Equivalent sp² atomic orbitals (trigonal)

2a

1s Orbital of hydrogen atom sp³ Atomic orbitals of carbon atom 4 Bonding σ molecular orbitals

C + 4 H → CH₄

1b

5 Bonding σ molecular orbitals Bonding π molecular orbital

2b

Methane Hydrogen phosphate Ammonium ion

1c

Alkene Carbonyl compound Aldimine

2c

A. Orbital hybridization and chemical bonding

	Formate	1,3-Butadiene	Benzene
π Molecular orbitals			
Formula			

B. Resonance

Molecular Structure

The physical and chemical behavior of molecules is largely determined by their constitution. Many of their features can, therefore, be predicted from structural formulae alone. The predictable properties of molecules include chemical reactivity, size, shape, and — to a certain extent — conformation, i.e., the preferred spatial arrangement of their atoms in solution. Some data that form the basis of such predictions are summarized here and on the facing page. Other diagrams illustrate how molecular models are represented in this book. L-Dihydroxyphenyl-alanine (L-dopa), an intermediate in catecholamine biosynthesis, is used as an example (see p. 322).

A. Bond length ○

Covalent radii are useful for estimating the distances between atoms within a molecule. The distance between singly bonded atoms is approximately equal to the sum of their covalent radii. The value for the corresponding double bond is usually 12–20% less than this. Nowadays, atomic radii and distances between atoms are given in picometers (pm, 1 pm = 10^{-12} m). An older unit, the angstrom (Å, 1 Å = 100 pm) is now outmoded.

B. Bond polarity ○

Atoms differ in their **electronegativity**, i.e., in their tendency to take up extra electrons. This property depends on their position in the periodic table. The values given here are on a scale between 2 and 4 (the higher the value, the more electronegative the atom). When two atoms with very different electronegativities are bound to one another, the bonding electrons are drawn toward the more electronegative atom, and the bond is said to be **polar**. One measure of the polarity of a bond is its **dipole moment**. Oxygen is the most electronegative of the biochemically important elements, with C=O double bonds being especially highly polar. The resulting partial positive charge on the carbonyl carbon strongly facilitates the *nucleophilic substitu-*

tion of carbonyl compounds, a reaction frequently encountered in metabolism (see p. 18).

C. Hydrogen bonds ●

The hydrogen bond, a special type of noncovalent bond, is of outstanding importance in biochemistry. In hydrogen bonds, hydrogen atoms of OH, NH or SH groups (so-called hydrogen bond **donors**) interact with free electrons of **acceptor** atoms (for example, O, N or S). The bonding energies of such bonds (10–40 kJ · mol^{-1}) are much lower than those of covalent bonds (> 400 kJ · mol^{-1}). However, as hydrogen bonds can be very numerous in macromolecules, they play a key role in the stabilization of these molecules (see pp. 66, 80). L-Dopa, for instance, can form two intramolecular hydrogen bonds. They are shown as dashed lines in the **ball-and-stick model** of the molecule.

D. Effective atomic radii ○

The size of an atom or ion is determined by the shape of its electron shell. However, this shell does not have a defined surface and, therefore, the effective size of an atom is usually given by its **van der Waals radius**. Such radii are derived from the energetically most favorable distance between two atoms *not directly bonded to one another*. This distance corresponds to the sum of the van der Waals radii of the respective atoms. Here the energy function determined by the forces of attraction and repulsion reaches a minumum. A good way to illustrate the actual shape and size of a molecule is by a space-filling **van der Waals model** in which each atom is shown as section of a sphere with the appropriate radius.

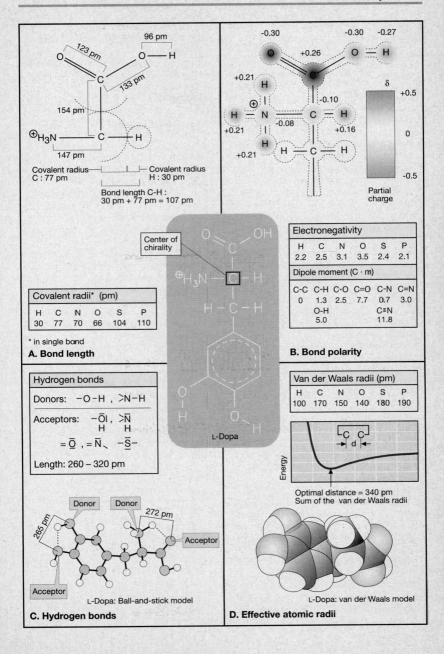

A. Bond length

96 pm
123 pm
133 pm
154 pm
147 pm

Covalent radius—C : 77 pm
Covalent radius—H : 30 pm
Bond length C–H :
30 pm + 77 pm = 107 pm

Covalent radii* (pm)

H	C	N	O	S	P
30	77	70	66	104	110

* in single bond

B. Bond polarity

−0.30   −0.30   −0.27
+0.26
+0.21
−0.10
+0.21   −0.08   +0.16
+0.21

δ
+0.5
0
−0.5
Partial charge

Electronegativity

H	C	N	O	S	P
2.2	2.5	3.1	3.5	2.4	2.1

Dipole moment (C · m)

C-C	C-H	C-O	C=O	C-N	C=N
0	1.3	2.5	7.7	0.7	3.0

O-H	C≡N
5.0	11.8

Center of chirality

L-Dopa

C. Hydrogen bonds

Hydrogen bonds

Donors: $-\overline{O}-H$, $>N-H$

Acceptors: $-\overline{O}I$, $>\overline{N}$
                    H   H

= $\overline{\underline{O}}$, = $\overline{N}$, $-\overline{\underline{S}}-$

Length: 260 – 320 pm

Donor   Donor
272 pm
265 pm
Acceptor
Acceptor

L-Dopa: Ball-and-stick model

D. Effective atomic radii

Van der Waals radii (pm)

H	C	N	O	S	P
100	170	150	140	180	190

Energy

Optimal distance = 340 pm
Sum of the van der Waals radii

L-Dopa: van der Waals model

Biomolecules

A. Important classes of compounds ●

Most biomolecules are compounds made up of the non-metals oxygen (O), hydrogen (H), nitrogen (N), sulfur (S) and phosphorus (P). Many of these compounds are derived from simple hydrides like H_2O, NH_3, and H_2S, respectively. In biological systems, phosphorus is found almost exclusively in derivatives of phosphoric acid, H_3PO_4.

When one or more of the hydrogen atoms of a non-metal hydride are replaced with another group, R, e.g., an alkyl group, compounds of the type $R-XH_{n-1}$, $R-XH_{n-2}-R'$ etc., are obtained. In this formal way, alcohols (R-OH) and ethers (R-O-R') are derived from water (H_2O), primary (R-NH_2), secondary (R-NH-R') and tertiary (R-N-R'R'') amines are obtained from ammonia, and thiols (R-SH) arise from hydrogen sulfide. Many organic compounds contain polar groups, such as -OH and -NH_2. As such groups are much more reactive than the hydrocarbon backbone to which they are attached, they are referred to as **functional groups**.

New functional groups can arise as a result of **oxidation** of the compounds mentioned above. For example, the oxidation of a **thiol** yields a **disulfide** (R-S-S-R'). The oxidation of a primary alcohol (R-CH_2-OH) gives rise to an **aldehyde** (R-C(O)-H), while further oxidation produces a **carboxylic acid**. In contrast, the oxidation of a secondary alcohol yields a **ketone**. The carbonyl group (C=O) is a structural element typical of aldehydes, ketones, undissociated carboxylic acids, and carboxylic acid derivatives.

Reaction of an alcohol with the carbonyl group of an aldehyde yields a **hemiacetal** (R-O- C(H)OH-R'). The cyclic forms of sugars are well-known examples of hemiacetals (see p. 30). The oxidation of hemiacetals produces carboxylic acid esters.

The **carboxylic acids** and their derivatives are very important compounds in biology. Formally, such derivatives are obtained from the acid by exchange of an OH group for another group. In reality, they are formed by nucleophilic substitutions of "activated" in-termediate compounds with the release of water. This is the way in which **carboxylic acid esters** (R-COO-R') are produced from carboxylic acids and alcohols. Fats (see p. 46) are a good example of carboxylic acid esters. A **thioester** (R-S-CO-R') is formed in an analogous way from a carboxylic acid and a thiol. Thioesters play an important role in carboxylic acid metabolism. The best known compound of this type is acetyl-coenzyme A (see p. 50).

Carboxylic acids and primary amines react to form **carboxylic acid amides** (R-NH-CO-R'). The amino acid building blocks of peptides and proteins are linked by carboxylic acid amide bonds, which are therefore also known as *peptide bonds* (see p. 64).

Phosphoric acid, H_3PO_4, is a tribasic acid, i.e., it contains three hydroxyl groups able to donate H^+ ions. One of these three groups is fully dissociated under normal physiological conditions, while the other two can react with alcohols. The resulting products are phosphoric acid monoesters and diesters. **Phosphoric acid monoesters** are found in carbohydrate metabolism, whereas **phosphoric acid diester** groups occur in phospholipids (see p. 48) and nucleic acids (see p. 78).

Compounds of one acid with another are referred to as **acid anhydrides**. A large amount of energy is required for the formation of an acid-anhydride bond, and therefore *phosphoric anhydride bonds* play a central role in the storage and release of chemical energy in the cell (see p. 114). **Mixed anhydrides** between carboxylic acids and phosphoric acid are also very important "energy-rich" intermediates in cellular metabolism.

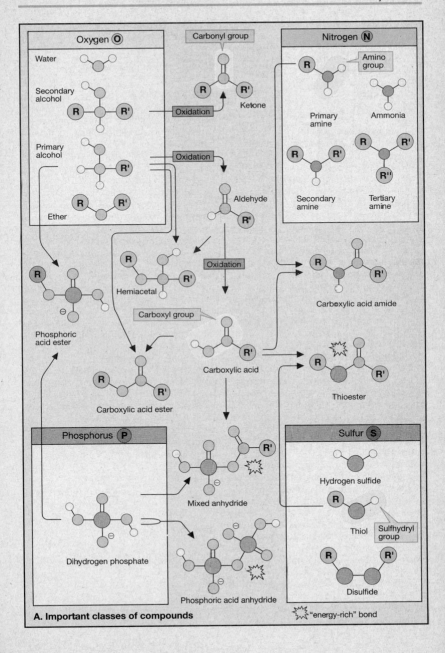

A. Important classes of compounds

Energetics

In order to understand the processes involved in energy storage and conversion in living cells, it is useful first to recall the physical basis for these processes.

A. Forms of work ●

There is essentially no difference between work and energy. Both are measured in **joule** (J, 1 J = 1 N · m). An outdated unit is the **calorie** (cal, 1 cal = 4.187 J). *Energy is defined as the ability of a system to perform work.* There are many different forms of energy, e.g., mechanical, chemical, and radiant energy.

In general, a system is capable of performing work when its components are moving along a potential gradient. This abstract definition is best understood by an example involving mechanical work (**1**). Due to the earth's gravitational pull, the mechanical potential energy of an object is the greater the further the object is away from the center of the earth. Between an elevated and a lower-lying point, there exists a **potential difference** (ΔP). In a waterfall, the water spontaneously follows this potential gradient and, in doing so, is able to perform work, e.g., turn a mill. Work and energy comprise two quantities, an **intensity factor**, which is a measure of the potential difference, i.e., the "driving force" of the process (here it is the height difference) and a **capacity factor**, which is a measure of the quantity of the substance being transported (here it is the weight of the water). In the case of electrical work (**2**), the intensity factor is the voltage, i.e., the electrical potential difference between the source of the electrical current and "ground", while the capacity factor is the amount of charge that is flowing.

Chemical work and chemical energy are defined in an entirely analogous way. Certain compounds or combinations of compounds exist at high chemical potential. When such substances react with one another, the result is products of lower potential. The chemical potential difference (the "driving force" of the reaction) is referred to as the **free-energy change** (ΔG). In the case of chemical work, the capacity factor is the amount of reacting substance (in mol).

B. Energetics and the occurrence of processes ●

Everyday experience shows that water never flows *spontaneously* uphill. Whether a particular process will occur spontaneously or not depends on whether the potential difference between the final and the initial state is positive or negative (ΔP = P_2-P_1). If P_2 is smaller than P1, then ΔP will be negative, and the process will take place and perform work. Processes of this type are called **exergonic** (**1**). If there is no potential difference, then the system is in **equilibrium** (**2**). In the case of **endergonic** processes, ΔP is positive (**3**). Such processes do not proceed spontaneously.

In order to take place, endergonic processes are dependent on **energetic coupling**. This effect can be illustrated by a mechanical analogy (**4**): When two masses M_1 and M_2 are connected by a rope, M_1 will move upward despite the fact that this is an endergonic process. The sum of both potential differences (ΔP_{eff} = ΔP_1 + ΔP_2) is the determining factor in coupled processes. When ΔP_{eff} is negative, the entire process can proceed.

By energetic coupling, it is possible to convert different forms of work and energy into one another. For example, in a flashlight, an exergonic chemical reaction provides an electrical voltage that can then be used for the endergonic generation of light energy. In muscles (see p. 310), chemical energy is converted into mechanical work and heat energy.

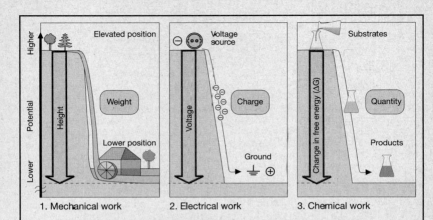

1. Mechanical work   2. Electrical work   3. Chemical work

Energy is the ability of a system to perform work

$J = Joule = N \cdot m, \; 1 \; cal = 4.187 \; J$

Form of work	Intensity factor ⟶	Unit	Capacity factor	Unit	Work = ⟹ · ▢	Unit
Mechanical	Height	m	Weight	$J \cdot m^{-1}$	Height · Weight	J
Electrical	Voltage	$V = J \cdot C^{-1}$	Charge	C	Voltage · Charge	J
Chemical	Free-energy change ΔG	$J \cdot mol^{-1}$	Quantity	mol	ΔG · Quantity	J

A. Forms of work

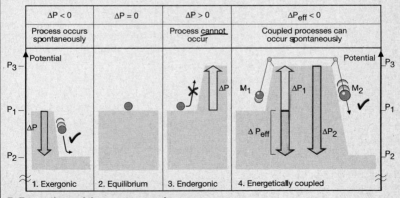

$\Delta P < 0$	$\Delta P = 0$	$\Delta P > 0$	$\Delta P_{eff} < 0$
Process occurs spontaneously		Process **cannot** occur	Coupled processes can occur spontaneously

1. Exergonic   2. Equilibrium   3. Endergonic   4. Energetically coupled

B. Energetics and the occurrence of processes

Free Energy

On the previous page, the free-energy change (ΔG) is defined as a measure of the "driving force" behind a chemical reaction. However, ΔG is not constant, but varies, depending on the nature of the reactants, their concentrations, and the reaction conditions (e.g., temperature). In order to predict whether a reaction is feasible under a given set of conditions and, if so, how much work it will perform, it is necessary to consider the parameters that determine the magnitude of ΔG.

A. Standard free-energy change ◑

As an example, let us discuss the oxidation of pyruvate to lactate, a reaction in which the coenzyme NAD^+ serves as the electron acceptor. Here, we are only interested in the conditions under which this reaction can occur spontaneously. This requires that ΔG be negative (see p. 10). The concentration dependence of ΔG results from the fact that it is the sum of two terms, a standard free-energy change ΔG^o and a concentration term ΔG_c. The latter term also includes the reaction temperature (T). ΔG^o is the free energy change measured when all reactants, including H^+, are present at a concentration of 1 M. Under these conditions, $\Delta G_c = 0$ and thus $\Delta G = \Delta G^o$. For biochemical reactions, it is customary to give the free-energy change at pH 7 ($\Delta G^{o\prime}$). In the case of lactate oxidation, both $\Delta G^{o\prime}$ and ΔG^o are positive, and therefore the oxidation of lactate is not possible under standard conditions.

B. Concentration dependence of ΔG ◑

Even reactions with unfavorable standard free- energy changes can take place when extreme reactant concentrations are chosen. Fig. **B** shows this for lactate oxidation at pH 7 and 37 °C. The spheres represent the relative concentrations of the reactants. Clearly, the *concentration ratios* of pyruvate to lactate and NAD^+ to NADH are the key factors determining whether the reaction will occur. When lactate and NAD^+ are in large excess, the value of ΔG_c becomes so negative that ΔG, despite the positive value of ΔG^o, also becomes negative, i.e., the reaction can proceed. At a specific concentration ratio, $-\Delta G_c$ becomes equal to ΔG^o and thus $\Delta G = 0$. This is the **equilibrium state** of the reaction, where there is no net interconversion of the reactants.

As the concentration ratios of the reactants, rather than absolute concentrations, are the key factors determining the progress of a reaction, it is possible to keep energetically unfavorable reactions running by continuously removing the products. This effect is frequently observed in metabolism, where most reactions are embedded in metabolic pathways. In gluconeogenesis, for example, the pyruvate that is derived from lactate is immediately converted to oxaloacetate, and, at the same time, NAD^+ is regenerated from NADH by the respiratory chain (see p. 158). It should be noted that the equations shown in Fig. **A** do not apply to such **steady-state conditions**.

C. ΔG^o and equilibrium state ◑

The standard free-energy change also provides information about the equilibrium state of a reaction. Here this is shown for the simple reversible interconversion of two compounds, A (red spheres) and B (green spheres). In general, the more *negative* the ΔG^o value, the *higher* the equilibrium constant $K_{eq} = [B]_{eq} / [A]_{eq}$. For example, for the strongly exergonic hydrolysis of ATP (see p. 114), the equilibrium constant is $K_{eq} = 5 \cdot 10^6 \, mol \cdot l^{-1}$. This reaction, therefore, goes almost to completion i.e., the equilibrium lies far to the side of the products ADP and P_i.

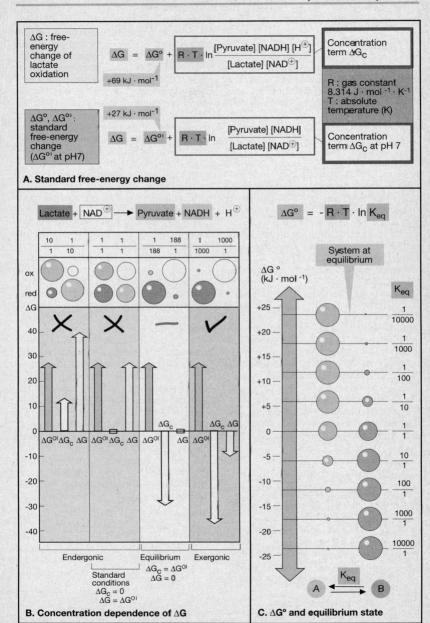

A. Standard free-energy change

B. Concentration dependence of ΔG

C. ΔG^o and equilibrium state

Enthalpy and Entropy

The free-energy change (ΔG) of a reaction is dependent on a range of parameters, e.g., the concentrations of the reactants and the temperature (see p. 12). Here we discuss two further parameters that are associated with molecular changes occurring during the reaction.

A. Heat of reaction and calorimetry ◐

All chemical reactions involve heat exchange. Reactions that release heat are called **exothermic** and those that consume heat are called **endothermic**. Heat exchange is measured as the enthalpy change ΔH (*heat of reaction*), which corresponds to the heat exchange at constant pressure. In the case of exothermic reactions, the system loses heat, and ΔH is negative. When the reaction is endothermic, the system gains heat, and ΔH becomes positive.

For many reactions, ΔH and ΔG are similar in magnitude (see for example **B1**). This fact is utilized to estimate the caloric content of foods. In living organisms, nutrients are usually oxidized by oxygen to CO_2 and H_2O (see p. 104). The maximum amount of chemical work supplied by a particular foodstuff (i.e., ΔG for the oxidation of the utilizable constituents) can be estimated by burning a weighed amount in a **calorimeter** in an oxygen atmosphere. The heat of the reaction will raise the water temperature in the calorimeter, and it can thus be calculated from the change in temperature.

B. Enthalpy and entropy ◐

The heat of reaction ΔH and the free-energy change ΔG are not always of the same magnitude. In fact, there are reactions that occur spontaneously ($\Delta G < 0$) even though they are endothermic ($\Delta H > 0$). This is because the change in the *degree of order* of the system also strongly affects the progress of a reaction. This change is measured as the **entropy change** (ΔS).

The smaller the degree of order of a system, the greater its entropy. Thus, when a process leads to a lower degree of order (everyday experience teaches us that this is the normal case), ΔS is positive for that process. An increase of the order in a system ($\Delta S < 0$) always requires an input of energy. Both of these statements are consequences of a fundamental natural law, the **second law of thermodynamics**. The quantitative relationship between the changes of enthalpy, entropy, and free energy is described by the so-called Gibbs-Helmholtz equation $\Delta G = \Delta H - T \cdot \Delta S$. The following examples will help explain these relationships.

In the so-called **knall-gas reaction (1)**, gaseous oxygen and gaseous hydrogen react to form liquid water. This reaction, like many redox reactions (see p. 28), is strongly exothermic ($\Delta H << 0$). However, during the reaction, the degree of order increases. The total number of molecules declines by 1/3, and a more highly ordered liquid is formed from freely moving gas molecules. As a result of the increase in the degree of order ($\Delta S < 0$), the term $-T \cdot \Delta S$ becomes positive. However, this is more than compensated for by the increase in enthalpy, and the reaction is still strongly exergonic ($\Delta G << 0$).

In the case of the **dissolution of salt** in water (**2**), ΔH is positive, i.e., the temperature of both the container and the water decreases. Nevertheless, the process still occurs spontaneously, due to the fact that there is a decrease in the degree of order in the system. In the undissolved state, the Na^+ and Cl^- ions are rigidly fixed in a crystal lattice. In solution, they move about independently and in random directions. The decrease in order ($\Delta S > 0$) leads to a negative $-T \cdot \Delta S$ term, which compensates for the positive ΔH term and results in a negative ΔG term overall. Processes of this type are said to be *entropy-driven*.

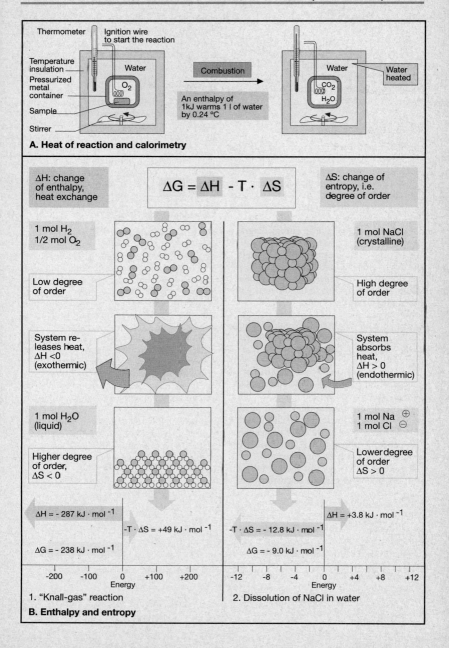

Thermometer   Ignition wire to start the reaction

Temperature insulation

Pressurized metal container

Sample

Stirrer

Water

O_2

Combustion

An enthalpy of 1kJ warms 1 l of water by 0.24 °C

Water   Water heated

CO_2 H_2O

A. Heat of reaction and calorimetry

ΔH: change of enthalpy, heat exchange

ΔS: change of entropy, i.e. degree of order

$$\Delta G = \Delta H - T \cdot \Delta S$$

1 mol H_2
1/2 mol O_2

Low degree of order

1 mol NaCl (crystalline)

High degree of order

System releases heat, $\Delta H < 0$ (exothermic)

System absorbs heat, $\Delta H > 0$ (endothermic)

1 mol H_2O (liquid)

Higher degree of order, $\Delta S < 0$

1 mol Na $\oplus$
1 mol Cl $\ominus$

Lower degree of order $\Delta S > 0$

$\Delta H = -287$ kJ $\cdot$ mol^{-1}

$-T \cdot \Delta S = +49$ kJ $\cdot$ mol^{-1}

$\Delta G = -238$ kJ $\cdot$ mol^{-1}

$\Delta H = +3.8$ kJ $\cdot$ mol^{-1}

$-T \cdot \Delta S = -12.8$ kJ $\cdot$ mol^{-1}

$\Delta G = -9.0$ kJ $\cdot$ mol^{-1}

-200   -100   0   +100   +200
Energy

-12   -8   -4   0   +4   +8   +12
Energy

1. "Knall-gas" reaction

2. Dissolution of NaCl in water

B. Enthalpy and entropy

Reaction Kinetics

The free-energy change, ΔG, tells us whether or not a particular reaction can proceed under a given set of conditions, and how much work it can perform (see p. 12). However, it does not tell us anything about the rate of the reaction, i.e., its kinetics.

A. Activation energy ◑

Most reactions involving organic substances (except for acid-base reactions, see p. 24) proceed only very slowly—regardless of the value of ΔG. The main reason for this is that the molecules involved (the reactants) must first attain a certain minimum energy before they can enter the reaction. This is best understood with the help of an **energy diagram** (**1**) of the simplest possible reaction A → B. The *reactant* A and the *product* B each attain a particular chemical potential (P_e and P_p, respectively). The free-energy change of the reaction, ΔG, corresponds to the difference between these two potentials. In order to be converted to B, A must first overcome a potential energy barrier, the peak of which, P_a, lies far above P_e. The potential difference $P_a - P_e$ is called the **activation energy (E_a)** of the reaction.

The fact that A can be converted to B at all is because the potential P_e only represents the average potential of all the molecules. Some of them occasionally attain much higher potentials, e.g., as the result of collisions with other molecules. When the increase in energy thus gained is greater than E_a, they will overcome the barrier and be converted to B. Figs. **2** and **3** show the energy distribution for such an assembly of molecules as calculated from a simple model. $\Delta n/n$ is the fraction of molecules that have attained or exceeded the energy E (in $kJ \cdot mol^{-1}$). At 27 °C, for example, approximately 10% of the molecules have energies $> 6\ kJ \cdot mol^{-1}$. The activation energies of chemical reactions are typically much higher. The shape of the energy function at energies around 50 $kJ \cdot mol^{-1}$ is shown in Fig. **3**. Statistically, at 27 °C only two out of 10^9 molecules have attained this energy. At 37 °C, it is already four. This explains the "Q_{10} law",

an empirically-derived rule of thumb that states that the rate of a biological process approximately doubles with an increase in temperature of 10 °C.

B. Reaction rate ◑

The rate of a chemical reaction is determined by following the change in concentration of one of its reactants or products, over time. In the example shown here, 3 mmol of reactant are consumed and 3 mmol of product are formed per liter of solution per second. This corresponds to a reaction rate of $v = 3\ mM \cdot s^{-1} = 3 \cdot 10^{-3}\ mol \cdot l^{-1} \cdot s^{-1}$.

C. Reaction order ○

Reaction rates are influenced not only by the activation energy and the temperature, but also by the concentrations of the reactants. When there is only one substrate, A (**1**), v is proportional to [A] (the concentration of A) and the reaction is a first-order one. When two substrates, A and B, react with one another (**2**), the reaction is said to be a second-order one. In this case, the rate v is proportional to the *product* of the concentrations of the substrates. The factors k and k' are called *rate constants*. They depend on the type of reaction and the reaction conditions.

Only the kinetics of simple irreversible reactions are discussed here. More complicated cases, e.g., reversible or multi-step reactions can usually be broken down to simple-order partial reactions, which can be described using the appropriate equations (for an example, see the Michaelis-Menten kinetics, p. 90).

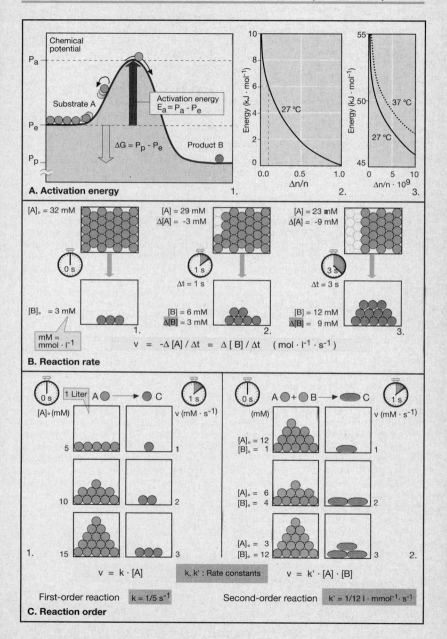

A. Activation energy

Chemical potential

P_a

Substrate A

Activation energy
$E_a = P_a - P_e$

P_e

$\Delta G = P_p - P_e$

Product B

P_p

Energy (kJ · mol^{-1})

27 °C

$\Delta n/n$

37 °C

27 °C

$\Delta n/n \cdot 10^9$

B. Reaction rate

$[A]_0 = 32$ mM

$[A] = 29$ mM
$\Delta[A] = -3$ mM

$[A] = 23$ mM
$\Delta[A] = -9$ mM

0 s

1 s

$\Delta t = 1$ s

3 s

$\Delta t = 3$ s

$[B]_0 = 3$ mM

$[B] = 6$ mM
$\Delta[B] = 3$ mM

$[B] = 12$ mM
$\Delta[B] = 9$ mM

mM =
mmol · l^{-1}

$$v = -\Delta[A]/\Delta t = \Delta[B]/\Delta t \quad (\text{mol} \cdot \text{l}^{-1} \cdot \text{s}^{-1})$$

C. Reaction order

0 s

1 Liter A → C

1 s

$[A]_0$ (mM)

v (mM · s^{-1})

5

1

10

2

15

3

$$v = k \cdot [A]$$

k, k' : Rate constants

First-order reaction $k = 1/5 \ \text{s}^{-1}$

0 s

A + B → C

1 s

(mM)

v (mM · s^{-1})

$[A]_0 = 12$
$[B]_0 = 1$

1

$[A]_0 = 6$
$[B]_0 = 4$

2

$[A]_0 = 3$
$[B]_0 = 12$

3

$$v = k' \cdot [A] \cdot [B]$$

Second-order reaction $k' = 1/12 \ \text{l} \cdot \text{mmol}^{-1} \cdot \text{s}^{-1}$

Catalysis

Catalysts are substances that increase the rate of a chemical reaction without themselves being consumed in the process. The principal catalysts of living cells are the *enzymes* (see p. 86). Only a small number of reactions are catalyzed by RNA molecules ("*ribozymes*," see p. 224).

A. Catalysis: principle ◑

The reason for the slow rates of most reactions involving organic substances is the high energy barrier (activation energy, see p. 16), which the reactants have to overcome before they can interact. In aqueous solutions, a large fraction of the activation energy is required to remove the hydration shells surrounding the reactants. During the course of the reaction, there is often a transient loss of resonance stabilization (see p. 4), which also requires an input of energy. The highest point on the reaction coordinate usually corresponds to the energetically unfavorable **transition state** (**1**). A catalyst creates a new pathway for the reaction (**2**). Whenever the highest transition state of the catalyzed pathway has a lower activation energy than that of the uncatalyzed sequence, the reaction will proceed more rapidly along the catalyzed pathway, even when the number of intermediates is greater. Since reactants and products are the same in both routes, the catalyst has no effect on the free-energy change of the reaction, ΔG. Catalysts—including enzymes—cannot, in principle, change the equilibrium state of a reaction (see p. 12).

The statement that "a catalyst decreases the activation energy of a reaction" is not strictly correct, because the reaction that occurs in the presence of a catalyst is different from the uncatalyzed one. The activation energy of the pathway followed in the presence of the catalyst, however, is lower than that of the uncatalyzed reaction.

B. Catalysis of ester hydrolysis by imidazole ○

As a simple example, let us consider the hydrolysis of a carboxylic acid ester. The **uncat-alyzed reaction** (above) involves a simple *nucleophilic substitution*. The oxygen atom of a water molecule serves as the nucleophile. Its site of attack is the carbonyl carbon of the ester, which, due to the strong polarizing effect of doubly bound oxygen, carries a partial positive charge (**1**; see p. 6). In the first step, an unstable tetrahedral *transition state* is produced by addition of water to the carbonyl group. In the second step, an alcohol molecule is eliminated, and the anion of the free carboxylic acid is formed (**3**). Most substitution reactions of biochemical interest proceed by similar addition-elimination mechanisms.

Even though its ΔG value is negative, ester hydrolysis in pure water is very slow. This is, because water is a very weak nucleophile. At alkaline pH values, the reaction proceeds much more rapidly, because of the presence of strongly nucleophilic OH^- ions (see p. 24). The rate of reaction can also be increased at neutral pH by adding bases, such as imidazole. The **imidazole-catalyzed reaction** (below) comprises *two* substitution steps. In the first, the catalyst itself acts as a nucleophile and *N*-acylimidazole is formed as a relatively stable intermediate, Z (**4**). In the second step, this intermediate is hydrolyzed to yield the carboxylic acid anion, as in the uncatalyzed reaction. Simultaneously, the catalyst is released in unaltered form. The schematic energy diagram (**5**) shows that the activation energy of each of the two partial reactions is lower than that of the uncatalyzed reaction. This is why imidazole increases the rate of the ester hydrolysis. Just as we see with enzymes (see p. 90), the rate of the imidazole-catalyzed reaction is proportional to the concentration of the catalyst.

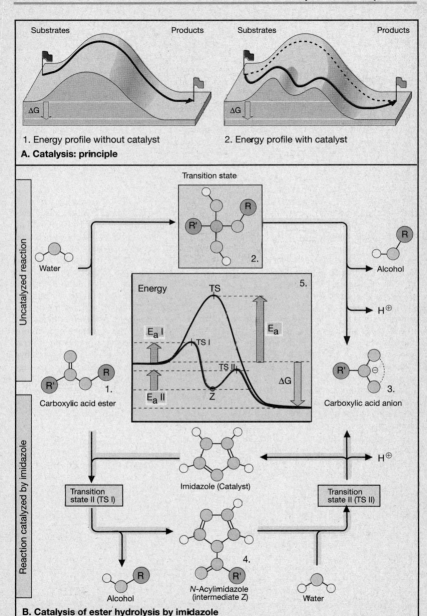

A. Catalysis: principle

1. Energy profile without catalyst

2. Energy profile with catalyst

Substrates   Products   Substrates   Products

ΔG

Uncatalyzed reaction

Transition state

R'   R

2.

Water

Alcohol

H$^\oplus$

Energy   5.

TS

E_a I   TS I

E_a

TS II

ΔG

E_a II   Z

R'   R

1.

Carboxylic acid ester

R'

⊖

3.

Carboxylic acid anion

Reaction catalyzed by imidazole

Imidazole (Catalyst)

H$^\oplus$

Transition
state II (TS I)

Transition
state II (TS II)

Alcohol

N-Acylimidazole
(intermediate Z)

4.

R'   R

Water

B. Catalysis of ester hydrolysis by imidazole

Water as Solvent

Life as we know it has evolved in water and is still dependent on it. The physical and chemical properties of water are, therefore, of fundamental importance to all living things.

A. Water and methane ◑

The special nature of water (H_2O) becomes apparent when we compare it with methane (CH_4). Both molecules are of similar mass and size. Nevertheless, the boiling point of water is 250 °C above that of methane. As a result of this, water is liquid at temperatures on the earth's surface, whereas methane is gaseous. The high boiling point of water is due to its high heat of vaporization, which, in turn, is caused by the uneven distribution of electrons within the molecule. Two corners of the tetrahedrally-shaped water molecule are occupied by unshared electrons (green), and the other two by hydrogen atoms. As a result of this, the H-O-H bond is bent. In addition, the O-H bonds are polarized due to the high electronegativity (see p. 6) of oxygen. One region of the molecule carries a partial negative charge (δ) of about -0.6 units, whereas the other is positively charged. The spatial separation of positive and negative charge gives the molecule the properties of an **electrical dipole**. Water molecules are attracted to one another like magnets, and are connected by hydrogen bonds (see **B**). When liquid water vaporizes, a large amount of energy has to be expended to disrupt these interactions. Methane molecules are not dipolar, and thus interact with one another only weakly. This is why liquid methane vaporizes at very low temperatures.

B. Structure of water and ice ◑

The dipolar nature of water molecules favors the formation of **hydrogen bonds** (see p. 6). Each molecule can act either as a donor or an acceptor in H bond formation, and thus many molecules in liquid water are connected by H bonds, which are constantly being broken and reformed. Tetrahedral networks of molecules, so-called water "clusters," often

arise. As the temperature decreases, the proportion of water clusters increases until the water begins to crystallize. Under normal atmospheric pressure, this occurs at 0 °C. In **ice**, most of the water molecules are fixed in a hexagonal lattice (**3**). Since the distance between the individual molecules in the frozen state is, on average, greater than in the liquid state, the density of ice is lower than that of liquid water. This fact is of great ecological importance. It means, for example, that in winter water bodies first form ice on the surface and seldom freeze to the bottom.

C. Hydration ◑

Unlike most other liquids, water is an excellent **solvent for ions**. In the electrical field of ions, the dipolar water molecules orient themselves in a regular fashion corresponding to the charge of the ion. In this way, the water molecules form **hydration shells**, and shield the central ion from oppositely charged ions. Water has a dielectric constant of 78, i.e., the electrostatic attraction between ions is reduced to 1/78. In the inner hydration spheres of ions, the water molecules are more or less immobilized and follow the central ion. Neutral molecules containing several hydroxyl groups, such as glycerol (left) or sugars, are also readily soluble in water, because they can form H bonds with the solvent.

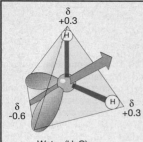

H₂O		CH₄
18	Molecular mass	16
+100 °C	Boiling point	-162 °C
41	Heat of vaporization (kJ · mol⁻¹)	8
6.2	Dipole moment (10⁻³⁰ C · m)	0

Water (H₂O)

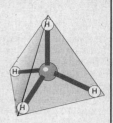

Methane (CH₄)

A. Water and methane

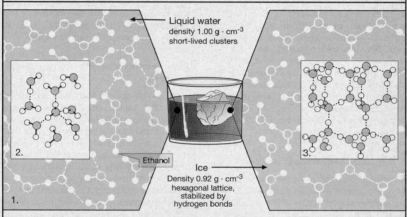

Liquid water
density 1.00 g · cm⁻³
short-lived clusters

2.

Ethanol

Ice
Density 0.92 g · cm⁻³
hexagonal lattice,
stabilized by
hydrogen bonds

1.

3.

B. Structure of water and ice

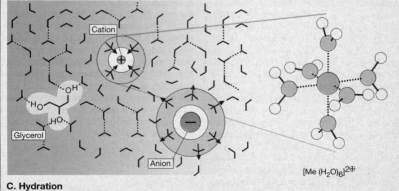

Cation

Glycerol

Anion

[Me (H₂O)₆]²⊕

C. Hydration

Hydrophobic Interactions

Water is an excellent solvent for ions and compounds containing polar bonds (see p. 20). Such compounds are referred to as **polar** or **hydrophilic** ("water-loving"). In contrast, substances, which are composed mainly of hydrocarbon structures dissolve only poorly in water. Such substances are said to be **apolar** or **lipophilic** ("lipid-loving").

A. Water solubility of fatty acids ○

The effects of polar and apolar groups on the water-solubility of organic compounds are illustrated by the solubility of various fatty acids. Under physiological conditions, the carboxyl groups of these acids are ionized, and therefore well hydrated. Nevertheless, there is a marked decrease in solubility with increasing length of the apolar hydrocarbon tail of the fatty acid. Acids with more than 10 carbons are more or less insoluble in water. As a consequence, they have to be transported in protein-bound form in the body (see p. 144).

B. Dissolution of methane in water ○

A better understanding of the poor solubility of hydrocarbons in water can be gained by considering the energetics of the process (see p. 10). Fig. 1 shows the data for the simplest of these compounds, methane. The dissolution of gaseous methane in water is clearly exothermic ($\Delta H^o < 0$). Nevertheless, the free-energy change (ΔG^o) is positive, because the entropy term, $-T \cdot \Delta S^o$, is strongly positive. Consequently, the entropy change of the process (ΔS^o) must be negative, i.e., a solution of methane in water has a higher degree of order than either water or gaseous methane alone. One reason for this is that the methane molecules are less mobile when surrounded by water. More importantly, however, apolar molecules cause the water to form cage-like "clathrate" structures, which — as in ice — are stabilized by hydrogen bonds. Thus, dissolution of methane increases the degree of order of the water phase. The greater the surface area of the region of contact between water and the apolar phase, the greater the increase in the degree of order.

C. The "oil drop effect" ○

The well-known spontaneous separation of a mixture of oil and water into two distinct phases reduces the energetically unfavorable formation of clathrate structures. A larger drop has a smaller surface area than several smaller drops of the same volume. Thus, phase separation decreases the area of contact between water and oil, and consequently also the extent of clathrate formation. The entropy change, ΔS, during phase separation, therefore, is positive, and the negative $-T \cdot \Delta S$ makes the process exergonic ($\Delta G < 0$). In other words, it proceeds spontaneously.

D. Arrangements of amphipathic substances in water ●

Molecules that contain both polar and apolar groups are called **amphipathic** or amphiphilic. Some examples include soaps (see p. 46), phospholipids (see p. 48), and bile acids (see p. 288). As a result of the "oil drop effect," amphipathic substances in water tend to orient themselves so as to minimize the area of contact between the apolar regions of the molecule and water. On the surface, they form **monolayers** in which the polar "head groups" are facing toward the water. Soap bubbles consist of lipid **bilayers** with the apolar parts facing outward and a thin layer of water in the interior. When surrounded by water, amphipathic compounds form **micelles** with their head groups facing toward the outside. By a back-to-back arrangement of two monolayers, extended **bilayers** arise. Most biological membranes are assembled according to this principle (see p. 202). Hollow membrane sacks are called **vesicles.** Such structures play a crucial role in the transport of substances within the cells and in body fluids (see pp. 210, 254).

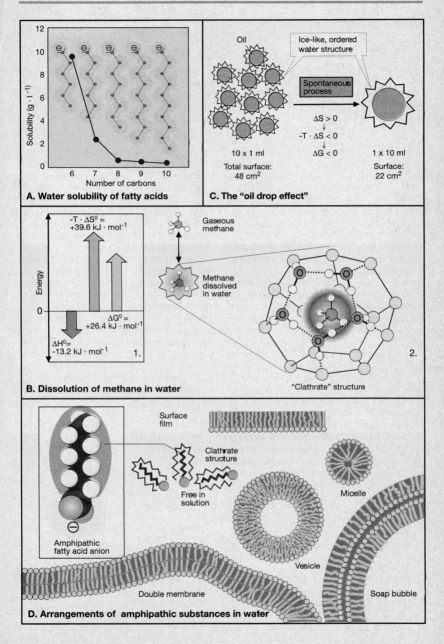

A. Water solubility of fatty acids

C. The "oil drop effect"

Oil

Ice-like, ordered water structure

Spontaneous process

$\Delta S > 0$

$-T \cdot \Delta S < 0$

$\Delta G < 0$

10 x 1 ml
Total surface: 48 cm^2

1 x 10 ml
Surface: 22 cm^2

B. Dissolution of methane in water

$-T \cdot \Delta S^0 = +39.6 \ kJ \cdot mol^{-1}$

$\Delta G^0 = +26.4 \ kJ \cdot mol^{-1}$

$\Delta H^0 = -13.2 \ kJ \cdot mol^{-1}$

1.

Gaseous methane

Methane dissolved in water

"Clathrate" structure

2.

D. Arrangements of amphipathic substances in water

Surface film

Clathrate structure

Free in solution

Micelle

Vesicle

Amphipathic fatty acid anion

Double membrane

Soap bubble

Acids and Bases

A. Acids and bases ●

There are several different definitions for acids and bases. In general, however, **acids** are defined as substances that can *donate* hydrogen ions (protons), whereas **bases** are compounds that *accept* protons. Water enhances the acidic or basic properties of dissolved substances, because water itself can act as an acid or a base. For example, when hydrogen chloride (HCl) is in aqueous solution, it donates protons to the solvent (**1**). This results in the formation of chloride ions (Cl^-) and protonated water molecules (hydronium ions, H_3O^+, usually simply referred to as H^+). The proton exchange between HCl and water is virtually quantitative, i.e., in water HCl behaves as a very strong acid. In methanol, HCl donates fewer protons; in this environment, it is a much weaker acid.

Bases, such as **ammonia** (NH_3) abstract protons from water molecules. As a result of this, **hydroxyl ions** (OH^-) and positively charged **ammonium ions** (NH_4^+, **3**) are formed. Hydronium and hydroxyl ions, like other ions, exist in water in hydrated rather than free form (see p. 20). These hydrated forms include $[H_5O_2]^+$ and $[H_3O_2]^-$ (**4**, **5**).

Acid-base reactions always involve **conjugate acid-base pairs** (see below). The stronger the acid or base, the weaker the conjugate base or acid, respectively. For example, the very weak base chloride ion is paired with the very strongly acidic hydrogen chloride (**1**). The weakly acidic ammonium ion is conjugated with the moderately strong base ammonia (**3**). When water is acting as a weak acid, the hydroxyl ion is the extremely strong conjugate base. In contrast, when water acts as a weak base, the hydronium ion is the very strong conjugate acid.

The equilibrium constant K_{eq} for the acid-base reaction between H_2O molecules (**2**) is small:

$$K_{eq} = [H^+] \cdot [OH^-]/[H_2O] = 2 \cdot 10^{-16} \text{ mol} \cdot l^{-1}$$
(at 25 °C)

The term $[H_2O]$, therefore, is practically constant. Substituting 55 mol $\cdot$ l^{-1} for $[H_2O]$ yields

$$K_w = [H^+] \cdot [OH^-] = 1 \cdot 10^{-14} \text{ mol} \cdot l^{-1}$$

Thus, even when other acids or bases are added to an aqueous solution, there is no change in the product $[H^+] \cdot [OH^-]$, the *ion product* of water. In pure H_2O at 25 °C, the concentrations of H^+ and OH^- ions are both $1 \cdot 10^{-7}$ mol $\cdot$ l^{-1}.

B. pH and pK_a ●

The dissociation of an acid is described by the equation HB $\leftrightarrow$ H^+ + B^- (**1**). The equilibrium constant for this process is called the *acid dissociation constant*, K_a. From the definition of K_a (**2**), it follows that the ratio $[HB]/[B^-]$ is proportional to $[H^+]$ (**3**). For every H^+ concentration, there is a unique acid/conjugate base concentration ratio. This relationship can be illustrated by a **titration curve** (**4**).

The H^+ concentration in an aqueous solution is usually given as the **pH value**, i.e., the negative logarithm of the H^+ concentration. For example, an H^+ concentration of $1 \cdot 10^{-4}$ mol $\cdot$ l^{-1} corresponds to a pH value of 4. Pure water at 25 °C has a pH value of 7. The K_a value of an acid can be determined from its titration curve. According to formula **3**, $[H^+] = K_a$ when $[HB] = [B]$, i.e., at the point of inflection of the curve. The pH value at this point corresponds to the negative logarithm of K_a, the so-called *pK_a value* of the acid. The stronger an acid, the lower its pK_a value.

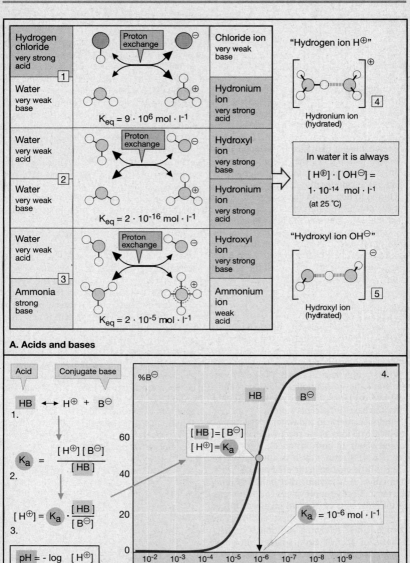

A. Acids and bases

B. pH and pK$_a$

Buffer Systems

A. pH values in the body ●

The pH within cells and in extracellular fluids is maintained at a relatively constant level. In the blood, the pH normally only varies between 7.35 and 7.45. This corresponds to a maximum change in the H^+ concentration of about 30% (see p. 256). The pH of the cytoplasm, with values ranging from 7.0 to 7.3, is somewhat lower than that of the blood. In lysosomes (see p. 212), the proton concentration is more than a hundred times higher than in the cytoplasm.

In the lumen of the digestive tract, which for the organism is part of the outside world, and in body secretions, the pH is much more variable. Extreme values are found in the stomach (around 2) and in the small intestine (> 8). Since the kidneys excrete both acids and bases (see p. 300), the pH of the urine is highly variable (4.8–7.5).

B. Buffers ●

Short-term pH changes in living organisms are counteracted by **buffer systems**. Such systems are mixtures of a weak acid (HB) with its conjugate base (B^-) or a weak base with its conjugate acid (see p. 24). Buffer systems are able to neutralize both hydronium and hydroxyl ions. In the case of hydronium ions, the base (B^-) accepts a proton giving rise to undissociated acid and water. In the case of the hydroxyl ions, these react with HB to give B^- and water. In both instances, it is mainly the [HB]/[B^-] ratio, which is affected, while there is little change in the pH. From the titration curve, it is obvious that buffer systems are most effective at pH values corresponding to the pK_a value of the acid involved. This is where the curve is steepest and the pH change (ΔpH), following addition of a given amount of acid or base, smallest. In other words, the **buffer capacity** is greatest at the pK_a value of the system.

C. Physiological buffer systems ◐

Many different buffer systems contribute to the buffering capacity of body fluids and the intracellular space. They can be classified as either high-molecular weight or low-molecular weight systems. The main high-molecular weight system involves **protein side chains** with acidic or basic properties. The most important of these residues are summarized in Fig. **1**. The pK_a values of protein-bound acids and bases are not constant, but depend on the environment of the respective residue in the protein. Thus, the buffering capacity of proteins spans almost the entire pH scale. At physiological pH values, the imidazole rings of histidine residues are most important, but carboxyl and amino groups also make substantial contributions (cf. p. 58).

The principal low-molecular weight buffer system in the blood is composed of **carbon dioxide** (CO_2), **water** and **hydrogen carbonate** (HCO_3^-). The pK_a value of this system (6.3) differs somewhat from the pH of the plasma. However, since CO_2 can be exhaled by the lungs, the CO_2/HCO_3^- system is nevertheless very efficient (see p. 256). A further buffer system in the blood consists of **dihydrogen phosphate** ($H_2PO_4^-$) and **hydrogen phosphate** (HPO_4^{2-}), which has a pK_a value of about 7. The main buffer system in the urine involves **ammonium ions** (NH_4^+) and **ammonia** (NH_3). Ammonia, a moderately strong base, can be synthesized in the kidneys from amino acids. After adaptation, the organism is able to neutralize large amounts of acids by this process (see p. 300).

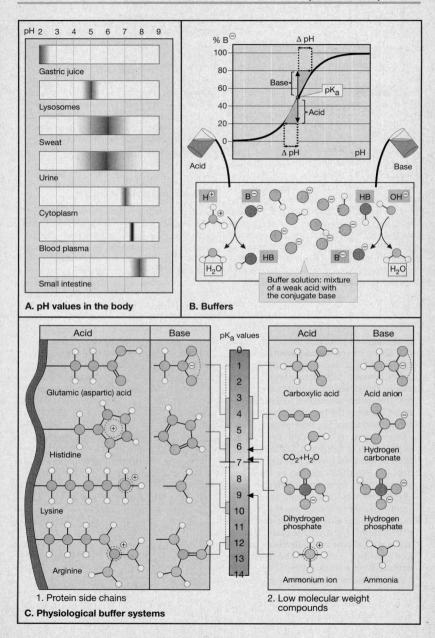

A. pH values in the body

B. Buffers

Buffer solution: mixture of a weak acid with the conjugate base

C. Physiological buffer systems

1. Protein side chains

2. Low molecular weight compounds

Redox Processes

A. Redox reactions ●

In redox reactions electrons are transferred from one reaction partner to another (**1**). Processes of this kind, just like acid-base reactions, always involve pairs of compounds. Such a pair is referred to as a **redox system** (**2**). The essential difference between the two components of a redox system is the number of electrons, which they contain. The more electron-rich component is called the **reduced form** of the system, while the other one is referred to as the **oxidized form**. In a redox reaction, the reduced form of one system (the **reducing agent**) donates electrons to the oxidized form of another one (the **oxidizing agent**). In the process, the reducing agent becomes oxidized and the oxidizing agent is reduced. Any given reducing agent can reduce only certain other redox systems. Such observations form the basis of **redox series** (**4**).

B. Determination of reduction potentials ○

The position of a particular redox system in a redox series, its **reduction potential**, can be determined by a device in which electron transfer between solutions of two different, spatially separated redox systems is monitored. The system is designed so that electron transfer between the two systems occurs via metallic conductors, i.e., the chemical energy of the reaction is converted to electrical energy (a flow of electrons). Technical applications of this principle include dry-cell batteries for electrical appliances and lead storage batteries for motor vehicles.

Replacement of the light bulb, or any other current-sink, with a voltage meter of high resistance stops the flow of current and allows the measurement of a voltage corresponding to the electrical potential difference (ΔE) between the two solutions. The **reduction potential** (E) of a redox system is defined as the voltage measured in this way against the system $2H^+/H_2$ (a "hydrogen electrode") as a reference. Reduction potentials can be either positive or negative, relative to this arbitrarily chosen zero reference potential. As E also depends on the concentrations of the reactants and the reaction conditions, a **standard potential**, E^0, has been defined for standard conditions and unit concentrations of all reactants. A corresponding value, $E^{0\prime}$, applies to a constant pH value of 7 (see p. 12). In a redox series (**A4**), the redox systems are arranged in order of increasing reduction potential. *Spontaneous* electron transfers are only possible when the reduction potential of the donor system is more negative than that of the acceptor system.

C. Redox scale and energetics ◑

For redox reactions, there is a simple linear relationship between the potential difference, ΔE (the intensity factor of electrical work, see p. 10), and the free-energy change, ΔG (**1**). Therefore, redox series allow predictions of whether a particular reaction will occur spontaneously and, if so, how much chemical work it can perform. Let us consider two examples: electron transfer from the $NAD^+/NADH$ system ($E^{0\prime} = -0.32$ V) to oxygen ($E^{0\prime} = +0.82$ V, $\Delta E^{0\prime} = +1.14$ V) is strongly exergonic. In fact, most of the chemical energy of the cell is obtained from this transfer (see p. 104). In contrast, the electron transfer from lactate to NAD^+ is not possible under standard conditions, because $\Delta E^{0\prime} < 0$. However, as discussed on p. 12, the concentration of the reactants affects the direction of the reaction (**2**). Thus, as the ratio [Pyr]/[Lac] increases, the reduction potential of the system also increases. If, at the same time, NAD^+ is in large excess over NADH, then $E\prime$ for this redox pair will be low enough for $\Delta E\prime$ to become positive. As $\Delta G\prime$ then becomes negative, the reaction will proceed spontaneously.

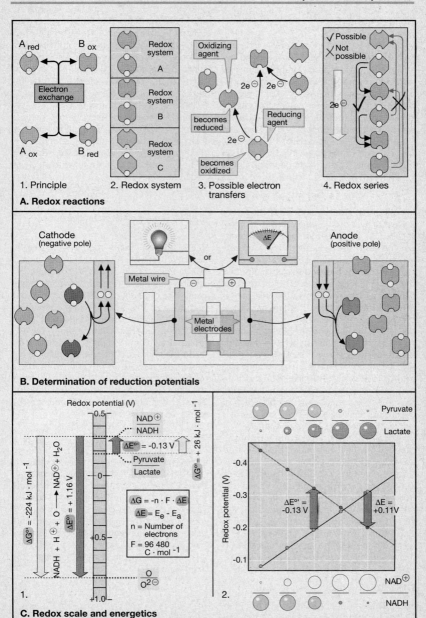

A. Redox reactions

1. Principle

A_{red} ⟷ B_{ox}

Electron exchange

A_{ox} ⟷ B_{red}

2. Redox system

Redox system A

Redox system B

Redox system C

3. Possible electron transfers

Oxidizing agent

becomes reduced

$2e^{\ominus}$ $2e^{\ominus}$

$2e^{\ominus}$

Reducing agent

becomes oxidized

4. Redox series

✓ Possible
✗ Not possible

$2e^{\ominus}$

B. Determination of reduction potentials

Cathode (negative pole)

Anode (positive pole)

ΔE

Metal wire

or

Metal electrodes

C. Redox scale and energetics

1.

Redox potential (V)

−0.5

$NAD^{\oplus}$
$NADH$

$\Delta E^{o'} = -0.13\ V$

Pyruvate
Lactate

$\Delta G^{o'} = +26\ kJ \cdot mol^{-1}$

$\Delta G^{o} = -224\ kJ \cdot mol^{-1}$

$NADH + H^{\oplus} + O \longrightarrow NAD^{\oplus} + H_2O$

$\Delta E^{o} = +1.16\ V$

0

+0.5

$\Delta G = -n \cdot F \cdot \Delta E$
$\Delta E = E_e - E_a$
n = Number of electrons
$F = 96\ 480\ C \cdot mol^{-1}$

$\dfrac{O}{O^{2\ominus}}$

+1.0

2.

Pyruvate
Lactate

Redox potential (V)

−0.4

−0.3

−0.2

−0.1

$\Delta E^{o'} = -0.13\ V$

$\Delta E = +0.11V$

$NAD^{\oplus}$

$NADH$

Stereochemistry

The **carbohydrates** are naturally occurring carbonyl compounds containing several hydroxyl groups. The carbohydrates include the **sugars (monosaccharides)** and their polymers, the **oligosaccharides** and **polysaccharides**. The monosaccharides are further divided into two distinct families — the **aldoses,** in which the carbonyl group is present at the end of the carbon chain as an aldehyde function, and the **ketoses,** in which the carbonyl group is located within the chain.

A. Aldoses of the D-series: stereochemistry ◑

The simplest **aldose** is glyceraldehyde (**1**). Glyceraldehyde has a *chiral center* (see p. 56) at C-2 and therefore exists in two **enantiomeric forms** (D- and L-glyceraldehyde). The chiral center is best represented using a so-called **Fischer projection formula.** Fig. **1a-c** uses glyceraldehyde as an example to show how the Fischer projection formula (**1c**) is derived from the three-dimensional model (**1a**): First of all, the tetrahedron is rotated until the most highly oxidized group (in this case the aldehyde function) is at the top. This is also the point where the numbering of the carbons begins. Then it is further rotated until the line connecting the first and third carbons (red) is in the plane of the page (**1b**). Using this type of representation, the OH group of D-glyceraldehyde is on the right, and that of L-glyceraldehyde (not shown) on the left.

Most naturally occurring monosaccharides are derived from the **triose** D-glyceraldehyde, and thus belong to the D-series. Formally, the other D-aldoses are obtained by introducing further chiral centers, of the structure H-C-OH or HO-C-H, between C-2 and the aldehyde group of D-glyceraldehyde. This yields two different D-tetroses, four D-pentoses, eight D-hexoses, and so on. All of these aldoses have the same configuration as D-glyceraldehyde at the penultimate carbon.

B. Hemiacetal formation ◑

In aqueous solution at neutral pH, less than 0.1% of sugar molecules actually contain free aldehyde functions. This is, because certain OH groups of the sugar molecule tend to react with the aldehyde group of the same molecule to produce a cyclic structure called a **hemiacetal** (see p. 8), which has an additional chiral center at C-1. In the case of aldohexoses, it is usually the hydroxyl group at C-5, which is involved in this reaction. This results in a 6-membered ring with 5 carbons and one oxygen, which resembles pyran. Sugars containing this type of ring structure are, therefore, referred to as **pyranoses.** When the hydroxyl group at C-4 reacts with the aldehyde group, a 5-membered furan-like ring is formed, giving rise to a **furanose** (see p. 34).

C. α-D-Glucopyranose: stereochemistry, Haworth projection ○

The hemiacetal forms of sugars are usually depicted using the so-called *Haworth projection formula* (**5**). To do this, C-1 through C-5 and the bridging oxygen are drawn in a single plane. The CH_2OH group then projects up out of this plane, and the OH groups at the chiral carbons are either above or below the plane depending on their configuration. The OH groups drawn on the right in the Fischer projection formula appear below the plane of the ring in the Haworth projection, while those located on the left appear above the plane of the ring. Figs. 1–5 illustrate the relationship between the Fischer projection and the Haworth perspective formula.

D. Conformation of α-D-glucopyranose ○

The Haworth projection does not take into account the fact that the rings of monosaccharides are not planar. In its most stable state, the pyran ring of α-D-glucopyranose assumes the so-called *chair conformation* (see p. 52). Most of the OH groups lie in the same plane as the ring (**equatorial** or **e** position). The only exception is the anomeric OH group at C-1, which, in the α anomer, adopts the **axial** position.

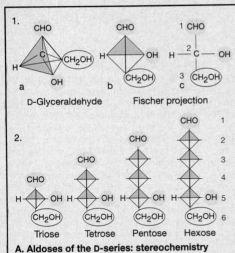

1.

D-Glyceraldehyde   Fischer projection

a   b   c

2.

Triose   Tetrose   Pentose   Hexose

A. Aldoses of the D-series: stereochemistry

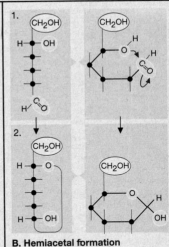

1.

2.

B. Hemiacetal formation

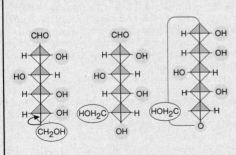

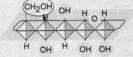

4. Tilted

1. Fischer
   projection

2. Following rotation
   around C-4/5

3. Hemi-
   acetal

5. Haworth representation

C. α-D-Glucopyranose: stereochemistry, Haworth projection

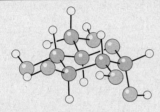

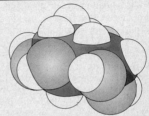

1. Ball-and-stick model

2. Van der Waals model

D. Conformation of α-D-glucopyranose

Sugar Derivatives

A. Reactions of the monosaccharides ◑

In the cyclic form — as opposed to the open-chain form — aldoses have a chiral center at C-1 (see p. 30). The two different enantiomers arising as a result of this fact are referred to as **anomers**. In β anomers, the OH group at C-1 (the anomeric OH group) and the CH_2OH group lie on the same side of the ring. In the α anomer, they are on opposite sides. The interconversion of anomers is known as **mutarotation** (see **B**).

Condensation of the anomeric OH group of a sugar with an alcohol, with elimation of water, yields an **O**-glycoside (**1**, here we have α-O-methylglucoside). Oligosaccharides and polysaccharides also contain O-glycosidic bonds. Reaction of the anomeric OH group with an amino group yields an **N**-glycoside (not shown). N-glycosidic bonds occur in nucleotides (see p. 78) and in glycoproteins (see p. 40). Like free hydroxyl groups, glycosidic bonds can adopt either the α or the β configuration.

Reduction of the anomeric center of glucose (**2**) produces the sugar alcohol **sorbitol** (a *glycitol*). Oxidation at C-1 (**3**) gives the intramolecular ester (lactone) of **gluconic acid** (a *glyconic acid*). Oxidation at C-6 (**5**) yields **glucuronic acid** (a *glycuronic acid*). Glucuronic acid plays an important role in biotransformations in the liver (see p. 290).

In alkaline solutions, glucose is in equilibrium with the ketohexose D-fructose and the aldohexose D-mannose (**4**). The only difference between glucose and mannose is the configuration at C-2. Such pairs of sugars are referred to as **epimers**, and their interconversion is called **epimerization**.

The hydroxyl groups of monosaccharides can form esters with acids. In metabolism, phosphoric acid esters, such as **glucose 6-phosphate** (**6**) are especially important.

Treatment of sugars with strong acids or bases yields degradation products that can be used for the detection of carbohydrates. The so-called reduction assays are based on **reductones** formed in strongly alkaline solutions (**7**). With acids, hydroxymethyl derivatives of furfural (furaldehyde) are obtained, which can be detected by color reactions (**8**).

B. Polarimetry, mutarotation ○

Sugar solutions can be analyzed by polarimetry, a method that depends on the interaction between chiral centers and plane-polarized light, i.e., light oscillating in only one plane. It is produced by passing normal light through a special filter (a **polarizer**). A second polarizing filter (the **analyzer**) is then placed behind the first. Light can only pass through when the polarizer and the analyzer are in phase with one another. In that case, the field of view seen through the analyzer appears bright (**1**). Solutions of chiral substances rotate the plane of polarized light either to the left or to the right (see p. 56). When such a solution is placed between the polarizer and the analyzer, the field of view appears darker (**2**). The angle of rotation, α, is estimated by turning the analyzer until the field of view becomes bright again (**3**). The **optical rotation** caused by a particular solution depends on the type of chiral compound, its concentration, and the pathlength through the solution.

The α and β anomers of glucose (**A**) can be obtained in pure form using special procedures. Initially, a 1 molar solution of α-D-glucose shows a rotation of +112°, whereas a corresponding solution of β-D-glucose has a value of +19°. These rotations change spontaneously, however, and after some time reach the same endpoint of +52° (**4**). The reason is that, in solution, **mutarotation** (**A**) establishes an equilibrium between the α and β forms, which — regardless of the starting conditions — contains 62% β anomer and 38% α anomer.

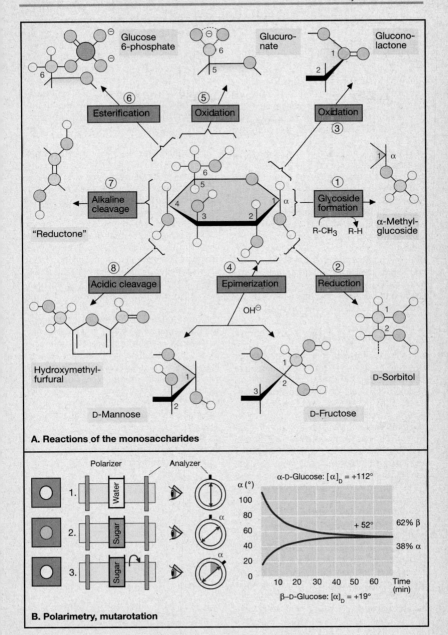

A. Reactions of the monosaccharides

Glucose 6-phosphate
Glucuronate
Gluconolactone

⑥ Esterification
⑤ Oxidation
③ Oxidation
① Glycoside formation

R-CH₃   R-H

α-Methylglucoside

⑦ Alkaline cleavage
"Reductone"

⑧ Acidic cleavage
④ Epimerization
OH⊖
② Reduction

Hydroxymethylfurfural

D-Mannose
D-Fructose
D-Sorbitol

B. Polarimetry, mutarotation

Polarizer   Analyzer

1. Water
2. Sugar
3. Sugar

α-D-Glucose: $[\alpha]_D = +112°$

+ 52°   62% β
38% α

10 20 30 40 50 60   Time (min)

β–D-Glucose: $[\alpha]_D = +19°$

Monosaccharides

A. Monosaccharides: nomenclature, ring forms ◑

The nomenclature of the carbohydrates takes into account not only the type of compound involved, but also its stereochemistry and ring form. The nomenclature rules for sugars and glycosides are illustrated here with two examples.

The name of the sugar is always preceded by an indication as to whether it belongs to the D or the L-series. In addition, the prefix α or β denotes the configuration of the anomeric center. Most sugars exist in the form of pyranose or furanose rings as a result of the formation of intramolecular hemiacetals (see p. 30). Thus, the *ring form* is also part of the name of a carbohydrate. The names of free monosaccharides are characterized by the '-ose' ending, whereas *glycoside* names end with '-oside'. With glycosides, the name of the substituent attached to the anomeric OH group precedes the name of the sugar.

B. Important monosaccharides ◑

Only the most important of the large number of naturally occurring monosaccharides are listed here. Their full names, as well as the official abbreviations, are given. In aqueous solution, most *free* sugars (hexoses as well as pentoses) exist predominantly in the *pyranose form*. Such solutions always contain both anomers (α and β) in various ratios. To account for this fact, the anomeric OH groups are drawn projecting out sideways from the plane of the ring rather than above and below it (the latter would indicate the β and α forms, respectively).

An important **aldopentose (1)** is D-ribose, which is widely distributed in nature as a constituent of RNA and of nucleotide coenzymes. In these compounds, ribose always exists in the furanose form (see p. 79). Like ribose, D-xylose and L-arabinose are rarely found as free sugars. However, they constitute the building blocks of polysaccharides in plant cell walls (see p. 36).

The most abundant **aldohexose (1)** is D-glucose. A large percentage of the total biomass is accounted for by glucose polymers, predominantly cellulose (see p. 38). Free D-glucose is found in fruits ("grape sugar") and as "blood sugar" in the blood of higher animals. D-galactose, a component of milk sugar, is also an important part of the human diet. Both D-mannose and D-galactose are regularly found in glycolipids and glycoproteins (see p. 40).

Phosphoric acid esters of the **ketopentose** D-ribulose (**2**) are intermediates in the hexose monophosphate pathway (see p. 142) and in photosynthesis (see p. 120). The most important **ketohexose** is D-fructose. In free form, it is present in fruit juices ("fruit sugar") and in honey. In bound form, it is found in sucrose and plant polysaccharides (inulin).

In the **deoxyaldoses (3)**, an OH group is replaced with a hydrogen atom. The illustration shows *2-deoxy-D-ribose*, a constituent of DNA (see p. 80), which is reduced at C-2, and *L-fucose*, a sugar from the L-series, which is reduced at C-6.

The acetylated **amino sugars (4)** N-acetyl-D-glucosamine and N-acetyl-D-galactosamine, as well as N-acetylneuraminic acid (sialic acid, **5**) are characteristic constituents of the carbohydrate part of glycoproteins (see p. 40), while the **acidic monosaccharides**, D-glucuronic acid, D-galacturonic acid, and L-iduronic acid, are typical building blocks of the glycosaminoglycans found in connective tissues (see p. 314).

Sugar alcohols (6), such as *sorbitol* and *mannitol* are of little importance for the metabolism of healthy animals.

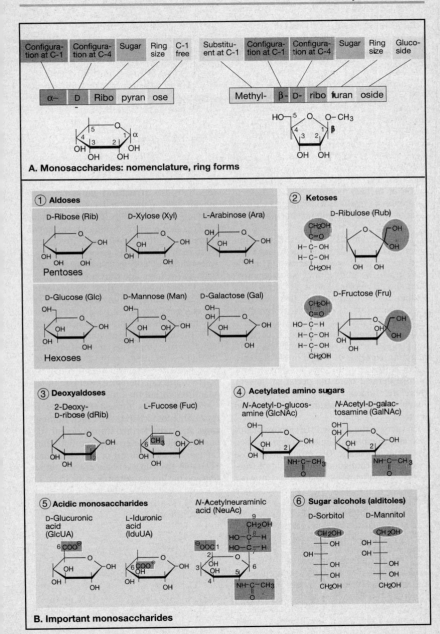

A. Monosaccharides: nomenclature, ring forms

① Aldoses

D-Ribose (Rib)   D-Xylose (Xyl)   L-Arabinose (Ara)

Pentoses

D-Glucose (Glc)   D-Mannose (Man)   D-Galactose (Gal)

Hexoses

② Ketoses

D-Ribulose (Rub)

D-Fructose (Fru)

③ Deoxyaldoses

2-Deoxy-D-ribose (dRib)   L-Fucose (Fuc)

④ Acetylated amino sugars

N-Acetyl-D-glucos-amine (GlcNAc)   N-Acetyl-D-galac-tosamine (GalNAc)

⑤ Acidic monosaccharides

D-Glucuronic acid (GlcUA)   L-Iduronic acid (IduUA)   N-Acetylneuraminic acid (NeuAc)

⑥ Sugar alcohols (alditoles)

D-Sorbitol   D-Mannitol

B. Important monosaccharides

Disaccharides and Polysaccharides

A. Disaccharides ○

The formation of a glycosidic bond between the anomeric hydroxyl group of one monosaccharide and an OH group of another one yields a disaccharide. As with all glycosides, the glycosidic bond does not allow *mutarotation*. In nature, the formation of glycosidic bonds is enzyme-catalyzed, and thus stereospecific. Naturally occurring disaccharides, therefore, exist in only one of the two possible configurations (α or β). The addition of further monosaccharides leads to the formation of **oligosaccharides** (3–20 residues) and ultimately **polysaccharides** (up to 10^5 residues).

There are two dissacharides that are important in human nutrition, sucrose and lactose. In the plant world, **sucrose** (**1**) is the form in which carbohydrates are transported throughout the plant, and it also serves as a soluble carbohydrate reserve. It is valued by man because of its intensely sweet taste. Plants with a particularly high sucrose content, such as sugar cane and sugar beet, are the main sources of sucrose for human consumption. *Honey* is produced in the digestive tracts of bees by enzymatic hydrolysis of sucrose derived from the nectar of flowers. Therefore, honey consists of an approximately equimolar mixture of glucose and fructose.

In sucrose, the anomeric OH groups of glucose and fructose are linked to one another. Since the glycosidic OH group of glucose adopts the α configuration and that of fructose the β configuration, the result is an α1↔2β bond. This unusual linkage results in a compact, egg-shaped molecule.

Lactose ("*milk sugar*") is the principal carbohydrate in the milk of mammals. Cow's milk contains 4.5% lactose, while human milk contains up to 7.5%. In lactose, the anomeric OH group of galactose is linked via a β1 → 4 bond to a glucose residue. As a consequence, the lactose molecule is elongated, and both of its pyran rings lie in the same plane.

It is not clear why galactose should be found in mammalian milk. Galactose formed by hydrolytic cleavage of lactose in the intestine is immediately converted into glucose (an epimer of galactose) in the liver (see p. 282). Disturbances of this process can cause serious diseases, due to the accumulation of free galactose in the body. One example of this is a hereditary metabolic disorder known as *galactosemia*.

B. Important polysaccharides ○

Polysaccharides are ubiquitous in nature. They can be classified into three separate groups, based on their different functions. The first, the **structural polysaccharides**, provide cells, organs, and whole organisms with mechanical stability. This group includes *cellulose* (see p. 38) and *chitin*, the main constituent of the shells of insects and crustaceans. The second group is made up of **water-binding polysaccharides**, which are strongly hydrated and thus prevent cells and tissues from drying out. These compounds include carbohydrates from algae (e.g., *agarose*, *carrageenan*), which are used in the laboratory for the production of gels (agar-agar). Finally, the **reserve polysaccharides** serve as a carbohydrate store that releases monosaccharides as required. Due to their polymeric nature, reserve carbohydrates are osmotically inactive, and can therefore be stored in large quantities within the cell.

Polysaccharides formed from only one type of monosaccharide are called **homoglycans**, and those formed from different sugar building blocks **heteroglycans**. Both forms can exist as either linear or branched chains. The table gives an overview of the composition and the linkage patterns of the important glycans. Some of the polysaccharides shown here are discussed in more detail elsewhere (e.g., pp. 38 and 160).

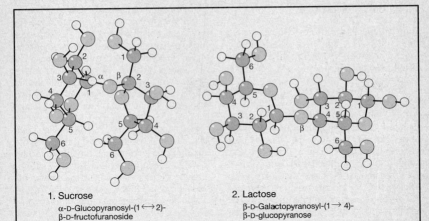

1. Sucrose

α-D-Glucopyranosyl-(1 ⟷ 2)-
β-D-fructofuranoside

2. Lactose

β-D-Galactopyranosyl-(1 → 4)-
β-D-glucopyranose

A. Disaccharides

Poly-saccharide	Mono-saccharide 1	Mono-saccharide 2	Linkage	Branch-ing	Occurrence	Function
Bacteria						
Murein	D-GlcNAc	D-MurNAc[1]	β1 → 4	—	Cell wall	SC
Dextran	D-Glc	—	α1 → 6	α1→3	Slime	WB
Plants						
Agarose	D-Gal	L-aGal[2]	β1 → 4	β1→3	Red algae (agar)	WB
Carrageenan	D-Gal	—	β1 → 3	α1→4	Red algae	WB
Cellulose	D-Glc	—	β1 → 4	—	Cell wall	SC
Xyloglucan	D-Glc	D-Xyl (D-Gal, L-Fuc)	β1 → 4	β1→6 (β1→2)	Cell wall (Hemicellulose)	SC SC
Arabinan	L-Ara	—	α1 → 5	α1→3	Cell wall (pectin)	SC
Amylose	D-Glc	—	α1 → 4	—	Amyloplasts	RC
Amylopectin	D-Glc	—	α1 → 4	α1→6	Amyloplasts	RC
Inulin	D-Fru	—	β2 → 1	—	Storage cells	RC
Animals						
Chitin	D-GlcNAc	—	β1 → 4	—	Insects, crabs	SK
Glycogen	D-Glc	—	α1 → 4	α1 → 6	Liver, muscle	RK
Hyaluronic acid	D-GlcUA	D-GlcNAc	β1 → 4 β1 → 3	—	Connective tissue	SK,WB

SC= structural carbohydrate, RC = reserve carbohydrate,
WB = water-binding carbohydrate; [1] *N*-acetylmuramic acid, [2] 3,6-anhydrogalactose

B. Important polysaccharides

Plant Polysaccharides

Among the polysaccharides, two glucose polymers of plant origin are of special importance. These are the $\beta1 \rightarrow 4$ linked polymer **cellulose** and the mostly $\alpha1 \rightarrow 4$ linked **starch**.

A. Cellulose ○

Cellulose, a linear homoglycan of $\beta1 \rightarrow 4$-linked glucose residues, is the *most abundant organic substance* in nature. Plant cell walls contain 40–50% cellulose, and cotton fibers are 98% cellulose. Cellulose molecules consist of more than 10^4 glucose residues (mass $1–2 \cdot 10^6$ Da) and can attain lengths of 6–8 µm. Naturally occurring cellulose has high *mechanical stability* and is highly *resistant* to chemical and enzymatic hydrolysis. These properties are due to the conformation of the molecules and their supramolecular organization. The unbranched $\beta1 \rightarrow 4$ linkage results in linear chains that are stabilized by hydrogen bonds within the chain and between neighboring chains (**1**). During their synthesis, 50–100 cellulose molecules associate to form an **elementary fibril** with a diameter of approximately 4 nm. About 20 such elementary fibrils then form a **microfibril**, which is readily visible with the electron microscope (**2**).

Cellulose microfibrils constitute the framework of the primary wall of young plant cells (**3**). There they form a complex network with other polysaccharides (**3**). The linking polysaccharides include **hemicellulose**, which is a mixture of predominantly neutral heteroglycans (xylans, xyloglucans, arabinogalactans, etc.). Hemicellulose associates with the cellulose fibrils via non-covalent interactions. These complexes are connected by neutral and acidic **pectins**, which typically contain galacturonic acid. Finally, a collagen-related protein, extensin, is also involved in the formation of primary walls.

B. Starch ◑

Starch, a reserve polysaccharide widely distributed in plants, is *the most important carbohydrate in the human diet.* In plants, starch is present in chloroplasts of leaves, as well as in fruits, seeds, and tubers. The starch content is especially high in cereal grains (up to 75% of the dry weight), potato tubers (approximately 65%), and in other plant storage organs. In such tissues, starch is found in the form of microscopically small granules in special organelles, *amyloplasts*. Starch granules are virtually insoluble in cold water, but swell dramatically when the water is heated. Approximately 15–25% of the starch goes into solution in colloidal form when the mixture is boiled for some time. This portion is called amylose ("soluble starch"). The remaining material, amylopectin, remains insoluble even after long periods of boiling.

Amylose is composed of *unbranched* $\alpha1 \rightarrow 4$-linked chains of 200–300 glucose residues. Due the α configuration at C-1, amylose molecules form *helices* with 6–8 residues per turn (**1**). The blue color of soluble starch upon addition of iodine indicates the presence of such helices. The iodine molecules line up inside the amylose helix and, in this largely nonaqueous environment, take on a blue color. In contrast, highly branched polysaccharides, such as amylopectin or glycogen turn brown or reddish-brown in the presence of iodine.

Unlike amylose, **amylopectin** is *branched*. On average, one in 20 to 25 glucose residues is the origin of a side chain that is attached via a $\alpha1 \rightarrow 6$ bond. This leads to an extended tree-like structure, which, like amylose, contains only *one* reducing end, i.e., one *free* anomeric OH-group. Amylopectin molecules can contain hundreds of thousands of glucose residues, with molecular masses exceeding 10^8 Da.

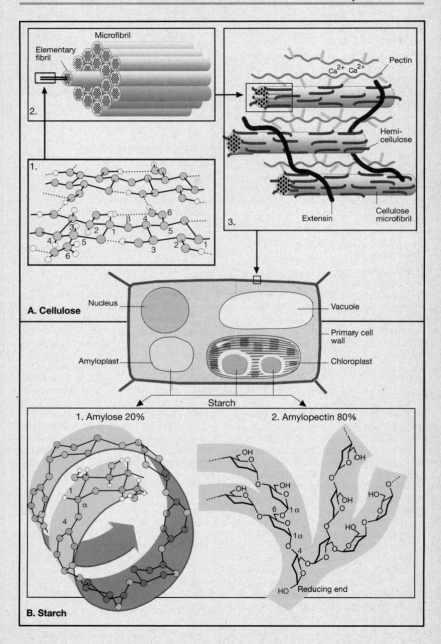

A. Cellulose

1.

2.

Elementary fibril

Microfibril

3.

Pectin

Ca^{2+} Ca^{2+}

Hemi-cellulose

Extensin

Cellulose microfibril

Nucleus

Vacuole

Primary cell wall

Amyloplast

Chloroplast

Starch

B. Starch

1. Amylose 20%

2. Amylopectin 80%

Reducing end

Glycoproteins and Glycosamino-glycans

Many proteins on the outer surface of the plasma membrane and most secreted proteins carry covalently linked oligosaccharide residues, which are added after translation in the endoplasmic reticulum and Golgi complex (see p. 210). In contrast, cytoplasmic proteins are rarely glycosylated. **Glycoproteins** may contain more than 50% carbohydrate, but, in general, the protein fraction predominates. The **glycosaminoglycans**, a group of acidic heteropolysaccharides, are the main constituents of proteoglycans and, as such, are important structural elements of the connective tissue matrix (see p. 314).

A. Glycoproteins: oligosaccharide of immunoglobulin G ○

As an example of the carbohydrate part of a glycoprotein, we show here the structure of the oligosaccharide chain of immunoglobulin G. This carbohydrate is *N-glycosidically* linked to the carboxamide group of an asparagine residue within the F_c part of the protein (see p. 268). Other glycoproteins contain *O-glycosidic* linkages between the carbohydrate and serine or threonine residues of the protein. This linkage (not shown) is less common than the *N-glycosidic* one.

The present oligosaccharide contains a T-shaped *core structure* consisting of two *N-acetylglucosamines* and three *mannose* residues (violet). Such core structures are found in all *N*-linked oligosaccharides. In addition, the structure contains two further *N*-acetylglucosamine residues, a *fucose* residue and a *galactose* residue. Glycoproteins exhibit many different types of branching. Here we not only have β1→ 4 , but also a β1→ 2, an α1→ 6, and an α1→ 3 linkage. During glycosylation in the endoplasmic reticulum, the apoprotein is first connected with an oligosaccharide consisting of the core structure described above and six further mannose residues and three terminal glucose residues (see p. 210). These additional residues are subsequently removed and, in part, replaced by other sugars. In this way, oligosaccharides of the present type ("**complex type**") are synthesized. Glycoproteins of the complex type often contain *N*-acetylneuraminic acid residues at the end of the oligosaccharide structure, which give it a negative charge. Another type of glycoprotein ("**mannose-rich type**") arises when only some of the additional sugars are removed from the primary product, and no new residues are added.

B. Glycosaminoglycans: hyaluronic acid ○

The glycosaminoglycans are acidic heteropolysaccharides made up from an *amino sugar* and either *glucuronic acid* or *iduronic acid*. Most polysaccharides belonging to this group are esterified with sulfuric acid to differing extents, which further enhances their acidic nature. Glycosaminoglycans are found in the extracellular matrix everywhere in the body, either in free form or as constituents of proteoglycans (see p. 314).

Hyaluronic acid, a simple, unesterified glycosaminoglycan, is assembled from disaccharide units in which **N-acetylglucosamine** and **glucuronic acid** are alternately β1 → 4 and β1→ 3- linked (**1**). Due to the β1→ 3 linkages, hyaluronic acid molecules, which may contain several thousand monosaccharide residues, adopt a helical conformation. The outward-facing hydrophilic carboxylate groups of the glucuronic acid residues bind Ca^{2+} ions (**2**). The strong hydration of these groups enables hyaluronic acid and other glycosaminoglycans to bind water up to 10,000 times their own volume to yield gels. Thus, the vitreous humor of the eye contains approximately 1% hyaluronic acid and 98% water.

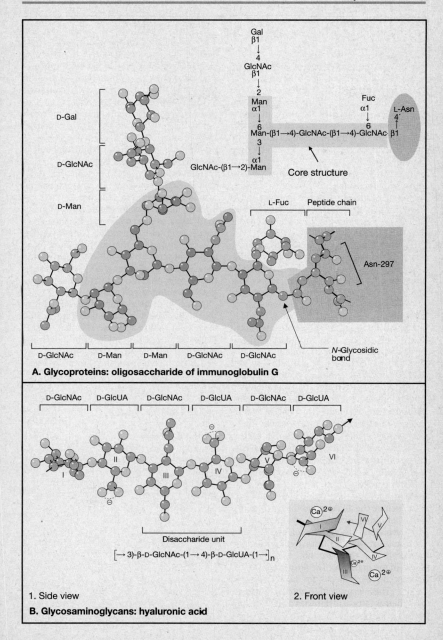

Gal
β1
↓
4
GlcNAc
β1
↓
2
Man
α1
↓
6
Man-(β1→4)-GlcNAc-(β1→4)-GlcNAc- β1
3
↑
α1
GlcNAc-(β1→2)-Man

Fuc
α1
↓
6

L-Asn
4'
↑

Core structure

D-Gal

D-GlcNAc

D-Man

L-Fuc   Peptide chain

Asn-297

D-GlcNAc   D-Man   D-Man   D-GlcNAc   D-GlcNAc

N-Glycosidic
bond

A. Glycoproteins: oligosaccharide of immunoglobulin G

D-GlcNAc   D-GlcUA   D-GlcNAc   D-GlcUA   D-GlcNAc   D-GlcUA

I   II   III   IV   V   VI

Disaccharide unit
$$[→ 3)-β-D-GlcNAc-(1→ 4)-β-D-GlcUA-(1→]_n$$

1. Side view

$Ca^{2⊕}$

I   VI   V
II   IV
III

$Ca^{2⊕}$ $Ca^{2⊕}$

2. Front view

B. Glycosaminoglycans: hyaluronic acid

Lipids: Overview

A. Classification ❍

The lipids are a large group of substances of biological origin which dissolve well in organic solvents, such as methanol, acetone, chloroform, and benzene (*definition of lipids*). On the other hand, they are either insoluble or only sparingly soluble in water. Their low water solubility is due to the small proportion of polarizing atoms (O, N, S, P), which they contain (see p. 6).

Lipids can be classified as either *hydrolyzable*, i.e., able to undergo hydrolytic cleavage, or *non-hydrolyzable*. Here, we mention only a few examples of the many lipids known. The individual classes of lipids will be discussed in more detail in the next few pages.

Hydrolyzable lipids (building blocks shown in parentheses). The *simple esters* include the fats (glycerol + three fatty acids), the waxes (fatty alcohol + fatty acid) and the sterol esters (sterol + fatty acid). The *phospholipids* are complex esters containing phosphate residues. This lipid family includes the phosphatidic acids (glycerol + two fatty acids + phosphate), the phosphatides (glycerol + two fatty acids + phosphate + alcohol) and the sphingolipids (sphingosine + fatty acid + phosphate + amino alcohol). Sphingosine instead of glycerol and a carbohydrate moiety are the common feature of the *glycolipids*. Cerebrosides (fatty acid + sphingosine + one sugar) and gangliosides (fatty acid + sphingosine + several different sugars, including neuraminic acid) are representatives of this group. The building blocks of the hydrolyzable lipids are linked to one another by ester bonds.

Non-hydrolyzable lipids. The *hydrocarbons* include the alkanes and the carotenoids. The *lipid alcohols* comprise alkanols and cyclic sterols, such as cholesterol, and steroids, such as estradiol and testosterone. Like the hydrocarbons, they are not easily broken down in the organism. The *fatty acids* are important building blocks for many different lipids. The *eicosanoids*, which can be classified as fatty acid derivatives, also belong to this group (see p. 354).

B. Biological roles ●

1. In the diet, lipids are an important *source of energy*. In quantitative terms, they are the principal *energy reserve* of the body. It is predominantly the fats stored as lipid droplets throughout the cells of the body, which serve as **metabolic fuel**. Their components are oxidized in the mitochondria to water and carbon dioxide. In the process, ATP is produced in large amounts (see p. 144).

2. Amphipathic lipids are **building blocks** for cellular membranes (see p. 202). Typical membrane lipids are phospholipids, glycolipids, and cholesterol. In contrast, neutral fats do not occur in membranes.

3. Lipids are excellent **insulators**. In mammals, they are found in the subcutaneous tissue, and also surround various organs where they serve as thermal insulators. As an important constituent of cell membranes, they are also involved in the electrical insulation of cells, and thus facilitate the formation of the membrane potential.

Some lipids have adopted special roles in the body. Fat-soluble vitamins (see p. 330) and essential fatty acids are essential constituents of the diet. Steroids, eicosanoids and retinoic acid function as signaling molecules, i.e., as hormones, mediators, or growth factors (see p. 340).

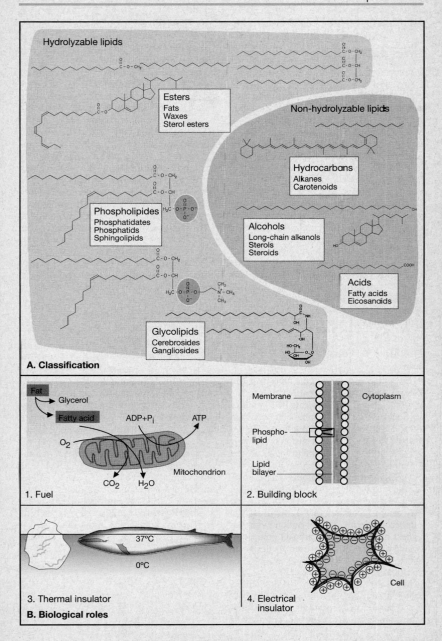

Hydrolyzable lipids

Esters
Fats
Waxes
Sterol esters

Non-hydrolyzable lipids

Hydrocarbons
Alkanes
Carotenoids

Phospholipides
Phosphatidates
Phosphatids
Sphingolipids

Alcohols
Long-chain alkanols
Sterols
Steroids

Acids
Fatty acids
Eicosanoids

Glycolipids
Cerebrosides
Gangliosides

A. Classification

Fat → Glycerol
→ Fatty acid

ADP+P_i → ATP

O_2

CO_2 H_2O Mitochondrion

1. Fuel

Membrane   Cytoplasm

Phospho-
lipid

Lipid
bilayer

2. Building block

37°C

0°C

3. Thermal insulator

Cell

4. Electrical
insulator

B. Biological roles

Fatty Acids

A. Fatty acids ●

Carboxylic acids with long hydrocarbon chains are called fatty acids. They are present in all organisms as building blocks of fats and membrane lipids. In these compounds, they are esterified with alcohols, e.g., with glycerol, sphingosine, or cholesterol. Small amounts also exist in unesterified form. The latter are referred to as *free fatty acids* (FFA).

The table lists most of the aliphatic monocarboxylic acids that are found in plants and animals. In higher plants and animals, unbranched, long-chain fatty acids with either 16 or 18 carbons are the most common. Almost without exception, naturally-occurring fatty acids have an even number of carbon atoms. This is a result of their biosynthesis from C_2 building blocks (see p. 144). Some fatty acids contain double bonds, and are therefore *unsaturated*. Both *cis* and *trans* isomers are possible, but it is usually only the *cis* form, which is found in natural lipids. Branched fatty acids only occur in bacteria.

B. Structure of oleate ◑

The full representation of a fatty acid (**1**) shows an elongated molecule, which, except for the carboxyl group, is entirely apolar. The molecule shown here is oleate, the anion of oleic acid, which has a double bond between C-9 and C-10. In order to simplify the illustration of fatty acids (**2**), it is customary to neglect the hydrogen atoms and to show only the carbon backbone.

The number of carbon atoms and the number and location of the double bonds are often given in parentheses after the name of the fatty acid, e.g., oleate (18:1;9). As usual, the numbering begins at the carbon with the highest oxidation state (carboxyl group = C-1), and Greek letters are used to designate each individual carbon (α = C-2; β = C-3; ω = last carbon).

In unsaturated fatty acids, the first double bond is usually located between C-9 and C-10. In polyunsaturated fatty acids, the individual double bonds are separated from one another by 3 carbons, i.e., the double bonds are *isolated*. Fatty acids with *conjugated* double bonds, which are separated by only two carbons, are very rare.

In saturated fatty acids, each single C-C bond in the hydrocarbon tail has complete freedom of rotation. As a result of this, these molecules are very flexible and tend to adopt a more or less linear conformation. In contrast, unsaturated fatty acids have kinks in the aliphatic chains, because of the *cis* configuration of their double bonds. Therefore, unsaturated fatty acids interfere with the regular arrangement of lipid molecules, leading to lower melting points and increased fluidity of membranes. For the same reason, fats containing a high amount of polyunsaturated fatty acids are liquid at room temperature (e.g., plant oils). Hydrogenation, i.e., saturation of the double bonds, causes them to solidify. In general, the melting points of fatty acids increase with increasing chain length, and decrease with increasing numbers of double bonds.

Further Information

In animals, polyunsaturated C_{20}-fatty acids are essential precursors for the biosynthesis of eicosanoids (**essential fatty acids,** see p. 354). This is, because animals have only a limited ability to introduce double bonds into fatty acids. Mammals are unable to form double bonds beyond C-9, and are therefore dependent on the presence of polyunsaturated fatty acids in their diet. The most important essential fatty acid is the C_{20} compound **arachidonic acid** (20:4;5,8,11,14). As the mammalian metabolism is able to extend the hydrocarbon chains of fatty acids, the two shorter C_{18} acids **linoleic acid** (18:2;9,12,15) and **linolenic acid** (18:3;9,12,15) can substitute for arachidonic acid in the diet.

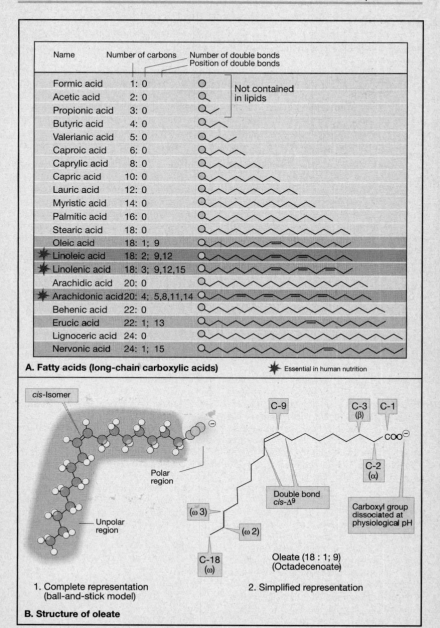

Name	Number of carbons	Number of double bonds / Position of double bonds	
Formic acid	1: 0		Not contained in lipids
Acetic acid	2: 0		
Propionic acid	3: 0		
Butyric acid	4: 0		
Valerianic acid	5: 0		
Caproic acid	6: 0		
Caprylic acid	8: 0		
Capric acid	10: 0		
Lauric acid	12: 0		
Myristic acid	14: 0		
Palmitic acid	16: 0		
Stearic acid	18: 0		
Oleic acid	18: 1; 9		
✳ Linoleic acid	18: 2; 9,12		
✳ Linolenic acid	18: 3; 9,12,15		
Arachidic acid	20: 0		
✳ Arachidonic acid	20: 4; 5,8,11,14		
Behenic acid	22: 0		
Erucic acid	22: 1; 13		
Lignoceric acid	24: 0		
Nervonic acid	24: 1; 15		

A. Fatty acids (long-chain carboxylic acids)   ✳ Essential in human nutrition

cis-Isomer

Polar region

Unpolar region

C-9   C-3 (β)   C-1

COO⁻

Double bond *cis*-Δ⁹

C-2 (α)

Carboxyl group dissociated at physiological pH

(ω 3)

(ω 2)

C-18 (ω)

Oleate (18 : 1; 9)
(Octadecenoate)

1. Complete representation (ball-and-stick model)

2. Simplified representation

B. Structure of oleate

Fats

Fats, phospholipids, and glycolipids are structurally very similar.

A. Structure of fats ●

Fats are esters of three **fatty acids** with the trivalent alcohol **glycerol**. When just one fatty acid is esterified with glycerol, the product is referred to as **monoacylglycerol** (fatty acid residue = acyl residue). Esterification with further fatty acids gives **diacylglycerols**, and ultimately **triacylglycerols** (triglycerides), the actual fat. Since fats are uncharged, they are also referred to as *neutral fats*. The carbons of glycerol in fats are not equivalent. They are distinguished by their "*sn*" number, where *sn* stands for "stereo-specific numbering."

The fatty acid residues of a fat molecule are inhomogeneous with respect to their chain length and the number of double bonds they contain. Fats extracted from biological materials are always *mixtures* of very similar compounds, which differ only in their fatty acid residues. Nutritional fats often contain palmitic, stearic, oleic and linoleic acids. The monounsaturated oleic acid is usually bound at position *sn*-C-2 of the glycerol.

Triacylglycerols are apolar and very flexible molecules. They can undergo rotation around the C-C single bonds, thereby adopting many different conformations. The van der Waals models show two possible conformations of tristearoylglycerol. The three apolar carbon chains are responsible for the lipophilic nature of the molecule (see p. 22), which makes it insoluble in water.

B. Biological functions ●

Nutritional fats are important *energy storage compounds*. In a balanced diet, they should supply 30–35% of the energy requirement of humans. However, this is not their only function. Fats also serve as *vehicle for fat-soluble vitamins* (see p. 330) and as the *source of the essential polyunsaturated fatty acids* linoleic, linolenic, and arachidonic acid (see p. 354).

Most of the fatty acids in animal fats are saturated. Plant fats (with the exception of coconut fat) usually contain a larger proportion unsaturated fatty acids. Therefore, they are often *oils* (= fluid fats). Plant fats can be converted from oils into solid fats by chemical hydrogenation of their double bonds. This is referred to as "hardening of oils." In this way, for example, margarine is made from plant oils.

Body fats are the most important **energy reserve** in animals. They are used as a source of carbon atoms for biosynthetic reactions. Most importantly, they supply acetyl-CoA. Fats are found in many cells in the form of small droplets. *Adipocytes* (fat cells) are specialized for fat storage. Their storage fats meet the energy requirements of a human being for two to three months. We have already mentioned the role of lipids as thermal and electrical **insulators**. They also serve as a mechanical **padding,** e.g., in the subcutaneous tissue and around organs.

Further information ◑

Fat hydrolysis: In the test tube ("*in vitro*"), fats can be broken down by alkaline hydrolysis ("*saponification*") to glycerol and the salts of the fatty acids ("soaps"). Soaps have been made in this way for centuries. Their amphipathic nature makes them well suited for washing. Soaps dissolve in water by forming micelles (see p. 22). By entering into fat droplets they emulsify fats, and thus facilitate their removal by water.

Within the organism ("*in vivo*"), fat cleavage is catalyzed by *lipases* of differing specificities. For example, the degradation of nutritional fats in the small intestine is catalyzed by pancreatic lipase (see p. 248), which cleaves preferentially at *sn*-C-1 and *sn*-C-3.

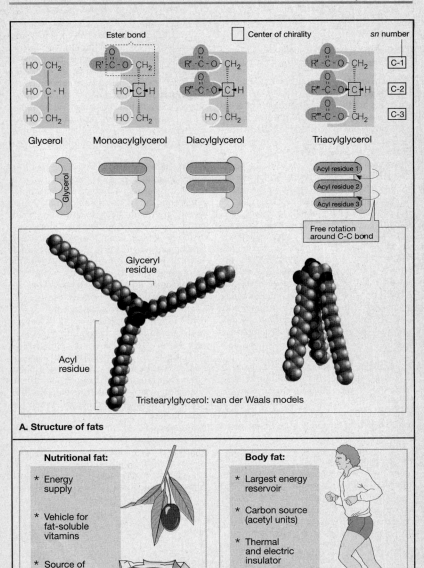

Ester bond   Center of chirality   *sn* number

HO - CH₂   Glycerol
HO - C - H
HO - CH₂

Monoacylglycerol

Diacylglycerol

Triacylglycerol — C-1, C-2, C-3

Acyl residue 1
Acyl residue 2
Acyl residue 3

Free rotation around C-C bond

Glyceryl residue

Acyl residue

Tristearylglycerol: van der Waals models

A. Structure of fats

Nutritional fat:

* Energy supply
* Vehicle for fat-soluble vitamins
* Source of essential fatty acids

Body fat:

* Largest energy reservoir
* Carbon source (acetyl units)
* Thermal and electric insulator
* Padding

B. Biological functions

Phospholipids and Glycolipids

A. Structure of fats, phospholipids and glycolipids ◑

As discussed on p. 46, the neutral fats are esters of glycerol with three fatty acids. **Phospholipids** are the main constituents of cellular membranes (see p. 202–5). Their common feature is a phosphate residue, which is esterified with the hydroxyl group at *sn*-C-3 of glycerol. The presence of this residue confers at least one negative charge to phospholipids. **Phosphatidic acids** (phosphatidates), the simplest phospholipids, are phosphate esters of diacylglycerol. Phosphatidates can be formed from phospholipids by phospholipases present in bee and snake venoms. In addition, they are important intermediates in the biosynthesis of fats and phospholipids (see p. 168). All other phospholipids are derived from phosphatidic acids by esterification of the phosphate group with the hydroxyl group of an amino alcohol (*choline, ethanolamine,* or *serine*) or a sugar alcohol (*myo-inositol*). The illustration shows phosphatidylcholine as a typical example of a phospholipid.

The result of linking two phosphatidyl residues with glycerol is **cardiolipin**, a lipid found only in the inner mitochondrial membrane (not shown). **Lysophospholipids** arise from phospholipids by cleavage of an acyl residue. They do not normally occur in mammalian metabolism.

Phosphatidylcholine (lecithin) is the most abundant phospholipid in membranes. **Phosphatidylethanolamine** (cephalin) has an ethanol-amine residue instead of choline, and **phosphatidylserine** a serine residue. **Phosphatidylinositol** contains the sugar-like alcohol *myo*-inositol. A more highly phosphorylated derivative of this phospholipid, phosphoinositol 4,5-bisphosphate, is a special constituent of membranes, which, by enzymatic cleavage, can give rise to the two *second messengers*, diacylglycerol (DAG) and inositol 1,4,5-trisphosphate (InsP$_3$, see p. 352).

In addition to the negative charge at the phosphate residue, some phospholipids also carry further charges. In phosphatidylcholine and phosphatidylethanolamine, the nitrogen atom of the amino alcohol is positively charged, making these two phosphatides electrically neutral. In contrast, phosphatidylserine (one additional positive and one additional negative charge) and phosphatidylinositol (no additional charge) have a negative net charge, because of the phosphate residue.

Sphingophospholipids, which are found in large quantities in the brain and nervous tissue, are somewhat different from the other phospholipids. In these compounds, *sphingosine*, an amino alcohol with a long side chain, replaces glycerol and one of the acyl residues. Amide bond formation between sphingosine and a fatty acid yields **ceramide** (**3**), the precursor of the sphingolipids. In **sphingomyelin** (**2**), the most important sphingolipid, the ceramide part of the molecule carries an additional phosphate residue with a choline group attached to it.

Glycolipids are present in all tissues on the outer surface of the plasma membrane (see p. 204). These lipids are composed of sphingosine, a fatty acid and a sugar or oligosaccharide residue, while the phosphate group typical of phospholipids is absent. Simple representatives include **galactosylceramides** and **glucosylceramides**. When the sugar moiety of glycolipids is esterified with sulfuric acid, they are referred to as **sulfatides**. **Gangliosides** are the most complex glycolipids. They constitute a large family of membrane lipids with receptor function. A characteristic component of the gangliosides is *N*-acetylneuraminic acid (sialic acid).

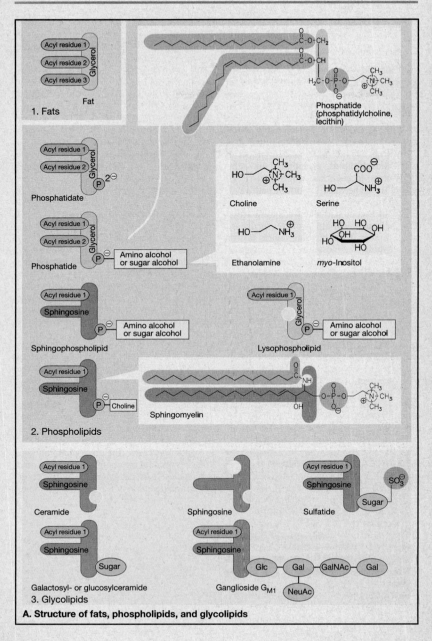

1. Fats

Acyl residue 1 / Acyl residue 2 / Acyl residue 3 — Glycerol
Fat

C-O-CH₂
C-O-CH
H₂C-O-P-O-CH₂...CH₃ N⁺ CH₃ CH₃

Phosphatide
(phosphatidylcholine, lecithin)

2. Phospholipids

Acyl residue 1 / Acyl residue 2 — Glycerol — P 2⊖
Phosphatidate

Acyl residue 1 / Acyl residue 2 — Glycerol — P ⊖ — Amino alcohol or sugar alcohol
Phosphatide

CH₃
HO N⁺ CH₃
CH₃
Choline

COO⊖
HO NH₃⁺
Serine

HO NH₃⁺
Ethanolamine

HO HO OH HO HO
myo-Inositol

Acyl residue 1 — Sphingosine — P ⊖ — Amino alcohol or sugar alcohol
Sphingophospholipid

Acyl residue 1 — Glycerol — P ⊖ — Amino alcohol or sugar alcohol
Lysophospholipid

Acyl residue 1 — Sphingosine — P — Choline
Sphingomyelin

C=O NH
O P O
CH₃ N⁺ CH₃ CH₃
OH ⊖

3. Glycolipids

Acyl residue 1 — Sphingosine
Ceramide

Sphingosine

Acyl residue 1 — Sphingosine — Sugar — SO₃⊖
Sulfatide

Acyl residue 1 — Sphingosine — Sugar
Galactosyl- or glucosylceramide

Acyl residue 1 — Sphingosine — Glc — Gal — GalNAc — Gal
NeuAc
Ganglioside G_{M1}

A. Structure of fats, phospholipids, and glycolipids

Isoprenoids

A. Acetyl-CoA as building block for lipids ●

Animals and plants contain a wide range of different lipids, which, despite their great diversity, are biogenetically related, i.e., they are all derived from **acetyl-CoA,** the "*activated acetic acid.*"

1. One major pathway leads from acetyl-CoA to fatty acids (see p. 144). Their CoA-derivatives are the basic building blocks for **fats**, **phospholipids**, **glycolipids,** and other derivatives. In quantitative terms, fatty acid synthesis is the most important pathway in animals and most plants.

2. The second pathway leads from acetyl-CoA to isopentenyl diphosphate ("*active isoprene*"), the basic building block for the **isoprenoids**. The biosynthesis of isopentenyl diphosphate is discussed in conjunction with cholesterol biosynthesis (see p. 170).

B. Isoprenoids ○

Formally, all isoprenoids are derived from **isoprene** (2-methyl-1,3-butadiene), a branched unsaturated alkane with five carbons. Plants and animals use **isopentenyl diphosphate** for the synthesis of linear and cyclic oligomers and polymers. Only a small subset of the large number of known isoprenoids is shown here. Each is listed along with the number of isoprene units, which it contains ($I = n$).

From activated isoprene, the main pathway leads, via dimerization, to activated **geraniol** ($I = 2$) and activated **farnesol** ($I = 3$). At this point, it divides in two. One branch involves the extension of farnesol, leading to chains with increasing numbers of isoprene units, e.g., **phytol** ($I = 4$), **dolichol** ($I = 14–24$) up to **rubber** ($I = 700–5000$). The other pathway involves a "head-to-head" linkage between two farnesol residues (see p. 170), giving rise to **squalene** ($I = 6$), which, in turn, is converted to **cholesterol** ($I = 6$) and the **steroids**.

The ability to synthesize particular isoprenoids is often limited to a few plant and animal species. For example, only the rubber tree (*Hevea brasiliensis*) and several related species can synthesize rubber. Some of the isoprenoids have essential roles in metabolism, but cannot be synthesized by animals. This group includes *vitamins* A, D, E, and K. Vitamin D, because of its structure and function, is now usually classified as a steroid hormone (see p. 304).

Isoprene metabolism in plants is very complex. Plants can synthesize a variety of fragrant substances and volatile oils from isoprenoids. **Menthol** ($I = 2$), **campher** ($I = 2$) and **citronellol** ($I = 2$) are shown as examples. These C_{10} compounds are also called *terpenes*. In an analogous way, compounds consisting of three isoprene units ($I = 3$) are classified as *sesquiterpenes,* and the steroids ($I = 6$) as *triterpenes*.

A very important group of isoprenoids comprises compounds with hormone and signaling functions e.g., **steroid hormones** ($I = 6$) and **retinoic acid** ($I = 3$) in vertebrates, and **juvenile hormone** ($I = 3$) in arthropods. Some plant hormones also have the isoprenoid structure, e.g., the *cytokinins, abscisic acid,* and *brassinosteroids acid*, to name but a few.

Polyisoprene chains sometimes serve as "lipid anchors" to fix molecules in membranes. For instance, proteins can be bound to membranes by isoprenylation (see p. 214). Coenzymes with isoprenoid anchors of various lengths include *ubiquinone* (*coenzyme Q*; $I = 6–10$), *plastoquinone* ($I = 9$) and *menaquinone* (vitamin K; $I = 4–6$). Chlorophyll bears a phytyl residue ($I = 4$).

In some cases, the isoprenoid building block is used for the chemical modification of molecules. N^6-*isopentenyl-AMP*, which occurs as a modified base in tRNA, is one example.

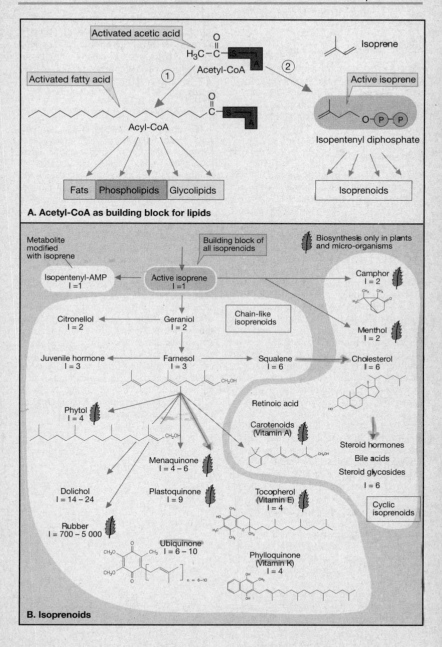

A. Acetyl-CoA as building block for lipids

B. Isoprenoids

Steroid Structure

A. Steroid building blocks ◑

Steroids are characterized by a core structure known as **gonane,** which consists of four saturated fused rings. Many steroids, such as **cholestane**, the basic building block of the sterols, carry a side chain attached to one end of the steroid core.

B. 3D structure ○

The four different rings of the steroids are designated A, B, C, and D. They are not flat, but puckered, due to the tetrahedrally arranged C-C bonds. Three major **ring conformations,** "chair," "boat," and "twisted" (not shown) are possible. The **chair** and **boat** conformations are the most common. Five-membered rings frequently adopt a conformation referred to as an **"envelope"**. While individual aliphatic rings are easily converted from one conformation to another at ambient temperature, this is not possible with steroids. Substituents of the steroid core lie either approximately in the same plane as the ring (e = **equatorial**) or perpendicular to it (a = **axial**). In three-dimensional representations, substituents pointing toward the observer are indicated by an unbroken line (β position), while bonds pointing into the plane of the page are indicated by a broken line (α position). The so-called "angular" methyl groups at C-10 and C-13 of the steroids always adopt the β position. Neighboring rings can lie in the same plane (*trans*; **2**) or at an angle to one another (*cis*; **1**). This depends on the positions of the substituents of the shared ring carbons, which can be arranged either *cis* or *trans* to the angular methyl group at C-10. The substituents at the points of intersection of the individual rings are usually in *trans* position. As a whole, the core of most steroids is more or less planar, and looks like a flat disk. The only exceptions to this are the ecdysteroids, the bile acids (A:B *cis*), the cardiac glycosides, and certain toad toxins.

A more realistic impression of the three-dimensional structure of the steroids is given by the ball-and-stick-model of **cholesterol**

(3). The four rings form a rigid scaffolding to which the mobile side chain is attached. Steroids are relatively apolar. Polar groups, e.g., hydroxyl groups, give them amphipathic properties. This is especially pronounced with the bile acids (see p. 288).

C. Thin-layer chromatography ◑

Thin-layer chromatography (TLC) is an efficient, predominantly analytical, technique for the rapid separation of lipids and other small molecules (amino acids, nucleotides, vitamins, etc). The sample to be analyzed is applied to a plate made of glass, aluminum, or plastic, which is covered with a thin layer of silica gel **(1)**. The plate is then placed in a chromatography tank, that contains a solvent at the bottom. The solvent moves up the plate drawn by capillary forces **(2)**. The substances in the sample move with the solvent. The speed at which they move is determined by their distribution between the *stationary phase*, i.e., the hydrophilic silica, and the *mobile phase*, i.e., the hydrophobic solvent. The chromatography is stopped shortly before the solvent has reached the top of the plate. Once the solvent has evaporated, the substances can be made visible using appropriate physical procedures (e.g., ultraviolet light) or chemical reactions **(3)**. The migration of a particular substance is described by its R_f value. Unknown compounds can be identified by comparison with reference substances.

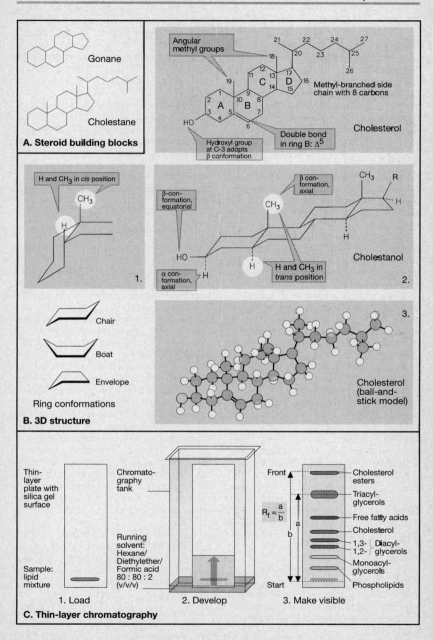

A. Steroid building blocks

Gonane

Cholestane

Angular methyl groups

Methyl-branched side chain with 8 carbons

Cholesterol

Hydroxyl group at C-3 adopts β conformation

Double bond in ring B: Δ^5

B. 3D structure

H and CH$_3$ in *cis* position

β-conformation, equatorial

β conformation, axial

α conformation, axial

H and CH$_3$ in *trans* position

Cholestanol

1.   2.

Chair

Boat

Envelope

Ring conformations

Cholesterol (ball-and-stick model)

3.

C. Thin-layer chromatography

Thin-layer plate with silica gel surface

Chromatography tank

Running solvent: Hexane/ Diethylether/ Formic acid 80 : 80 : 2 (v/v/v)

Sample: lipid mixture

$R_f = \dfrac{a}{b}$

Front

Start

Cholesterol esters

Triacyl-glycerols

Free fatty acids

Cholesterol

1,3- 1,2- Diacyl-glycerols

Monoacyl-glycerols

Phospholipids

1. Load   2. Develop   3. Make visible

Classes of Steroids

The three most important families of steroids are the **sterols**, the **bile acids**, and the **steroid hormones**. Further compounds with steroid structure, noteworthy for their pharmacological effects, are especially abundant in plants. These are the steroid alkaloids, the digitalis glycosides ("cardiac glycosides,") and the saponines.

A. Sterols ◗

Sterols are *steroid alcohols*. They bear a hydroxyl group in β position at C-3 of the A ring and one or more double bonds in ring B and in the side chain; further oxygen atoms, such as in carbonyl and carboxyl groups, are absent. The most important sterol in animals is **cholesterol**. In plants and micro-organisms, cholesterol is replaced by a range of closely related sterols, e.g., **ergosterol**, β-**sitosterol**, and **stigmasterol**.

Cholesterol is present in all animal tissues, being particularly abundant in nervous tissue. It is a major constituent of cellular *membranes*, where it contributes to the fluidity of the membrane (see p. 204). The storage and transport forms of cholesterol are its *esters* with *fatty acids*. In plasma *lipoproteins*, cholesterol and cholesterol esters are associated with other lipids (see p. 254). Cholesterol is a normal constituent of the *bile* and is, therefore, found in many *gallstones*. Its biosynthesis, metabolism and transport are discussed elsewhere (see pp. 170 and 284).

Cholesterol significantly contributes to the development of *arteriosclerosis,* where alterations in the walls of the arteries, e.g., calcification, may occur as a result of elevated plasma cholesterol levels. A diet rich in foods of plant origin is low in cholesterol, whereas foods derived from animals, especially egg yolk, red meat, liver, and brain, contain much higher amounts.

B. Bile acids ◗

Bile acids are synthesized from cholesterol in the liver (see p. 288). Their structures are therefore derived from that of cholesterol. Characteristic for the bile acids is a side-chain shortened by three carbons in which the last carbon atom is oxidized to a carboxyl group. The double bond in ring B is reduced (ring A:B, *cis*!, see p. 52). Up to three hydroxyl groups (in α position) are found in the steroid core at positions 3, 7, and 12.

Bile acids increase the solubility of cholesterol and promote the digestion of lipids in the intestine (see p. 246). **Cholic acid** and **chenodeoxycholic acid** are *primary bile acids*, that are formed in the liver. Their dehydroxylation at C-7 by micro-organisms from the intestinal flora gives rise to the *secondary bile acids* **lithocholic acid** and **deoxycholic acid.**

C. Steroid hormones ◗

The conversion of cholesterol to steroid hormones is of minor importance quantitatively, but of major importance in a physiological sense. The steroid hormones are a group of lipophilic signal molecules, which regulate metabolism, growth and reproduction (see p. 346). The steroid hormones of vertebrate animals are **progesterone, cortisol, aldosterone, testosterone, estradiol,** and **calcitriol**. With the exception of calcitriol, they have only a short side chain, consisting of two carbons or none at all. Characteristic for most of them is an oxo group at C-3 conjugated with a double bond between C-4 and C-5 of ring A. Differences occur in rings C and D. Estradiol is aromatic in ring A, and its hydroxyl group at C-3 is, therefore, phenolic in character. Calcitriol differs from the other steroid hormones of vertebrate animals, but still contains the complete carbon backbone of cholesterol. Due to the light-dependent opening of ring B during its biosynthesis, it is a so-called "secosteroid" (steroid with an open ring).

Ecdysone is the steroid hormone of the arthropods. It represents an early form of the steroid hormones found in higher organisms, such as man. Steroid hormones with signaling functions also occur in plants.

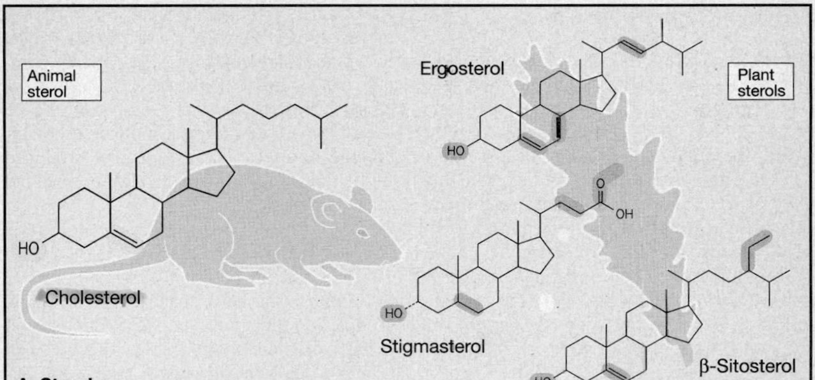
Animal sterol
Cholesterol
HO
Ergosterol
HO
Plant sterols
Stigmasterol
HO
OH
O
β-Sitosterol
HO

A. Sterols

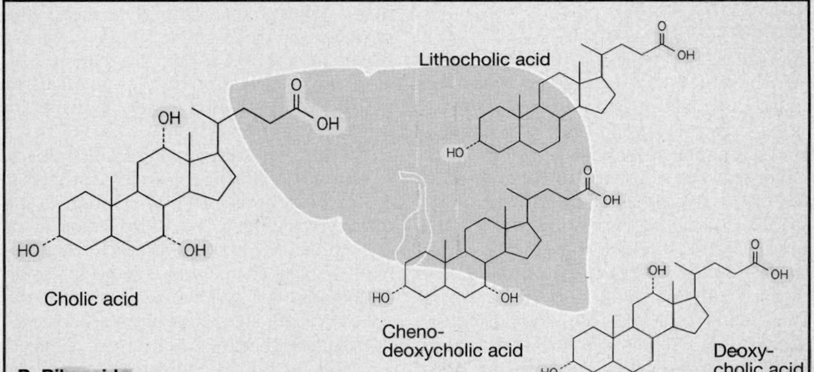
Cholic acid
OH
HO
OH
O
OH
Lithocholic acid
HO
Cheno-
deoxycholic acid
Deoxy-
cholic acid

B. Bile acids

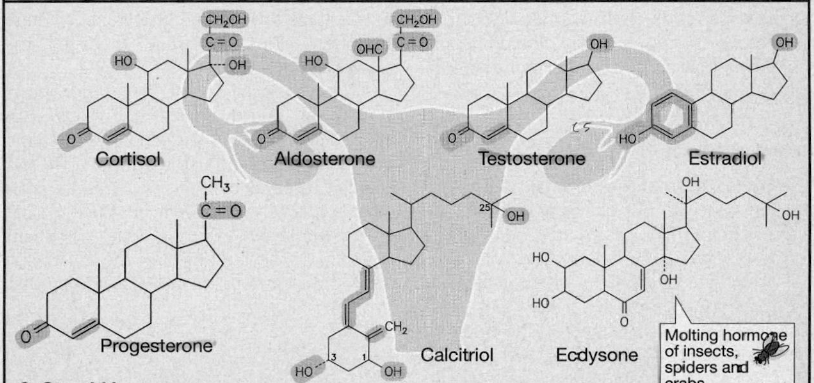
Cortisol
CH_2OH
C=O
HO
OH
O
Aldosterone
OHC
Testosterone
OH
Estradiol
Progesterone
CH_3
C=O
Calcitriol
25
OH
CH_2
3
1
Ecdysone
Molting hormone of insects, spiders and crabs

C. Steroid hormones

Amino Acids: Properties

The **amino acids** (*2-aminocarboxylic acids*) are the building blocks of peptides and proteins. Only 20 of the naturally occurring amino acids (the **proteinogenic amino acids**) are found in all proteins, because they are the only ones that are contained in the genetic code (see p. 226). There are many other amino acids that are derived from enzymatic conversions of proteinogenic amino acids or formed by *posttranslational modification* within a polypeptide (see p. 210). On the following pages, we provide an overview of the proteinogenic amino acids.

A. Stereochemistry ◑

In all amino acids except glycine, carbon C-2 (also called α carbon, or C_α) carries four different substituents, a *carboxylate group*, an *amino group*, a *hydrogen*, and a *side chain* (referred to as the R group here), which differs from one amino acid to the next. Thus the α-carbon constitutes a **center of chirality**, i.e., a site where two different configurations are possible, which are mirror images of one another. The interconversion of such mirror-image isomers, or **enantiomers**, requires a chemical reaction. Almost all of the amino acids found in nature are **L enantiomers** (left). In the *Fischer projection formula* (see p. 30), the carboxylate group of the L form is shown at the top, the amino group at the left, the hydrogen at the right and the side chain at the bottom. Two amino acids, threonine and isoleucine, contain a second center of chirality at C-3 (see p. 58).

B. Optical activity ○

The chemical properties of enantiomers are almost identical, but they differ from one another in their *optical activity*. The L- and D-forms of enantiomers rotate the plane of polarized light to the same extent, but in *opposite directions* (see p. 32; here L- and D-histidine are used as an example). The chemical synthesis of amino acids is not usually stereospecific, and therefore yields **racemic mixtures**, i.e., mixtures containing both enantiomers in equal amounts. Since the rotations caused by the L- and D-forms cancel one another, racemic forms are optically inactive. Only L-amino acids are active in the organism, and this is why amino acids to be used for parenteral application or other biomedical applications are usually produced with the help of genetically engineered micro-organisms.

C. Dissociation curve of histidine ◑

All amino acids have at least two ionizable groups, and thus their net charge depends on the pH of their environment. The carboxyl groups at C_α have pK_a values (see p. 24) between 1.8 and 2.8, and thus are more acidic than simple monocarboxylic acids. The pK_a values of the α-amino groups vary between 8.8 and 10.6. Acidic and basic amino acids also carry ionizable groups in their side chain. The pK_a values of these side chains are listed on page 59.

Let us consider the amino acid **histidine** as an example of the pH dependence of the net charge of an amino acid. In addition to the carboxyl group and the amino group at C_α (p K_a values 1.8 and 9.2, respectively), histidine also has an imidazole residue in its side chain with a pK_a value of 6.0. Therefore, the net charge (the sum of the positive and negative charges) changes from +2 to -1 as the pH increases. At pH 7.6, the net charge is zero, even though the molecule contains two almost completely ionized groups under these conditions. This pH value is called the **isoelectric point**.

At its isoelectric point, an amino acid is said to be **zwitterionic**, i.e., it has both anionic *and* cationic properties. Most amino acids are zwitterionic at neutral pH. The net charge carried by peptides and proteins is almost exclusively determined by the ionizable groups in the side chains of their constituent amino acids, because most of the α-carboxyl and α-amino groups are linked to one another by peptide bonds.

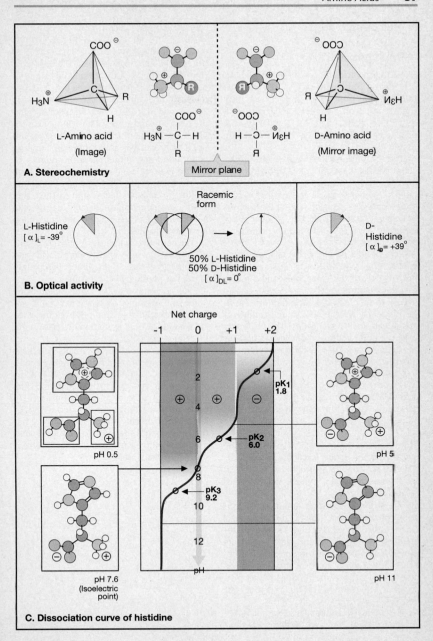

COO⊖

H₃N⊕ —C— R

H

L-Amino acid
(Image)

COO⊖
H₃N⊕ —C— H
R

Mirror plane

⊖OOC
H —C— ИεⁿH⊕
Я

⊖OOC

Я —C— ⊕ИεⁿH

H

D-Amino acid
(Mirror image)

A. Stereochemistry

L-Histidine
$[\alpha]_L = -39°$

Racemic
form

50% L-Histidine
50% D-Histidine
$[\alpha]_{DL} = 0°$

D-
Histidine
$[\alpha]_D = +39°$

B. Optical activity

Net charge

-1   0   +1   +2

pH 0.5

pH 7.6
(Isoelectric
point)

⊕   ⊕   ⊖

2

4

6

8

10

12

pH

pK₁
1.8

pK₂
6.0

pK₃
9.2

pH 5

pH 11

C. Dissociation curve of histidine

Proteinogenic Amino Acids

A. The proteinogenic amino acids ◐

The 20 amino acids that are incorporated into proteins by *translation* (see p. 228–231) are referred to as proteinogenic. Their classification is based on the chemical **structure** of their side chains and their **polarity** (see p. 22), which defines their behavior in proteins.

The illustration shown here takes into account both of these criteria. Starting from the outside and working in, the square block contains a) the *abbreviation* of the name of each amino acid, consisting of the first three letters of its name; b) the *single letter symbol*, introduced to save space during the electronic processing of sequence data; c) a *value for the polarity* of the side-chain (see below); and d) *membership of the structural classes* I-VII. The polarity of each side chain is also indicated by the color of the sector. This increases from yellow to green to blue. In addition, the illustration shows the *structural formulae* of the side chains (the R groups), the *full names* of the amino acids, and the pK_a *values* for those side chains that are ionizable. Here, the polarity of the side chain is expressed as the free energy change, ΔG, (in kJ · mol^{-1}, see p. 12) for the transfer of the respective amino acid side chain from ethanol, a rather apolar solvent, to water. The more negative the ΔG value, the more *polar* the side chain.

The **aliphatic** amino acids (class I) include *glycine, alanine, valine, leucine,* and *isoleucine.* They do not contain heteroatoms (N, O, or S) in their side chain, which makes them strongly apolar.

The **sulfur-containing** amino acids, *cysteine* and *methionine* (class II), are also apolar (in the case of Cys this applies to the undissociated state only). Cysteine plays an important role in the stabilization of proteins, because of its ability to form disulfide bonds (see p. 8). Two cysteine residues linked by a disulfide bridge are referred to as *cystine* (not shown).

The **aromatic** amino acids (class III) have resonance-stabilized ring systems in their side chain. *Phenylalanine* is the only member of this group, that is strongly apolar. *Tyrosine* and *tryptophan* are intermediate, and histidine is, in fact, strongly polar. The imidazole ring of *histidine* is easily protonated at neutral or weakly acidic pH values (see p. 56). Therefore, histidine can also be classified as a basic amino acid. Tyrosine with its phenolic hydroxyl group and tryptophan strongly absorb light at wavelengths between 250 and 300 nm. This property is used for the photometric determination of protein concentrations.

The **neutral** amino acids (class IV) have hydroxyl groups (*serine, threonine*) or amide groups (*asparagine, glutamine*) in their side chains. Despite their non-ionic nature, the amide groups of asparagine and glutamine are markedly polar.

The carboxylate groups in the side chains of the **acidic** amino acids *aspartic acid* and *glutamic acid* (class V) are almost completely ionized at physiological pH values.

The side chains of the **basic** amino acids *lysine* and *arginine*, on the other hand, are fully protonated at neutral pH. Arginine, with its guanido group, is extremely basic, and therefore very polar.

Proline (VII) is a special case. Its side chain forms a five-membered ring including C_α and the α-amino group. Therefore, proline is actually an imino acid rather than an amino acid. The nitrogen atom of proline is only weakly basic, and therefore not protonated at physiological pH. The ring structure of proline has the consequence that, within peptide chains, proline residues cause bends that interrupt secondary structures.

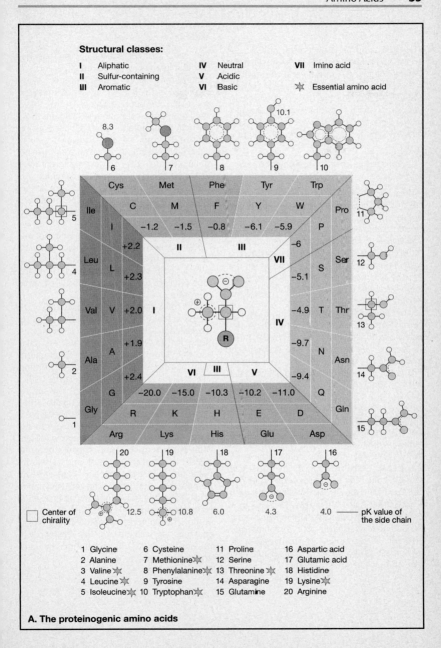

Structural classes:

I Aliphatic	**IV** Neutral	**VII** Imino acid
II Sulfur-containing	**V** Acidic	
III Aromatic	**VI** Basic	Essential amino acid

Cys Met Phe Tyr Trp

C M F Y W Pro
I −1.2 −1.5 −0.8 −6.1 −5.9 P
Ile +2.2 −6
II **III** −5.1
Leu L +2.3 **VII** S Ser
Val V +2.0 **I** −4.9 T Thr
Ala A +1.9 **IV** −9.7
G +2.4 −9.4 N Asn
VI **III** **V**
Gly −20.0 −15.0 −10.3 −10.2 −11.0 Q Gln

R K H E D
Arg Lys His Glu Asp

Center of chirality

12.5 10.8 6.0 4.3 4.0 —— pK value of the side chain

1 Glycine	6 Cysteine	11 Proline	16 Aspartic acid
2 Alanine	7 Methionine	12 Serine	17 Glutamic acid
3 Valine	8 Phenylalanine	13 Threonine	18 Histidine
4 Leucine	9 Tyrosine	14 Asparagine	19 Lysine
5 Isoleucine	10 Tryptophan	15 Glutamine	20 Arginine

A. The proteinogenic amino acids

Amino Acid Analysis

The separation and analysis of amino acids and amino acid derivatives is employed in the determination of the amino acid composition of proteins, in the analysis of peptide sequences, and for the diagnosis of disturbances in amino acid and protein metabolism. Two important procedures for the analysis of amino acids are presented here.

A. Ion-exchange chromatography of free amino acids ○

Ion-exchange chromatography is based on *electrostatic interactions* between ions of opposite charge, one type of which are covalently bound to a solid support matrix. The matrix with its covalently bound ions is referred to as an **ion-exchange resin**. The resin binds ions of opposite charge, which can then be selectively removed (*eluted*) by treatment with solutions of higher ionic strength or different pH.

   Amino acid analysis by ion-exchange chromatography utilizes synthetic resins, bearing sulfonate ($-SO_3^-$) groups (red), as the *stationary phase*. These are completely ionized, i.e., negatively charged, over almost the entire pH range. Prior to the analysis, the resin is poured into a column, and then washed with a buffer containing Na^+. The sulfonate residues bind Na^+ ions (blue). When an amino acid solution adjusted to pH 2 is loaded onto the column (**1a**), the amino acids (green), which are positively charged at this pH, displace Na^+ ions, and bind to the resin themselves. Since amino acids are uncharged at their isoelectric point (see p. 56), they can be eluted from the column by washing it with a buffer that has a higher pH (**1b**). This procedure is shown schematically in Fig. **3** using the amino acids aspartic acid, threonine, and histidine as examples. An examination of their dissociation curves (**2**) explains why they elute in the order they do.

B. Partition chromatography of PTC-amino acids ○

The separation of substances by partition chromatography is based on differences in polarity rather than charge differences. When a column is filled with an apolar *stationary phase* and a mixture of apolar substances is loaded onto the column, they will bind to the resin by hydrophobic interactions. Washing with a mixture of polar solvents (*mobile phase*) will result in migration of the substances through the column at different rates, depending on their polarity. Hydrophilic substances, which form only weak interactions with the stationary phase, will be eluted first and the more hydrophobic later.

The original version of partition chromatography was based on a hydrophilic stationary phase and a lipophilic mobile phase (see, e. g., p. 53). The version described here is, therefore, referred to as **reversed-phase chromatography**.

   The first step is the *derivatization* (**1**) of the amino acids in the mixture with phenylisothiocyanate. The PTC-amino acids formed have two advantages — they are less polar, and they can be easily detected photometrically, because of their ultraviolet absorption. The stationary phase consists of small silica gel particles (diameter 5–10 μm), the surface of which is covered with long hydrocarbon chains. Finer particles improve the separation, but they also make it necessary to use high pressure on the columns. Thus, **high-performance liquid chromatography (HPLC, 2)** employs columns and capillaries made of stainless steel and high-pressure pumps. For the separation of PTC-amino acids, buffers with a continuously increasing content of acetonitrile (CH_3CN) are used for elution. The composition of the mobile phase, and thus the resolution of the substances to be separated (**3**), is controlled by programmable gradient mixers.

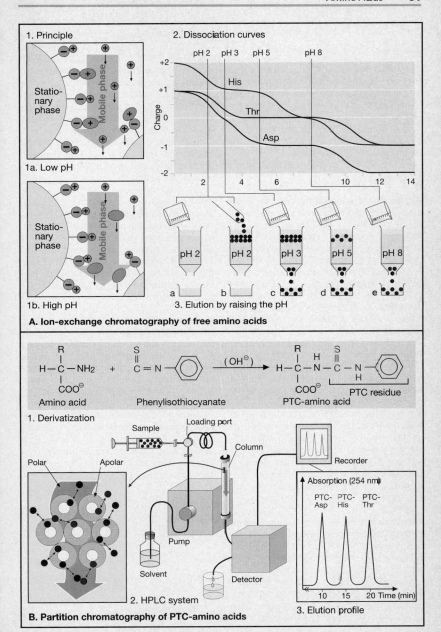

A. Ion-exchange chromatography of free amino acids

1. Principle

Stationary phase

Mobile phase

1a. Low pH

Stationary phase

Mobile phase

1b. High pH

2. Dissociation curves

pH 2 pH 3 pH 5 pH 8

Charge

+2
+1
0
-1
-2

His

Thr

Asp

2 4 6 10 12 14

3. Elution by raising the pH

pH 2   pH 2   pH 3   pH 5   pH 8

a   b   c   d   e

B. Partition chromatography of PTC-amino acids

$$H-\overset{\displaystyle R}{\underset{\displaystyle COO^{\ominus}}{C}}-NH_2 \quad + \quad \overset{\displaystyle S}{\underset{}{C}}=N-\bigcirc \quad \xrightarrow{(OH^{\ominus})} \quad H-\overset{\displaystyle R}{\underset{\displaystyle COO^{\ominus}}{C}}-\overset{H}{\underset{}{N}}-\overset{\displaystyle S}{\underset{H}{C}}-N-\bigcirc$$

Amino acid   Phenylisothiocyanate   PTC-amino acid   PTC residue

1. Derivatization

Sample   Loading port   Column   Recorder

Polar   Apolar

Pump

Solvent   Detector

2. HPLC system

Absorption (254 nm)

PTC-Asp PTC-His PTC-Thr

10   15   20 Time (min)

3. Elution profile

Functions of Proteins

A. Proteins ◗

Every organism contains thousands of different proteins with a variety functions. In order to give an impression of their variety, highly simplified illustrations of the structures of a few intra- and extracellular proteins are shown here at a magnification of about 1.5 million. Most of the molecules depicted are discussed in detail elsewhere.

The functions of proteins can be classified as follows:

Establishment and maintenance of structure. The structural proteins are responsible for the *mechanical stability* of organs and tissues. A small part of a **collagen** molecule is shown as an example (right). The complete molecule has dimensions of 1.5 ×300 nm, and would span three pages at the magnification used here. **Histones** (upper left) also belong to the structural proteins. They are components of chromatin, where they organize the arrangement of DNA in the nucleosomes (see p. 218).

Transport. A well-known transport protein **hemoglobin,** which is found in the erythrocytes (lower left) and is essential for the transport of oxygen and carbon dioxide between the lungs and other tissues (see p. 258). Many other proteins with transport functions occur in the blood plasma. For example, **serum albumin** (middle) transports free fatty acids, bilirubin, steroid hormones and many drugs (see p. 252). **Ion channels** and other integral membrane proteins (see p. 206) facilitate the transport of ions and metabolites across biological membranes.

Protection and defense. The immune system protects the body from pathogenic agents and foreign substances. An important component of this system is **immunoglobulin G** (lower left, see p. 266). It is shown here bound to an erythrocyte by complex formation with surface glycolipids. **Fibrinogen** (upper right), the soluble precursor of fibrin, is important for blood cogulation (see p. 262).

Control and regulation. In biochemical signal transduction pathways, proteins function not only as signal substances (proteohormones, see p. 342), but also as hormone receptors. The complex between the hormone **insulin** and the **insulin receptor** is shown as an example (middle). DNA binding proteins are central to the regulation of intermediary metabolism and the differentiation of cells, tissues, and organs. The structure and function of the **catabolite activator protein** (above left) and similar bacterial repressor proteins have been particularly well characterized (see p. 108).

Catalysis. With several thousand representatives known, the **enzymes** (see p. 86) form an especially large group of proteins. The smallest enzymes have molecular masses of 10–15 kDa. Intermediate-sized enzymes, such as **lactate dehydrogenase** (above) are around 100 kDa, whereas the largest, including **glutamine synthetase** with its 12 subunits have masses of more than 500 kDa.

Movement. The interaction between actin and myosin is responsible for muscle contraction and cell movement (see p. 306). **Myosin** (right), with a length of over 150 nm, is among the largest proteins known, while **actin** filaments arise due to the polymerization of relatively small protein subunits.

Storage. In the seeds of plants, special **storage proteins** are found, which are also important for the nutrition of man (not shown). In animals, the *muscle proteins* constitute a nutrient reserve that can be mobilized in an emergency.

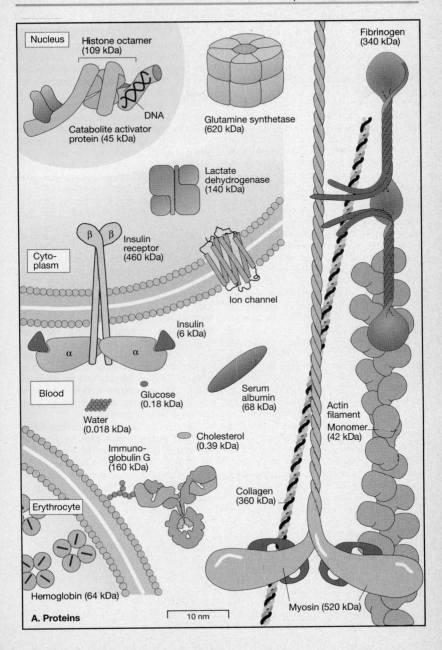

A. Proteins

Nucleus

Histone octamer
(109 kDa)

DNA

Catabolite activator
protein (45 kDa)

Glutamine synthetase
(620 kDa)

Lactate
dehydrogenase
(140 kDa)

β β

Cyto-
plasm

Insulin
receptor
(460 kDa)

Ion channel

α   α

Insulin
(6 kDa)

Blood

Glucose
(0.18 kDa)

Water
(0.018 kDa)

Serum
albumin
(68 kDa)

Cholesterol
(0.39 kDa)

Immuno-
globulin G
(160 kDa)

Erythrocyte

Collagen
(360 kDa)

Hemoglobin (64 kDa)

Fibrinogen
(340 kDa)

Actin
filament
Monomer
(42 kDa)

Myosin (520 kDa)

10 nm

The Peptide Bond

Amino acids, the building blocks of peptides and proteins, are linked to one another by *amide bonds* (see p. 8) between the carboxyl group of one molecule and the amino group of the next. Since they are found in peptides, such bonds are also called **peptide bonds**. Many of the properties of peptides and proteins can be explained by properties of the peptide bond. Thus, a more detailed consideration of the chemistry and stereochemistry of this type of bond will help us to understand protein structure.

A. Peptide bond ◖

The linking of two amino acids gives rise to a **dipeptide** (**1c**). In dipeptides, one of the two residues (the *N-terminal* amino acid) contains a free amino group, and the other one (the *C-terminal* amino acid) a free carboxylate group. The naming and representation of peptides always begins at the left with the *N*-terminal amino acid. During the *chemical* synthesis of peptides (**1**), the formation of the peptide bond is only allowed to occur once the amino and carboxyl groups, which are not supposed to react have been blocked by **protecting groups** (X,Y) (**1a, b**). Otherwise, in the case shown, not only the desired product alanine-glycine (Ala-Gly), but also Gly-Ala, Gly-Gly, and Ala-Ala would be formed. To increase its reactivity, the carboxyl group is first reacted with a group (Z) that can be easily displaced by the amino group as a result of nucleophilic attack. Nowadays, peptide synthesis is an automated process, with special instruments linking the amino acids in the appropriate order. The *biosynthesis* of the peptide bond is discussed on p. 228–230.

Like all amide bonds, the peptide bond is stabilized by **resonance** (**2**), i.e., the delocalization of electrons between the carbonyl oxygen and the amide nitrogen. As a result of this, the peptide bond is *planar*. Large amounts of energy are required to drive rotation about the C-N bond, which is therefore more or less *unable to rotate freely*. The plane in which the 6 atoms of the peptide bond lie is highlighted in light blue here and on the following pages.

B. Conformations of the peptide chain ○

With the exception of the terminal residues, every amino acid in a peptide is involved in two peptide bonds (one with the preceding and one with the succeeding residue). Because rotation about the C-N bond is hindered, rotations are restricted to the N-C_α and the C_α-C bonds. Their state of rotation is characterized by the **dihedral angles** ϕ (**phi**) and ψ (**psi**). The angle ϕ describes rotation about the N-C_α bond, and the angle ψ describes the rotation about C_α-C. For steric reasons, not all combinations of the angles ϕ and ψ are possible — a fact that has important implications for peptide structure. Possible conformations are best illustrated by a so-called ϕ/ψ diagram (**1**).

Most combinations of ϕ and ψ are sterically forbidden (red region). For example, the combination $\phi = 0°$ and $\psi = 180°$ (**4**) would place the two carbonyl oxygen atoms less than 115 pm apart, i.e., at a distance much lower than the sum of their van der Waals radii (see p. 6). Similarly, in the case of $\phi = 180°$, $\psi = 0°$ (**5**), the two NH hydrogens would collide. Only combinations within the green regions are sterically feasible (e.g., **2** and **3**). The diagram also indicates where the most important of the *secondary structures* of peptides are located (details will be discussed later). The conformations corresponding to the yellow regions are energetically less favorable, but still possible.

The ϕ/ψ diagram (1) was developed from modeling studies of small peptides. However, the actual conformations of most of the amino acids in proteins also fall into the "allowed" region. The corresponding data for the small protein insulin (see p. 70) is shown by black dots in **1**.

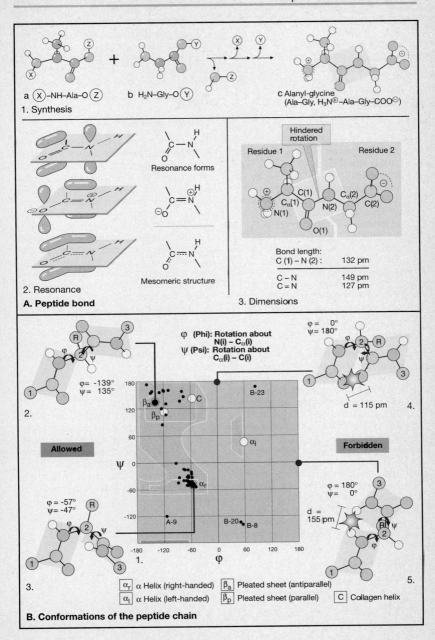

a (X)–NH–Ala–O (Z)   **b** H₂N–Gly–O (Y)   **c** Alanyl-glycine
1. Synthesis   (Ala–Gly, H₃N⊕–Ala–Gly–COO⊖)

Resonance forms

Mesomeric structure

2. Resonance

A. Peptide bond

Hindered rotation

Residue 1   Residue 2

C(1)
C_α(1)   C_α(2)   C(2)
N(1)   N(2)
O(1)

Bond length:

C (1) – N (2) :	132 pm
C – N	149 pm
C = N	127 pm

3. Dimensions

φ (Phi): Rotation about N(i) – C_α(i)
ψ (Psi): Rotation about C_α(i) – C(i)

2. φ = −139°   ψ = 135°

φ = 0°   ψ = 180°   **4.**   d = 115 pm

β_a
β_p
C
B-23
α_l
α_r
A-9   B-20 B-8

Allowed

Forbidden

φ = 180°   ψ = 0°   d = 155 pm   **5.**

3. φ = −57°   ψ = −47°

1. φ

α_r	α Helix (right-handed)	β_a	Pleated sheet (antiparallel)
α_l	α Helix (left-handed)	β_p	Pleated sheet (parallel)	C	Collagen helix

B. Conformations of the peptide chain

Secondary Structures

In naturally occurring proteins, certain combinations of the dihedral angles ϕ and ψ are much more common than others (see p. 64). When several successive residues conform to a given combination, defined **secondary structures** arise, which are stabilized by hydrogen bonds either within the peptide chain or between neighboring chains. When a large part of a protein takes on a defined secondary structure, the protein often forms mechanically stable filaments or fibers. Such **structural proteins** (see p. 68) usually have characteristic amino acid compositions.

Let us first discuss the most important secondary structural elements. The illustrations of the structures of the peptide chains are highly simplified. The α carbons of each chain are numbered, while the side chains are not shown. In order to accentuate the course of the peptide backbone, the planes of the peptide bonds are highlighted in blue. The bond angles ϕ and ψ for the structures shown here also are included in diagram B1 on p. 64.

A. α Helix ◑

The right-handed α helix (α_R) is one of the most common secondary structures. In this conformation, the peptide chain is wound like a screw. Each turn of the screw (screw axis in orange) encompasses approximately 3.6 amino acid residues. The **pitch** of the screw (i.e., the smallest distance between equivalent points) is 0.54 nm. α Helices are stabilized by almost linear *hydrogen bonds* (red dots) between the NH and CO groups, which are 4 residues apart from one another in the sequence. In long α helices, most amino acid residues enter into two H bonds. Interestingly, the mirror image of the α_R helix, the left-handed helix (α_L), is rarely found in nature, even though it is energetically feasible.

B. Collagen helix ◑

Another type of helix occurs in the collagens, which are important constituents of the connective-tissue matrix (see p. 312). The collagen helix is left-handed and, with a pitch of 0.96 nm and 3.3 residues per turn, it is steeper than the α helix. In contrast to the α helix, H bonds are not possible *within* the collagen helix. However, the conformation is stabilized by the association of three left-handed helices to form a right-handed *collagen triple helix* (see pp. 68 and 312).

C. Pleated-sheet structures ◑

Two further, largely extended conformations of the peptide chain are known as **pleated-sheet structures**, because the peptide planes are arranged like a regularly folded sheet of paper (see also p. 69). Once again, H bonds can only form between *neighboring* chains. When the two peptide chains run in opposite directions (**1**), the structure is referred to as an **antiparallel pleated sheet** (β_a). When they run in the same direction, we have a **parallel pleated sheet** (β_p) (**2**). In both cases, the C_α-atoms occupy the highest and lowest points of the structure, and the side chains point alternately straight up or straight down.

Energetically, the β_a structure, with its almost linear H bonds, is more favorable. In extended pleated sheets, the single strands of the sheet are usually twisted with respect to each other rather than running in parallel (see p. 73, B).

D. β Turns ◑

β Turns are frequently found at sites where the peptide chain abruptly reverses direction. In this case, four amino acid residues are involved in a bend, which mediates a 180° reversal in the direction of the chain. Two arrangements (*type I* and *type II*) are possible. Both are stabilized by H bonds between residues 1 and 4.

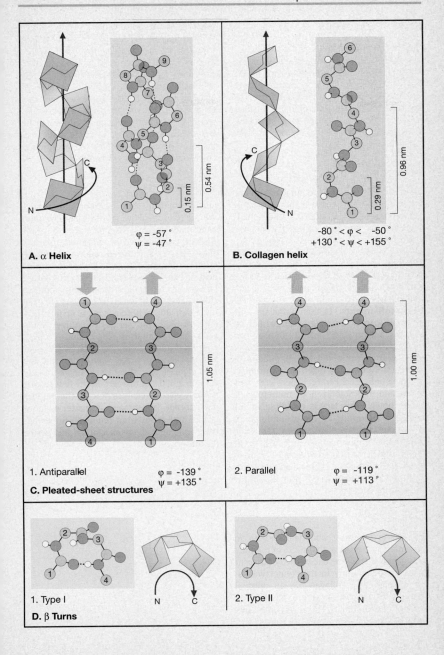

$\varphi = -57°$
$\psi = -47°$

A. α Helix

$-80° < \varphi < -50°$
$+130° < \psi < +155°$

B. Collagen helix

1. Antiparallel $\varphi = -139°$
    $\psi = +135°$

C. Pleated-sheet structures

2. Parallel $\varphi = -119°$
    $\psi = +113°$

1. Type I

D. β Turns

N   C

2. Type II

N   C

Structural Proteins

The structural proteins provide extracellular structures with mechanical stability, or constitute parts of the cytoskeleton (see p. 190). Most of these proteins contain a high percentage of a particular secondary structure (see p. 66). Since certain amino acid residues are especially common in such structures, structural proteins are usually rich in a few particular amino acids.

A. α Keratin ○

α Keratin is a structural protein made up predominantly of α helices. It is the most abundant protein in hair (wool), feathers, nails, claws and the hooves of animals. It is also a constituent of the cytoskeleton (see p. 190), where it appears in so-called **intermediary filaments**.

In the keratins, four right-handed α-helical peptide chains get together to form a left-handed **superhelix**. Eight of these superhelical **protofilaments** (**1**) then associate with one another to form an intermediary filament with a diameter of 10 nm (**2**). Similar filaments are found in hair. In a single wool fiber with a diameter of about 20 μm, millions of filaments are bundled together within dead cells. The individual keratin helices are cross-linked, and thus stabilized, by numerous disulfide bonds. This fact is exploited in the *perming of hair*. Initially, the disulfide bonds of hair keratin are disrupted by reduction with thiol compounds (see p. 8). Then the hair is done in the desired new style and dried with heat. In the process, new disulfide bonds are formed by oxidation, which maintain the hairstyle for some time.

B. Collagen ◑

Collagen is the most abundant protein in mammals, constituting about 25% of the total body protein. It is present in many different forms, predominantly in the connective tissues (see p. 314). Collagen has an unusual amino acid compostion. Approximately one third of the amino acids are *glycine* (Gly), about 10% *proline* (Pro), and 10% *hydroxypro-*
line (Hyp). The latter amino acid is formed during collagen biosynthesis as a result of *posttranslational modification* (see p. 312). The reason for the unusual amino acid composition is that collagen exists almost entirely as a **triple helix** made up of three single collagen helices (**1**). In such triple helices, every third residue lies on the inside (**3**) where, for steric reasons, only glycine residues can fit. Only a small segment of a triple helix is illustrated here. The complete collagen molecule is approximately 300 nm long.

C. Silk fibroin ○

Silk is produced from the cocoons of silkworms (larvae of the butterfly *Bombyx mori* and related species). The main protein in silk, **fibroin**, is composed of antiparallel *pleated sheet structures* arranged one on top of the other in numerous layers (**1**). Since the amino acid side chains in pleated sheets point either straight up or straight down, only compact side chains fit between the layers. In fact, more than 80% of fibroin is made up from glycine, alanine, and serine, the three amino acids with the shortest side chains. A typical repetitive amino acid sequence is (*Gly-Ala-Gly-Ala-Gly-Ser*)$_n$. It has been found that the individual pleated sheet layers lie alternately 0.35nm and 0.57 nm apart. In the first case, it is solely glycine residues (R = -H) that oppose one another. The somewhat greater distance of 0.57 nm arises from repulsion forces between the side chains of alanine and serine residues (**2**).

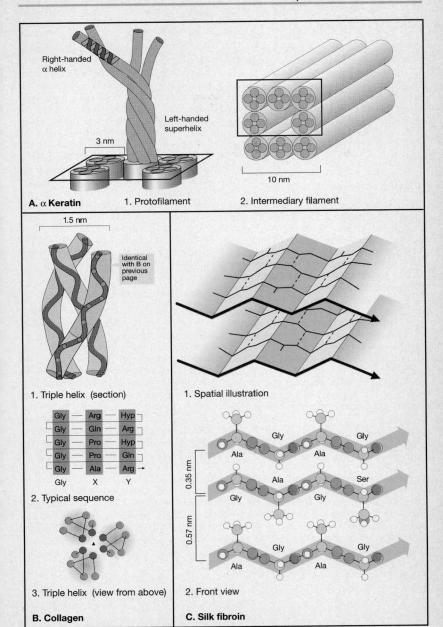

A. α Keratin

Right-handed α helix

Left-handed superhelix

3 nm

1. Protofilament

10 nm

2. Intermediary filament

1.5 nm

Identical with B on previous page

1. Triple helix (section)

Gly	—	Arg	—	Hyp
Gly	—	Gln	—	Arg
Gly	—	Pro	—	Hyp
Gly	—	Pro	—	Gln
Gly	—	Ala	—	Arg	→

Gly   X   Y

2. Typical sequence

3. Triple helix (view from above)

B. Collagen

1. Spatial illustration

Ala   Gly   Gly

0.35 nm

Ala   Gly   Ser

Gly   Gly

0.57 nm

Ala   Gly   Gly

Ala

2. Front view

C. Silk fibroin

Globular Proteins

Soluble proteins are, in general, more complex in structure than the fibrous, insoluble structural proteins. They are more or less spherical (globular) in shape. Such **globular proteins** have a well-defined three-dimensional structure (*conformation*), which is essential for their biological function. Here we discuss their structure using the smallest proteohormone insulin (see p. 342) as an example.

A. Insulin: primary structure ○

The primary structure of a protein is its **amino acid sequence**. Insulin was the first protein the amino acid sequence of which was fully elucidated (*F. Sanger* and colleagues, 1952). In its biologically active form, the insulin molecule consists of two polypeptide chains (**A chain** and **B chain**) held together by disulfide bonds (yellow; in the illustration, the A chain is light brown and the B chain is dark brown). Another disulfide bond is located within the A chain. Insulin is synthesized in the pancreas as a precursor, **proinsulin**, in which the *C*-terminal amino acid of the B chain and the *N*-terminal amino acid of the A chain are connected by an additional 33 amino acids. This **C peptide** (open circles) is proteolytically removed once the disulfide bridges have been correctly formed (see p. 164).

B. Secondary structure ○

Secondary structure is defined as regions of the peptide chain with a defined conformation stabilized by H bonds. In most globular proteins, α **helices** and β **pleated sheets** are found at the same time. In addition, there are **disordered regions** of the chain. A further structural element typical of globular proteins is the β **Turn**, a segment made up of four amino acids in which the peptide chain reverses direction. β turns usually occur at transitions between different structural elements.

C. Tertiary structure ○

The three-dimensionally folded, biologically active conformation of a protein is referred to as its tertiary structure. The conformations of hundreds of proteins are known today. Almost all were determined using **X-ray crystallography**. This complicated procedure depends upon the diffraction of X-rays by ordered protein crystals. From these data, the distribution of the electron density in the crystal can be calculated with the help of computers to determine the structure of the molecule. The analysis of insulin (*D. Hodgkin* and colleagues, 1971) showed that the molecule contains two short α helices in the A chain and a longer one in the B chain. The *N* terminus of the A chain and the *C* terminus of the B chain lie very close to one another. The tertiary structure of proinsulin is not yet known.

D. Quaternary structure ○

Individual protein molecules often assemble to form symmetrical complexes held together by non-covalent interactions. Such complexes are referred to as **oligomers,** and the individual components (usually 2–12) are called subunits or **monomers**. Insulin also has a quaternary structure. In the blood, it is present, in part, as a *dimer* with a two-fold axis of symmetry (**1**). In addition, there is also a Zn^{2+}-stabilized *hexamer* consisting of 6 subunits (**2**), which is the form in which insulin is stored in the pancreas (see also p. 164). The histidine residues at position B-10 in each of the 6 subunits are involved in complex formation with Zn^{2+}. In each case, three histidine residues and three water molecules are coordinated with one Zn^{2+} ion in an octahedral fashion. The insulin hexamer contains two identical complexes of this type (see p. 76).

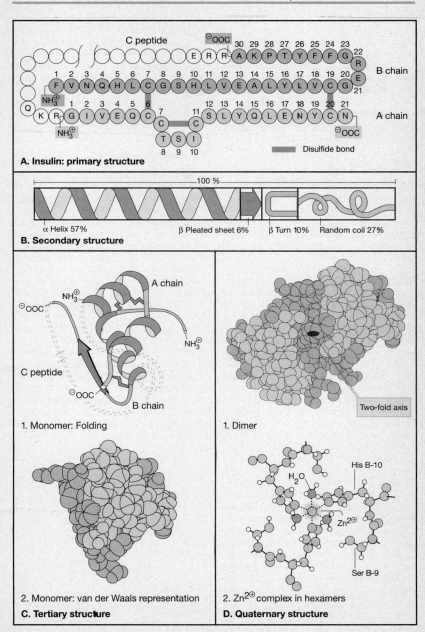

A. Insulin: primary structure

C peptide

B chain

A chain

Disulfide bond

B. Secondary structure

100 %

α Helix 57% β Pleated sheet 6% β Turn 10% Random coil 27%

1. Monomer: Folding

A chain

C peptide

B chain

1. Dimer

Two-fold axis

2. Monomer: van der Waals representation

C. Tertiary structure

His B-10

H_2O

Zn^{2+}

Ser B-9

2. Zn^{2+} complex in hexamers

D. Quaternary structure

Protein Folding

Many proteins spontaneously take on their secondary, tertiary, and quaternary structures. Apparently, all of the information required for formation of the biologically active ("*native*") conformation of proteins is encoded in their amino acid sequence. It is an important goal of biochemistry to better understand the rules that govern protein folding and its role in biological activity.

A. Protein folding ◐

The folding of proteins to yield their native conformations is favored under physiological conditions. Loss of the native conformation (**denaturation**) occurs at extreme pH values, at high temperatures, or as a result of treatment with organic solvents, detergents, and other denaturing agents.

The interactions, which stabilize protein conformation include hydrogen bonds, disulfide bridges, electrostatic interactions and complex formation with metal ions. Interactions of these types are illustrated on the following page using insulin as an example. A further stabilizing factor of outstanding importance is the "*hydrophobic effect.*" As discussed on p. 22, a mixture of an apolar substance with water will separate into two distinct phases in such a way that the area of contact between the two phases is at a minimum (the "oil drop effect"). In an analogous fashion, soluble proteins in water fold so that the majority of the apolar amino acid side chains lie within the molecule, whereas the polar side chains face toward the solvent (**1**). This behavior is well illustrated by the conformation of the insulin molecule (see p. 76).

Our understanding of the **energetics** of protein folding (**2**) is still incomplete. Here we present a highly simplified model. The major factor working *against* folding is the large increase in the degree of order of the molecule. As a result, the change of *conformational entropy* ΔS_{conf} is negative and the entropy term -T · ΔS_{conf} is strongly positive (violet). On the other hand, there is a *stabilizing* influence due to formation of covalent and non-covalent bonds inside the folded protein. Thus, the change of *folding enthalpy,* ΔH_{fold}, is negative (orange). A third factor is the entropy change of the system due to the hydrophobic effect. During folding, the degree of order of the *water* decreases, i.e., ΔS_{ap} is positive and therefore -T · ΔS_{ap} is negative (violet). When the sum of ΔH_{fold}, and -T · ΔS_{ap} is greater than -T . ΔS_{conf}, ΔG_{fold} becomes negative (blue), and spontaneous folding of the protein into a stable conformation occurs.

B. Protein folding: examples ○

Owing to X-ray crystallography (see p. 70), we now know the native conformations of many proteins. The folding of the majority of large globular proteins is characterized by a number of characteristic **folding patterns**, which are seen again and again. Some examples are shown here (α helices in red, pleated sheet strands in green). A predominantly helical organization of a protein, such as in **myoglobin** (**1**, heme in yellow) is rare. Usually, pleated sheet and helical elements exist side by side as in **flavodoxin,** a small flavoprotein serving a redox function (**2**, FMN in yellow). Here, a fan-shaped pleated sheet made up of five parallel strands forms the nucleus of the molecule. It is surrounded by four α helices on the outside. The **immunoglobulins** (see p. 164) consist of several similar *domains* (independently folded partial structures) in which two antiparallel pleated sheets form a barrel-like structure. The C_H2 domains (**3**) shown here carry an oligosaccharide, which is illustrated in detail on p. 40.

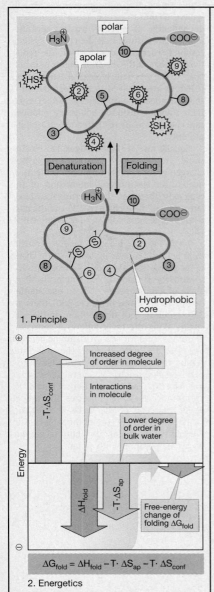

1. Principle

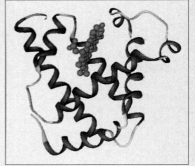

1. Myoglobin

2. Flavodoxin

In the principle diagram:

polar
$H_3\overset{\oplus}{N}$
apolar
10
$COO^{\ominus}$
1 HS
2
5
6
9
8
3
4
SH 7

Denaturation ⇄ Folding

$H_3\overset{\oplus}{N}$
10
$COO^{\ominus}$
9
1 S
7 S
8
6
4
2
3
5

Hydrophobic core

2. Energetics

⊕

Increased degree of order in molecule

Interactions in molecule

Lower degree of order in bulk water

Energy

$-T\cdot\Delta S_{conf}$

ΔH_{fold} $-T\cdot\Delta S_{ap}$

Free-energy change of folding ΔG_{fold}

⊖

$$\Delta G_{fold} = \Delta H_{fold} - T\cdot\Delta S_{ap} - T\cdot\Delta S_{conf}$$

2. Energetics

A. Protein folding

3. Immunoglobulin G: C_H2 domains

B. Protein folding: examples

Insulin: Structure

Here we illustrate some of the typical elements of protein structure, taking insulin (see p. 70) as an example.

A. Structural elements ○

The partial structures shown here account for almost half of the molecule. Their locations in the complete structure (**1**) are highlighted using various colors. As on the previous pages, the hydrogen atoms, which cannot be localized by X-ray crystallography, are omitted.

The only β **pleated** sheet structure in insulin (**2**) is unusual in that it is only found in dimers and hexamers. There two antiparallel strands are formed by corresponding residues (Phe B-24 to Tyr B-26) from *neighboring* subunits (see p. 70). As always, the side chains in the β pleated sheets project alternately above and below the plane of the peptide. The four H bonds shown, as well as hydrophobic interactions between the aromatic side chains, help to hold the monomers together.

The apex of the wedge-shaped insulin molecule (**3**) is very important for the binding of insulin to its receptor. As a result, its conformation is especially well stabilized. This part of the molecule is formed by the intersection of the C-terminal end of the A chain with the transition between the helical and extended segments of the B chain. Residues Cys B-19 to Gly B-23 form *two* intertwined β **turns** with H bonds between the carbonyl groups of Cys-19 and Gly-20 and the NH groups of each of the residues. From the φ/ψ diagram on p. 64 it is obvious that the conformations of the peptide chain at glycine residues B-20 and B-23 fall into the "forbidden" region. It is only the distortion of the conformation in this section, which makes the tight bend of the B chain possible. The energy necessary for the distortion is derived from more favorable interactions elsewhere in the molecule.

There are two other bonds in this region, which link the A and B chains. One of the three **disulfide bonds** forms a covalent bridge between cysteines B-19 and A-20. This bond is formed in proinsulin (see p. 164), which, in contrast to mature insulin, folds independently of already existing disulfide bonds. This ensures that the correct cysteine residues react with one another. The reductive cleavage of the disulfide bonds of insulin leads to the loss of its biological activity. Such reactions are involved in the degradation of the hormone *in vivo* (see p. 348). Despite their stability, disulfide bonds are relatively rare and are usually restricted to extracellular proteins.

Equally rare are **salt bridges**, i.e., electrostatic interactions between oppositely charged groups *on the surface* of proteins. In insulin, an electrostatic interaction of this type exists between the C-terminal carboxylate group of the A chain (A-21) and the positively charged guanido group of arginine B-22 (**3**).

The longest α **helix** in insulin stretches from Ser B-9 to Cys B-19 (**4**). The side chains of the amino acid residues in α helices project toward the outside of the molecule at an angle of approximately 100° to one another (in one turn of the α helix, i.e., 360°, there are 3.6 residues). The distribution of the side chains along the helix follows a consistent pattern: except for Leu-17, all amino acids on one side are *polar*, and on the other side *apolar*. This is apparent when the distribution of the residues is observed in the direction of the helix axis. This type of **"helical wheel" diagram** (**5**) clearly emphasizes the *amphipathic character* of the helix. The residues facing in toward the core of the molecule are predominantly apolar, and those facing outwards are predominantly polar.

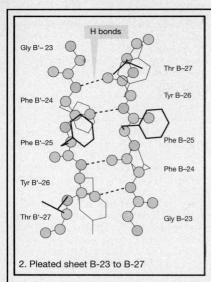

H bonds

Gly B'–23

Thr B–27

Phe B'–24

Tyr B–26

Phe B'–25

Phe B–25

Phe B–24

Tyr B'–26

Thr B'–27

Gly B–23

2. Pleated sheet B-23 to B-27

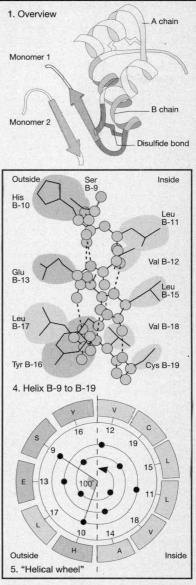

1. Overview

A chain

Monomer 1

B chain

Monomer 2

Disulfide bond

Outside   Ser   Inside
                           B-9
His
B-10   Leu
                                                  B-11

Glu   Val B-12
B-13
                                                  Leu
                                                  B-15
Leu
B-17   Val B-18

Tyr B-16   Cys B-19

4. Helix B-9 to B-19

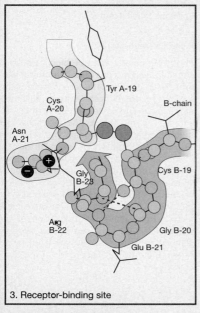

Tyr A-19

Cys
A-20   B-chain

Asn
A-21

                                 Cys B-19
Gly
B-23

Arg   Gly B-20
B-22
              Glu B-21

3. Receptor-binding site

5. "Helical wheel"

Outside   Y   V
      S   C
         16   12
      9   19
                         15
   E 13   100°   11
      17   18
         10   14   V
   L   L
Outside   H   A   Inside

A. Insulin: structural elements

Molecular Models: Insulin

The illustration on the adjacent page shows models of the peptide hormone **insulin**. A detailed discussion of the structure, biosynthesis, and function of this molecule can be found elsewhere (pp. 71, 77, and 165).

Large quantities of insulin are required for the treatment of *diabetes* (see p. 164). In Germany alone, the yearly demand is more than 500 kg. Until recently, the hormone has predominantly been isolated from the pancreas of slaughtered animals, i.e., by a complicated and expensive procedure. In addition, animal insulins, although their amino acid sequences differ only slightly from that of human insulin, can trigger antibody formation after long periods of use. The effectiveness of treatment then declines. Furthermore, serious *hypersensitivity reactions* can arise. Some years ago, therefore, several pharmaceutical companies started producing **human insulin** in genetically engineered bacteria (see p. 240).

The illustration shows models of **pig insulin**, which differs from human insulin in only one amino residue (instead of Ala-30, the human hormone contains a threonine residue). Fig. **A** highlights important structural elements of the insulin monomer. Fig. **B** shows, from two different angles, the folding pattern of the *hexameric* zinc complex.

A. Insulin (monomer) ○

The insulin monomer, with its 51 amino acid residues and a molecular mass of 5.5 kDa, is only about half as big as the smallest enzyme. Nevertheless, it has the typical properties of a globular protein. In solution, the molecule maintains a stable **tertiary structure**, which is essential for its function as a signal substance. In the *van der Waals model* shown on the left, the atoms of the A chain are highlighted in a light color and those of the B chain in a darker color. The molecule is almost cone-shaped. The tip of the cone is formed by the B-chain, which undergoes a reversal in direction at this point.

In the second model (middle), all side chains of *polar* amino acids (see p. 58) are highlighted in green, while *apolar* ones are shown in in *yellow and orange*. This model emphasizes the importance of the **hydrophobic effect** in protein folding (see p. 72). Most of the highly apolar residues (yellow) are located on the inside of the molecule, whereas the hydrophilic groups are located on the surface. There is a seeming contradiction to this rule since, in the insulin monomer, several distinctly apolar side chains (orange) are exposed to the solvent. However, all of these residues are involved in hydrophobic interactions that stabilize the insulin dimer (see p. 84) and the hexamer (**B**).

In the third model (right), the residues shown in red are those that are located on the surface of the molecule, and, at the same time, are **invariant** (red) or almost invariant (orange) in all known insulin sequences. Amino acid residues that do not change in the course of evolution may be assumed to be essential for the function of the protein. In the case of insulin, most of the invariant residues are located on one side of the molecule. They are probably involved in the binding of the hormone to its receptor.

B. Insulin (hexamer) ○

Prior to its release into the blood, insulin is stored in the form of a **zinc-containing hexamer** in the B cells of the islets of Langerhans (see p. 164). Here, these 6-subunit oligomers are shown in simplified form using ribbon diagrams. As in Fig. **A**, the A chains are highlighted in a light color and the B chains in a darker shade. As discussed on p. 70, the hexamers are stabilized by two zinc complexes involving the histidine residues in position B-10 of all 6 monomers. The hexamer (molecular mass 33 kDa) has a *threefold axis* of symmetry. This is best seen when the molecule is viewed from above (left). Rotation of the structure by 90° (right) reveals that two Zn^{2+} centers (Zn^{2+} ions shown in light blue) are arranged on top of each other.

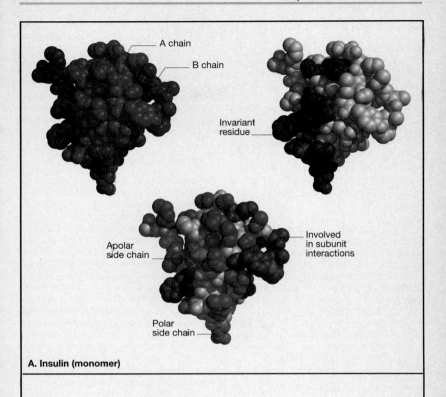

A chain

B chain

Invariant
residue

Apolar
side chain

Involved
in subunit
interactions

Polar
side chain

A. Insulin (monomer)

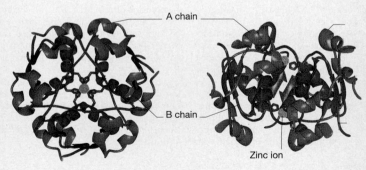

A chain

B chain

Zinc ion

Front view   Rotated by 90°

B. Insulin (hexamer)

Nucleobases, Nucleotides, RNA

The nucleic acids play a central role in the storage and expression of genetic information (see p. 218). They can be divided into two major classes: **deoxyribonucleic acid (DNA)**, which functions solely in information storage, and the **ribonucleic acids (RNAs)**, which are involved in almost every step of gene expression and protein biosynthesis. All nucleic acids are made up from **nucleotides**. These in turn consist of a *base*, a *sugar*, and a *phosphate residue*. DNA and RNA differ from one another in the type of the sugar and in one of the bases, which they contain.

A. Bases ◑

The bases in nucleic acids are *heterocyclic* compounds derived from either **pyrimidine** or **purine**. Five of these bases are found in all organisms. The purine bases **adenine** (Ade) and **guanine** (Gua) and the pyrimidine base **cytosine** (Cyt) are present in both RNA and DNA. In contrast, **uracil** (Ura) is only found in RNA. In DNA, uracil is replaced by **thymine** (Thy), the 5-methyl derivative of uracil. Many other modified bases occur in tRNAs (see p. 226) and in other types of RNA.

B. Nucleosides, nucleotides ◑

Linkage of a nucleic acid base to ribose or 2-deoxyribose (see p. 34) yields a **nucleoside**. In this way, for instance, the nucleoside *adenosine* (abbreviation: A) is formed from adenine and ribose. The corresponding derivatives of the other bases are called *guanosine* (G), *uridine* (U), *thymidine* (T) and *cytidine* (C). When the sugar component is 2-deoxyribose, the product is a **deoxyribonucleoside**, e.g., *2′-deoxyadenosine* (dA). In the cell, the 5′-OH-group of the sugar component of the nucleoside is usually esterified with phosphoric acid. Thus, adenosine gives rise to **adenosine-5′-monophosphate** (AMP) and dA to dAMP. When the 5′-phosphate residue is linked via an acid-anhydride bond to further phosphate residues, the resulting molecules are nucleoside diphosphates and triphosphates, e.g., ADP and ATP, which are key coenzymes in energy metabolism (see p. 114). All nucleoside phosphates are grouped together under the term *nucleotide*.

In nucleosides and nucleotides, the pentose residues are present in the furanose form (see p. 35). The sugars and bases are linked by a *N-glycosidic bonds* between C-1 of the sugar and either N-9 of the purine ring or N-1 of the pyrimidine ring. This bond always adopts the β configuration.

C. Oligonucleotides, polynucleotides ◑

Nucleotides can be linked to one another via acid-anhydride bonds between their phosphate residues. The resulting products are dinucleotides with phosphoric acid-anhydride structure (not shown). This group includes the coenzymes NAD+, NADP+, FAD (see p. 100) and coenzyme A (see p. 102). When, by contrast, the phosphate residue of a nucleotide reacts with the 3′-OH-group of a second nucleotide, the result is a dinucleotide with a phosphoric acid diester structure. Such dinucleotides carry a free phosphate residue at the 5′ end and a free OH group at the 3′ end. This allows them to be extended by additional mononucleotides by the formation of further phosphoric diester bonds. This is the way in which **oligonucleotides**, and ultimately **polynucleotides**, are synthesized.

Polynucleotides consisting of ribonucleotide building blocks are called **ribonucleic acid (RNA)**, while those made up of deoxyribonucleotide monomers are called **deoxyribonucleic acid (DNA**, see p. 80). The structures of the oligonucleotides and polynucleotides are written in the 5′→ 3′ direction from left to right, using the abbreviations for the nucleoside building blocks. The position of phosphate residue is also sometimes indicated by a "p". Thus, the structure of the RNA shown in **C** may be abbreviated as 5′-pApU.....-3′.

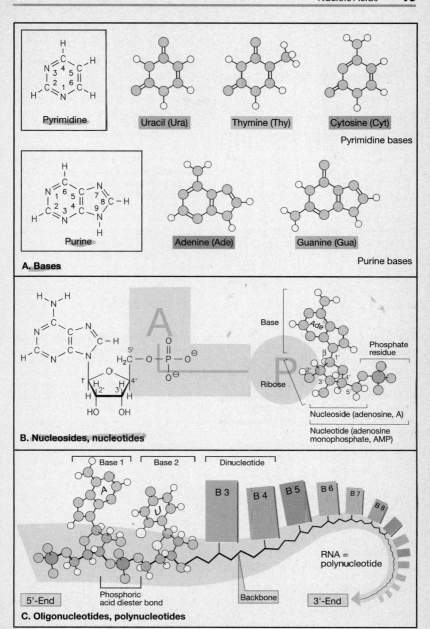

A. Bases

Pyrimidine

Uracil (Ura)   Thymine (Thy)   Cytosine (Cyt)

Pyrimidine bases

Purine

Adenine (Ade)   Guanine (Gua)

Purine bases

B. Nucleosides, nucleotides

Base

Ribose

Phosphate residue

Nucleoside (adenosine, A)

Nucleotide (adenosine monophosphate, AMP)

C. Oligonucleotides, polynucleotides

Base 1   Base 2   Dinucleotide

B 3   B 4   B 5   B 6   B 7   B 8

RNA = polynucleotide

5'-End

Phosphoric acid diester bond

Backbone

3'-End

DNA

The biological functions of the nucleic acids are based mainly on the ability of the nucleic acid bases to enter into specific interactions with one another.

A. Base pairing in DNA ◑

The first indication of the existence of such interactions came from the observation that DNA from all sources contains approximately equal amounts of adenine and thymine. The same is true for guanine and cytosine. In contrast, the ratio of the *sum* of Ade and Thy to the *sum* of Gua and Cyt varies from one organism to the next. This constant base ratio and other findings were explained by a model formulated by *J. Watson* and *F. Crick* in 1953. Intact DNA consists of *two* polydeoxynucleotide molecules ("*strands*"). Each base in each strand is linked by H bonds to a complementary base in the other strand. Thus, adenine is complementary to thymine, and guanine is complementary to cytosine. In other words, every **base pair** involves one purine and one pyrimidine base.

The complementarity of A with T, and of G with C, can be understood by considering the H bonds that are possible between the different bases. Potential donors (see p. 6) are amino groups (Ade, Cyt, Gua) and the ring NH groups. Possible acceptors are carbonyl oxygen atoms (Thy, Cyt, Gua) and ring nitrogen atoms. Thus, in A-T pairs, *two* and in G-C pairs, *three* linear and thus very stable H bonds can be formed. In RNA, uracil behaves like thymine during base pairing.

B. DNA (B conformation) ◑

In DNA, the base pairings shown in Fig. **A** can extend over millions of bases. However, they are not feasible unless the two strands have opposite polarity, i.e., when they run in opposing directions. In addition, the two strands must be intertwined to form a **double helix**. RNA cannot form extended double helices, because of steric hindrance due to the 2′-OH-groups of the ribose residues. Base pairing in RNA, therefore, is limited to short regions. As a consequence, the overall structure of RNA molecules is much less organized than that of DNA.

The conformation of DNA, which predominates within the cell (the so-called **B-DNA**) is illustrated schematically in Fig. **1** and by van der Waals models on p. 85. In the schematic diagram, the deoxyribose-phosphate "backbone" is shown here as a ribbon. The bases (indicated by lines) are located on the inside of the double helix. Their aromatic rings are stacked 0.34 nm apart from one another and almost at 90° with respect to the axis of the helix. Each base is rotated at an angle of 35° with respect to the previous one. Therefore, a full turn of the double helix (360°) contains approximately 10 base pairs (abbreviation: bp), and the *pitch of the helix* is about 3.4 nm. The inside of the DNA double helix is apolar, while its surface is polar and negatively charged, because of the sugar and phosphate residues of the backbone. There are two depressions referred to as the "minor groove" and the "major groove" that lie between the strands for the whole length of the DNA molecule.

The two strands are held together by non-covalent interactions only, and therefore native DNA can be separated into the individual strands simply by heating (**denaturation**). The original double helix is spontaneously reformed as a result of base pairing (**renaturation**) during slow cooling of complementary single strands. Denaturation and renaturation of DNA play an important role in genetic engineering (see p. 236).

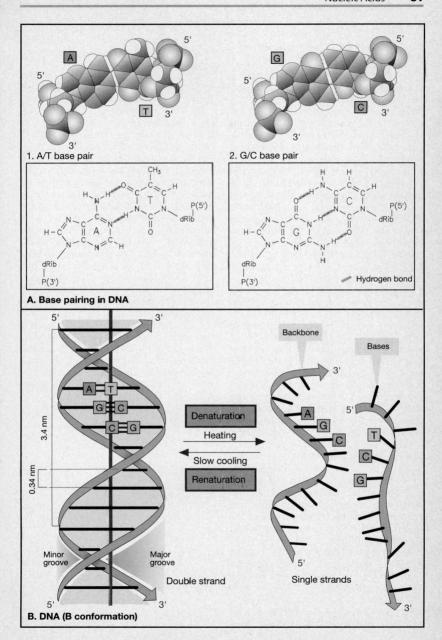

1. A/T base pair

2. G/C base pair

Hydrogen bond

A. Base pairing in DNA

Backbone

Bases

Denaturation

Heating

Slow cooling

Renaturation

3.4 nm

0.34 nm

Minor groove

Major groove

Double strand

Single strands

B. DNA (B conformation)

Functions

A. Nucleic acids: functions ●

The nucleic acids are of central importance in the storage and processing of genetic information. A key feature that allows them to perform this function is their ability to enter into specific base pairing (see p. 80). The following is a brief overview of the functions of the nucleic acids. More details are provided elsewhere (see p. 218–230).

Individual regions of DNA with specific functions are referred to as **genes**. Living organisms possess thousands of genes that code for proteins, i.e., genes that store information determining the amino acid sequences of these proteins. Each amino acid residue is encoded in the DNA sequence by a code word (**codon**) consisting of three sequential bases. For example, one of the codons for the amino acid phenylalanine (Phe) is *TTC*.

Expression of a gene, i.e., the synthesis of the protein which the gene encodes, requires that the sequence information in the DNA be converted into an amino acid sequence. As DNA itself does not take any part in protein synthesis, the information it contains must first be transferred from the nucleus to the site of protein synthesis, i.e., to ribosomes in the cytoplasm. This is achieved by copying the relevant part of the gene into **messenger RNA (mRNA)** by a process called **transcription** (**1**). The sequence of the resulting mRNA is complementary to that of the *coding strand* of the DNA. Since RNA contains uracil instead of thymine, the triplet *AAG*, the respective Phe codon on the coding strand of DNA, gives rise to the mRNA codon *UUC*.

In eukaryotes, mRNAs formed by transcription must first be modified before they can leave the nucleus (**2**). Several smaller RNA molecules, referred to as **small nuclear RNA (snRNA)**, are involved in this process, which is known as *mRNA maturation*. The functions of snRNAs are discussed on p. 224. The mature mRNA enters the cytoplasm where it binds to a **ribosome**. In addition to many different proteins, the ribosomes also contain several RNA molecules of various lengths, the so-called **ribosomal RNA (rRNA)**. These rRNAs are the most abundant RNAs in the cell. As compared with mRNAs, they are are very long-lived. rRNAs serve as structural elements of the ribosomes. In addition, they are involved in the binding of mRNAs to the ribosome. Once again, base pairing plays an important role in the process.

The actual information transfer takes place when the codons of mRNA interact with a further RNA species, **transfer RNA (tRNA)**. tRNAs, which occur in many different types, are responsible for positioning the correct amino acid residue at the ribosome, as dictated by the sequence information in the mRNA. A tRNA carrying a phenylalanine residue at its 3′-end is designated *Phe-tRNA^{Phe}*. Somewhere near the center of this molecule, it contains a base triplet *GAA*, the so-called **anticodon**, which is complementary to the mRNA codon *UUC*. The anticodon of the yeast tRNA shown here and on pp. 84 and 226 contains 2′-O-methylguanosine (Gm) instead of guanosine (G). However, Gm behaves exactly like guanine during base pairing. When the codon *UUC* appears at the ribosome, the anticodon associated with Phe-tRNA^{Phe} binds to the mRNA, bringing the phenylalanine residue bound at the other end of the molecule into a position where it can accept the growing polypeptide chain from a neighboring tRNA (**4**). With this step, the information transfer is completed.

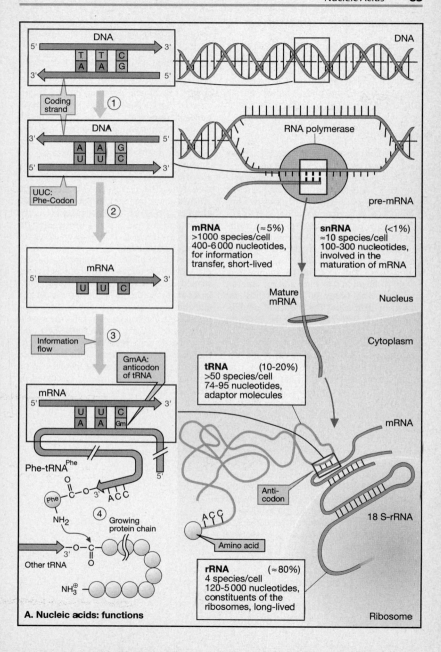

A. Nucleic acids: functions

DNA

5' T T C 3'
    A A G
3'   5'

① Coding strand

DNA

3'   5'
    A A G
5' U U C 3'

UUC: Phe-Codon

②

mRNA

5' U U C 3'

③ Information flow

GmAA: anticodon of tRNA

mRNA

5' U U C 3'
    A A Gm

Phe-tRNA^Phe

Phe—C—O—3' A C C
    ‖
    O
NH₂

④ Growing protein chain

Other tRNA

—O—C—O
     ‖
     O

NH₃⊕

DNA

RNA polymerase

pre-mRNA

mRNA   (≈5%)
>1000 species/cell
400–6 000 nucleotides,
for information
transfer, short-lived

snRNA   (<1%)
≈10 species/cell
100–300 nucleotides,
involved in the
maturation of mRNA

Mature mRNA

Nucleus

Cytoplasm

tRNA   (10–20%)
>50 species/cell
74–95 nucleotides,
adaptor molecules

mRNA

Anti-codon

18 S-rRNA

A C C

Amino acid

rRNA   (≈80%)
4 species/cell
120–5 000 nucleotides,
constituents of the
ribosomes, long-lived

Ribosome

Molecular Models: DNA, tRNA

The facing page contains van der Waals models (cf. p. 6) of two small nucleic acid molecules. In Fig. **A**, a **DNA molecule** consisting of 17 nucleotide pairs is depicted, while Fig. **B** shows the phenylalanine-specific **tRNA** from yeast (75 nucleotides).

A. B-DNA ○

Investigations of synthetic DNA molecules have shown that DNA can adopt several different conformations. By far the most common is the **B conformation** shown here. As discussed on p. 80, DNA consists of two opposing polydeoxyribonucleotide strands intertwined with one another to form a *right-handed double helix*. The "**backbone**" of such a strand is formed by alternating deoxyribose and phosphate residues linked by phosphoric acid-diester bonds. Here, the backbone of one strand is shown in dark blue, and the other in mid-blue. The **bases** of both strands are shown in light blue. The model of double-standed DNA (left) shows how the bases on the inside of the molecule are stacked almost at right angles to the axis of the helix. Two **grooves** of different widths are found between the backbones of the two strands. The wider *major groove* is visible above and below, and the *minor groove* in the middle. DNA-binding proteins usually interact with the more accessible bases in the region of the major groove (see pp. 108 and 346). The model showing only one strand of the molecule (right) emphasizes the spatial relationship between the backbone and the bases. In all models, the topmost adenine is highlighted in yellow and the complementary thymine in orange. In the axial views (middle), these bases are especially obvious.

Under certain conditions, DNA can adopt the so-called **A-conformation** (not shown). In this state, the double helix is still right-handed, but the bases are no longer arranged at right angles to the axis of the helix, as in the B-form. In the **Z-conformation** (not shown), which can occur within GC-rich regions of B-DNA, the organization of the nucleotides is completely different. Here the helix is *left-handed*, and the backbone adopts a characteristic *zigzag* conformation (hence "Z-DNA").

B. Phe-tRNA^Phe ○

RNA molecules are unable to form extended double helices, and are therefore less highly ordered than DNA molecules. A special tRNA molecule from yeast is shown here as an example. Its nucleotide sequence, and the general principles of tRNA structure, are discussed on p. 226. As in Fig. **A**, the bases are shown in light blue, and the backbone, composed of ribose- and phosphate residues, in dark blue. The corresponding amino acid, phenylalanine (orange), is bound to the 3′-terminal adenosine of the molecule. As the amino acid residue was not included in the measured structure, it has been added by "molecular modelling".

From the models, it is apparent that the tRNA is tightly folded into a compact, cone-shaped structure. As in DNA, most of the bases are found on the inside of the molecule, and the more polar backbone is located on the outside. An exception to this rule are the three bases making up the **anticodon** (yellow), which must interact with mRNA, and therefore have to be located on the surface. Most of the other bases are involved in *intramolecular* base pairings. As discussed on p. 226, some of these pairings do not conform to the standard principle (A with U, and G with C).

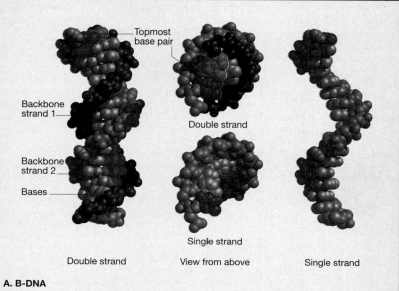

Topmost base pair

Backbone strand 1

Backbone strand 2

Bases

Double strand

Double strand

View from above

Single strand

Single strand

A. B-DNA

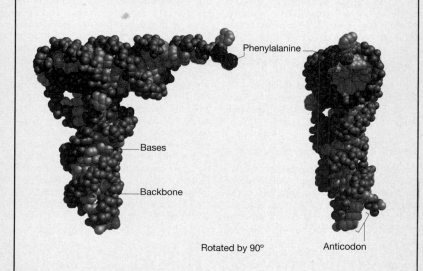

Phenylalanine

Bases

Backbone

Rotated by 90°

Anticodon

B. Phe-tRNA^Phe

Enzyme Classification

Enzymes are **biological catalysts**, i.e., substances of biological origin that accelerate chemical reactions. The orderly progression of metabolic pathways would be impossible if it were not for the genetically determined suite of enzymes found in all cells. It is the existence of these enzymes that establishes coordinated series of reactions (**metabolic pathways**, see p. 104). Most mechanisms that control metabolic processes involve enzymes, and thus regulation of their activities allows the metabolism to adapt to changing conditions (see p. 106). Almost all enzymes are **proteins**. However, there are also catalytically active nucleic acids, the *"ribozymes"* (see p. 224).

A. Enzymatic activity ●

The **activity** of an enzyme, a measure of its catalytic action, is obtained by determining the *increase in reaction rate* under defined conditions, i.e., the difference (violet) between the turnover of the catalyzed (orange) and the uncatalyzed reactions (yellow) within a particular period of time. Normally, reaction rates are expressed as *change in concentration per unit time* (mol · l^{-1} · s^{-1} see p. 16). Since the catalytic activity of an enzyme is independent of the volume, *substrate turnover per unit time* is commonly used for enzyme-catalyzed reactions, with the unit **katal** (kat, mol turnover · s^{-1}). A second customary unit is the **international unit** (**U**, μmol · min^{-1}, 1 U = 16.7 nkat).

B. Reaction and substrate specificity ●

The action of most enzymes is very *specific*. This specificity not only applies to the type of reaction being catalyzed (**reaction specificity**), but also to the nature of the reactants ("substrates") that are involved (**substrate specificity**). Fig. B uses a bond-breaking enzyme as an example. Highly specific enzymes (type A, top) catalyze the cleavage of only one type of bond, provided the structure of the substrate is the correct one. Other enzymes (type B, middle) have a narrow reac-

tion specificity, but a broad substrate specificity. Enzymes of type C (low reaction specificity *and* low substrate specificity, bottom) are very rare.

C. The enzyme classes ◐

Today, approximately 2000 different enzymes are known. A system of *classification* has been developed that takes into account both the *reaction specificity* and the *substrate specificity* of the enzymes. All enzymes are entered in the "Enzyme Catalogue" under a four-digit number, the **EC number** (see p. 378 and subsequent pages). The first digit indicates membership of one of the 6 **major classes**. The next two indicate subclasses and sub-subclasses. The last digit indicates where the enzyme belongs in the sub-subclass. Thus, lactate dehydrogenase (see p. 94) has the EC number *1.1.1.27* (Class 1, oxidoreductases; subclass 1.1, CH-OH group as electron *donor*; sub-subclass 1.1.1, NAD(P)$^+$ as electron *acceptor*).

Each of the six major classes contains enzymes with the same reaction specificity. The **oxidoreductases** (class 1), catalyze the transfer of reducing equivalents from one redox system to another. The **transferases** (class 2) catalyze the transfer of other functional groups from one substrate to the other. Most oxidoreductases and transferases require coenzymes (see p. 100). The **hydrolases** (class 3) are also involved in group transfer, but the acceptor is always a water molecule. **Lyases** (class 4), sometimes also called "synthases", catalyze reactions involving either the removal or the formation of a double bond. The **isomerases** (class 5) move groups within a molecule without changing the gross composition of the substrate. The ligation reactions catalyzed by **ligases** ("synthetases," class 6) are energy-dependent. As such, they are always coupled to the hydrolysis of nucleoside triphosphates.

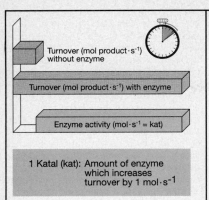

Turnover (mol product·s⁻¹) without enzyme

Turnover (mol product·s⁻¹) with enzyme

Enzyme activity (mol·s⁻¹ = kat)

1 Katal (kat): Amount of enzyme which increases turnover by 1 mol·s⁻¹

A. Enzymatic activity

B. Reaction and substrate specificity

	Reaction specificity	Substrate specificity
Ⓐ	High	High
Ⓑ	High	Low
Ⓒ	Low	Low

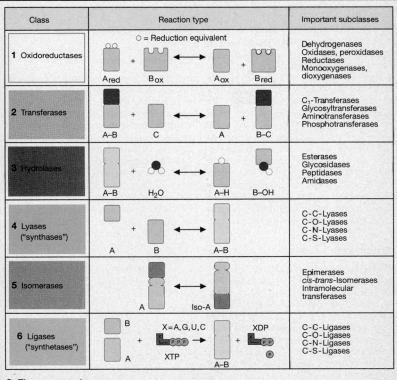

Class	Reaction type	Important subclasses
1 Oxidoreductases	$\bigcirc$ = Reduction equivalent — $A_{red} + B_{ox} \leftrightarrow A_{ox} + B_{red}$	Dehydrogenases Oxidases, peroxidases Reductases Monooxygenases, dioxygenases
2 Transferases	$A{-}B + C \leftrightarrow A + B{-}C$	C_1-Transferases Glycosyltransferases Aminotransferases Phosphotransferases
3 Hydrolases	$A{-}B + H_2O \leftrightarrow A{-}H + B{-}OH$	Esterases Glycosidases Peptidases Amidases
4 Lyases ("synthases")	$A + B \leftrightarrow A{-}B$	C-C-Lyases C-O-Lyases C-N-Lyases C-S-Lyases
5 Isomerases	$A \leftrightarrow Iso{-}A$	Epimerases cis-trans-Isomerases Intramolecular transferases
6 Ligases ("synthetases")	$B + A + XTP \xrightarrow{X=A,G,U,C} A{-}B + XDP + \text{\textcircled{P}}$	C-C-Ligases C-O-Ligases C-N-Ligases C-S-Ligases

C. The enzyme classes

Enzyme Catalysis

Enzymes are extremely effective **catalysts**. Some enzymes increase the rate of the reaction they catalyze by a factor of 10^{12} or more. In order to better understand the mechanism of enzymatic catalysis, we first consider the course of an uncatalyzed reaction.

A. Uncatalyzed reaction ○

We can take the reaction $A + B \rightarrow C + D$ as an example. In solution, the **reactants** A and B are surrounded by a shell of water molecules (*hydration shell*), and they are moving in random directions as the result of Brownian motion. A and B can only interact with one another when they collide in a favorable orientation — a circumstance that rarely occurs. Prior to conversion to the products C + D, the resulting **collision complex** must proceed through a **transition state**, the formation of which usually requires a large amount of **activation energy**, E_a (see p. 16). Since only a few of the A-B complexes have this amount of energy, formation of a productive transition state is even less likely than that of the collision complex. In solution, a large proportion of the activation energy is required for the *removal of the hydration shells* surrounding A and B. However, *charge displacements* and other *chemical processes* within the reactants also play a role. Because of these restrictions, product formation is rare in the absence of a catalyst and the reaction rate is low, even when the reaction is thermodynamically feasible, i.e., $\Delta G^o < 0$.

B. Enzyme-catalyzed reaction ○

Enzymes are able to specifically bind the reactants (their substrates) at the **active site**. This enables them to position the substrates in an *optimal orientation* for the formation of the transition state (**1–3**). The closer **proximity** and **optimized orientation** of the reactants dramatically increases the likelihood of the formation of *productive* A-B complexes. In addition, binding of the substrates at the active center of the enzyme results in removal of their hydration shells. The **exclusion of water** means that the reactants experience very different conditions when they are associated with the enzyme than if they had been free in solution (**3–5**). A third factor is the **stabilization of the transition state** as a result of interactions between the amino acid residues of the enzyme protein and the substrate (**4**). This results in an even greater decrease in the activation energy required for the formation of the transition state. In addition, many enzymes bring about the transfer of specific groups to and from the substrate during catalysis. *Proton transfer*, or **acid-base catalysis**, is especially common. Enzyme-facilitated acid-base catalysis is much more effective than the exchange of protons between acids and bases in solution. Formation of covalent bonds between certain chemical groups of the substrate and amino acid residues of the enzyme is also common. This is referred to as **covalent catalysis**.

C. Principles of enzyme catalysis ○

It is difficult to provide quantitative estimates of the contributions of the individual catalytic effects to the overall catalytic activity of an enzyme. Nevertheless, it is thought that the **stabilization of the transition state** is the most important factor, i.e., it is not the tight binding of the substrate that is important, but rather the binding of the transition state. This conclusion is supported by the very high affinity of many enzymes for analogs of the transition state. A simple mechanical analogy will help to explain this fact (right). In order to transfer the metal balls (the reactants) from EA (the substrate state) via the higher energy transition state to EP (the product state), the magnet (the catalyst) must be orientated so that its force of attraction acts on the transition state (bottom) rather than on EA (top).

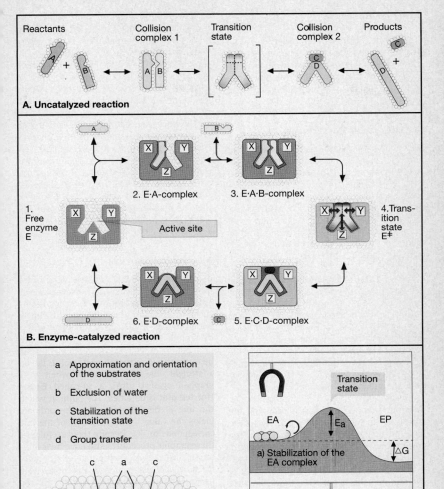

A. Uncatalyzed reaction

Reactants — Collision complex 1 — Transition state — Collision complex 2 — Products

B. Enzyme-catalyzed reaction

1. Free enzyme E
2. E·A-complex
3. E·A·B-complex
4. Transition state E‡
5. E·C·D-complex
6. E·D-complex

Active site

C. Principles of enzyme catalysis

a Approximation and orientation of the substrates

b Exclusion of water

c Stabilization of the transition state

d Group transfer

Transition state

a) Stabilization of the EA complex

EA E_a EP ΔG

b) Stabilization of the transition state

EA   EP

Enzyme Kinetics

The kinetics of an enzyme-catalyzed reaction (i.e., the dependence of its rate on the reaction conditions) is mainly determined by the *properties of the catalyst*. Thus, it is more complex than that of an uncatalyzed reaction (see p. 16).

A. The Michaelis-Menten model ◐

The complete mathematical analysis of enzyme-catalyzed reactions leads to complicated equations that are not suitable for practical applications. However, a simple model, developed in 1913 by *L. Michaelis* and *M. Menten*, is usually sufficient. It explains the characteristic hyperbolic dependence of enzyme activity on substrate concentration and also provides constants that describe the catalytic capacity of the enzyme in quantitative terms.

The *Michaelis-Menten model* assumes that the substrate A binds to the enzyme E and, once bound, reacts much more rapidly to form the product B than in the absence of the enzyme. Thus, the rate constant k_{cat} (2) is much larger than k, the corresponding constant for the uncatalyzed reaction; k_{cat}, the so-called **turnover number,** is the number of substrate molecules converted to product per enzyme molecule per second.

According to the Michaelis-Menten model, the most important factor determining the activity of an enzyme is the proportion of the total enzyme molecules, which exist in the form of the EA complex, i.e., the ratio $[EA]/[E]_t$ (3). For the sake of simplicity, the model assumes that E, A, and EA are all in chemical equilibrium (this is not strictly true, because of the constant removal of EA). Thus, for the dissociation of the EA complex, the law of mass action yields $[E] \cdot [A]/[EA] = K_m$. Given that $[E]_t = [E] + [EA]$, it follows that $[EA]/[E]_t = [A]/(K_m + [A])$. Now, as $v = k_{cat} [A]$ (2), we finally arrive at the **Michaelis-Menten equation** (4).

The equation contains two terms that do not depend on the substrate concentration: the term $k_{cat} \cdot [E]_t$ corresponds to the **maximum velocity V** that is attained at high substrate concentrations (when $[A] > K_m$, v approximately equals $k_{cat} \cdot [E]_t$), i.e., $k_{cat} = V/[E]_t$.

The **Michaelis constant K_m**, on the other hand, corresponds to the substrate concentration at which v is half of the maximum velocity V (when $v = V/2$, $[A]/(K_m + [A]) = 1/2$, i.e., $K_m = [A]$). An enzyme with a high affinity for its substrate has a low K_m value and *vice versa*.

The Michaelis-Menten model is based on rather unrealistic assumptions: first, that the conversion of EA to E + B is irreversible, second, that E, A, and EA are in equilibrium, and third, that there are no forms of the enzyme other than E and EA. Only when these assumptions all apply does K_m correspond to the dissociation constant of the EA complex, and k_{cat} to the rate constant of the reaction EA → E + B.

B. Determination of V and K_m ○

In principle, V and K_m can be determined from a plot of v against [A]. However, since v approaches V asymptotically with increasing substrate concentration [A], it is difficult to obtain reliable values for V (and thus of K_m) by simple extrapolation. To overcome this problem, the Michaelis-Menten equation can be transformed so that the data points fall on a *straight* line. For example, when v is graphed against v/[A], a so-called **Eadie-Hofstee plot** is obtained. The point at which the line of best fit intersects the ordinate yields the value V and the slope of the line corresponds to $-K_m$. Despite their usefulness, these graphic procedures also have their limitations. Nowadays, kinetic data can be analyzed much more quickly and objectively using computer programs written for the purpose.

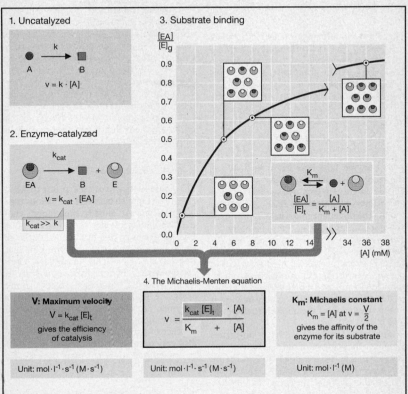

1. Uncatalyzed

$$v = k \cdot [A]$$

2. Enzyme-catalyzed

$$v = k_{cat} \cdot [EA]$$

$$k_{cat} \gg k$$

3. Substrate binding

$$\frac{[EA]}{[E]_t} = \frac{[A]}{K_m + [A]}$$

4. The Michaelis-Menten equation

V: Maximum velocity

$$V = k_{cat} [E]_t$$

gives the efficiency of catalysis

Unit: $mol \cdot l^{-1} \cdot s^{-1}$ ($M \cdot s^{-1}$)

$$v = \frac{k_{cat} [E]_t \cdot [A]}{K_m + [A]}$$

Unit: $mol \cdot l^{-1} \cdot s^{-1}$ ($M \cdot s^{-1}$)

K_m: Michaelis constant

$$K_m = [A] \text{ at } v = \frac{V}{2}$$

gives the affinity of the enzyme for its substrate

Unit: $mol \cdot l^{-1}$ (M)

A. The Michaelis-Menten model

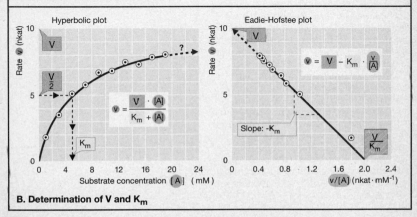

Hyperbolic plot

$$v = \frac{V \cdot [A]}{K_m + [A]}$$

Eadie-Hofstee plot

$$v = V - K_m \cdot \frac{v}{[A]}$$

Slope: $-K_m$

Substrate concentration [A] (mM)

$v/[A]$ (nkat · mM^{-1})

B. Determination of V and K_m

Inhibitors

There are numerous substances that affect metabolic processes by modulating the activity of specific enzymes. Of particular importance are **enzyme inhibitors**. Many drugs, either natural or synthetic in origin, act as enzyme inhibitors. Inhibitory metabolites are also involved in enzyme regulation (see p. 106)

A. Types of inhibitors ○

Most enzyme inhibitors act reversibly, i.e., they do not bring about any irreversible change in the enzyme. However, there are also *irreversible* inhibitors that permanently modify the target enzyme. The mechanism of action of an inhibitor, i.e., its type of inhibition, can be determined by comparing the kinetics (see p. 90) of the inhibited and uninhibited reactions (see **B**). In this way, it is possible to distinguish **competitive** inhibitors (left) from **non-competitive** inhibitors (right). **Allosteric** inhibition is very important in the regulation of metabolism (below; see also p. 112).

Substrate analogs (**2**) have properties similar to those of one of the substrates of the target enzyme. They are bound to the enzyme like a substrate, but cannot be converted to the product(s). Thus, they *reversibly* block a certain fraction of the enzyme molecules present. Because the substrate and the inhibitor compete with one another for the same binding site on the enzyme, this type of inhibition is referred to as competitive. **Analogs of the transition state** (**3**) usually also act in competitive fashion.

When an inhibitor interacts with a functional group that is essential for enzyme activity, but does not affect binding of the substrate, then the inhibition is **non-competitive** (right). In this case, the Michaelis constant remains unchanged, but the concentration of functional enzyme, and thus the maximal rate V, decrease (see p. 90). Non-competitive inhibitors often act *irreversibly* by modifying functional groups of the target enzyme (**4**).

"Suicide substrates" (**5**) are substrate analogs that contain an additional reactive group. Initially, they bind reversibly, and then they form a covalent bond with the active center of the enzyme (for an example, see p. 233C). Inhibition by such compounds is *non-competitive*.

Allosteric inhibitors bind to a separate binding site outside the active center (**6**). This results in a *conformational change* of the enzyme protein that indirectly decreases its activity (see also p. 112).

B. Kinetics of inhibition ○

The *Eadie-Hofstee plot* (see p. 90) makes it easy to distinguish between the various types of inhibition. As already mentioned, **competitive** inhibitors only affect K_m, but not V. Thus, the lines of best fit obtained in the absence and presence of a competitive inhibitor intersect on the ordinate. In contrast, a **non-competitive** inhibitor gives a line with the same slope (K_m unchanged), while it intersects the ordinate lower down. **Allosteric** enzymes, i.e., those that can exist in various conformations, depending on the presence or absence of particular ligands (see p. 112), are also easy to recognize, because they give rise to curved Eadie-Hofstee plots (not shown).

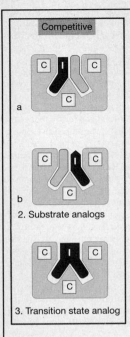

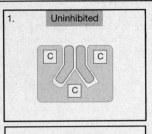

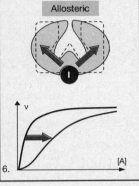

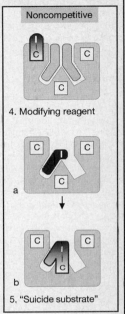

Competitive

a

b
2. Substrate analogs

3. Transition state analog

1. **Uninhibited**

Allosteric

6.

Noncompetitive

4. Modifying reagent

a

↓

b
5. "Suicide substrate"

A. Types of inhibitors

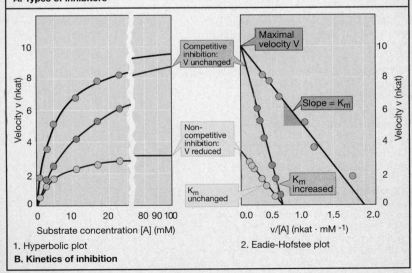

Competitive inhibition: V unchanged

Non-competitive inhibition: V reduced

K_m unchanged

Maximal velocity V

Slope = K_m

K_m increased

1. Hyperbolic plot

2. Eadie-Hofstee plot

B. Kinetics of inhibition

Lactate Dehydrogenase: Structure

In the following, we discuss **lactate dehydrogenase** (LDH, *EC 1.1.1.27*) as an example of the structure and function of an enzyme.

A. Structure ○

The active form of lactate dehydrogenase (mass 144 kDa) is a **homotetramer**, i.e., it is made up of four identical subunits (**1**). Each subunit is formed by a single peptide chain of 334 amino acids (36 kDa). In the tetramer, the subunits occupy equivalent positions; each monomer harbors an active site.

In mammals, LDH has two *different* types of subunits (H and M) with somewhat different amino acid sequences. These subunits associate randomly to form tetramers. This random association gives rise to five different **isozymes** of LDH. In heart muscle tissue, tetramers made up of H subunits predominate (hence the designation H), whereas in the liver and skeletal muscle, M monomers are most common.

The **active site** of LDH adopts different conformations in the absence or presence of ligands. In Fig. 2 we show a part of one subunit to which the substrate (lactate, red) and the coenzyme (NAD⁺, yellow) are bound. In this state, a loop encompassing amino acid residues 98–111 (magenta) is closed over the active site. The role of this loop in catalysis, as well as the functions of three further amino acid residues (Arg-171, Arg-109, and His-195; green), are discussed in more detail on the next page.

B. Pyridine nucleotide coenzymes ◑

All dehydrogenases require *coenzymes* for the transfer of reducing equivalents. The coenzymes are usually dinucleotides (see p. 100), i.e., they consist of two nucleoside 5'-monophosphates linked by a phosphoric acid-anhydride bond. LDH requires **nicotinamide adenine dinucleotide**, abbreviated **NAD⁺**. NAD⁺ is made up of 5'-AMP and a nucleotide containing *nicotinamide* (see p. 333) as the base. Another coenzyme with a very similar structure is **NADP⁺** (NAD⁺ phosphate).

In this molecule, the 3'-OH group of the ribose residue linked to adenine is esterified with phosphoric acid (**1**). Despite their structural similarity, NAD⁺ and NADP⁺ play very different roles in metabolism (see p. 104).

It is only the *nicotinamide ring* of the pyridine nucleotide coenzymes that is involved in redox reactions (**2**). Nicotinamide is the acid amide of pyridine-3-carboxylic acid (*nicotinic acid*). In the oxidized form (above), the ring is aromatic, and carries a positive charge. For this reason, the coenzyme in the oxidized state is designated NAD⁺. During the *oxidation* of lactate by LDH, two hydrogen atoms (i.e., two electrons and two protons) are removed from the substrate (middle). However, it is only a **hydride ion** (H⁻, one proton with two electrons) that is transferred to NAD⁺. The target of the hydride transfer is the carbon in *para* position to the ring nitrogen. At this site, an alicyclic CH_2 group is formed, the double bonds are rearranged, and the positive charge disappears (below).

The proton released during the hydride transfer to NAD⁺ must not be ignored and, therefore, the correct designation for the reduced coenzyme is *not NADH₂* but **NADH + H⁺**.

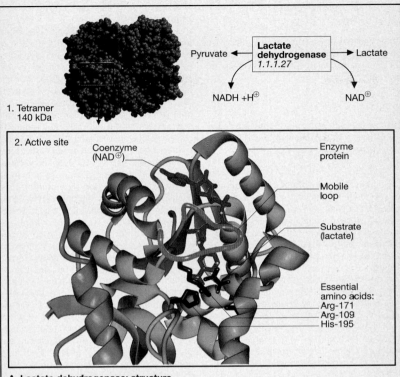

Pyruvate ◄─── **Lactate dehydrogenase** *1.1.1.27* ───► Lactate

NADH +H$^{\oplus}$ NAD$^{\oplus}$

1. Tetramer 140 kDa

2. Active site

Coenzyme (NAD$^{\oplus}$)

Enzyme protein

Mobile loop

Substrate (lactate)

Essential amino acids:
Arg-171
Arg-109
His-195

A. Lactate dehydrogenase: structure

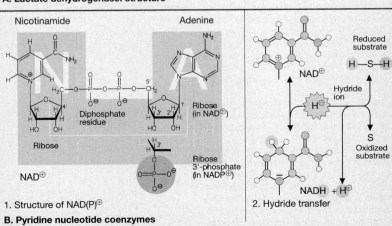

Nicotinamide   Adenine

Diphosphate residue

Ribose

Ribose (in NAD$^{\oplus}$)

Ribose 3'-phosphate (in NADP$^{\oplus}$)

NAD$^{\oplus}$

1. Structure of NAD(P)$^{\oplus}$

Reduced substrate

NAD$^{\oplus}$

H —S—H

Hydride ion

H$^{\ominus}$

S Oxidized substrate

NADH + H$^{\oplus}$

2. Hydride transfer

B. Pyridine nucleotide coenzymes

Lactate Dehydrogenase: Mechanism

The principles of enzyme catalysis discussed on p. 88 can be illustrated using the reaction mechanism of lactate dehydrogenase as an example.

A. Catalytic Cycle ○

LDH catalyzes the transfer of reducing equivalents (see p. 100) either from NADH to pyruvate or from lactate to NAD⁺.

Pyruvate + NADH + H⁺ ↔ Lactate + NAD⁺

The equilibrium state of the reaction strongly favors lactate *formation* (see p. 12). However, in the presence of high concentrations of lactate and NAD⁺, the oxidation of lactate to pyruvate is also possible. LDH catalyzes the reaction in both directions, but — like all enzymes — does not affect the chemical equilibrium.

Since the reaction is reversible, the catalytic process can be represented as a closed loop. Six "snapshots" from the **catalytic cycle** of LDH are shown. Catalytic intermediates like those included here are extremely short-lived and therefore difficult to detect. Their existence has been deduced indirectly from a large number of experimental findings.

Many amino acid residues play a role in the **active center** of LDH (see preceding page). Either they are involved in the binding of the substrate and coenzyme, or they take part directly in one of the steps in the catalytic cycle. Only the three most important side chains are shown here. The positively charged guanidinium group of **arginine-171** binds the carboxylate group of the substrate by electrostatic interactions. The imidazole group of **histidine-195** is involved in acid-base catalysis, and the side chain of **arginine-109** is important for the stabilization of the transition state. The two essential arginine residues are permanently protonated during catalysis, while His-195 alters its charge. In addition, the illustration shows the **peptide loop** already mentioned on p. 94 (magenta). Subsequent to the binding of the substrate and coenzyme, the loop, by closing the active site, serves to exclude water during the electron transfer

Now let us consider the partial reactions involved in the LDH-catalyzed reduction of pyruvate by LDH. His-195 is protonated in the free enzyme (**1**). Thus, this form of the enzyme is designated E . H⁺. The coenzyme NADH binds first (**2**), followed by pyruvate (**3**). It is important that the carbonyl group of the pyruvate and the nicotinamide ring of the coenzyme attain the optimal orientation with respect to one another, and that this orientation becomes fixed (*proximity and orientation of substrates*). The 98–111 loop now closes over the active site. This causes a marked decrease in polarity, and thus promotes the formation of the transition state (**4**) (*water exclusion*). In the transition state, a hydride ion, H⁻ (see p. 94B), is transferred from the coenzyme to the carbonyl carbon (*group transfer*). The transiently formed negative charge on the oxygen is stabilized by electrostatic interaction with Arg-109 (*stabilization of the transition state*). Simultaneously, a proton from His-195 is transferred to this oxygen atom (*group transfer*), giving rise to lactate and NAD⁺ as enzyme-bound products (**5**). After this, the loop opens, lactate dissociates from the enzyme, and the temporarily uncharged imidazole-group of His-195 binds a proton from the surrounding water (**6**). Finally, the oxidized coenzyme NAD⁺ is released, and the enzyme returns to its original state (**1**).

The same steps occur during the oxidation of lactate to pyruvate, but in the opposite direction.

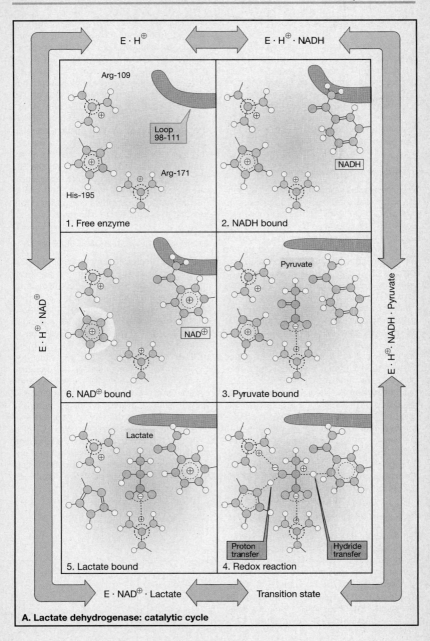

A. Lactate dehydrogenase: catalytic cycle

Enzymatic Analysis

Enzymes play an important role in *biochemical analyses*. Even minute quantities of an enzyme can be detected in biological material, e.g., body fluids, by measuring its activity. Enzymes are also used as reagents for the determination of metabolite concentrations, e.g., blood glucose levels (**C**). Most procedures involving enzymatic analysis employ absorption photometry.

A. Principle of spectrophotometry ◑

Many molecules *absorb* light in the visible or ultraviolet region of the spectrum. This property can be used to determine the concentration of these molecules. The amount of light absorbed depends on the type and concentration of the substance being analyzed, and on the wavelength of the light being used. Therefore, it is important to use **monochromatic light**, i.e., light of a defined wavelength, derived from white light with the help of a *monochromator*. Monochromatic light with an intensity I_0 is passed through a rectangular cell made of glass or quartz (a *cuvette*) containing a solution of the absorbing substance. The intensity I of the light transmitted through the solution is measured using a *detector*. Clearly, intensity I will be less than I_0. The **absorption A** (often also referred to as the *extinction*) of a solution is defined as the *negative logarithm of the ratio I/I_0*. The **Beer-Lambert law** states that A is proportional to the concentration, c, of the absorbing substance and the length of the light path through the solution, d. As stated above, the **absorption coefficient** ε is dependent on the type of substance and the wavelength.

B. Assay of lactate dehydrogenase activity ◑

The determination of the activity of lactate dehydrogenase (LDH) is based on the fact that the reduced coenzyme NADH + H+ absorbs light at 340 nm, whereas NAD+ does not.

Absorption spectra (i.e., plots of A vs. wavelength) for the substrate and the coenzyme of the LDH reaction are shown in Fig. **1**. The difference in the absorption be-

havior of NAD+ and NADH between 300 and 400 nm is due to the changes in the nicotinamide ring resulting from oxidation or reduction (see p. 94). For the determination of LDH activity, a solution containing lactate and NAD+ is added to a cuvette, and absorption is monitored at a *constant* wavelength of 340 nm. The uncatalyzed LDH reaction is extremely slow and, therefore, it is only following addition of the enzyme that measurable quantities of NADH are formed and the absorption increases. According to the Beer-Lambert law, the rate of increase in the absorption $\Delta A/\Delta t$ is proportional to the reaction rate $\Delta c/\Delta t$. Thus, from this ratio, and taking into account the absorption coefficient ε, it is possible to calculate LDH activity.

C. Enzymatic determination of glucose ○

Most biomolecules do not absorb light in the visible region of the spectrum. In addition, they are usually present as mixtures with other compounds. Both of these problems can be circumvented by using an appropriate enzyme to selectively convert the metabolite to be determined into a colored substance, with an absorption that can then be measured. A frequently used procedure for the determination of glucose in the monitoring of blood glucose levels involves two successive reactions. In the first step, a glucose-specific *glucose oxidase* produces hydrogen peroxide (H_2O_2). In the second step, a *peroxidase* catalyzes the conversion by H_2O_2 of a colorless precursor to a green dye. When all of the glucose in the sample has been used up, the final concentration of the dye ([Green dye]$_\infty$) is the same as the original glucose concentration ([Glucose]$_0$).

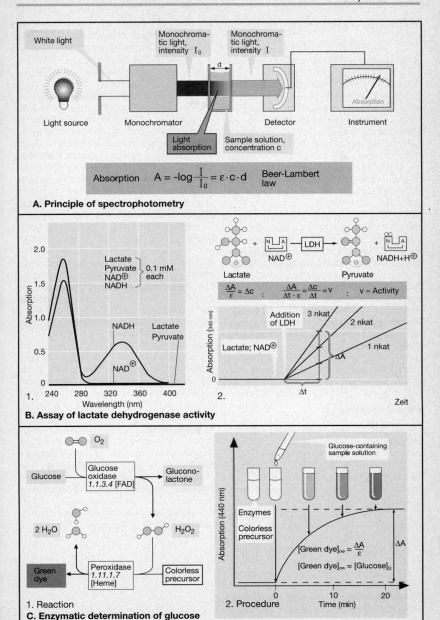

Absorption $A = -\log \dfrac{I}{I_0} = \varepsilon \cdot c \cdot d$ Beer-Lambert law

A. Principle of spectrophotometry

$\dfrac{\Delta A}{\varepsilon} = \Delta c$; $\dfrac{\Delta A}{\Delta t \cdot \varepsilon} = \dfrac{\Delta c}{\Delta t} = v$; $v \approx$ Activity

B. Assay of lactate dehydrogenase activity

$[\text{Green dye}]_\infty = \dfrac{\Delta A}{\varepsilon}$

$[\text{Green dye}]_\infty = [\text{Glucose}]_0$

1. Reaction
2. Procedure

C. Enzymatic determination of glucose

Redox Coenzyme

A. Coenzymes: definitions ●

Many enzyme-catalyzed reactions involve the transfer of electrons or groups of atoms from one substrate to another. Such reactions always involve "helper" molecules, which act as temporary acceptors of the group being transferred. Helper molecules of this type are called **coenzymes**. Since they are not catalytically active themselves, the term "*cosubstrate*" would be more appropriate. Enzymes are usually highly specific for their substrates, whereas coenzymes cooperate with many different enzymes of differing substrate specificity.

A distinction is made between soluble coenzymes and prosthetic groups, based on the nature of their interaction with the enzyme. During a reaction, **soluble coenzymes (1)** are bound like substrates, undergo a chemical change, and are then released. A second, independent reaction is required for the regeneration of the original form of the coenzyme. In contrast, **prosthetic groups (2)** are coenzymes that are *tightly bound to the enzyme*, and remain associated with it during the reaction. The group bound by the coenzyme is later transferred to another substrate of the *same* enzyme (not shown in Fig. 2).

B. Redox coenzymes ◗

All oxidoreductases (see p. 86) require coenzymes. The most important of these redox coenzymes are shown here. They can act either in soluble form (S) or as prosthetic groups (P). Redox reactions often involve the transfer of one or two protons, together with the electrons, from one molecule to another. Together, these are referred to as **reducing equivalents**. The standard reduction potential, E^o, of a prosthetic group (see p. 28) can vary markedly depending on its environment within a particular enzyme.

The usual coenzymes associated with dehydrogenases are the pyridine nucleotides **NAD+ and NADP+ (1)**. They transfer hydride ions ($2e^-$ and $1 H^+$, see p. 94) and are always acting in soluble form. NAD+ carries reducing equivalents from catabolic pathways to the respiratory chain and thereby plays a role in energy metabolism. In contrast, NADP+ is the most important reductant in biosynthetic processes (see p. 104).

The flavin coenzymes **FMN** and **FAD (2)** are usually tightly bound to the enzyme. The redox-active group of both coenzymes is *flavin* (isoalloxazine), a structure made up of three fused rings that can become reduced by accepting up to two electrons and two protons. FMN carries the phosphorylated sugar alcohol *ribitol* at the flavin ring. FAD consists of FMN linked to AMP. Both coenzymes perform similar functions. They are found in *dehydrogenases, oxidases*, and *monooxygenases*.

In **lipoic acid (3)**, an intramolecular *disulfide bond* is the redox center. The active lipoic acid is covalently bound to a lysine residue (R′) of the corresponding enzyme (**lipoamide**). Lipoic acid is involved primarily in the decarboxylation of 2-oxoacids (see p. 124). The peptide coenzyme **glutathione** also contains a disulfide bond (see p. 260).

The role of **ubiquinone (coenzyme Q, 4)** in the transfer of reducing equivalents in the respiratory chain is discussed on p. 130. During reduction, the *quinone* form is converted into the *hydroquinone (quinol)* form. Similar quinone/hydroquinone systems are found in photosynthesis (see p. 118). *Vitamins E and K* (see p. 330) also belong to this class of redox systems.

Heme groups (5) are important redox cofactors in the *respiratory chain* (see p. 130), in *photosynthesis* (see p. 118), in *monooxygenases* (see p. 292) and *peroxidases*, and other processes. Unlike in hemoglobin, the iron ion undergoes a change in valence here. The illustration shows the heme of cytochrome c, which is covalently bound to two cysteine residues of the protein (R_2).

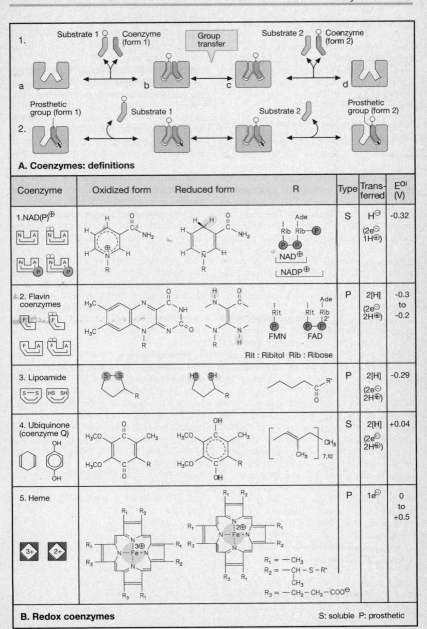

A. Coenzymes: definitions

Coenzyme	Oxidized form	Reduced form	R	Type	Transferred	E^{0I} (V)
1. NAD(P)$^{\oplus}$				S	H$^{\ominus}$ (2e$^{\ominus}$ 1H$^{\oplus}$)	-0.32
2. Flavin coenzymes			Rit : Ribitol Rib : Ribose	P	2[H] (2e$^{\ominus}$ 2H$^{\oplus}$)	-0.3 to -0.2
3. Lipoamide				P	2[H] (2e$^{\ominus}$ 2H$^{\oplus}$)	-0.29
4. Ubiquinone (coenzyme Q)				S	2[H] (2e$^{\ominus}$ 2H$^{\oplus}$)	+0.04
5. Heme			$R_1 = -CH_3$ $R_2 = -CH-S-R'$ $\qquad CH_3$ $R_3 = -CH_2-CH_2-COO^{\ominus}$	P	1e$^{\ominus}$	0 to +0.5

B. Redox coenzymes   S: soluble P: prosthetic

Group-Transferring Coenzymes

On the previous page the redox coenzymes are summarized. Here we consider coenzymes involved in group transfer reactions. Cobalamine (coenzyme B_{12}) is discussed on p. 334.

A. Group-transferring coenzymes ◑

The **nucleoside phosphates (1)** not only act as precursors for nucleic acids, but also as coenzymes. They are involved in *energy conservation* and, in addition, they provide the driving force for endergonic processes by energetic coupling (see p. 114). Metabolic intermediates can be made more reactive ("activated") as a result of the attachment of phosphate-containing residues. For example, the attachment of nucleoside diphosphate residues generates activated precursors for polysaccharide and lipid synthesis (see pp. 160 and 168). Reactions catalyzed by *ligases* also depend on nucleoside triphosphates.

Acyl residues are usually activated by transfer to **coenzyme A (2)**. Coenzyme A is basically *pantetheine* linked to to 3′-phospho-ADP by a phosphoric ester bond. Pantetheine consists of three components connected by amide bonds: *pantoinic acid*, β*-alanine* and *cysteamine*. The latter two components are biogenic amines formed by the decarboxylation of aspartate and cysteine, respectively (see p. 154). **Thioesters**, such as acetyl-CoA, are produced by the reaction of the thiol group of the cysteamine residue with different carboxylic acids. As the reaction is strongly endergonic, it is always coupled to exergonic processes. Thioesters are activated forms of the carboxylic acids, because their acyl residues are readily transferred to other molecules. This principle is frequently applied in intermediary metabolism.

Thiamine diphosphate (3) is capable of activating aldehydes or ketones, producing *hydroxyalkyl groups*, which are then passed on to other molecules. This type of group transfer is important in the transketolase reaction, for example (see p. 142). Hydroxyalkyl residues are also involved in the decarboxyla-

tion of oxoacids. Here, they are either released as aldehydes or transferred to lipoamide, as in the case of the 2-oxoacid dehydrogenases (see p. 124)

Pyridoxal phosphate (4) is the most important coenzyme in amino acid metabolism. Its role in *transamination* reactions is discussed in detail on p. 152. Many other reactions involving amino acids, such as decarboxylations and dehydratations, also require pyridoxal phosphate. The aldehyde form of pyridoxal phosphate shown here is not usually found free in nature. In the absence of substrates, its aldehyde group is attached to the ε-amino group of a lysine residue of the enzyme as an *aldimine* ("Schiff's base").

The carboxylases contain **biotin (5)** as coenzyme. It is bound via the carboxyl group to the side chain of a lysine residue of the enzyme. When ATP is present, biotin reacts with hydrogen carbonate (HCO_3^-) to form *N-carboxybiotin*. From this activated derivative, carbon dioxide can then be transferred to other molecules. Examples of biotin-dependent reactions are the synthesis of malonyl-CoA from acetyl-CoA (see p. 144), and the formation of oxaloacetate from pyruvate (see p. 158)

Tetrahydrofolate (THF, **6**) is a coenzyme that transfers one-carbon (C_1) residues of different oxidation states. THF is synthesized from the vitamin *folic acid* (see p. 332) by repeated hydrogenation of the pterin ring. The C_1 units are bound to N-5, N-10, or both. In the most important derivatives, the C_1 residue has the oxidation state of a) a carboxylic acid in N^{10}-*formyl-THF*, b) an aldehyde in N^5,N^{10}-*methylene-THF*, and c) an alcohol in N^5-*methyl-THF*. C_1 units transferred by THF play a role in the synthesis of purine nucleotides (see p. 174), dTMP (see p. 176), and methionine (see p. 375).

Coenzyme (symbol)	Free form	Charged form	Group(s) transferred	Important enzymes
1. Nucleoside phosphates P P P	Ade, Gua, Cyt, Ura Bases B	$H_2C-O-P-P-P$ a | b | c Nucleoside monophosphate (a) diphosphate (b) triphosphate (c)	P B-Rib B-Rib-P B-Rib-$P$$P$	Phospho-transferases Nucleotidyl-transferases (2.7.n.n) Ligases (6.n.n.n)
2. Coenzyme A A	Pantoinic acid | β-Ala | Cysteamine H_3C CH_3 ... OH ... SH P P—Rib—Ade P 3'	O S C R	Acyl residues	Acyltrans-ferases (2.3.n.n) CoA trans-ferases (2.8.3.n)
3. Thiamine diphosphate TPP	H_3C N NH_2 H S O—$P$$P$ N^+ CH_3 CH_3	R $HO-C-H$ S N^+ CH_3	Hydroxy-alkyl residues	Decarboxy-lases (4.1.1.n) Oxoacid de-hydrogenases (1.2.4. n) Transketolase (2.2.1.1)
4. Pyridoxal phosphate PLP	O C H HO H_3C N^+ H P	$NH_3^{⊕}$ CH_2 HO H_3C N^+ H P	Amino group Amino acid residues	Transaminases (2.6.1.n) Many lyases (4.n.n.n)
5. Biotin B	O HN NH H H S $COO^⊖$	O O $^⊖O$ C N NH H H S	[CO_2]	Carboxylases (6.4.1.n)
6. Tetra-hydrofolate THF	H_2N N H H HN 5 CH_2 O N H $HN-R$ 10 $R = $ O—N—L-Glu H	5 H N CH_2 H N N $H-C-O$ 10 a 5 H N CH_2 H_2C N 10 b 5 H N CH_2 H_3C $HN-$ 10 c	C_1 groups a Formyl- b Methylen- c Methyl-	C_1 trans-ferases (2.1.n.n)

A. Group-transferring coenzymes

Intermediary Metabolism

In every cell there are hundreds of chemical reactions occurring, which, together, are referred to as **metabolism**. The compounds involved are called **metabolites**. Outside of the cell, these reactions would occur only very slowly and without any specific direction. By contrast, within cells, the existence of specific **enzymes** (see p. 86) establishes series of reactions with a high rate of flux in a particular direction, i.e., **metabolic pathways**.

A. Intermediary metabolism: Overview ●

The central metabolic pathways are common to most cells and organisms. These pathways, which serve in the synthesis, degradation, and interconversion of the most important metabolites, as well as in energy conservation, are referred to as **intermediary metabolism**. Here we present a very simplified overview of this branch of metabolism.

In order to survive, cells are constantly in need of organic and inorganic *nutrients,* as well as *chemical energy* derived primarily from ATP (see below). According to the way in which these needs are satisfied, organisms cans be divided into two groups—autotrophic and heterotrophic. The **autotrophs**, which include the plants and many micro-organisms, are capable of synthesizing organic compounds from inorganic precursors (CO_2). An autotrophic lifestyle is, for example, made possible by **photosynthesis** (see p. 118).

In contrast, the **heterotrophs**, e.g., the animals, are dependent on a constant supply of organic compounds in their diet. Most of these nutrients (proteins, carbohydrates, nucleic acids and lipids) cannot be utilized directly; they first have to be broken down into simpler compounds by degradative or **catabolic pathways** (red arrows). The metabolites that arise (together they are sometimes referred to as the "*metabolite pool*") are then catabolized further to yield energy, or else they are used in **anabolic pathways** (blue arrows) for the synthesis of more complex molecules. Although there are numerous metabolites in the "pool" we show only *pyruvate, acetyl-CoA* and *glycerol,* three particularly important molecules that are links between the metabolism of proteins, carbohydrates and lipids. The metabolite pool also includes the intermediates of the citric acid cycle (6). This cyclic pathway plays both catabolic and anabolic roles, i.e., it is **amphibolic** (see p. 128). The waste products from the degradation of organic substances by animals are *carbon dioxide* (CO_2), *water* (H_2O), and *ammonia* (NH_3). NH_3 is incorporated into *urea,* and it is in this form that it is excreted (see p. 286).

The most important storage form of energy in all cells is **adenosine triphosphate** (ATP, see p. 114). The synthesis of ATP is an energy-requiring process, i.e., it is *endergonic.* The cleavage of ATP to yield ADP and phosphate, on the other hand, releases a large amount of energy, it is strongly *exergonic.* By the process of **energetic coupling** (see p. 10), the exergonic hydrolysis of ATP can be used to drive energy-dependent (endergonic) processes. For example, most anabolic pathways, as well as movement and transport processes, are energy-dependent.

The most important pathway for the synthesis of ATP is **oxidative phosphorylation** (see p. 130). In this process, electrons are transferred from reduced coenzymes, produced by catabolic pathways, to oxygen. This exergonic process is used indirectly for the synthesis of ATP. Under *anaerobic* conditions, i.e., in the absence of oxygen, most organisms satisfy their needs with ATP formed in glycolysis (3). This less efficient alternative for the synthesis of ATP is referred to as **fermentation** (see p. 136). Only NADH can be used for oxidative phosphorylation, while NADPH, chemically a very similar coenzyme, serves as reducing agent in anabolic pathways. NADPH is derived predominantly from the hexose monophosphate pathway (HMP, 1).

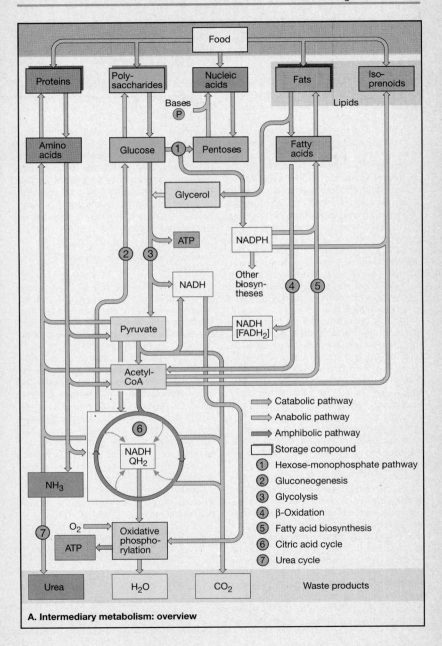

A. Intermediary metabolism: overview

Regulatory Mechanisms

A. Fundamental mechanisms of metabolic regulation ◑

The activities of the metabolic pathways (see p. 104) are constantly being monitored and adjusted so that the synthesis and degradation of metabolites satisfy the prevailing physiological requirements. Here we present an overview of the various mechanisms involved.

One important factor is the **availability of precursors** (metabolite A). The availability of a precursor increases with increasing activity of the metabolic pathways responsible for its synthesis (**1**), and decreases with increasing activity of pathways that consume it (**2**). Transport processes can also restrict the availability of A. In many cases, it is the **availability of coenzymes** that limits the flux through a pathway (**3**). When the coenzyme is regenerated via a second independent pathway, the rate of this second process can limit that of the first.

The flux through a metabolic pathway is determined primarily by the activities of the **enzymes** involved. For the purposes of metabolic regulation, it is sufficient to change the activity of the enzyme catalyzing the slowest step in the reaction chain. Such **key enzymes**, which are subject to various regulatory mechanisms, are found in most metabolic pathways.

The activity of key enzymes is regulated at three independent levels (**4–6**). The first is the expression of the enzyme protein. Intervention at this level is usually mediated by alterations in the synthesis of the mRNA encoding the enzyme, i.e., transcription of the particular gene is affected. This mechanism, which is referred to as **transcriptional control** (**4**, see p. 108) depends on *regulatory proteins* that directly interact with DNA. The activity of these proteins is, in turn, affected by hormones or metabolites. Stimulation of enzyme synthesis as a result of transcriptional regulation is referred to as **induction**. When enyzme synthesis is inhibited or prevented, this is referred to as **repression**. Induction and repression are relatively slow

to take effect, whereas enzyme **interconversion (5)** is much quicker. In the latter case, the enzyme is already present, but in an inactive form. When the need arises, the inactive enzyme is converted into its catalytically active form by an activating enzyme. This usually occurs in response to a hormonal signal. Most interconversions involve either ATP-dependent phosphorylation of the enzyme or phosphatase-dependent dephosphorylation of it (see p. 110).

Finally, the activity of key enzymes can be directly regulated by **ligands** (**6**). The most common example of this is the inhibition of a key enzyme by the immediate products of the catalyzed reaction, by the end products of the pathway ("*feedback*" inhibition), or by metabolites from an entirely different pathway.

B. Isosteric and allosteric enzyme regulation ◑

The regulatory mechanisms described above affect the kinetics of the enzymes involved in different ways. There are both **isosteric** and **allosteric processes**. In the first case (**1**), the substrate saturation curve for the enzyme (see p. 90) is hyperbolic, and its activity is mainly determined by changes in *substrate concentration* (**1a**, substrate red circles), by increased or decreased *enzyme synthesis* (**1b**) or by *inhibitors* (**1c**, inhibitors open circles). Allosteric enzymes (**2**), on the other hand, are always oligomers, and characteristically have *sigmoidal substrate saturation curves* (**2a**). Many allosteric enzymes exist in two conformations (see p. 112), a less active *T form* (left) and a more active *R form* (right). The binding of allosteric activators (open circles) increases the fraction of molecules in the R state (**2b**), whereas allosteric inhibitors (open rectangles) usually stabilize the T form (**2c**).

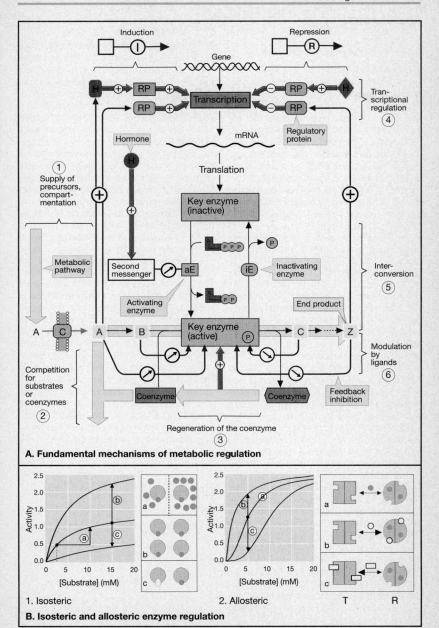

A. Fundamental mechanisms of metabolic regulation

1. Isosteric

2. Allosteric

B. Isosteric and allosteric enzyme regulation

Transcriptional Control

A. Functions of regulatory proteins ◑

In all cells, gene expression is directed by *regulatory proteins* that bind to particular DNA sequences and thus activate or inhibit the transcription of specific genes (see p. 224). The effects of these regulatory proteins are usually reversible and often *ligand-dependent*. Although new DNA binding proteins are constantly being discovered, our knowledge of them is still rather limited. Depending on the mechanism involved, various different terms are used to refer to the DNA binding proteins and the DNA sequences to which they bind. These designations are not always strictly defined. Here, we present only a few general terms. A regulatory protein that affects gene transcription is referred to as a **transcription factor**. When it inhibits transcription, it is called a **repressor,** and when it stimulates transcription, it is called an **inducer**. DNA sequences to which regulatory proteins bind are referred to as **control elements**. In prokaryotes, control elements serving as binding sites for RNA polymerases are called **promoters**, whereas repressor binding sequences are usually called **operators**.

The regulatory proteins known to date can be classified into four different groups, based on their mechanisms of action. **Negative gene regulation**, i.e., the *switching off* of a given gene, is brought about by repressors. Some repressors only bind to DNA in the absence of specific ligands (**1a**). In the presence of the ligand, the repressor-ligand complex is unable to remain bound to the DNA, and the promoter region becomes available for the binding of RNA polymerase (**1b**). In other cases, the *ligand-free* repressor cannot bind and, consequently, transcription is inhibited in the presence of the ligand (**2a, 2b**). Similarly, there are two different types of **positive gene regulation**. If it is only the ligand-free inducer that binds, then transcription is inhibited by the appropriate ligand (**3**). On the other hand, many inducers only become active when they have bound a ligand (**4**). One example of this is the receptors for steroid hormones (see p. 346).

B. Lactose operon ○

Let us consider the **lactose operon** of the bacterium *E. coli* as an example. This is a DNA sequence that is subject to both negative and positive control. The operon contains *structural genes* for three enzymes that are required for the utilization of lactose, as well as *control elements* that serve in the regulation of the operon. Since, in the cell, lactose is converted to glucose, it is unnecessary to express the genes of the lactose operon when glucose is already available. In fact, the genes are only transcribed when *glucose is absent* and *lactose is present* (**3**). This is achieved by the interaction of two regulatory proteins. In the absence of lactose, the **lac** *repressor* blocks the promoter region (**2**). When lactose is available, it is converted into **allolactose,** which binds to the repressor and detaches it from the operator (**3**). However, this is still not sufficient for the transcription of the structural genes. In order for the RNA polymerase to bind, an *inducer*, the **catabolite activator protein (CAP)**, is required, which only binds to DNA when it is present as a complex with **cAMP** (see p. 352). In *E. coli*, cAMP is a signal indicating nutrient deficiency and, as such, is only synthesized in the absence of glucose.

Fig. **5** illustrates the interaction between the CAP · cAMP complex and DNA. Each subunit of the dimeric inducer protein (yellow and orange) binds one molecule of cAMP (red). Contact with the DNA (blue) is mediated by two "recognition helices" that interact with the major groove of the DNA.

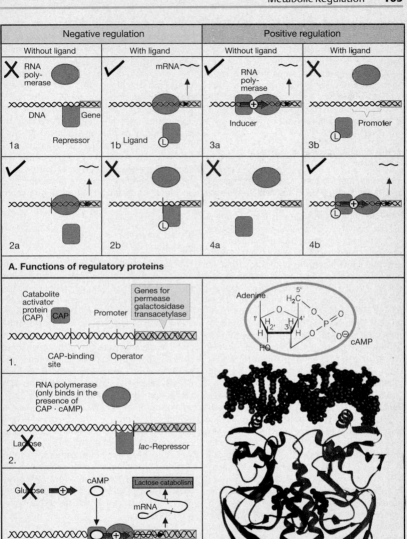

Negative regulation		Positive regulation	
Without ligand	With ligand	Without ligand	With ligand

A. Functions of regulatory proteins

B. Lactose operon

Hormonal Control

The enzyme-catalyzed activation and inactivation of key enzymes in intermediary metabolism is referred to as **interconversion**. Such processes are often initiated by hormones. Here, we show interconversion processes that are involved in the regulation of glycogen metabolism in the *liver*.

A. Hormonal control of glycogen metabolism ◑

In the liver and in muscles, glycogen serves as a carbohydrate store from which glucose phosphates are rapidly mobilized (see p. 160). The rate of glycogen synthesis is determined by the activity of *glycogen synthase* (lower right), while the degradation is catalyzed by *glycogen phosphorylase* (lower left). Both enzymes function on the surface of insoluble glycogen particles, where they reside in active or inactive form, depending on the metabolic state of the cell. In hunger or stress ("fight or flight"), the organism needs more glucose. In such situations, the hormones **epinephrine** and **glucagon** are released. They activate glycogen degradation and, at the same time, inhibit glycogen synthesis. Epinephrine acts on the muscles and the liver, whereas glucagon only affects the liver.

Both hormones bind to their respective **receptors** on the plasma membrane (**1**, see p. 350) Mediated by G proteins, the receptors activate *adenylate cyclase* (**2**), which catalyzes the synthesis of **cAMP** from ATP. The level of this "**second messenger**" is also influenced by a *cAMP phosphodiesterase*, which inactivates cAMP by catalyzing its hydrolysis to AMP. cAMP now binds to a further enzyme, *protein kinase A*, and activates it (**4**). Protein kinase A has two different points of attack. Firstly, it converts the active form of *glycogen synthetase* into the inactive D-form by *phosphorylation* with ATP as coenzyme, and thereby stops the synthesis of glycogen (**5**). Secondly, it activates, also by phosphorylation, a further protein kinase, *phosphorylase kinase* (**6**). Finally, active phosphorylase kinase phosphorylates the inactive b form of *glycogen phosphorylase*, converting it into the

active a form (**7**). This leads to the release of glucose 1-phosphate from glycogen (**8**), which can enter into glycolysis after conversion to glucose 6-phosphate (**9**). In the liver, free glucose is also formed, which ultimately turns up in the blood (**10**).

When the level of cAMP falls again, *phosphoprotein phosphatases* become predominant, which dephosphorylate the phosphoproteins described above (**11**). As a consequence, glycogen degradation is stopped, and glycogen synthesis is initiated again. These processes all occur within a few seconds. Therefore, glycogen metabolism can be rapidly adjusted to suit changing conditions.

B. Interconversion of glycogen phosphorylase ○

The structural changes that accompany the interconversion of *glycogen phosphorylase* were elucidated by X-ray crystallography (see p. 70). They are shown here schematically. The enzyme is a *dimer* with twofold symmetry. Each subunit has an active site that lies inside the protein and, in the b form, is difficult for the substrate to reach. Interconversion begins with the *phosphorylation* of a serine residue (Ser-14) near the *N*-terminus of each subunit. Arginine residues from adjacent subunits then bind to the newly introduced phosphate groups. This triggers a *conformational change* in the dimer, which, in turn, increases the affinity of the enzyme for the allosteric activator AMP.

The effect of AMP binding and of the conformational change on the active centers results in a higher activity of the a form. Following the removal of the phosphate residues, the enzyme spontaneously returns to the b conformation.

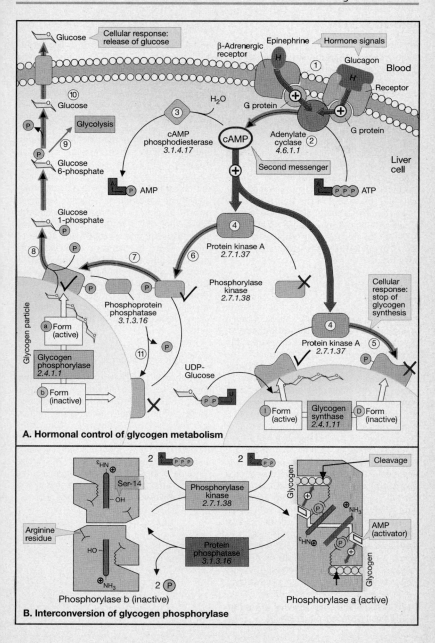

A. Hormonal control of glycogen metabolism

B. Interconversion of glycogen phosphorylase

Allosteric Regulation

As an example of allosteric enzyme regulation, we discuss here the regulation of *aspartate carbamoyltransferase* (**ACTase**, *2.1.3.2*), a key enzyme of pyrimidine biosynthesis.

A. Aspartate carbamoyltransferase: reaction ○

ACTase catalyzes the transfer of a carbamoyl residue from carbamoyl phosphate to the amino group of L-aspartate. The *N*-carbamoyl-L-aspartate thus formed already contains all of the atoms of the later pyrimidine ring (see p. 174). The ACTase of the bacterium *E. coli* is inhibited by cytidine triphosphate (CTP), an end product of the anabolic metabolism of pyrimidines, and is activated by ATP, a precursor.

B. Kinetics ○

In contrast to the kinetics of isosteric enzymes (see p. 90), ACTase shows S-shaped or **sigmoidal** substrate **saturation curves**. The slope of such a curve is a measure of the affinity of the enzyme for its substrate. In the case of allosteric enzymes, the slope initially increases with increasing substrate concentration [A]. At a higher [A], it decreases again, because most of the substrate binding sites are already occupied. Since the affinity of allosteric enzymes changes with [A], it cannot be described by a constant, such as the Michaelis constant, K_m (see p. 90). Instead, one gives [A] at half-maximal rate, i.e., $[A]_{0.5}$. The sigmoidal character of the curve is described by the **Hill coefficient**, h. For isosteric systems, h = 1 and h increases with increasing sigmoidicity. **Allosteric effectors**, such as CTP and ATP in the case of ACTase, not only change $[A]_{0.5}$, but also h. The inhibitor CTP leads to a *right-hand shift* of the curve (II), with an increase in $[A]_{0.5}$ and h. ATP, on the other hand, causes a *left-hand shift,* decreasing $[A]_{0.5}$ as well as h (III). At an aspartate concentration of 5 mM, 0.5 mM CTP is already sufficient to reduce the activity of ACTase by half. Activation by ATP is most effective at a low [A].

C. R and T conformations ○

As a rule, allosteric enzymes are *oligomers* of 2 -12 subunits. ACTase consists of six catalytic (blue) and six regulatory subunits (yellow). The latter bind the allosteric effectors CTP and ATP. Two conformations of ACTase exist simultaneously, a less active **T conformation** ("tense") and a more active **R conformation** ("relaxed"). Substrates and effectors affect the equilibrium between the two conformations, and thereby give rise to sigmoidal saturation behavior. With increasing aspartate concentration, the equilibrium is progressively shifted toward the more active R conformation. The R conformation is further stabilized by ATP binding to the regulatory subunits. The binding of CTP to the same sites, on the other hand, promotes conversion to the T conformation. In the case of ACTase, the structural differences between the R and T states are especially dramatic. During the T → R conversion, the catalytic subunits separate from one another by 1.2 nm, and they also rotate about the axis of symmetry. In contrast, there is little change in the folding of the subunits.

D. Structure of a dimer ○

The subunits of ACTase each consist of two *domains*, i.e., independently folded partial structures. The *N*-terminal domain of the regulatory subunit (right) mediates the interaction with CTP or ATP (green). A second, Zn^{2+}-containing domain of this monomer (Zn^{2+} shown in light blue) mediates contacts to the neighboring catalytic subunits. The active center is located between the domains of the catalytic subunit. In the structure shown, it is occupied by two substrate analogs (red).

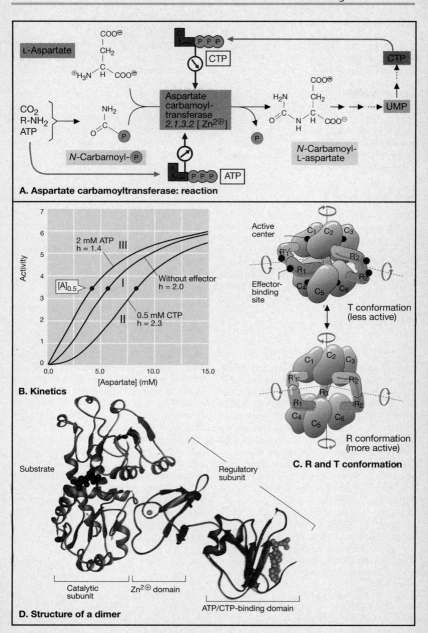

A. Aspartate carbamoyltransferase: reaction

B. Kinetics

C. R and T conformation

D. Structure of a dimer

ATP

The coenzyme **adenosine triphosphate (ATP)** is the most important *storage form of chemical energy*. Its cleavage is strongly exergonic. The chemical work (ΔG) provided by this reaction drives endergonic processes, such as biosyntheses, movement and transport by way of *energetic coupling* (see p. 10). Although the other nucleoside triphosphate coenzymes (GTP, CTP, and UTP) are similar to ATP chemically, these compounds perform other tasks in metabolism.

ATP: structure ◑

In ATP, a chain of three phosphate residues is linked to the 5'-hydroxyl group of adenosine (see p. 78). These phosphate residues are designated α, β, and γ. The α phosphate is bound to ribose by a phosphoric acid diester bond. The linkages between the three phosphate residues, on the other hand, involve much more labile **phosphoric acid anhydride bonds**. At physiological pH values, ATP carries four negative charges.

B. Phosphoric acid anhydride bonds ○

The formula shown in Fig. **A,** with its single and double bonds within the phosphate residues, is not an accurate representation of the actual charge distribution. In ATP, the partial negative charges at the oxygen atoms of all three phosphate residues are approximately equal, while the phosphorus atoms are centers of positive charge. One reason for the instability of phosphoric anhydride bonds is the *repulsion between the negatively-charged oxygen atoms*, which is partially relieved by the removal of a phosphate residue. Such reactions are, therefore, strongly exergonic (**C**). In addition, the hydrolysis of ATP to ADP creates a free phosphate anion, which is better hydrated and more effectively resonance-stabilized than in ATP. This also contributes to the strongly exergonic character of ATP hydrolysis.

C. Free energy of hydrolysis ◑

Under standard conditions, the free-energy change $\Delta G^{o'}$ (see p. 12) of hydrolysis of phosphoric acid anhydride bonds in ATP amounts to -30 to -35 kJ · mol^{-1}. Which one of the anhydride bonds of ATP is cleaved, has only a minor effect on ΔG^o. Even the hydrolysis of free diphosphate still yields more than -30 kJ · mol^{-1}. By contrast, the cleavage of the ester bond between the ribose and the α-phosphate residue provides only -9 kJ · mol^{-1}.

In the cell, the actual free-energy change of ATP hydrolysis ($\Delta G'$) is even greater, because the concentrations of ATP, ADP and P_i are much lower than under standard conditions, and ATP is present in excess over ADP (cf. p. 12). The pH value and the concentration of Mg^{2+} also affect the magnitude of $\Delta G'$. The *physiological energy yield* of ATP hydrolysis is probably as negative as -50 kJ · mol^{-1}.

D. Energetic coupling ◑

In order to utilize the chemical energy of ATP hydrolysis for endergonic processes, it is not sufficient simply to let the two reactions occur next to one another. In this case, ATP hydrolysis would just provide heat without affecting the endergonic process. In order to achieve energetic coupling, the two reactions have to be carried out so that a *common intermediate* arises. Here this is shown using the *glutamine synthetase* reaction as an example. Initially, the γ-phosphate residue of ATP is transferred to glutamate to yield an "energy-rich" *mixed acid anhydride* (**a**). In the second step (**b**), the phosphate residue of this compound is displaced by NH_3 and glutamine and free phosphate are produced. The $\Delta G^{o'}$ of the complete reaction (-16 kJ · mol^{-1}) corresponds to the sum of the free-energy changes of the two partial reactions.

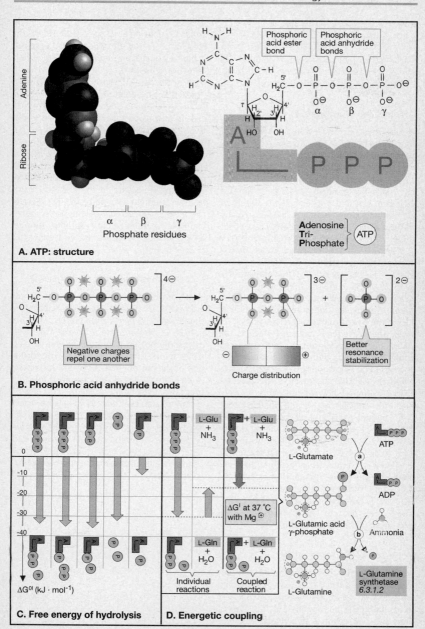

A. ATP: structure

B. Phosphoric acid anhydride bonds

Negative charges repel one another

Charge distribution

Better resonance stabilization

C. Free energy of hydrolysis

ΔG^{0I} (kJ · mol^{-1})

Individual reactions

D. Energetic coupling

Coupled reaction

ΔG^{I} at 37 °C with Mg$^{\oplus}$

L-Glutamate

L-Glutamic acid γ-phosphate

L-Glutamine

L-Glutamine synthetase 6.3.1.2

ATP

ADP

Ammonia

Phosphoric acid ester bond

Phosphoric acid anhydride bonds

Adenine

Ribose

α   β   γ
Phosphate residues

Adenosine **T**ri-**P**hosphate ATP

Energy Conservation at Membranes

The synthesis of ATP from ADP and P_i is strongly endergonic. It can only take place when linked to other, exergonic processes.

A. Forms of ATP synthesis ◑

During the course of evolution, two important pathways for ATP synthesis have arisen, which are basically similar in all cells.

1. The first (and most important) mechanism takes advantage of the energy stored in *electrochemical gradients* (see **B**) to synthesize ATP from ADP and P_i. The energy required for the formation of such gradients is usually derived from a respiratory chain, i.e., a series of *redox reactions*. This mechanism of ATP synthesis is therefore referred to as "**oxidative phosphorylation.**" The ATP is synthesized by *H^+-transporting ATP synthases* (see p. 132), which utilize the energy stored in the gradient. In eukaryotes, oxidative phosphorylation only occurs under aerobic conditions, i.e., in the presence of oxygen (see p. 136).

2. The second mechanism (which is the older one in evolutionary terms) occurs under both aerobic and anaerobic conditions. It is based on the coupling of ATP synthesis to the **hydrolysis of "energy-rich" metabolites**. In the example shown here, ATP is synthesized by transferring a phosphate residue from *creatine phosphate*, an energy reserve of the muscles, to ADP. The free-energy change $\Delta G^{o\prime}$ for the coupled process equals the sum of the changes ΔG^o for the partial reactions.

There are only a few metabolic reactions that are able to raise inorganic phosphate (P_i) to a chemical potential high enough to allow ATP formation. Such processes are collectively referred to as "**substrate level phosphorylation**" (see p. 140).

B. Electrochemical gradient ◑

Electrically charged atoms and molecules (ions) require special transport proteins (ion channels or ion "pumps") to allow them to cross biological membranes. The transport of ions by an ion pump is an *active* process (see p. 206), resulting in the creation of an unequal distribution of ions on either side of the membrane, i.e., an **electrochemical gradient** is built up that stores electrical energy like a capacitor.

In most cells, ATP synthesis utilizes **proton gradients**. The amount of energy stored in such a gradient is dependent on the difference in the proton concentration between the two sides of the membrane, i.e., the *pH difference, ΔpH*. The unequal distribution of positive and negative charges, the *membrane potential, $\Delta\psi$*, is also an important factor. Both quantities contribute to the **proton motive force Δp**, a measure of the chemical work, ΔG, the gradient can perform. The electrochemical gradient across the inner mitochondrial membrane, for example, supplies approximately 24 kJ per mol of transported H^+ (see p. 130).

C. Energy conservation at membranes ◑

There are different ways of creating proton gradients. An unusual example is bacteriorhodopsin (**1**), a *light-driven proton pump* (see p. 122). In photosynthesis (see p. 118), *plastoquinone* transports protons as well as electrons across the membrane (**2**). The formation of the proton gradient by the *respiratory chain* (see p. 130) is also coupled to redox processes. *Cytochrome-c oxidase* (complex IV) is shown as an example (**3**). In all three cases, the gradient is utilized for ATP synthesis by *ATP synthase* (**4**). ATP synthases consist of two components, a proton channel (F_o) and a protein complex (F_1), which uses the proton flow across the membrane for the synthesis of ATP (see p. 132).

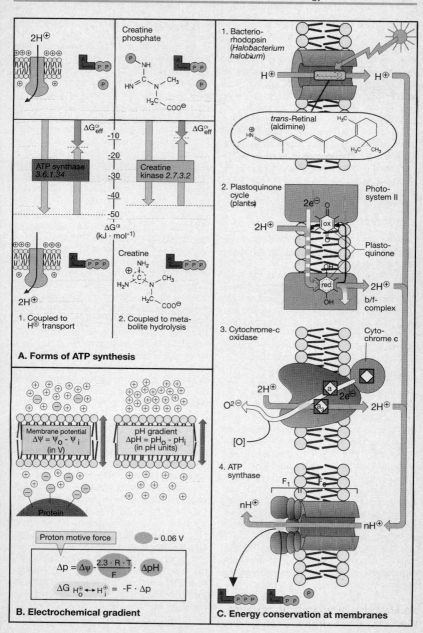

A. Forms of ATP synthesis

Creatine phosphate

$\Delta G^{\circ i}_{eff}$

ATP synthase 3.6.1.34

Creatine kinase 2.7.3.2

-10
-20
-30
-40
-50

$\Delta G^{\circ i}$ (kJ · mol^{-1})

$2H^{\oplus}$

Creatine

1. Coupled to $H^{\oplus}$ transport

2. Coupled to metabolite hydrolysis

B. Electrochemical gradient

Membrane potential
$\Delta \Psi = \Psi_o - \Psi_i$
(in V)

pH gradient
$\Delta pH = pH_o - pH_i$
(in pH units)

Protein

Proton motive force   ≈ 0.06 V

$$\Delta p = \Delta \Psi - \frac{2.3 \cdot R \cdot T}{F} \cdot \Delta pH$$

$$\Delta G_{H_o^{\oplus} \leftrightarrow H_i^{\oplus}} = -F \cdot \Delta p$$

C. Energy conservation at membranes

1. Bacterio-rhodopsin (*Halobacterium halobium*)

$H^{\oplus}$ $H^{\oplus}$

trans-Retinal (aldimine)

2. Plastoquinone cycle (plants)

$2e^{\ominus}$

$2H^{\oplus}$

Photo-system II

ox

Plastoquinone

red

$2H^{\oplus}$

b/f-complex

3. Cytochrome-c oxidase

Cyto-chrome c

$2H^{\oplus}$

a

$2e^{\ominus}$

$O^{2\ominus}$

a_3

$2H^{\oplus}$

[O]

4. ATP synthase

F_1

$nH^{\oplus}$

$nH^{\oplus}$

Photosynthesis: Light Reactions

The primary and thus most important source of energy for all living things is sunlight. With the help of **photosynthesis**, light energy is used to drive the synthesis of organic substances from CO_2 and water. *Heterotrophic organisms* (e.g., animals), which are dependent on a supply of organic substances in their diet (see p. 104), benefit from the ability of *phototrophic organisms* (plants, algae and certain bacteria) to perform photosynthesis. Atmospheric *oxygen*, which is essential for the life of higher organisms is also derived from photosynthesis.

A. Photosynthesis: overview ◗

The chemical balance of photosynthesis is simple. One hexose molecule is formed from six molecules of CO_2 (right). The hydrogen atoms required for this process are derived from water, and molecular oxygen is formed as a byproduct (left). The requirement for light energy is due to the fact that water is a very poor reducing agent, and therefore unable to reduce CO_2.

The light-dependent reactions of photosynthesis, or simply **"light reactions,"** involve the splitting of H_2O molecules to give protons, electrons and oxygen atoms. The electrons undergo *excitation* by light energy, which raises them to an energy level sufficiently high to reduce $NADP^+$. The NADPH thus formed, in contrast to H_2O, is a reductant suitable for the "fixing" of CO_2, i.e., for its incorporation into organic compounds. The other products of the light reactions are ATP, which is also required for CO_2 fixation, and molecular oxygen. When NADPH, ATP, and the necessary enzymes are available, CO_2 fixation can occur in the dark, and the reactions involved are, therefore, referred to as the **"dark reactions."**

The excitation of the electrons is a complicated photochemical process involving **chlorophyll**, a Mg^{2+}-containing *tetrapyrrol* pigment with an extra phytol residue.

B. Light reactions ○

In green algae and higher plants, photosynthesis occurs in **chloroplasts**. These are organelles, which, like mitochondria, are surrounded by two membranes and contain their own DNA. The chloroplast *stroma*, i.e., the interior space, contains *thylakoids* or flattened membrane sacks stacked one on top of the other to form *grana*. The inside of the thylakoid is referred to as the *lumen*. The light reactions are catalyzed by enzymes located in the thylakoid membrane, whereas the dark reactions occur in the stroma. The light reactions, like the mitochondrial respiratory chain (see p. 130) bring about the stepwise transfer of electrons from one redox system to the next, i.e., they constitute an **electron transport chain**. However, the direction of transport is the *opposite* of that found in the mitochondrial electron transport chain. In the respiratory chain, the electrons are transferred from NADH to O_2 with the production of water and useable energy. In photosynthesis, electrons are transferred from water to $NADP^+$ with an *input of energy*. Thus, photosynthetic electron transport is energetically "uphill." It is only made possible by the excitation of electrons at two different sites due to the *absorption of light energy*. These two sites are referred to as **photosystems**, protein complexes containing multiple chlorophyll molecules and other pigments (see p. 120). A further component of the photosynthetic electron transport chain is the **cytochrome b/f complex**, an aggregate of integral membrane proteins containing two cytochromes (b_{563} and f). The ubiquinone-related **plastoquinone** and two soluble proteins, the copper-containing **plastocyanin** as well as **ferredoxin**, act as mobile electron carriers. At the end of the chain, an enzyme transfers the electrons from ferredoxin to $NADP^+$. Both photosystem II and the cytochrome b/f complex release protons into the lumen. Therefore, photosynthetic electron transport, like the repiratory chain, builds up an **electrochemical gradient** that is used for the synthesis of ATP by an *ATP synthase* located in the thylakoid membrane.

Cyclic photophosphorylation (pg 604 Stryer)
- takes place when NADPH is abundant & NADP is unavailable to accept electrons from ferredoxin.
- builds up proton gradient to drive ATP synthesis.

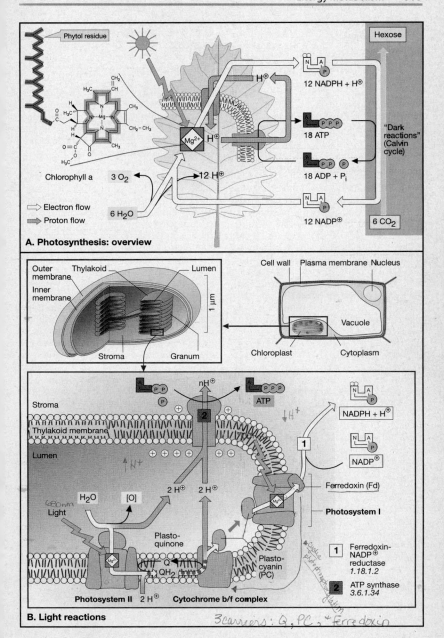

A. Photosynthesis: overview

Phytol residue

Chlorophyll a

⇨ Electron flow
⇨ Proton flow

Hexose

12 NADPH + H$^{\oplus}$

18 ATP

18 ADP + P$_i$

12 NADP$^{\oplus}$

"Dark reactions" (Calvin cycle)

H$^{\oplus}$

3 O$_2$

6 H$_2$O

12 H$^{\oplus}$

6 CO$_2$

Outer membrane
Thylakoid
Lumen
Inner membrane
Stroma
Granum
1 µm

Cell wall   Plasma membrane   Nucleus
Vacuole
Chloroplast   Cytoplasm

Stroma
Thylakoid membrane
Lumen

nH$^{\oplus}$
ATP
↓H^{+}
NADPH + H$^{\oplus}$
NADP$^{\oplus}$
2
1

H$_2$O
[O]
Light
680 nm
↑H^{+}

2 H$^{\oplus}$ 2 H$^{\oplus}$
Ferredoxin (Fd)
Photosystem I

Plastoquinone
Q
QH$_2$
Plastocyanin (PC)

| **1** | Ferredoxin-NADP$^{\oplus}$ reductase 1.18.1.2 |
| **2** | ATP synthase 3.6.1.34 |

Cyclic photophosphorylation

Photosystem II   2 H$^{\oplus}$ **Cytochrome b/f complex**

B. Light reactions

3 carriers: Q, PC, + ferredoxin

Photosynthesis: Dark Reactions

A. Photosystem II ○

The photosynthetic electron transport chain of plants starts at **photosystem II** (PS II, see p. 118). PS II consists of several protein subunits (brown) that contain bound **pigments**, i.e., molecules involved in the absorption and transfer of light energy. Only the most important pigments are shown in the schematic illustration of PS II (**1**), i.e., a special chlorophyll molecule (the *reaction center* P_{680}), a neighboring Mg^{2+}-free chlorophyll molecule (*pheophytin*) and two bound *plastoquinones* (Q_A and Q_B). A third quinone (Q_P) is not bound to PS II, but belongs to the plastoquinone "pool." The white arrows indicate the direction of electron flow from water to Q_P. Only about 1% of the chlorophyll molecules in PS II are *directly* involved in photochemical electron transfer. Most of them are associated with other pigments in so-called **light-harvesting complexes** (green). The energy of light quanta striking these complexes is funneled toward the reaction center where it can be made use of.

In Fig. 2, photosynthetic electron transport at PS II is separated into the individual steps involved. Light energy coming from the light-harvesting complexes raises an electron of the reaction center chlorophyll molecule to an excited state (a). The excited electron is immediately passed on to the neighboring pheophytin, leaving behind an "electron-hole" in the reaction center, i.e., a positively charged P_{680} radical (b). This hole is then filled by an electron removed from an H_2O molecule by the water-splitting complex (b). The excited electron is further transferred from pheophytin via Q_A to Q_B, converting the latter to a *semiquinone radical* (c). Q_B is then reduced to *hydroquinone* by means of a second excited electron and finally exchanged for an oxidized quinone (Q_P) from the plastoquinone pool.

B. Redox series ○

From the *standard reduction potentials* $E^{0\prime}$ (see p. 28) of the most important redox systems involved in the light reactions, it is clear why *two* excitation processes are required to transfer electrons from H_2O to $NADP^+$.

When $NADP^+$ is unavailable, photosynthetic electron transport can still be used for the synthesis of ATP. During **cyclic photophosphorylation**, electrons return from ferredoxin (Fd), via the plastoquinone pool, to the b/f complex. This type of electron transport does not produce any NADPH, but it does lead to the formation of the H^+ gradient required for the synthesis of ATP.

C. Calvin cycle ○

The illustration shows a greatly simplified version of the synthesis of hexoses from CO_2. A more complete scheme is found on p. 367. The actual **CO_2 fixation**, i.e., the incorporation of CO_2 into an organic compound, is catalyzed by *ribulose-bisphosphate carboxylase* (abbrev. "rubisco"). Rubisco is thought to be the most abundant enzyme on earth. It converts **ribulose 1,5-bisphosphate**, CO_2 and water into two molecules of **3-phosphoglycerate**. These are then converted, via **1,3-bisphosphoglycerate** and 3-phosphoglycerate, to **glyceraldehyde 3-phosphate**. In this way, 12 glyceraldehyde 3-phosphate are synthesized from 6 CO_2. Two molecules of this intermediate product are used for the synthesis of **glucose 6-phosphate** by reactions of gluconeogenesis (below right). From the remaining 10 molecules, 6 molecules ribulose 1,5-bisphosphate are regenerated, with which the cycle can start over again. In the Calvin cycle, ATP is required for the phosphorylation of 3-phosphoglycerate and ribulose 5-phosphate. NADPH, the second product of the light reactions, is consumed in the reduction of 1,3-bisphosphoglycerate to glyceraldehyde 3-phosphate.

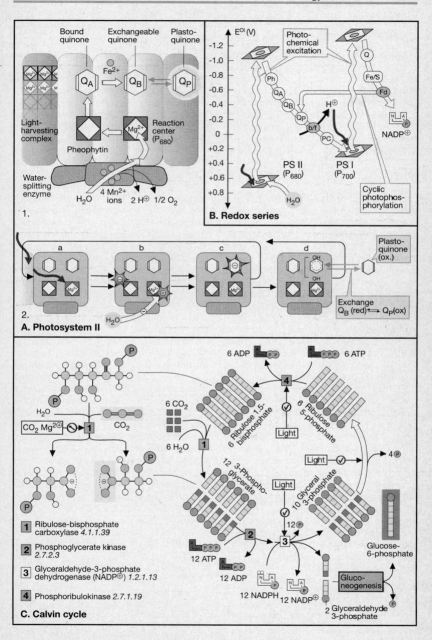

1.

B. Redox series

2.

A. Photosystem II

1 Ribulose-bisphosphate
carboxylase *4.1.1.39*

2 Phosphoglycerate kinase
2.7.2.3

3 Glyceraldehyde-3-phosphate
dehydrogenase (NADP⊕) *1.2.1.13*

4 Phosphoribulokinase *2.7.1.19*

C. Calvin cycle

Molecular Models: Bacterio-rhodopsin, Reaction Center

The illustration shows **bacteriorhodopsin** from the bacterium *Halobacterium halobium* and the **photosystem** of the purple bacterium *Rhodopseudomonas viridis*. These are two of the few transmembrane proteins whose structure is known in detail. Both have been used as model systems to investigate the mechanisms underlying energy metabolism. The structures shown are highly simplified.

A. Bacteriorhodopsin ○

Bacteria of the species *H. halobium* can grow at extremely high salt concentrations, e.g., in salt lakes. Their plasma membrane contains a protein that is similar to the rhodopsin of the eye (see p. 324), and is therefore called **bacteriorhodopsin**. This protein is unusual in that it can use light directly for the formation of an electrochemical gradient (see p. 116). As in the case of sight, this process is based on a light-induced *cis-trans* isomerization of retinal.

Most of the protein part of bacteriorhodopsin consists of seven α helices (shown in blue), that span the membrane and form a flattened hollow cylinder (the remainder of the molecule and amino acid side chains are not shown). Within the cylinder, the aldehyde group of the retinal molecule (orange) is covalently linked to the amino group of a lysine residue (red). In the dark, the aldimine group is protonated, and the double bond of retinal adopts the *trans* configuration (see p. 116). Upon exposure to light, retinal is converted to the 13-*cis* form and the aldimine group releases its proton, which is transferred to the outside via two aspartate residues of the protein (light blue). With the return of retinal to the all-*trans* form, the aldimine group is protonated again. The necessary proton is derived from within the cell (above), from where it is transferred via another aspartate residue (green).

B. Reaction center of *Rhodopseudomonas viridis* ○

The photosystem of the purple bacterium *Rhodopseudomonas viridis* is similar in structure to the photosystem II of higher plants. Thus, the elucidation of the structure of the bacterial system by J. Deisenhofer, H. Michel, and R. Huber, was a key step toward our understanding of the light reactions of photosynthesis (see p. 118).

Only the membrane-spanning part of the complex is shown in the model. The approximate location within the membrane is indicated by four schematized lipid molecules (left and right). Eleven *transmembrane helices*, associated with three different subunits (shown in various shades of blue), surround a space in the interior of the membrane that is filled with a chain of *pigments* (some pigments not directly involved in electron transport have been omitted). The principle of photosynthetic electron transport is discussed in more detail on p. 120.

In *R. viridis*, light energy is absorbed by two adjacent chlorophyll molecules known as the "special pair" (green, Mg^{2+} red). Both molecules have an absorption maximum at 870 nm. Therefore, the bacterial reaction center is designated P_{870}. Following light absorption, the excited electron is transferred, within a few picoseconds (1 ps = 10^{-12} s), from the reaction center to a neighboring magnesium-free pheophytin molecule (orange). From there it reaches the tightly bound quinone Q_A (above left, yellow) within about 200 ps. At the same time, the electron hole in the special pair is filled. After another 0.2 ms, the excited electron arrives at the exchangeable *quinone* Q_B (above, right). With this molecule, it eventually leaves the photosystem.

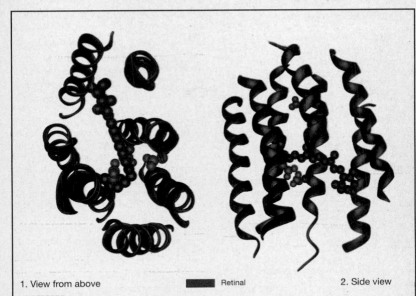

1. View from above   ▬ Retinal   2. Side view

A. Bacteriorhodopsin

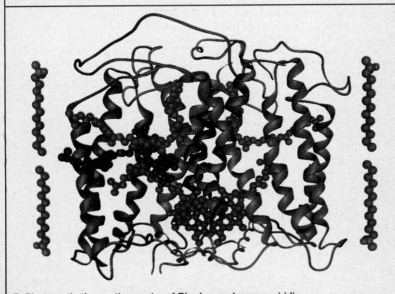

B. Photosynthetic reaction center of *Rhodopseudomonas viridis*

Oxoacid Dehydrogenases

The oxoacid dehydrogenases of intermediary metabolism are *multienzyme complexes* catalyzing complicated reactions that result in the oxidative decarboxylation of 2-oxoacids and the transfer of the acyl residue thus produced to coenzyme A. NAD+ acts as the final electron acceptor for these enzymes. Thiamine diphosphate, lipoamide, and FAD are also involved. The oxoacid dehydrogenases include a) the **pyruvate dehydrogenase complex** (PDH, pyruvate → acetyl-CoA), b) the **2-oxoglutarate dehydrogenase complex** of the citric acid cycle (ODH, 2-oxoglutarate → succinyl-CoA), and c) the **branched-chain dehydrogenase complex**, which is involved in the catabolism of valine, leucine and isoleucine (see p. 371). Here we take the PDH complex as an example.

A. Pyruvate dehydrogenase: reaction ◑

Three different enzymes [E1-E3] are involved in the pyruvate dehydrogenase reaction. *Pyruvate dehydrogenase* [E1] catalyzes the decarboxylation of pyruvate, the transfer of the resulting hydroxyethyl residue to **thiamine diphosphate** (TPP, **1a**), and *oxidation* of the hydroxyethyl group to yield an acetyl residue. This acetyl moiety and the reducing equivalents obtained in the first step are then transferred to **lipoamide** (**1b**). A second enzyme, *dihydrolipoamide acetyltransferase* [E2], transfers the acetyl residue from lipoamide to **coenzyme A** (**2**), leaving dihydrolipoamide to be re-oxidized by a third enzyme, *dihydrolipoamide dehydrogenase*. The products of this last reaction are the disulfide form of lipoamide and **NADH + H+ (3)**. The electrons are transferred to soluble NAD+ via **FAD** and a catalytically active disulfide bond of the E3 subunit.

The five different coenzymes of the reaction are associated in different ways with the various protein components of the complex. Thiamine diphosphate is bound noncovalently to E1, whereas lipoamide is covalently bound to a lysine residue of E2. FAD is tightly associated with E3 as a *prosthetic group* (see p. 100), while NAD+ and coenzyme A are *soluble coenzymes*, and therefore only temporarily interact with the complex.

B. PDH complex of *E. coli* ○

The PDH complex of the bacterium *Escherichia coli* has been particularly well characterized. It has a molecular mass of $5.3 \cdot 10^6$ and a diameter of more than 30 nm, making it larger than a ribosome. The complex consists of a total of 60 polypeptides. The almost cube-shaped core of the complex is made up of 24 molecules of E2 (8 trimers). Each of the six surfaces of the cube is occupied by a dimer (possibly a tetramer) of the E3 component, whereas each of the twelve edges of the cube is occupied by dimers of E1 molecules. The other oxoacid dehydrogenases have similar structures, except that they may differ in the number of subunits and in their molecular masses.

The spatial organization of the different components of the complex is very important for the catalytic process. As a result of its linkage to a lysine side chain of E2, the lipoic acid coenzyme is very mobile. The *lipoamide arm*, which is almost 1.4 nm long, swings backwards and forwards between E1 and E3 during catalysis (**3**). This enables lipoic acid to interact with thiamine diphosphate bound to E1, as well as with the soluble coenzyme A and the electron-accepting FAD associated with E3. The protein domain of the acetyl transferase, which binds the lipoic acid is very flexible, and thus further extends the reach of the lipoamide "arm".

The **regulation** of the animal PDH complex is discussed on p. 134.

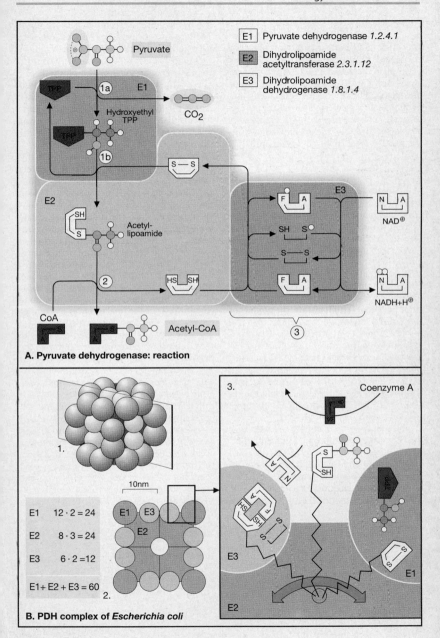

A. Pyruvate dehydrogenase: reaction

E1   Pyruvate dehydrogenase *1.2.4.1*

E2   Dihydrolipoamide acetyltransferase *2.3.1.12*

E3   Dihydrolipoamide dehydrogenase *1.8.1.4*

Pyruvate

Hydroxyethyl TPP

CO_2

Acetyl-lipoamide

CoA

Acetyl-CoA

$NAD^\oplus$

$NADH+H^\oplus$

B. PDH complex of *Escherichia coli*

10nm

E1 $12 \cdot 2 = 24$

E2 $8 \cdot 3 = 24$

E3 $6 \cdot 2 = 12$

E1+ E2 + E3 = 60

Coenzyme A

Citric Acid Cycle: Reactions

The **citric acid cycle** (also know as *tricarboxylic acid cycle*, or TCA cycle) is a metabolic pathway in the mitochondrial matrix that oxidizes acetyl residues (CH_3-CO-) to carbon dioxide (CO_2). The reducing equivalents thus obtained are transferred to NAD^+ or ubiquinone, and from there to the respiratory chain (see p. 130). Further metabolic functions of the cycle are discussed on p. 128.

A. Reactions ◑

Most of the acetyl-CoA that feeds the citric acid cycle with acetyl residues is derived from the β-*oxidation* of fatty acids (see p. 144) and from the *pyruvate dehydrogenase reaction*. Like the cycle itself, these processes take place in the mitochondrial matrix.

The oxidation of acetyl residues occurs via a series of intermediate steps. Initially, the residue is transferred to a carrier molecule, **oxaloacetate**, in a reaction catalyzed by *citrate synthase* [1]. It is **citrate**, the product of this reaction, that gave the cycle its name. In the next step [2], citrate undergoes isomerization to yield **isocitrate**. In the process, a hydroxyl group is shifted within the molecule. Unsaturated *aconitate* is produced as an enzyme-bound intermediate (not shown), and therefore the enzyme catalyzing the reaction [2] is called a*conitate hydratase* ("aconitase").

The properties of aconitase are such that the isomerization is absolutely *stereospecific*. Citrate is not chiral, but isocitrate has two centers of chirality (see p. 56). Thus there are four potential isomeric forms of isocitrate. However, in the citric acid cycle only one of these stereoisomers, *(2R,3S)*-isocitrate, is produced.

The next step is the first oxidative one. *Isocitrate dehydrogenase* [3] oxidizes the hydroxyl group of isocitrate to an oxo group. At the same time, a carboxyl group is released as CO_2, and **2-oxoglutarate** is formed. Reaction [4], the formation of **succinyl-CoA**, also involves an oxidation and a decarboxylation. It is catalyzed by *2-oxoglutarate dehydrogenase*, a multienzyme complex (oxoacid dehydro-

genases are discussed on the previous page). The subsequent cleavage of succinyl-CoA into **succinate** and coenzyme A catalyzed by *succinate-CoA ligase* ("thiokinase" [5]) is sufficiently exergonic to allow the synthesis of a phosphoric acid anhydride bond ("*substrate level phosphorylation*"). However, it is not ATP that is produced here, but rather **guanosine triphosphate (GTP)**. GTP is readily converted to ATP by a *nucleoside diphosphate kinase* (not shown).

Through the reactions described up to this point, the acetyl residue has been completely oxidized to CO_2. However, at the same time, the carrier molecule oxaloacetate has been reduced to succinate. Three further reactions of the cycle now regenerate oxaloacetate from succinate. Initially, *succinate dehydrogenase* [6] oxidizes succinate to **fumarate**. In contrast to the other enzymes of the cycle, succinate dehydrogenase is an integral protein of the inner mitochondrial membrane. Although the enzyme contains FAD as a prosthetic group (see p. 130), **ubiquinone** (coenzyme Q) is the final electron acceptor of the reaction. *Fumarate hydratase* ("fumarase", [7]) now catalyzes the addition of water to the double bond of fumarate resulting in the production of chiral **(2S)-malate**. In the last step of the cycle, malate is oxidized to oxaloacetate by *malate dehydrogenase* [8], once again with the production of $NADH + H^+$. With this reaction, the cycle is complete and can begin again from the start.

The outcome of the cycle is the production of 2 CO_2, 3 $NADH + H^+$ and one ubiquinol per acetyl residue. By oxidative phosphorylation, the cell gains 9 ATP from these reduced coenzymes (see p. 136). Together with the directly formed GTP, this yields a total of 10 ATP per acetyl residue.

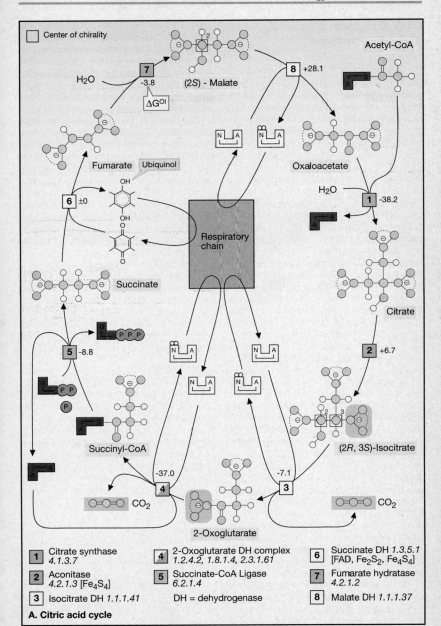

□ Center of chirality

H₂O

7 -3.8

ΔG^{0}

(2S) - Malate

8 +28.1

Acetyl-CoA

Fumarate   Ubiquinol

6 ±0

Respiratory chain

Oxaloacetate

H₂O

1 -38.2

Succinate

Citrate

5 -8.8

2 +6.7

P P P

P P

P

Succinyl-CoA

4 -37.0

3 -7.1

(2R, 3S)-Isocitrate

CO₂

2-Oxoglutarate

CO₂

1	Citrate synthase *4.1.3.7*	**4**	2-Oxoglutarate DH complex *1.2.4.2, 1.8.1.4, 2.3.1.61*	**6**	Succinate DH *1.3.5.1* [FAD, Fe₂S₂, Fe₄S₄]
2	Aconitase *4.2.1.3* [Fe₄S₄]	**5**	Succinate-CoA Ligase *6.2.1.4*	**7**	Fumarate hydratase *4.2.1.2*
3	Isocitrate DH *1.1.1.41*		DH = dehydrogenase	**8**	Malate DH *1.1.1.37*

A. Citric acid cycle

Citric Acid Cycle: Metabolic Functions

A. Metabolic functions ◑

The citric acid cycle (see p. 126) is the central pivot of intermediary metabolism. It has both catabolic and anabolic functions (see p. 104) — it is **amphibolic**. Many catabolic pathways generate citric acid cycle intermediates or metabolites, such as pyruvate and acetyl-CoA, which can be oxidized to CO_2 by the cycle as required. The reducing equivalents thus obtained are used for *oxidative phosphorylation*, i.e., for the synthesis of ATP.

The citric acid cycle also supplies important *precursors for anabolic pathways*. For instance, the precursors for the biosynthetic pathways for **glucose** (upper right), the **porphyrins** (lower left) and most of the **amino acids** are all derived from the cycle.

In addition to these functions, the cycle furnishes the **acetyl-CoA** required for the biosynthesis of fatty acids in the cytoplasm. Acetyl-CoA cannot cross the inner mitochondrial membrane as such, but is exported by a shuttle mechanism. First, acetyl-CoA is condensed with oxaloacetate to yield citrate, a reaction catalyzed by mitochondrial *citrate synthase*. Then citrate is exported from the mitochondria via an antiport system in exchange for malate (right; see also, p. 198). In the cytoplasm, citrate is cleaved again by an ATP-dependent *citrate lyase* [4] to regenerate acetyl-CoA and oxaloacetate. Oxaloacetate is reduced to malate by a cytoplasmic *malate dehydrogenase* [2]. Malate then reenters the mitochondria via the antiport mentioned above, or undergoes oxidative decarboxylation to yield pyruvate. One of the products of this reaction, which is catalyzed by the oxidoreductase *"malic enzyme"* [5], is NADPH + H^+, and therefore the reaction also stimulates fatty acid synthesis.

The intermediates of the citric acid cycle are present in the mitochondria in very small quantities. They are constantly being regenerated during the oxidation of acetyl residues and, therefore, their concentrations remain relatively constant over time. On the other hand, *anabolic* pathways constitute a major drain on the intermediates of the cycle. They would be rapidly depleted, and the cycle would stop, if it were not for the fact that certain metabolites enter the cycle at other points to replace those that have been consumed by anabolic processes. Reactions, which supply such intermediates are called **anaplerotic** (anaplerotic = replenishing).

For instance, the degradation of most amino acids is anaplerotic, because it produces either intermediates of the cycle or pyruvate (see p. 144). Gluconeogenesis, the *de novo* synthesis of glucose, is largely sustained by the degradation of amino acids. A key anaplerotic reaction in animals is the conversion of pyruvate to oxaloacetate. This ATP-dependent reaction is catalyzed by *pyruvate carboxylase* ([1], see p. 158). It allows pyruvate-yielding amino acids and lactate to be used for gluconeogenesis.

In contrast to pyruvate, acetyl-CoA is a non-anaplerotic metabolite in animals. It is completely oxidized to CO_2 in the citric acid cycle, and thus its carbon atoms are not available for biosynthetic pathways. As acetyl-CoA is the only product of fatty acid degradation, animal cells cannot convert fatty acids to glucose.

Additional information

Plant seeds often store large quantities of fats. Unlike animals, plants can utilize these fats for the synthesis of amino acids and carbohydrates. This occurs via an anaplerotic pathway known as the **glyoxylate cycle**, which is also found in bacteria. The glyoxylate cycle starts with the condensation of acetyl residues (derived from the degradation of fatty acids) to yield succinate, which then feeds into the citric acid cycle.

*Malate + pyruvate can be transported through into inner membrane

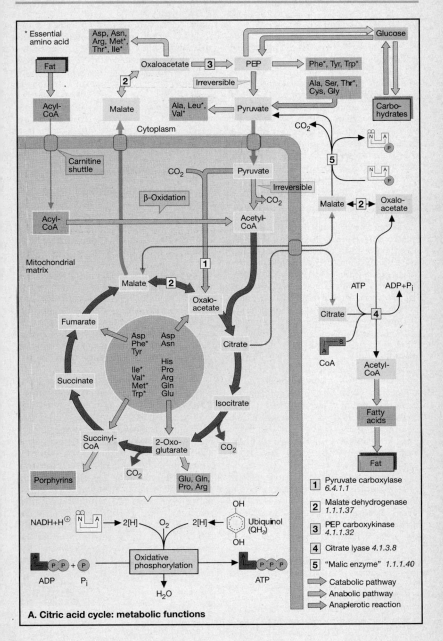

A. Citric acid cycle: metabolic functions

1 Pyruvate carboxylase 6.4.1.1
2 Malate dehydrogenase 1.1.1.37
3 PEP carboxykinase 4.1.1.32
4 Citrate lyase 4.1.3.8
5 "Malic enzyme" 1.1.1.40

Respiratory Chain

The **respiratory chain** is part of the process of oxidative phosphorylation (see p. 116). It catalyzes the transport of electrons from NADH or ubiquinol (QH_2) to molecular oxygen (O_2). The large differences in reduction potential between the donors (NADH + H+ and ubiquinol) and the acceptor (O_2) make this process *strongly exergonic* (see p. 28). Most of the energy released is used for the formation of an electrochemical gradient (see p. 116), which ultimately drives the synthesis of ATP by *ATP synthase*.

A. Components of the respiratory chain ○

The respiratory chain consists of three **protein complexes** (*complexes I, III, IV*) integrated in the inner mitochondrial membrane (see p. 196) and two mobile **carrier molecules** ubiquinone (coenzyme Q) and *cytochrome c*. Succinate dehydrogenase, which actually belongs to the citric acid cycle, may also be considered part of the chain. As such, it is designated *complex II. ATP synthase* (see p. 132) is sometimes included as *complex V*, although it does not take part in electron transport.

All of the complexes of the respiratory chain are made up of multiple polypeptide subunits, and contain a series of different protein-bound **redox coenzymes** (see p. 132). The latter include *flavins* (FMN or FAD in complexes I and II), *iron-sulfur centers* (Fe/S centers; I, II and III) and *heme groups* (II, III and IV). The detailed structures of most of the complexes are not yet known.

Various pathways feed electrons into the chain. **Complex I** transfers electrons from NADH to ubiquinone via FMN and several Fe/S centers. Reducing equivalents derived from the oxidation of succinate, acyl-CoA, choline, and other substrates are passed on to ubiquinone via a mitochondrial electron transfer protein (ETF, see p. 146). These reactions are catalyzed by **complex II** (*succinate dehydrogenase*) and other flavin-dependent *mitochondrial dehydrogenases*.

Ubiquinol transfers electrons to **complex III**, which in turn delivers them to the small,

soluble heme protein *cytochrome c*. Cytochrome c then passes on electrons to **complex IV** or *cytochrome c oxidase*, which catalyzes the final transfer of the electrons to *oxygen*. The strongly basic O^{2-} ion produced by the reduction of oxygen immediately binds two protons and is thereby converted to *water*. The electron transfer is energetically coupled to the formation of a **proton gradient** by complexes I, III and IV (see p. 116).

B. Organization ○

Proton transport via complexes I, III and IV occurs *vectorially* from the matrix to the intermembrane space. Thus during electron transport, the H+ concentration in the intramembrane space increases, i.e., the pH value there decreases. Although the number of transported protons is still not known with certainty, it is estimated that for each NADH oxidized (i.e., each electron pair transferred to oxygen) 10 protons are translocated from the mitochondrial matrix into the intermembrane space. In intact mitochondria, the protons can return to the matrix via *ATP synthase*. This is the structural basis of the **energetic coupling** of electron transport to ATP formation (see following page).

Although complexes I through V are all integrated into the inner mitochondrial membrane, they are not usually in direct contact with one another. The long apolar side chain of ubiquinone allows it to move freely in the membrane. In contrast, cytochrome c is readily soluble in water, and is therefore found on the *outer surface* of the inner membrane. NADH oxidation via complex I occurs on the *inside* of the membrane, i.e., in the matrix space where the most important sources of NADH, the citric acid cycle and β-oxidation, are located. O_2 reduction and ATP formation also occur on the inside of the membrane. The ATP formed is ultimately exported via an antiport in exchange for ADP (see p. 198).

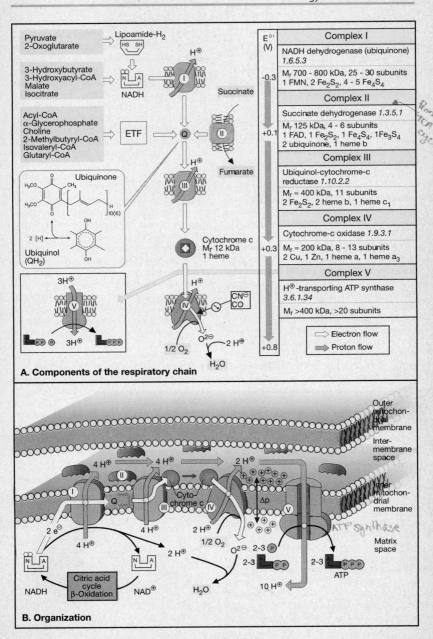

A. Components of the respiratory chain

Pyruvate
2-Oxoglutarate

3-Hydroxybutyrate
3-Hydroxyacyl-CoA
Malate
Isocitrate

Acyl-CoA
α-Glycerophosphate
Choline
2-Methylbutyryl-CoA
Isovaleryl-CoA
Glutaryl-CoA

Lipoamide-H_2
HS SH

$H^\oplus$

NADH

ETF

Succinate

Q

II

Fumarate

Ubiquinone

$H^\oplus$

III

Cytochrome c
M_r 12 kDa
1 heme

$3H^\oplus$

V

$3H^\oplus$

$H^\oplus$

IV

CN$^\ominus$
CO

$1/2\ O_2$ $O^{2\ominus}$ $2\ H^\oplus$

H_2O

Ubiquinol
(QH_2)

$2\ [H]$

$E^{\circ\prime}$ (V)	Complex I
-0.3	NADH dehydrogenase (ubiquinone) *1.6.5.3*
	M_r 700 - 800 kDa, 25 - 30 subunits 1 FMN, 2 Fe_2S_2, 4 - 5 Fe_4S_4
	Complex II
+0.1	Succinate dehydrogenase *1.3.5.1*
	M_r 125 kDa, 4 - 6 subunits 1 FAD, 1 Fe_2S_2, 1 Fe_4S_4, 1Fe_3S_4 2 ubiquinone, 1 heme b
	Complex III
	Ubiquinol-cytochrome-c reductase *1.10.2.2*
	$M_r \approx 400$ kDa, 11 subunits 2 Fe_2S_2, 2 heme b, 1 heme c_1
+0.3	**Complex IV**
	Cytochrome-c oxidase *1.9.3.1*
	$M_r \approx 200$ kDa, 8 - 13 subunits 2 Cu, 1 Zn, 1 heme a, 1 heme a_3
	Complex V
	$H^\oplus$-transporting ATP synthase *3.6.1.34*
	M_r >400 kDa, >20 subunits
+0.8	⇨ Electron flow ⇨ Proton flow

B. Organization

Outer mitochondrial membrane

Inter-membrane space

Inner mitochondrial membrane

Matrix space

ATP synthase

$4\ H^\oplus$ $4\ H^\oplus$ $2\ H^\oplus$

II

I

Q

III

'Cyto-chrome c'

IV

Δp

V

$2\ e^\ominus$

$4\ H^\oplus$

$4\ H^\oplus$

$2\ H^\oplus$

$2\ H^\oplus$

$1/2\ O_2$

$O^{2\ominus}$ 2-3 P

H_2O

$10\ H^\oplus$

2-3 P

2-3 ATP

Citric acid cycle
β-Oxidation

NADH   NAD$^\oplus$

ATP Synthesis

In the **respiratory chain**, electrons are transferred from NADH or ubiquinol (QH_2) to O_2. The energy thus obtained is used to build up a proton gradient across the inner mitochondrial membrane (see p. 130).

A. Redox systems of the respiratory chain ◑

The electrons donated by NADH are not transferred directly to O_2. Instead, they are passed on in a stepwise fashion. There are at least 12 intermediate redox systems involved, most of which are bound as **prosthetic groups** within the complexes I, III and IV. At first sight, the large number of coenzymes involved in respiratory electron transport comes as a surprise. However, as discussed on p. 28, the total *free-energy change ΔG* for redox reactions only depends on the difference in redox potential between the initial donor and the final acceptor. Thus, although the insertion of additional redox systems between NADH and O_2 does not change the overall ΔG of the reaction, it divides the total energy of about 200 kJ · mol⁻¹ into smaller, more amenable "packets" the size of which is given by the difference in the redox potentials of the respective intermediates. It is assumed that this division into packets is responsible for the surprisingly high energy yield (about 60%) achieved by the respiratory chain.

The illustration shows the important redox systems of the mitochondrial electron transport chain and their approximate reduction potentials. These potentials determine the path followed by the electrons. This is, because the members of a **redox series** must be organized in order of increasing redox potential if transport is to occur spontaneously (see p. 28).

In complex I, the electrons are passed from NADH to FMN (see p. 100) and then on to various different **iron/sulfur (Fe/S) centers**. These structures are stable only when located inside a protein. Depending on the type, Fe/S centers may contain 2 to 6 iron ions, which form complexes of differing composition with inorganic sulfide and with the SH groups of cysteine residues. The illustration shows a

schematic representation of the structure of a so-called Fe_4S_4 center (Fe blue, S yellow).

Several different **heme groups** are also involved in electron transport. The *type b hemes* are the same as those found in hemoglobin (see p. 180). *Heme c* is covalently bound to the protein (see p. 100), whereas the tetrapyrrol ring of *heme a* is isoprenylated (see p. 214) and carries a formyl group. In complex IV, a **copper ion** (Cu_B) and heme a_3 react directly with O_2. Coenzyme Q and cytochrome c are discussed on the previous page.

B. ATP synthase ◑

The synthesis of ATP is coupled to the flux of protons from the intermembrane space into the matrix (see p. 130). The *H^+-transporting ATP synthase* consists of two parts — a membrane-integrated *proton channel* (F_o) made up of at least 13 subunits, and a catalytic subunit (F_1) that protrudes out into the matrix. The "head" of the catalytic portion consists of 3 α and 3 β subunits. In between these, there are 3 active sites. The "stem" of the structure is made up of polypeptides belonging to the F_o part, as well as the γ, δ and ε subunits of the head.

The catalytic process can be divided into three phases through which the active sites cycle sequentially. First, ADP and P_i are bound (1), then the anhydride bond forms (2), and finally the product is released (3). Every time protons pass through the F_o channel and into the matrix, all three active sites change from their current status to the next. It is assumed that the energy of the proton transport is initially converted into a rotation of the γ subunit, which then changes the conformation of the α and β subunits in a cyclic fashion.

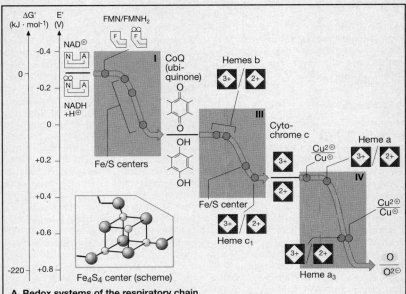

A. Redox systems of the respiratory chain

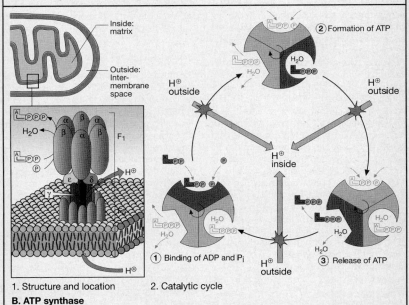

1. Structure and location   2. Catalytic cycle

B. ATP synthase

Regulation of Energy Metabolism I

Nutrient degradation and ATP synthesis have to be continually adapted to the ever-changing energy requirements of the cell. Here we discuss some of the mechanisms involved in this coordination process.

A. Respiratory control ◗

The primary reason why production and utilization of ATP have to be be coordinated with one another is the fact that the *total amount* of the coenzymes in the body is very small. The "caloric content" of a standard diet is about 12,000 kJ per day (see p. 320). Assuming a yield of 50%, this converts to 120 mol ATP, i.e., about 65 kg per day. However, the human body contains a total of only 3–4 g of free adenine nucleotides (AMP, ADP, and ATP). Thus, each ADP molecule must be phosphorylated to give ATP and then dephosphorylated again many thousands of times a day.

There is a simple regulatory mechanism that ensures that ATP synthesis is coordinated with ATP utilization. It is referred to as **respiratory control**, and is based on the fact that these two processes are **coupled** via shared coenzymes and other factors (1). If a cell does not use any ATP, there is hardly any ADP available in the mitochondria. Without ADP, *ATP synthase* (3) cannot dissipate the proton gradient across the inner mitochondrial membrane. This, in turn, inhibits electron transport in the respiratory chain (2), which means that NADH can no longer be reoxidized to NAD+. Finally, the resulting high NADH/NAD+ ratio inhibits the citric acid cycle (see **C**), thus slowing the degradation of the substrates SH_2 (1). Vice versa, high rates of ATP utilization stimulate nutrient degradation and oxidative phosphorylation by the same mechanism.

When the establishment of a proton gradient is prevented (2), substrate oxidation (1) and electron transport (2) proceed much more rapidly. However, instead of ATP, only heat is produced.

B. Uncouplers ○

Substances that cause the functional separation of oxidation and phosphorylation are referred to as uncouplers. They allow protons to be transported from the intermembrane space back into the mitochondrial matrix without the involvement of ATP synthase. Uncoupling can simply result from **mechanical damage** to the inner membrane (1), or to the action of substances, such as **2,4-dinitrophenol**, which pass protons through the membrane (2). A naturally occuring uncoupler is **thermogenin** (3), a ligand-gated proton channel (see p. 208) in the mitochondria of the so-called *brown fat*. This tissue is found, for example, in newborns or in hibernating animals, where its sole purpose is the generation of heat. When the temperature drops, norepinephrine activates the *hormone-sensitive lipase* (see p. 144). The result is an increase in lipolysis leading to the production of large quantities of free fatty acids, which are then degraded by β-oxidation and the respiratory chain. Since fatty acids also open thermogenin for protons, their degradation becomes independent of ADP availability, i.e., it can occur at maximum velocity and the only product is heat (**A**).

C. Regulation of the citric acid cycle ◗

The most important factor in the regulation of the cycle is the **NADH/NAD+ ratio**. Not only *pyruvate dehydrogenase* and *oxoglutarate dehydrogenase* (see p. 124), but also *citrate synthase* and *isocitrate dehydrogenase* are inhibited by NADH. With the exception of isocitrate dehydrogenase, these enzymes are also subject to **product inhibition** by acetyl-CoA, succinyl-CoA, and citrate, respectively. **Interconversion processes** (see p. 110) are also important. Here they are illustrated using the pyruvate dehydrogenase complex as an example (above). The inactivating *protein kinase* [1a] is inhibited by the substrate pyruvate and activated by the product acetyl-CoA. The corresponding *protein phosphatase* [1b] is activated by Ca^{2+}. Ca^{2+} also stimulates *isocitrate dehydrogenase* [3] and the *ODH complex* [4] — an effect especially important during muscle contraction.

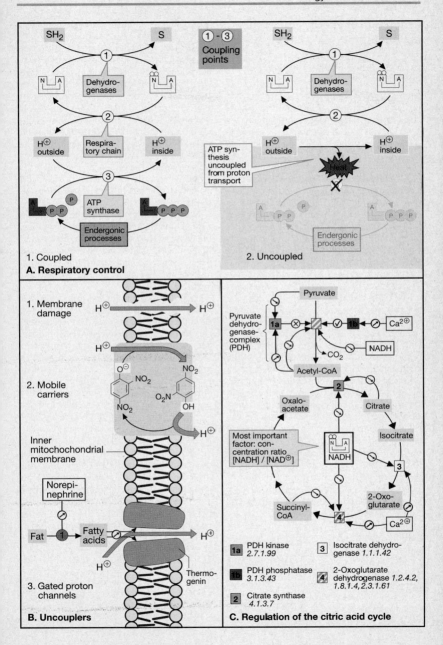

A. Respiratory control

1. Coupled

(1) - (3) Coupling points

2. Uncoupled

ATP synthesis uncoupled from proton transport

B. Uncouplers

1. Membrane damage

2. Mobile carriers

Inner mitochochondrial membrane

Norepi-nephrine

Fat → Fatty acids

Thermo-genin

3. Gated proton channels

C. Regulation of the citric acid cycle

Pyruvate

Pyruvate dehydro-genase-complex (PDH)

1a   1b   Ca²⁺

CO₂   NADH

Acetyl-CoA

2

Oxalo-acetate   Citrate

Isocitrate

Most important factor: concentration ratio [NADH] / [NAD⁺]

NADH

3

2-Oxo-glutarate

Ca²⁺

Succinyl-CoA

4

1a PDH kinase 2.7.1.99

1b PDH phosphatase 3.1.3.43

2 Citrate synthase 4.1.3.7

3 Isocitrate dehydro-genase 1.1.1.42

4 2-Oxoglutarate dehydrogenase 1.2.4.2, 1.8.1.4, 2.3.1.61

Regulation of Energy Metabolism II

A. Aerobic and anaerobic oxidation of glucose ◑

Under **aerobic conditions (1)**, i.e., in the presence of oxygen, ATP is derived almost solely from oxidative phosphorylation (see p. 104). Under these conditions, all nutrients (carbohydrates, lipids and amino acids) can be degraded by cellular "respiration," i.e., completely broken down by oxidative processes. **Fatty acids** enter the mitochondria with the help of carnitine (see p. 196). There they are degraded to CoA-bound acetyl residues by β-oxidation (see p. 140). **Glucose** is initially converted to pyruvate by glycolysis (see p. 140) in the cytoplasm. Under aerobic conditions, pyruvate is then transported into the mitochondrial matrix, where it is oxidatively decarboxylated by the pyruvate dehydrogenase complex (see p. 124) to yield acetyl-CoA. The reducing equivalents (2 NADH + H^+ per glucose) enter the mitochondrial matrix via the malate shuttle (see p. 198), where they are once again transferred to NAD^+. Acetyl residues from fatty acids and sugar degradation are oxidized to CO_2 in the citric acid cycle (see p. 126). The degradation of amino acids also yields acetyl residues or products that can directly enter the citric acid cycle (see p. 128, 154). The reducing equivalents stored in the reduced coenzymes are transferred to oxygen via the respiratory chain as required (see p. 130). In the process, chemical energy is released, which is used (via a proton gradient) for the synthesis of ATP (see p. 132). In the illustration, the flow of electrons is indicated by white arrows, whereas proton movement is shown by red arrows.

In the absence of oxygen, i.e., under **anaerobic conditions**, the picture is quite different. The electron acceptor for the respiratory chain is missing and, therefore, NADH + H^+ and QH_2 can no longer be reoxidized. As a result, not only ATP synthesis, but also almost every other aspect of metabolism in the mitochondrial matrix comes to a halt. One reason for this is the high NADH concentration,

which inhibits the citric acid cycle and the pyruvate dehydrogenase reaction (see p. 132). β-Oxidation and the malate shuttle also stop, because both are dependent on free NAD^+. Since amino acid degradation can no longer contribute to energy production either, the cell becomes totally dependent on ATP synthesized via the degradation of glucose by **glycolysis**. However, in order for this process to continue, the NADH + H^+ that is formed has to be continuously reoxidized. Since this can no longer occur in the mitochondria, animal cells dispose of these reducing equivalents by reducing pyruvate to lactate. A process, such as this is referred to as **fermentation** (see p. 138)

The table summarizes the steps in glucose degradation that contribute to ATP synthesis under anaerobic, as well as aerobic, conditions. Clearly, anaerobic glycolysis releases only a small proportion of the maximum amount of chemical energy theoretically available.

In order to estimate the number of ATP molecules formed in oxidative phosphorylation, it is necessary to know the **P/O quotient**, i.e., the molar ratio between ATP synthesized and water formed. During the transport of two electrons from NADH to oxygen, about 10 protons are translocated into the intermembrane space, while from ubiquinol (QH_2), it is only 6 (see p. 130). ATP synthase requires 3 protons for the synthesis of one ATP, and therefore the maximum possible P/O quotient is 3 (or 2 in the case of QH_2). However, it must be noted that the transport of phosphate into the matrix and the exchange of mitochondrial ATP for cytoplasmic ADP (see p. 198) also use protons from the intermembrane space. The actual P/O quotients are therefore about 2.5 for NADH oxidation, and 1.5 for the oxidation of QH_2. These values were used to calculate the data of the table.

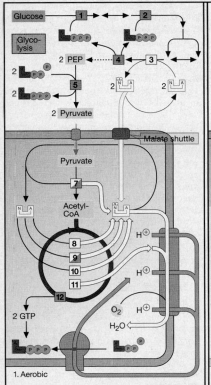

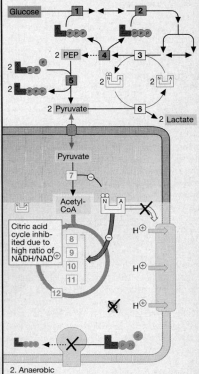

ATP	Coenzymes		Enzymes	Coenzymes	ATP
−1		−1 ATP	**1** Hexokinase	−1 ATP	−1
−2		−1 ATP	**2** 6-Phosphofructokinase	−1 ATP	−2
+3	+5 ATP ←	+2 NADH	**3** Glyceraldehyde-3 (P) DH	+2 NADH ←	−2
+5		+2 ATP	**4** Phosphoglycerate kinase	+2 ATP	0
+7		+2 ATP	**5** Pyruvate kinase	+2 ATP	+2
			6 Lactate dehydrogenase	−2 NADH ←	
+12	+5 ATP ←	+2 NADH	**7** Pyruvate dehydrogenase		
+17	+5 ATP ←	+2 NADH	**8** Isocitrate dehydrogenase		
+22	+5 ATP ←	+2 NADH	**9** Oxoglutarate dehydrogenase		
+27	+5 ATP ←	+2 NADH	**10** Malate dehydrogenase		
+30	+3 ATP ←	+2 QH$_2$	**11** Succinate dehydrogenase		
+32	+2 ATP ←	+2 GTP	**12** Succinate-CoA ligase		

Sum: 32 ATP/glucose   DH = dehydrogenase   **Sum: 2 ATP/glucose**

A. Aerobic and anaerobic oxidation of glucose

Fermentations

As discussed on p. 136, the degradation of glucose to pyruvate is the only way by which ATP can be synthesized in the *absence of oxygen*. The NADH formed must be reoxidized to NAD+ if glycolysis and, consequently, the synthesis of ATP are to continue. In animals, this is brought about by the reduction of pyruvate to lactate. In micro-organisms, there are many other pathways for the regeneration of NAD+. Processes of this type are referred to as **fermentations**. Microbial fermentation processes are often used for the production or preservation of foodstuffs and alcoholic beverages. All fermentation processes begin with pyruvate, and only occur under anaerobic conditions.

A. Lactic acid and propionic acid fermentation ○

Many milk products, such as clotted milk, yogurt, and cheese are made by *bacterial lactic acid fermentation* (**1**). The reaction is the same as in animals. Pyruvate, produced mainly by degradation of the disaccharide lactose (see p. 36) is reduced to **lactate** by *lactate dehydrogenase* [1] (see p. 94). Lactic acid fermentation also plays an important role in the production of sauerkraut and silage. These products usually keep for a long time, because the *decline in pH* that accompanies fermentation inhibits the growth of putrefying bacteria. It is bacteria of the genera *Lactobacillus* and *Streptococcus*, which are involved in the initial stages of the production of dairy products (**3**). The raw materials arising from their activity attain their final characteristics after further fermentation processes. For example, the characteristic taste of Swiss cheese develops during a subsequent propionic acid fermentation. In this process, bacteria of the genus *Propionibacterium* convert pyruvate to **propionate** via a complicated series of reactions (**2**).

B. Alcoholic fermentation ○

Alcoholic beverages arise from the fermentation of plant products with a high carbohydrate content.

Pyruvate, formed from glucose, is initially decarboxylated by *pyruvate decarboxylase* [2] to yield **acetaldehyde**. This is then reduced by *alcohol dehydrogenase* to **ethanol** [3], with the concomitant oxidation of NADH. It is **yeasts** (eucaryotic unicellular fungi) rather than bacteria, that are responsible for this type of fermentation (**3**). Yeasts are also used in baking. They produce CO_2 and ethanol, which raise the dough. Brewer's and baker's yeasts (*Saccharomyces cerevisiae*) are usually haploid and reproduce asexually by budding (**3**). They can live both aerobically and anaerobically. Wine is produced by yeasts that live on the grapes that are harvested for the purpose. The production of ethanol during alcoholic fermentation is dependent on the exclusion of oxygen. Therefore, dough is usually covered with a cloth during rising. For the same reason, fermentations leading to alcoholic beverages usually have to be carried out in airtight containers.

C. Beer brewing ○

Barley is the traditional starting material for the brewing of beer. Cereal grains are rich in starch, but hardly contain any *free* sugars. Therefore, the barley grains are first allowed to germinate so that *amylases* (see p. 245) can be synthesized. Careful heating of the germinated grain produces **malt**. This is then ground and soaked in warm water for some time. A large fraction of the starch is thereby degraded to the disaccharide maltose. The product (the *wort*) is then boiled, **yeast** and **hops** are added, and the mixture is allowed to ferment for several days. The addition of hops increases the keeping quality of the beer, and gives it a slightly bitter taste. Other compounds derived from the hops act as sedatives and diuretics.

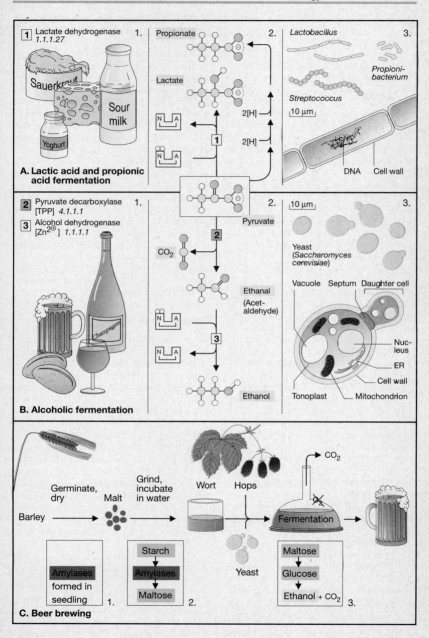

1 Lactate dehydrogenase *1.1.1.27*

1. Sauerkraut · Sour milk · Yoghurt

A. Lactic acid and propionic acid fermentation

2. Propionate · Lactate · 2[H] · 2[H]

N A · N A · **1**

3. *Lactobacillus* · *Propionibacterium* · *Streptococcus* · 10 µm · DNA · Cell wall

2 Pyruvate decarboxylase [TPP] *4.1.1.1*

3 Alcohol dehydrogenase [Zn²⁺] *1.1.1.1*

B. Alcoholic fermentation

1.

2. Pyruvate · **2** · CO_2 · Ethanal (Acetaldehyde) · N A · **3** · N A · Ethanol

3. 10 µm · Yeast (*Saccharomyces cerevisiae*) · Vacuole · Septum · Daughter cell · Nucleus · ER · Cell wall · Tonoplast · Mitochondrion

C. Beer brewing

Barley → Germinate, dry → Malt → Grind, incubate in water → Wort · Hops → Fermentation → CO_2 · Yeast

1. Amylases formed in seedling

2. Starch → Amylases → Maltose

3. Maltose → Glucose → Ethanol + CO_2

Glycolysis

A. Balance ●

Glycolysis is a catabolic pathway occurring in the cytoplasm of most cells, irrespective of whether they live aerobically or anaerobically. Glycolysis has a simple balance: for every glucose degraded, two molecules of pyruvate are formed. Under aerobic conditions, additional products are two *ATP* and two *NADH + H+* (**aerobic glycolysis**). Under anaerobic conditions, pyruvate undergoes further transformations to allow the regeneration of NAD+ (see p. 138). The result is *fermentation products*, such as lactate or ethanol (**anaerobic glycolysis**). In the absence of oxygen, glycolysis is the only pathway for the synthesis of ATP from ADP and inorganic phosphate (see p. 136).

B. Reactions ◑

Sugars are metabolized mainly as phosphate esters. Most animal cells obtain glucose from the blood. Once inside the cell, it is phosphorylated to **glucose 6-phosphate** in an ATP-dependent reaction catalyzed by *hexokinase* [1]. Following isomerization of glucose 6-phosphate into **fructose 6-phosphate** [2], another phosphorylation step leads to **fructose 1,6-bisphosphate**. *Phosphofructokinase* [3], the enzyme catalyzing this step, is a key regulatory enzyme in glycolysis (see p. 162). Up to this point, two ATP have been *consumed* per glucose molecule. Fructose 1,6-bisphosphate is subsequently cleaved by *aldolase* [4] to yield two phosphorylated C_3 compounds, **glyceraldehyde 3-phosphate** and **dihydroxyacetone 3-phosphate**, which are rapidly interconverted by *triose phosphate isomerase* [5]. Glyceraldehyde 3-phosphate is then oxidized in the presence of *glyceraldehyde-3-phosphate dehydrogenase* [6] to yield **1,3-bisphosphoglycerate** and NADH + H+. In this reaction, inorganic phosphate is incorporated into a high-energy bond (**substrate-level phosphorylation**). 1,3-Bisphosphoglycerate contains a *mixed acid-anhydride bond*, the cleavage of which is strongly exergonic (see p. 8). In the next step (catalyzed by *phos-*

phoglycerate kinase), the hydrolysis of this bond is energetically coupled to the formation of ATP. A further glycolytic intermediate whose hydrolysis is coupled to the synthesis of ATP is produced by isomerization of the 3-*phosphoglycerate* formed in reaction [7] to yield **2-phosphoglycerate** (enzyme: *phosphoglyceromutase*, [8]) followed by the elimination of water (enzyme: *phosphopyruvate hydratase*, "*enolase*," [9]). The product of this reaction is the phosphate ester of the *enol form* of pyruvate, which is, therefore, called **phosphoenolpyruvate** (PEP). In the last step of the pathway, *pyruvate kinase* [10] catalyzes the synthesis of *pyruvate* and ATP from PEP. The free energy-change upon hydrolysis of PEP is so high ($\Delta G^{0'}$ = -55 kJ · mol^{-1}) that the reaction is strongly exergonic, and therefore essentially irreversible.

On the minus side, glycolysis consumes two ATP per glucose for activation. On the plus side, however, 2 ATP are formed *per C_3 fragment*. In total, this represents a net gain of 2 mol ATP per mol of glucose.

C. Energy profile ○

The **energy balance** of metabolic pathways not only depends on the standard free-energy changes, $\Delta G^{0'}$, but also on the actual concentrations of the metabolites. Here we present an energy diagram showing the actual free-energy changes ΔG, (see p. 12) for the individual steps in glycolysis in erythrocytes (see p. 260). Only three reactions (1, 3, and 10), exhibit large free-energy changes. For these three reactions, the equilibrium lies far on the side of the products. The other steps, which are freely reversible, are also involved in glucose biosynthesis (*gluconeogenesis*), where they are catalyzed by exactly the same enzymes. This contrasts with the non-reversible steps 1, 3, and 10, which are bypassed by alternate enzymes in gluconeogenesis (see p. 158).

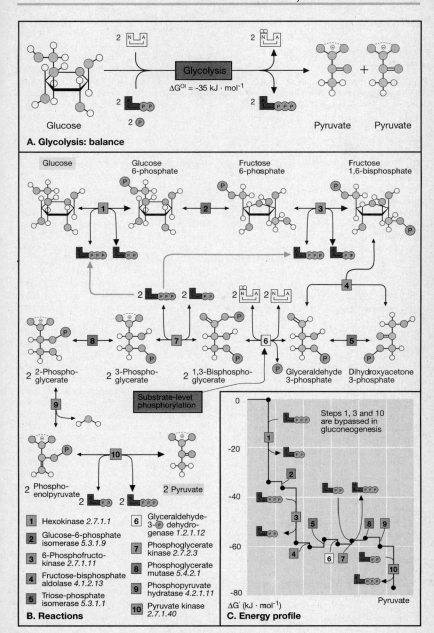

A. Glycolysis: balance

Glucose

Glycolysis

$\Delta G^{0l} = -35 \text{ kJ} \cdot \text{mol}^{-1}$

Pyruvate   Pyruvate

B. Reactions

Glucose

Glucose 6-phosphate

Fructose 6-phosphate

Fructose 1,6-bisphosphate

2 2-Phospho-glycerate

2 3-Phospho-glycerate

2 1,3-Bisphospho-glycerate

Glyceraldehyde 3-phosphate

Dihydroxyacetone 3-phosphate

Substrate-level phosphorylation

2 Phospho-enolpyruvate

2 Pyruvate

1	Hexokinase *2.7.1.1*	6	Glyceraldehyde-3-P dehydrogenase *1.2.1.12*
2	Glucose-6-phosphate isomerase *5.3.1.9*	7	Phosphoglycerate kinase *2.7.2.3*
3	6-Phosphofructokinase *2.7.1.11*	8	Phosphoglycerate mutase *5.4.2.1*
4	Fructose-bisphosphate aldolase *4.1.2.13*	9	Phosphopyruvate hydratase *4.2.1.11*
5	Triose-phosphate isomerase *5.3.1.1*	10	Pyruvate kinase *2.7.1.40*

C. Energy profile

Steps 1, 3 and 10 are bypassed in gluconeogenesis

Pyruvate

$\Delta G'$ (kJ · mol^{-1})

Hexose Monophosphate Pathway

A. Hexose monophosphate pathway: oxidative part ●

The hexose monophosphate pathway (HMP, also known as the *pentose phosphate cycle*) is an oxidative metabolic pathway located in the cytoplasm, which, like glycolysis, begins with glucose 6-phosphate. It supplies two important precursors for anabolic pathways: **NADPH + H+**, which is required for many biosynthetic pathways, and **ribose 5-phosphate**, a precursor in nucleic acid biosynthesis. The **oxidative segment** of the pathway converts glucose 6-phosphate to ribulose 5-phosphate. One CO_2 and 2 NADPH are produced in the process. The more complex **regenerative segment** of the HMP (**B**) may either convert a fraction of the pentose phosphates back to hexose phosphates, or it may channel them into glycolysis for breakdown. The routes chosen depend on the metabolic state of the cell. In most cases, however, less than 10% of glucose 6-phosphate is degraded via the HMP.

B. Reactions ◑

The oxidative segment of the HMP begins with *glucose-6-phosphate dehydrogenase* [1], which catalyzes the oxidation of **glucose 6-phosphate**, yielding NADPH + H+ and **6-phosphogluconolactone,** the intramolecular ester (*lactone*) of 6-phosphogluconate. A *hydrolase* [2] then cleaves the ester bond, exposing the carboxyl group of **6-phosphogluconate**. The last enzyme of the oxidative segment is *phosphogluconate dehydrogenase* [3], which releases the carboxylate group of 6-phosphogluconate as CO_2. At the same time, the hydroxyl group at C-3 is oxidized to an oxo group. Thus, besides a second molecule of NADPH, the reaction also produces the ketopentose **ribulose 5-phosphate**. This is converted by an isomerase to ribose 5-phosphate, the precursor for nucleotide biosynthesis.

Only a simplified representation of the **regenerative part** of the HMP is shown here. A complete reaction scheme can be found on p. 367. The function of the regenerative seg-ment of the HMP is to match the net production of NADPH with the requirement for pentose phosphates. Normally, cells require much more NADPH than pentose phosphates. Under these conditions, the reactions shown convert 6 ribulose 5-phosphate to 5 fructose 6-phosphate, which, in turn, can regenerate 5 glucose 6-phosphate by isomerization. These could then be used by the oxidative part of the HMP to produce further NADPH. Repetition of this series of reactions will eventually result in the oxidation of one glucose 6-phosphate to give 6 CO_2. At the same time, 12 NADPH + H+ are produced without any net synthesis of pentose phosphates.

Two enzymes are especially important in the interconversion of the various sugar phosphates in the regenerative segment of the HMP (cf. p. 366). *Transaldolase* [5] transfers C_3 units from sedoheptulose-7-phosphate, a keto-sugar with 7 carbons, to the aldehyde group of glyceraldehyde 3-phosphate. Similarly, *transketolase* [4] transfers C_2 fragments from one sugar phosphate to another. The reactions of the regenerative segment are freely reversible, and can therefore be used to convert hexose phosphates into pentose phosphates. This may occur when the demand for pentose phosphates is high, e.g., during DNA replication in the S-phase of the cell cycle (see p. 356).

When not only NADPH, but also energy in the form of ATP is required, cells can channel the products of the regenerative part of the HMP (fructose 6-phosphate and glyceraldehyde 3-phosphate) into glycolysis. The pyruvate formed is further broken down to CO_2 and water via the citric acid cycle and the respiratory chain. In this way, 12 NADPH and 165 ATP are obtained from 6 glucose 6-phosphate.

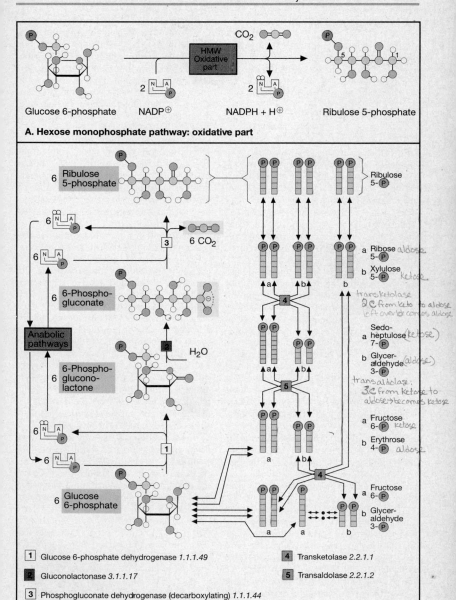

CO$_2$

HMW Oxidative part

2 NADP$^{\oplus}$

2 NADPH + H$^{\oplus}$

Glucose 6-phosphate   NADP$^{\oplus}$ NADPH + H$^{\oplus}$ Ribulose 5-phosphate

A. Hexose monophosphate pathway: oxidative part

6 Ribulose 5-phosphate

6 6 CO$_2$

3

6

6 6-Phospho-gluconate

Anabolic pathways

6 6-Phospho-glucono-lactone

2 H$_2$O

6

6

6

1

6 Glucose 6-phosphate

Ribulose 5-P

a Ribose 5-P   *aldose*
b Xylulose 5-P   *ketose*

transketolase 2 C from keto to aldose left over becomes aldose

4

a Sedo-heptulose 7-P (*ketose*)
b Glycer-aldehyde 3-P (*aldose*)

transaldolase: 3 C from ketose to aldose → becomes ketose

5

a Fructose 6-P *ketose*
b Erythrose 4-P *aldose*

4

a Fructose 6-P
b Glycer-aldehyde 3-P

1 Glucose 6-phosphate dehydrogenase *1.1.1.49*

2 Gluconolactonase *3.1.1.17*

3 Phosphogluconate dehydrogenase (decarboxylating) *1.1.1.44*

4 Transketolase *2.2.1.1*

5 Transaldolase *2.2.1.2*

B. Reactions

Fat Metabolism: Overview

Fats (triacylglycerols) are the most important energy reserve in animals. They are mostly stored in the cells of the adipose tissue, the *adipocytes*. There they are constantly being degraded and resynthesized (see p. 46 and 166).

A. Fat metabolism ⟩

Fat metabolism in the adipose tissue

For the synthesis of fats (lipogenesis), triacylglycerols are supplied by the liver and the intestine in the form of lipoproteins (VLDL and chylomicrons). A *lipoprotein lipase* [1] located on the surface of the endothelial cells of blood capillaries releases fatty acids from these lipoproteins, which are taken up and reconverted to triacylglycerols by the adipocytes (see p. 248).

The degradation of fats in adipose tissues (lipolysis) is catalyzed by a *hormone-sensitive lipase* [2], which is subject to complex control by hormones. The amount of fatty acids released by the adipose tissue is dependent on the activity of this lipase, i.e., it is mainly this enzyme that regulates the plasma lipid level.

The fatty acids released by the adipose tissue are transported in the blood in unesterified form (**FFA,** free fatty acids). However, only short-chain fatty acids are water soluble and, therefore, the less soluble long-chain fatty acids are transported bound to **albumin**.

Degradation of fatty acids in the liver (left)

The tissues take up fatty acids from the blood to rebuild fats or to obtain energy from their oxidation. The metabolism of fatty acids is especially intensive in the cells of the liver (hepatocytes).

The most important process for the degradation of fatty acids is β-**oxidation** (see p. 142), which takes place in the mitochondria. To allow this, the fatty acids in the cytoplasm first become activated by binding to coenzyme A [3]. Then, with the help of a transport system (carnitine shuttle [4]; see p. 196), they enter the mitochondria where they are degraded by β-oxidation to **acetyl-CoA**. The resulting acetyl residues can be further oxidized to CO_2 by the citric acid cycle and the respiratory chain with the production of ATP. When the acetyl-CoA production exceeds the energy requirements of the hepatocytes they form **ketone bodies** (see p. 278), which serve to supply other tissues with energy. This is the case when the fatty acid supply in the blood plasma is high, e.g., during starvation or diabetes mellitus.

Fat synthesis in the liver (right)

The biosynthesis of fatty acids occurs in the cytoplasm, especially of the liver, adipose tissue, kidneys, lungs, and mammary glands. The most important supplier of the carbons is **glucose**. However, other precursors of acetyl-CoA, such as ketogenic amino acids, can also be used.

The first step is the carboxylation of **acetyl-CoA** to malonyl-CoA. This reaction is catalyzed by *acetyl-CoA carboxylase* [5], which is the key regulatory enzyme in fatty acid biosynthesis. The polymerization to fatty acids is catalyzed by *fatty acid synthase* [6]. This multi-functional enzyme (see p. 165) starts with one molecule of acetyl-CoA and extends it by adding malonyl groups in seven reaction cycles until palmitate is produced. One CO_2 molecule is released in every reaction cycle and, therefore, the fatty acid grows by two carbon units every round. NADPH + H$^+$, derived either from the *hexose monophosphate pathway* or from the *isocitrate dehydrogenase* and *malate dehydrogenase* reactions, is used as the reducing agent.

The elongation of the fatty acid by *fatty acid synthase* stops at C_{16}, and the product, **palmitate** (16:0), is released. Further reactions can then use palmitate as the precursor for the synthesis of unsaturated fatty acids and longer-chain fatty acids. The subsequent biosynthesis of fats requires activated fatty acids (acyl-CoA) and glycerol 3-phosphate (see p. 166). In order to supply the other tissues, the fats are packed into lipoprotein complexes (**VLDL**) by the hepatocytes, and released into the blood (see p. 248).

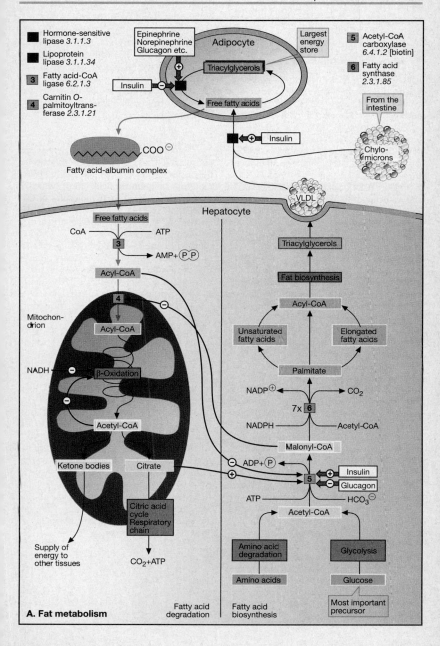

A. Fat metabolism

Degradation of Fatty Acids: β-Oxidation

A. Fatty acid degradation: β-oxidation ◑

Following their uptake by the cell, fatty acids are "activated" by conversion to their CoA derivatives, i.e., **acyl-CoAs** are formed. This requires the expense of two energy-rich anhydride bonds of **ATP** (see p. 140). The entry of activated fatty acids into the mitochondria is made possible by *carnitine*, which carries acyl residues through the inner membrane (see p. 190).

The degradation of the fatty acids occurs in the mitochondrial matrix through an oxidative reaction cycle in which C_2 units are successively released in the form of acetyl-CoA (*activated acetic acid*). Cleavage of the acetyl groups starts at the carboxyl end of the activated fatty acids between C-2 (α carbon) and C-3 (β carbon), and this is why this cyclic degradation process is called β-**oxidation**. It is closely linked to the citric acid cycle (see p. 126) and the respiratory chain, both spatially and functionally (see p. 130).

The first step of the pathway is the oxidation of an activated fatty acid (**acyl-CoA**) to yield an acid with a *trans* double bond (reaction **[1]**: *dehydrogenation*). In the process, two protons with their electrons are transferred from the enzyme [1] to an **electron-transferring flavoprotein (ETF)**. An *ETF dehydrogenase* [5] then passes on the reducing equivalents to ubiquinone (coenzyme Q), which is a component of the respiratory chain (see p. 130). The next step in fatty acid degradation is the addition of a **water** molecule to the double bond of the unsaturated fatty acid (reaction **[2]**: *hydration*). The third step oxidizes the hydroxyl group at C-3 to a carbonyl group (reaction **[3]**: *dehydrogenation*). In this case, the acceptor for the reducing equivalents is **NAD+**, which passes them on to the respiratory chain. In the fourth and last step of the cycle, the activated β-keto acid is cleaved by an *acyl transferase* (β-ketothiolase) in the presence of CoA (reaction **[4]**: *thioclastic cleavage*). The products are acetyl-CoA and an activated fatty acid with one less C_2 unit than the original acid.

For the complete degradation of long-chain fatty acids, the cycle must go through multiple rounds. For stearyl-CoA (C18:0), for example, eight cycles are required. The acetyl-CoA formed is condensed with oxaloacetate to yield citrate, which then enters the **citric acid cycle** (see p. 126). In the liver, an excess of acetyl-CoA leads to the synthesis of ketone bodies (see p. 278).

B. Energy balance ○

To calculate the energy balance of fatty acid degradation, we use as an example one molecule of **palmitic acid** (16:0), which is completely oxidized to 16 molecules of CO_2. In the first step, two energy-rich bonds (ATP) are used for the activation of the fatty acid to **palmitoyl-CoA**, which consists of eight C_2 units. Seven cycles of β-**oxidation** are then required. These release 7 molecules of reduced **ETF** and 7 molecules of **NADH + H+**. Both feed into the respiratory chain. The oxidation of ETF via ubiquinone yields 1.5 ATP, and that of NADH + H+ gives 2.5 (see p. 130). Thus, the β-oxidation of one palmitoyl residue yields 28 (7 × 4) ATP. The oxidation of each molecule of **acetyl-CoA** produces 10 ATP, which means that another 80 (8 ×10) ATP can be added. From the 28 + 80 ATP, we must subtract the two ATP that were required for the *activation* of the palmitic acid (see above). In total, therefore, 106 ATP are produced per molecule of palmitic acid. This corresponds to a free-energy gain of 3,300 kJ · mol^{-1} (106 · 30.5 kJ per mol ATP).

The amount of energy obtained from the degradation of fatty acids is greater than that from the degradation of carbohydrates (32 ATP per molecule of glucose) and proteins, even when the different sizes of the molecules are taken into account. Thus, fats are excellent energy storage compounds.

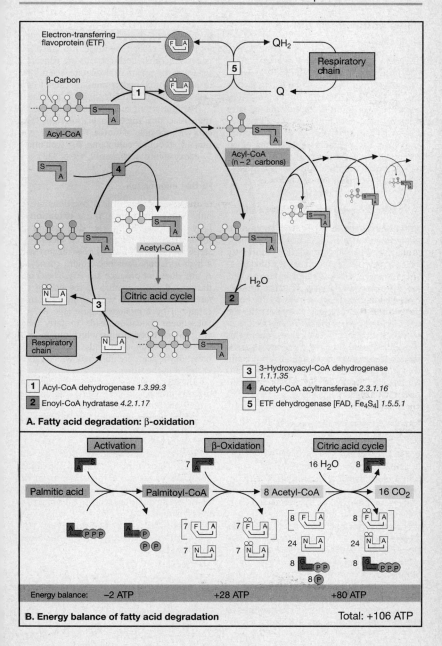

A. Fatty acid degradation: β-oxidation

Electron-transferring flavoprotein (ETF)

QH₂

Respiratory chain

β-Carbon

Acyl-CoA

Acyl-CoA (n − 2 carbons)

Acetyl-CoA

Citric acid cycle

H₂O

Respiratory chain

1 Acyl-CoA dehydrogenase 1.3.99.3

2 Enoyl-CoA hydratase 4.2.1.17

3 3-Hydroxyacyl-CoA dehydrogenase 1.1.1.35

4 Acetyl-CoA acyltransferase 2.3.1.16

5 ETF dehydrogenase [FAD, Fe₄S₄] 1.5.5.1

B. Energy balance of fatty acid degradation

Activation	β-Oxidation	Citric acid cycle
Palmitic acid → Palmitoyl-CoA	7 → 8 Acetyl-CoA	16 H₂O → 16 CO₂

Energy balance: −2 ATP | +28 ATP | +80 ATP

Total: +106 ATP

Minor Pathways of Fatty Acid Degradation

The main pathway for the degradation of fatty acids is β-oxidation (see p. 140). However, there are also other special pathways e.g., for the degradation of unsaturated fatty acids (see **A**), for the degradation of fatty acids with an uneven number of carbons (see **B**), as well as α and ω oxidation of fatty acids, and fatty acid degradation in peroxisomes. Although these pathways are quantitatively less important, their failure can nevertheless lead to severe disorders (see below).

A. Degradation of unsaturated fatty acids ○

Unsaturated fatty acids usually contain *cis* double bonds at positions 9 or 12, e.g., linoleic acid (18:2; 9,12). The degradation of this fatty acid occurs via β-oxidation of the CoA derivative until the C-9-*cis* double bond is reached. Since β-oxidation requires that the unsaturated intermediate carry a *trans* double bond, an *isomerase* [1] first has to convert the the 3,4-*cis* isomer into a 2-*trans* isomer before β-oxidation can continue. In cases where this conversion is not possible, the double bond is reduced with the help of NADPH+H⁺ [2], and then the degradation of the fatty acid continues with a shift of the double bond followed by β-oxidation as usual.

B. Degradation of odd-numbered fatty acids ○

These fatty acids are treated in the same way as the "normal" even-numbered acids, i.e., they are taken up by the cell, undergo ATP-dependent activation to acyl-CoA derivatives, and are transported into the mitochondria via the carnitine shuttle, where they are degraded by β-oxidation (see p. 140). The **propionyl-CoA** that is left behind is then carboxylated by *propionyl-CoA carboxylase* [3] to yield **methylmalonyl-CoA**, which, after isomerization (not shown, cf. p. 370), is converted to **succinyl-CoA** [4].

Succinyl-CoA is an intermediate of the citric acid cycle, and is therefore available for *gluconeogenesis*. Thus, some glucose can be synthesized from odd-numbered fatty acids. In contrast, the acetyl-CoA units produced by β-oxidation cannot be used for gluconeogenesis, because both carbons are converted to CO_2 on the way to oxaloacetate (see p. 128). In other words, higher animals cannot convert even-numbered fatty acids to glucose.

Two important coenzymes are involved in these reactions. The carboxylation step [3] occurs with the help of **biotin**. The mutase reaction [4] requires **coenzyme B₁₂** (cobamide, 5'-deoxyadenosylcobalamin, see p. 328).

Further information

α **and** ω **oxidation**. These reactions are of only minor importance. The α **oxidation** of fatty acids serves in the degradation of methyl-branched fatty acids. It occurs by stepwise removal of C_1 residues, begins with a hydroxylation, does not require coenzyme A, and does not produce any ATP. The ω **oxidation**, i.e., an oxidation of the end of the fatty acid, also starts with an hydroxylation catalyzed by a *monooxygenase* (see p. 292). Subsequent oxidation leads to fatty acids with two carboxyl groups, which can undergo β-oxidation from both ends until C_8 or C_6 dicarboxylic acids are obtained, which can be excreted in the urine.

An alternative form of β-oxidation takes place in *liver peroxisomes* (see p. 198), which are specialized for the degradation of particularly long fatty acids (n > 20). The degradation products are acetyl-CoA and H_2O_2. No ATP is formed.

Several rare diseases are associated with defects of enzymes involved in the minor pathways of fatty acid degradation. In the case of **Refsum syndrome**, the methyl-branched phytanic acid (obtained from plant foods) cannot be degraded by α oxidation. In the case of **Zellweger syndrome**, long-chain fatty acids cannot be degraded due to a peroxisomal defect.

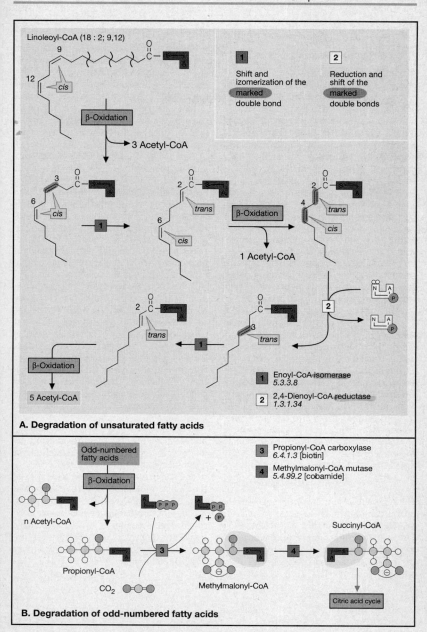

Linoleoyl-CoA (18 : 2; 9,12)

1
Shift and izomerization of the marked double bond

2
Reduction and shift of the marked double bonds

β-Oxidation

3 Acetyl-CoA

β-Oxidation

1 Acetyl-CoA

β-Oxidation

5 Acetyl-CoA

1 Enoyl-CoA isomerase
5.3.3.8

2 2,4-Dienoyl-CoA reductase
1.3.1.34

A. Degradation of unsaturated fatty acids

Odd-numbered fatty acids

3 Propionyl-CoA carboxylase
6.4.1.3 [biotin]

4 Methylmalonyl-CoA mutase
5.4.99.2 [cobamide]

β-Oxidation

n Acetyl-CoA

Propionyl-CoA

CO_2

Methylmalonyl-CoA

Succinyl-CoA

Citric acid cycle

B. Degradation of odd-numbered fatty acids

Nitrogen Balance, Peptidases

A. Protein metabolism: overview ●

In quantitative terms, proteins are the most important group of biomolecules (see p. 188). A person weighing 70 kg contains about 10 kg protein, most of it in the muscles. By contrast, the amounts of other nitrogen-containing compounds are low. Thus, the nitrogen balance of the organism is determined primarily by the metabolism of proteins. Several hormones, above all *testosteron* and *cortisol*, regulate protein synthesis and degradation.

The **nitrogen metabolism** of a healthy adult is, in general, balanced, i.e, the quantities of protein nitrogen taken in and excreted are approximately equal. It has been estimated that 300–400 g of protein are degraded to amino acids per day (*proteolysis*). On the other hand, approximately the same amount of amino acids is re-incorporated into proteins (*protein synthesis*). In order to balance the losses of nitrogen, at least 30 g of protein must be taken up with the diet each day (see p. 326). While this minimum amount is scarcely achievable in some countries, in industrial nations the protein content of the diet is usually much higher than necessary. Under these conditions, and given that amino acids cannot be stored, up to 100 g of excess amino acids per day are used for biosynthesis, or degraded in the liver. Most of the nitrogen released is converted to urea (see p. 152) and excreted as such in the urine. Small quantities of amino acids are also synthesized *de novo* from precursors.

B. Life span of proteins in the organism ◑

The high rate of turnover of proteins in the cells is due to the fact that most proteins are relatively *short-lived*. Their half-lives average 2–8 days. Structural proteins and hemoglobin of the erythrocytes are much more persistent. On the other hand, many *key enzymes* of intermediary metabolism, such as *HMG-CoA-reductase* or *ALA-synthase*, have half-lives of just a few hours.

C. Proteinases and peptidases ◑

Proteolysis, i.e., the complete degradation of proteins to free amino acids, requires a whole series of enzymes with differing specificities. *Proteinases* and *peptidases* are not only found in the digestive tract (see p. 244), but also within each cell. There they are in part located in the *lysosomes* (p. 212), but proteolytic activities are also found in the cytoplasm and other parts of the cell.

Based on the position of their cleavage site within the substrate molecule, proteolytic enzymes are divided into *endopeptidases* and *exopeptidases*. The **endopeptidases** or *proteinases* (1) catalyze the hydrolysis of peptide bonds within the peptide chain. They "recognize" and bind to short amino acid sequences, and then specifically hydrolyze bonds between particular residues. There are groups of proteinases with similar reaction mechanisms. In the case of the *serine proteinases*, for example, a serine residue of the enzyme is important for catalysis. In the case of the *cysteine proteinases*, it is a cysteine residue, and so on. In contrast to the endopeptidases, the **exopeptidases** (2) only attack peptides at their termini. They usually cleave off single amino acids one after the other. However, there are also enzymes that release dipeptides or tripeptides. Exopeptidases that attack peptides at the amino terminus chain are referred to as *aminopeptidases* and those that recognize the carboxy terminus are called *carboxypeptidases*. The *dipeptidases* only hydrolyze dipeptides. In order to completely degrade a protein, enzymes from each of the above-named groups are required.

Fig. **4** shows the structure of a complex of the serine proteinase **trypsin** (see p. 244) with a substrate peptide (blue). Trypsin specifically cleaves peptide bonds next to lysine or arginine residues (here it is the bond between a lysine and an alanine, arrow). Important functional residues of trypsin are highlighted in pink.

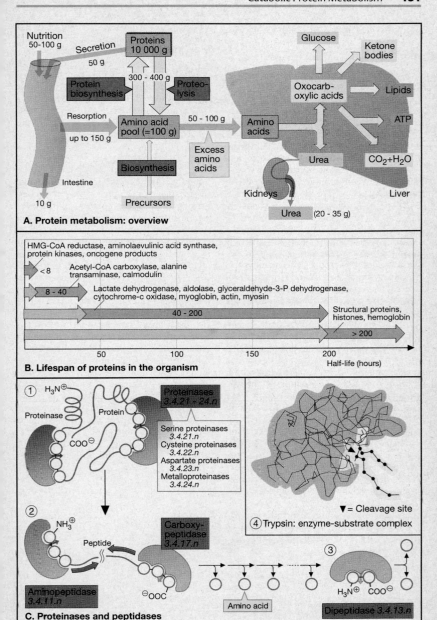

A. Protein metabolism: overview

Nutrition 50-100 g
Secretion 50 g
Proteins 10 000 g
Protein biosynthesis
Proteolysis
300 - 400 g
Resorption up to 150 g
Amino acid pool (=100 g)
50 - 100 g
Biosynthesis
Precursors
Excess amino acids
Intestine
10 g
Amino acids
Urea
Kidneys
Urea (20 - 35 g)
Glucose
Ketone bodies
Oxocarboxylic acids
Lipids
ATP
$CO_2 + H_2O$
Urea
Liver

B. Lifespan of proteins in the organism

HMG-CoA reductase, aminolaevulinic acid synthase, protein kinases, oncogene products
< 8
Acetyl-CoA carboxylase, alanine transaminase, calmodulin
8 - 40
Lactate dehydrogenase, aldolase, glyceraldehyde-3-P dehydrogenase, cytochrome-c oxidase, myoglobin, actin, myosin
40 - 200
Structural proteins, histones, hemoglobin
> 200
50   100   150   200
Half-life (hours)

C. Proteinases and peptidases

① $H_3N^{\oplus}$
Proteinase
Protein
$COO^{\ominus}$
Proteinases 3.4.21 - 24.n
Serine proteinases 3.4.21.n
Cysteine proteinases 3.4.22.n
Aspartate proteinases 3.4.23.n
Metalloproteinases 3.4.24.n

▼ = Cleavage site
④ Trypsin: enzyme-substrate complex

② $NH_3^{\oplus}$
Peptide
Carboxy-peptidase 3.4.17.n
Aminopeptidase 3.4.11.n
$^{\ominus}OOC$
Amino acid
③
$H_3N^{\oplus}$ $COO^{\ominus}$
Dipeptidase 3.4.13.n

Transamination and Deamination

A. Transamination and deamination ◐

During the degradation of proteins, amino nitrogen accumulates, which, in contrast to carbon, is not suitable for oxidative energy production. Therefore, those amino groups that cannot be re-used for biosynthesis are incorporated into urea (see p. 286) and excreted. Intermediates of NH_2 metabolism are amino groups bound to metabolites and coenzymes, and free ammonia (NH_3).

Among the NH_2 transfer reactions, **transaminations** (**1**) are especially important. They are catalyzed by *transaminases*, and are involved in both catabolic and anabolic amino acid metabolism. During transamination, the amino group of an amino acid (amino acid 1) is transferred to a 2-oxoacid (oxoacid 2). A 2-oxoacid is thereby produced from amino acid (**a**) and an amino acid from the original oxoacid (**b**). The amino group being transferred is taken over temporarily by enzyme-bound **pyridoxal phosphate** (PLP, **B**).

When the amino group of an amino acid is released as ammonia, this is referred to as **deamination** (**2**). There are different mechanisms for this. The removal of NH_3 from an amide group is known as **hydrolytic deamination**. A well-known example is the *glutaminase* reaction ([3], see **B**). In rare cases, NH_3 is eliminated and a double bond remains (**eliminating deamination**, not shown here). Of particular importance is **oxidative deamination** (**2**). In this reaction, the amino group is initially oxidized to an imino group (**a**), and the reducing equivalents are transferred to NAD^+ or $NADP^+$. In the second step, the imino group is cleaved by hydrolysis. Besides NH_3, a 2-oxoacid is thus produced, as in the case of transamination. This mechanism is seen in the *glutamate dehydrogenase* reaction, a process central to the amino acid metabolism of the liver (**C**).

B. Mechanism of transamination ○

In resting transaminases, the aldehyde group of PLP is covalently bound to a lysine residue of the enzyme (**1**). This type of compound, which is also found in the rhodopsins (see p. 324), is referred to an aldimine or "Schiff's base". During the reaction (**1**), amino acid 1 displaces the lysine residue and a new aldimine is formed (**2**). The C=N double bond is then shifted by isomerization. The ketimine (**3**) is hydrolyzed to yield the 2-oxoacid and pyridoxamine phosphate (**4**). In the second part of the reaction (**1**), the same steps occur *in the opposite direction*.

C. NH_3 metabolism of the liver ◐

Not only urea synthesis (see p. 286), but also the formation of the precursors NH_3 and aspartate, occur predominantly in the liver. Amino nitrogen, which accumulates in the tissues, is transported in the blood incorporated into **glutamine** (Gln) and **alanine** (Ala). In the liver, Gln is deaminated by *glutaminase* [3] to yield NH_3 and **glutamate** (Glu). The amino group of alanine is transferred to **2-oxoglutarate** (2-OG) by *alanine transaminase* [1]. Another glutamate is produced in this transamination (**A**). Ultimately, NH_3 is released from glutamate by oxidative deamination (**A**). This reaction is catalyzed by *glutamate dehydrogenase* [4], an enzyme typical of the liver. **Aspartate** (Asp), the second amino group donor in the urea cycle, also arises from glutamate. *Aspartate transaminase* [2], which is responsible for this reaction, like alanine transaminase, is present with high activity in the liver. However, both enzymes are found in other cells as well.

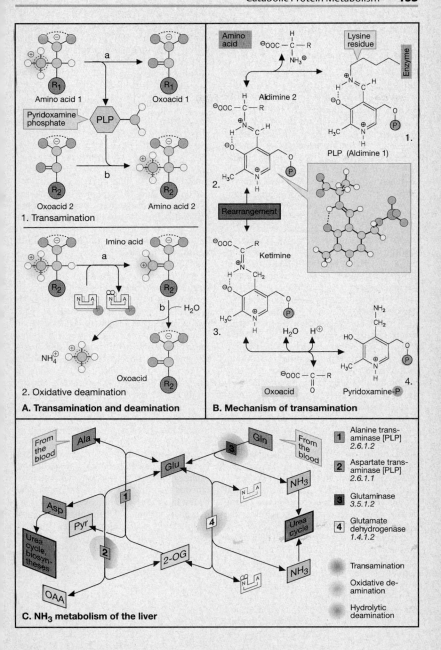

A. Transamination and deamination

1. Transamination

Amino acid 1 → Oxoacid 1
Pyridoxamine phosphate — PLP
Oxoacid 2 → Amino acid 2

2. Oxidative deamination

Imino acid
$NH_4^{\oplus}$
Oxoacid
H_2O

B. Mechanism of transamination

Amino acid

Lysine residue
Enzyme

Aldimine 2
$^{\ominus}OOC-C-R$

1. PLP (Aldimine 1)

2. Rearrangement

Ketimine
$^{\ominus}OOC$

3. H_2O $H^{\oplus}$

4. Pyridoxamine-P
$^{\ominus}OOC-C-R$
Oxoacid

C. NH₃ metabolism of the liver

From the blood — Ala
Gln — From the blood
Glu
Asp
Pyr
NH_3
2-OG
NH_3
OAA
Urea cycle, biosyntheses
Urea cycle

1 Alanine trans-aminase [PLP] 2.6.1.2
2 Aspartate trans-aminase [PLP] 2.6.1.1
3 Glutaminase 3.5.1.2
4 Glutamate dehydrogenase 1.4.1.2

Transamination
Oxidative de-amination
Hydrolytic deamination

Amino Acid Degradation

A. Amino acid degradation : overview ◑

The degradation of amino acids involves numerous metabolic pathways, which are presented here in an overview. Further details are summarized on pages 370 and 371.

The carbon skeletons of the 20 proteinogenic amino acids (see p. 59) yield just 7 different **degradation products** (highlighted in pink and light blue). Five of these metabolites (2-oxoglutarate, succinyl-CoA, fumarate, oxoalacetate, and pyruvate) are *precursors of gluconeogenesis* (see p. 154). The first four substances are already intermediates of the citric acid cycle; pyruvate may be converted to oxaloacetate by *pyruvate carboxylase* (green arrow), and thus made available to gluconeogenesis. Amino acids, the degradation of which supplies one of the five metabolites named above, are referred to as **glucogenic**. Except for two (lysine and leucine, see below), all proteinogenic amino acids are glucogenic, too.

Two further degradation products (acetoacetate and acetyl-CoA) cannot be converted to glucose in animal metabolism, because there is no way to transform them into precursors of gluconeogenesis. However, they can be used for the synthesis of ketone bodies, fatty acids, and isoprenoids. Amino acids that are broken down to acetyl-CoA or acetoacetate are, therefore, called **ketogenic.** Only leucine and lysine are purely ketogenic. Several other amino acids yield degradation products that are both glucogenic and ketogenic. This group includes phenylalanine, tyrosine, tryptophan, and isoleucine.

Different alternatives exist for the removal of the amino group (deamination) during amino acid degradation. Frequently the amino group is transferred to 2-oxoglutarate by *transamination* (see p. 152). The glutamate thus formed is then *oxidatively deaminated* (green) in a further step catalyzed by glutamate dehydrogenase. In this reaction, 2-oxoglutarate is regenerated, and free ammonia is produced. In animal metabolism, the latter is converted into urea and excreted (see p. 286). Ammonia may also originate from the hydrolysis of the amide groups of asparagine and glutamine (*hydrolytic deamination*, orange). A further conversion in which NH_3 is formed is the *eliminating deamination* (blue) of serine to pyruvate.

B. Biogenic amines ◑

Small quantities of certain amino acids are also degraded by *decarboxylation*. In general, this results in the production of monoamines, the so-called **biogenic amines**. Several compounds of this type are constituents of biomolecules. Thus, phospholipids (see p. 48) not only contain the amino acid serine, but also the corresponding biogenic amine, **ethanolamine**. **Cysteamine** and β-**alanine** are constituents of coenzyme A (see p. 102) and pantetheine (see p. 166). **Aminopropanol**, an amine formed from threonine, is a building block of vitamin B_{12} (see p. 334).

Other biogenic amines function as signal substances. An important *neurotransmitter* is γ-**aminobutyrate** (GABA, see p. 316), which is derived from glutamate. Other neurotransmitters are formed by decarboxylation of non-proteinogenic amino acids. The transmitter **dopamine** arises from 3,4-dihydroxyphenylalanine (Dopa) in this way. Dopamine is a precursor of the catecholamines epinephrine and norepinephrine (see p. 322). Defects in dopamine metabolism are involved in *Parkinson's disease.* **Serotonin** is a signaling substance with many sites of action. It is formed from tryptophan via the intermediate 5-hydroxytryptophan.

Many monoamines, e.g., the catecholamines, are inactivated by *(mono)amine oxidase* ("MAO"). The reaction involves simultaneous deamination and oxidation to aldehydes. Inhibitors of amine oxidase, therefore, play an important role in the pharmacological modification of neurotransmitter metabolism.

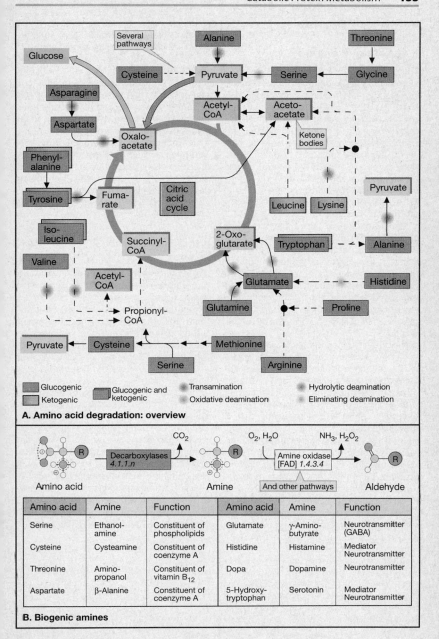

A. Amino acid degradation: overview

Legend:
- Glucogenic
- Ketogenic
- Glucogenic and ketogenic
- Transamination
- Oxidative deamination
- Hydrolytic deamination
- Eliminating deamination

Pathway labels visible: Glucose, Several pathways, Alanine, Threonine, Cysteine, Pyruvate, Serine, Glycine, Asparagine, Acetyl-CoA, Aceto-acetate, Aspartate, Oxalo-acetate, Ketone bodies, Phenyl-alanine, Pyruvate, Tyrosine, Fumarate, Citric acid cycle, Leucine, Lysine, Iso-leucine, Succinyl-CoA, 2-Oxo-glutarate, Tryptophan, Alanine, Valine, Acetyl-CoA, Glutamate, Histidine, Propionyl-CoA, Glutamine, Proline, Pyruvate, Cysteine, Methionine, Serine, Arginine

B. Biogenic amines

Reaction scheme: Amino acid →(Decarboxylases 4.1.1.n, CO_2)→ Amine →(Amine oxidase [FAD] 1.4.3.4, O_2, H_2O / NH_3, H_2O_2; And other pathways)→ Aldehyde

Amino acid	Amine	Function	Amino acid	Amine	Function
Serine	Ethanol-amine	Constituent of phospholipids	Glutamate	γ-Amino-butyrate	Neurotransmitter (GABA)
Cysteine	Cysteamine	Constituent of coenzyme A	Histidine	Histamine	Mediator Neurotransmitter
Threonine	Amino-propanol	Constituent of vitamin B_{12}	Dopa	Dopamine	Neurotransmitter
Aspartate	β-Alanine	Constituent of coenzyme A	5-Hydroxy-tryptophan	Serotonin	Mediator Neurotransmitter

Nucleotide Degradation

A. Degradation of nucleotides ◑

The nucleotides are among the most complex of metabolites. Their biosynthesis is lengthy, and requires a high energy input (see p. 174). It is, therefore, understandable that nucleotide building blocks are not completely degraded, but, for the most part, *recycled*. This especially holds for the purine bases adenine and guanine. In animals, up to about 90% of these bases are reconverted into nucleoside monophosphates by linkage with phosphoribosyl diphosphate (PRPP) (enzymes [1] and [2]). The pyrimidine bases are recycled to a much smaller extent.

The degradation of purine and pyrimidine nucleotides occurs by very different pathways. In man, the **purines** are degraded to uric acid, and excreted in this form. Thus, the purine ring remains intact in this process. In contrast, the **pyrimidine** rings of uracil, thymine, and cytosine are broken down into small fragments, which can be re-introduced into the metabolism or excreted without difficulty.

Guanosine monophosphate (GMP, **1**) is degraded in two steps, to the nucleoside *guanosine* and then to *guanine* (Gua). Guanine is converted, by deamination, into a further purine base, *xanthine*. In the most important degradative pathway for adenosine monophosphate (AMP), it is the nucleotide that is deaminated, and *inosine monophosphate* (IMP) arises. The purine base of IMP, *hypoxanthine*, is formed in a way similar to that observed for GMP. One and the same enzyme, *xanthine oxidase* [3], then converts hypoxanthine to xanthine and xanthine into uric acid. In each of these steps, an oxo group derived from molecular oxygen is introduced into the substrate. The other reaction product is hydrogen peroxide, which is toxic and must be removed by peroxidases.

Most mammals further degrade uric acid with ring opening to **allantoin** (enzyme: *uricase*). Allantoin is then excreted. In other animals, purine degradation continues further to allantoic acid or urea and glyoxylate. The primates, including man, are not capable of synthesizing allantoin. In their case, therefore, **uric acid** is the final product of purine catabolism. The same is true for reptiles and birds.

In the degradation of **pyrimidine nucleotides (2)**, the free bases *uracil* (Ura) and *thymine* (Thy, 5-methyluracil) are important intermediates. Both are further metabolized in similar ways. The pyrimidine ring is reduced and then hydrolytically cleaved. In the next step, β-**alanine** is produced as the degradation product of uracil, while β-**aminoisobutyrate** is formed from thymine. In the case of further degradation, β-alanine is cleaved to yield acetyl-CoA, CO_2 and NH_3. The ultimate degradation products of thymine, propionyl-CoA, CO_2, and NH_3, arise in an analogous way (not shown).

B. Hyperuricemia (gout) ○

The fact that, in man, purine degradation is already terminated with uric acid can lead to problems. In contrast to allantoin, uric acid is *poorly soluble* in water. When uric acid is formed in excessive amounts, or its utilization is disturbed, elevated concentrations in the blood (*hyperuricemia*) and, as a result, accumulation of uric acid crystals in the body can occur. The deposition of such crystals in joints is the cause of very painful attacks of **gout**.

Most cases of hyperuricemia can be traced back to defects in uric acid excretion via the kidneys (1). A high purine diet (e.g., meat) also may have unfavorable consequences (2). A rare, hereditary disease, the *Lesch-Nyhan syndrome*, is due to a defect in *hypoxanthine-phosphoribosyl transferase* (see Fig A, enzyme [1]). Under these circumstances, the impaired recycling of the purine bases leads to hyperuricemia and severe neurological disorders. Hyperuricemia can be treated with **allopurinol**, an inhibitor of xanthine oxidase.

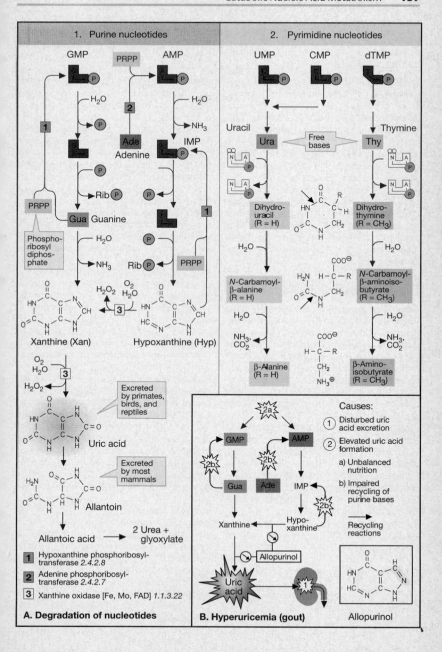

1. Purine nucleotides

GMP · PRPP · AMP

- Xanthine (Xan)
- Hypoxanthine (Hyp)
- Uric acid — Excreted by primates, birds, and reptiles
- Allantoin — Excreted by most mammals
- Allantoic acid → 2 Urea + glyoxylate

1 Hypoxanthine phosphoribosyl-transferase 2.4.2.8

2 Adenine phosphoribosyl-transferase 2.4.2.7

3 Xanthine oxidase [Fe, Mo, FAD] 1.1.3.22

2. Pyrimidine nucleotides

UMP · CMP · dTMP

Uracil (Ura) — Free bases — Thymine (Thy)

- Dihydro-uracil (R = H)
- Dihydro-thymine (R = CH₃)
- *N*-Carbamoyl-β-alanine (R = H)
- *N*-Carbamoyl-β-aminoiso-butyrate (R = CH₃)
- β-Alanine (R = H)
- β-Amino-isobutyrate (R = CH₃)

A. Degradation of nucleotides

B. Hyperuricemia (gout)

Causes:

① Disturbed uric acid excretion

② Elevated uric acid formation

a) Unbalanced nutrition

b) Impaired recycling of purine bases

→ Recycling reactions

Allopurinol

Gluconeogenesis

A. Carbohydrate metabolism during starvation ◑

Some tissues, such as the brain and the erythrocytes, are dependent on a constant supply of glucose. When the amount of carbohydrate taken up with the diet is insufficient, the concentration of glucose in the blood (the "blood glucose level") can, for a limited period of time, be maintained by *degradation of liver glycogen*. In Fig. **A**, data for starving rats are shown. The glycogen reserves are already depleted after one day, and the blood glucose level begins to fall. Not much later, however, it regains its original value, and the glycogen reserves of the liver also begin to recover. Both effects are due to the onset of *de novo* synthesis of glucose (**gluconeogenesis**).

B. Gluconeogenesis: overview ●

Gluconeogenesis occurs predominantly in the **liver**. The tubule cells of the **kidney** also have high gluconeogenetic activity (see p. 302). However, because of the much smaller mass of these cells, the contribution of the kidneys to the synthesis of glucose is only about 10% of the total. The main precursors of gluconeogenesis are **amino acids** derived from the muscles. Prolonged fasting, therefore, results in a massive degradation of muscle protein. A further important precursor is **lactate**, which is formed in erythrocytes and in muscles when oxygen is in short supply (see p. 310). **Glycerol** produced from the degradation of fats can also sustain gluconeogenesis. In contrast, the conversion of fatty acids to glucose is not possible in animal metabolism, because fatty acid degradation does not have anaplerotic effects (see p. 128). The human organism can synthesize several hundred grams of glucose per day by gluconeogenesis. *Cortisol, glucagon*, and *epinephrine* promote gluconeogenesis, whereas *insulin* inhibits it (see p. 162).

C. Reactions ◑

Many of the reactions of gluconeogenesis are catalyzed by enzymes that are also involved in glycolysis (see p. 140). Other enzymes, however, are *specific* for gluconeogenesis, and are only synthesized when they are required. It is only such enzymes, which are referred to in the table. Whilst glycolysis occurs entirely in the cytoplasm, gluconeogenesis also involves the *mitochondria* and the *endoplasmic reticulum*.

The first steps of the reaction chain occur in the *mitochondria*. The reason for this "detour" is the equilibrium state of the pyruvate kinase reaction (see p. 140). Even coupling to ATP hydrolysis would not be sufficient to convert pyruvate *directly* into phosphoenolpyruvate (PEP). In order to bypass this step, **pyruvate** is initially carboxylated to **oxaloacetate** in a biotin-dependent reaction catalyzed by *pyruvate carboxylase* [2]. Since oxaloacetate is an intermediate of the citric acid cycle, all amino acids whose degradation feeds the cycle, or, which supply pyruvate, can serve in the synthesis of glucose (*glucogenic amino acids*, see p. 154). The oxaloacetate formed by pyruvate carboxylase is reduced to **malate** or transaminated to **aspartate** in the mitochondria. Both metabolites can leave the mitochondria via the transport systems of the inner mitochondrial membrane (see p. 198). In the cytoplasm, oxaloacetate is reformed and then converted to **phosphoenolpyruvate** by a GTP-dependent *PEP carboxykinase* [3]. The succeeding steps up to **fructose 1,6-bisphosphate** are identical with those of glycolysis. Two gluconeogenesis-specific *phosphatases* then cleave off the phosphate residues of fructose 1,6-bisphosphate one after the other. In between these reactions lies the isomerization of **fructose 6-phosphate** to **glucose 6-phosphate** (also a glycolytic reaction). In the liver, *glucose 6-phosphatase* [5] is located in the endoplasmatic reticulum. From there, glucose is finally released into the blood.

Free glycerol is first phosphorylated to glycerol 3-phosphate, and then enters gluconeogenesis via dihydroxyacetone 3-phosphate.

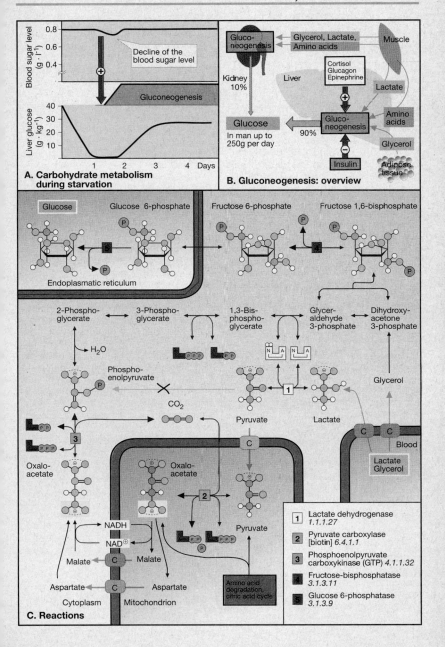

A. Carbohydrate metabolism during starvation

Blood sugar level (g · l⁻¹): 0.8, 0.6, 0.4

Decline of the blood sugar level

Gluconeogenesis

Liver glucose (g · kg⁻¹): 40, 30, 20, 10

1 2 3 4 Days

B. Gluconeogenesis: overview

Gluconeogenesis

Glycerol, Lactate, Amino acids — Muscle

Cortisol
Glucagon
Epinephrine

Kidney 10%

Liver

Lactate

Glucose
In man up to 250g per day

90%

Gluco-neogenesis

Amino acids

Glycerol

Insulin

Adipose tissue

C. Reactions

Glucose

Glucose 6-phosphate

Fructose 6-phosphate

Fructose 1,6-bisphosphate

Endoplasmatic reticulum

2-Phospho-glycerate

3-Phospho-glycerate

1,3-Bis-phospho-glycerate

Glycer-aldehyde 3-phosphate

Dihydroxy-acetone 3-phosphate

H_2O

Phospho-enolpyruvate

CO_2

Pyruvate

Lactate

Glycerol

Glycerol

Oxalo-acetate

Oxalo-acetate

Blood

Pyruvate

Lactate
Glycerol

NADH

NAD⊕

Malate

Malate

Aspartate

Aspartate

Amino acid degradation, citric acid cycle

Cytoplasm

Mitochondrion

1 Lactate dehydrogenase
   1.1.1.27

2 Pyruvate carboxylase
   [biotin] *6.4.1.1*

3 Phosphoenolpyruvate
   carboxykinase (GTP) *4.1.1.32*

4 Fructose-bisphosphatase
   3.1.3.11

5 Glucose 6-phosphatase
   3.1.3.9

Glycogen Metabolism

A. Glycogen balance ◐

In animals, glycogen (see p. 36) serves as a **carbohydrate reserve**, from which glucose phosphates and glucose can be released as required. The storage of glucose itself would not make sense, because high glucose concentrations within cells would make them strongly hypertonic, and thus would lead to the influx of water. Since the osmotic activity of solutes is dependent not on their mass, but rather on the *number* of molecules, polysaccharides are almost inactive osmotically.

The human body can store up to 450 g of glycogen, a third in the **liver** and almost all of the rest in the **muscles.** The glycogen content of the other organs is low. Liver glycogen serves, above all, in the *maintenance of the blood glucose level* in the postresorptive phase (see p. 280). The glycogen content of the liver is, therefore, very variable. It declines to almost zero in periods of starvation that last more than one day. Thereafter, gluconeogenesis (see p. 158) provides the organism with a supply of glucose. Muscle glycogen serves as an *energy reserve*. It is not involved in the maintenance of blood glucose, since muscles do not possess *glucose 6-phosphatase* and, therefore, cannot release glucose into the blood. Their glycogen content fluctuates less markedly than that of the liver.

B. Glycogen metabolism ◐

Animal glycogen, like plant amylopectin (see p. 38), is a *branched homopolymer of glucose*. Most glucose residues are linked by $\alpha 1 \rightarrow 4$ bonds. Every twelfth glucose, on the average, is connected to a further residue via a $\alpha 1 \rightarrow 6$ bond. These branches are extended by further $\alpha 1 \rightarrow 4$-linked residues. The result is insoluble, tree-like molecules of up to 50,000 glucose residues with masses of up to 10^7 Da.

Glycogen is never completely degraded. In general, only the non-reducing ends of the tree are shortened, or lengthened when glucose is abundant. The elongation of the chains is catalyzed by *glycogen synthase* [2]. Since the formation of glycosidic bonds between sugars is *endergonic*, an activated precursor, **UDPglucose**, is first synthesized by reaction of glucose 1-phosphate with UTP [1]. The transfer of a glucose residue from this intermediate to glycogen is then readily possible ($\Delta G < 0$). When the growing chain has reached a certain length (> 11 residues), an oligosaccharide of 6–7 residues is removed from the end by a *branching enzyme* [3] and then added to an internal part of the same or a neighboring chain in $\alpha 1 \rightarrow 6$ linkage. These branches are then further extended by glycogen synthase.

The branched structure of glycogen facilitates the rapid release of sugar residues. The most important degradative enzyme, *glycogen phosphorylase* [4], releases **glucose 1-phosphate** residues, one after the other, from non-reducing ends. The larger the number of these ends, the more phosphorylase molecules can attach at the same time. The formation of glucose 1-phosphate instead of glucose has the advantage that the liberated residue can be channeled into glycolysis or the HMP *without expending any ATP*. Due to the structure of glycogen phosphorylase, however, the degradation comes to a halt 4 residues before branchings (see p. 110). Two further enzymes, [5] and [6], overcome this blockage. Initially, a trisaccharide from a side chain is transferred to the end of the main chain. A *1,6-glucosidase* [6] then cleaves the remaining glucose residue and leaves behind an unbranched chain that is again accessible to phosphorylase.

The **regulation of glycogen metabolism** by interconversion and the role of hormones in these processes are discussed on pages 110 and 158.

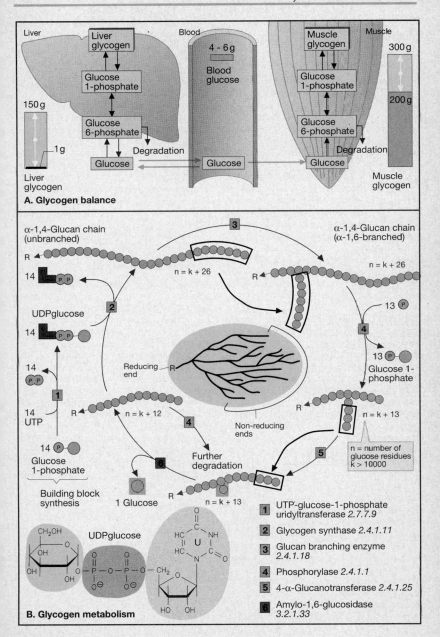

A. Glycogen balance

- Liver
- Liver glycogen
- Glucose 1-phosphate
- Glucose 6-phosphate
- Glucose
- Degradation
- 150 g
- 1 g
- Liver glycogen

- Blood
- 4 - 6 g
- Blood glucose
- Glucose

- Muscle
- Muscle glycogen
- Glucose 1-phosphate
- Glucose 6-phosphate
- Glucose
- Degradation
- 300 g
- 200 g
- Muscle glycogen

B. Glycogen metabolism

α-1,4-Glucan chain (unbranched)

α-1,4-Glucan chain (α-1,6-branched)

R

14

$n = k + 26$

R

$n = k + 26$

UDPglucose

14

13 P

4

14

P P

1

14 UTP

R

$n = k + 12$

4

13 P

Glucose 1-phosphate

Reducing end

R

Non-reducing ends

R

$n = k + 13$

14 P

Glucose 1-phosphate

Further degradation

5

$n = k + 13$

n = number of glucose residues
$k > 10000$

Building block synthesis

1 Glucose

R

$n = k + 13$

6

CH$_2$OH

UDPglucose

U

HO   OH

1   UTP-glucose-1-phosphate uridyltransferase *2.7.7.9*

2   Glycogen synthase *2.4.1.11*

3   Glucan branching enzyme *2.4.1.18*

4   Phosphorylase *2.4.1.1*

5   4-α-Glucanotransferase *2.4.1.25*

6   Amylo-1,6-glucosidase *3.2.1.33*

Regulation of Carbohydrate Metabolism

A. Regulation of carbohydrate metabolism ◑

In higher organisms, the metabolism of the carbohydrates is subject to complex regulatory mechanisms that involve *hormones, metabolites,* and *coenzymes.* The scheme shown here applies to the liver, which is the central site of carbohydrate metabolism (see p. 282). Some of the control mechanisms shown here are not effective in other tissues.

One of the most important tasks of the liver is to store excess glucose in the form of glycogen, and to release glucose from glycogen as required (*buffer function*). When the glycogen reserves are exhausted, the liver can provide glucose by *de novo* synthesis (*gluconeogenesis,* see p. 158). Moreover, it degrades glucose via glycolysis, as do all tissues. These functions must be coordinated with one another. So, for instance, it makes no sense for glycolysis and gluconeogenesis to occur at the same time, nor should glycogen synthesis and glycogen degradation proceed simultaneously. Such "futile cycles" are prevented by the fact that separate enzymes exist for the key steps of catabolic and anabolic pathways. These enzymes are specific for either the catabolic or the anabolic reaction, and are regulated differently. Only such **key enzymes** are shown here.

The hormones that influence carbohydrate metabolism include the peptides insulin and glucagon, the glucocorticoid cortisol, and the catecholamine epinephrine. **Insulin** stimulates, by *induction,* the expression of *glycogen synthase* [1], as well as of several other enzymes of glycolysis ([3], [5] and [7]). At the same time, insulin inhibits the synthesis of key enzymes of gluconeogenesis (*repression,* [4], [6], [8] and [9]). **Glucagon,** an opponent (antagonist) of insulin, has the opposite effect: It induces enzymes of gluconeogenesis ([4], [6], [8] and [9]) and represses *pyruvate kinase* [7], a key enzyme of glycolysis. Further effects of glucagon are based on the *interconversion* of enzymes. They are mediated by the second messenger cAMP (see p. 111). In this way, the synthesis of glycogen is inhibited [1],

while glycogenolysis is activated [2]. **Epinephrine** acts in a similar fashion. The inhibition of *pyruvate kinase* [7] by glucagon is also due to interconversion.

Cortisol, like glucagon, is an important inducer of key enzymes of gluconeogenesis ([4], [6], [8] and [9]). Because several enzymes involved in amino acid degradation are induced as well, cortisol also contributes to the provision of precursors for gluconeogenesis.

High concentrations of the **metabolites** ATP and citrate inhibit glycolysis by binding to *phosphofructokinase.* ATP also inhibits pyruvate kinase. Both metabolites are products of glucose degradation (*feedback inhibition*). Acetyl-CoA, an inhibitor of pyruvate kinase, functions in a similar way. In contrast, AMP, a *signal for ATP deficiency,* activates glycogen degradation and inhibits gluconeogenesis.

B. Fructose 2,6-bisphosphate ○

Fructose 2,6-bisphosphate is an important regulator of liver metabolism. This molecule, like fructose 1,6-bisphosphate, is formed from fructose 6-phosphate, but has a purely *regulatory function.* It stimulates glycolysis by activation of *phosphofructokinase* [5] and inhibits gluconeogenesis by inhibition of *fructose 1,6-bisphosphatase* [6]. Both the synthesis and degradation of fructose 2,6-bisphosphate are catalyzed by one and the same protein [10a,b]. In its unphosphorylated form, it catalyzes the formation of fructose 2,6-bisphosphate [10a]. Following phosphorylation by a cAMP-dependent *protein kinase,* it acts as a *phosphatase* [10b] and catalyzes the degradation of fructose 2,6-bisphosphate to fructose 6-phosphate, and thereby its inactivation.

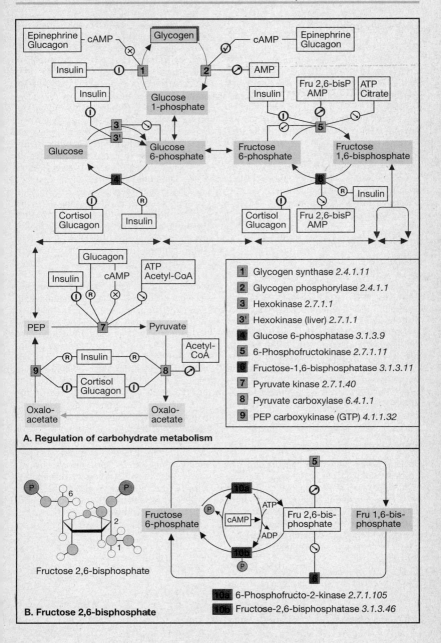

A. Regulation of carbohydrate metabolism

1	Glycogen synthase	*2.4.1.11*
2	Glycogen phosphorylase	*2.4.1.1*
3	Hexokinase	*2.7.1.1*
3'	Hexokinase (liver)	*2.7.1.1*
4	Glucose 6-phosphatase	*3.1.3.9*
5	6-Phosphofructokinase	*2.7.1.11*
6	Fructose-1,6-bisphosphatase	*3.1.3.11*
7	Pyruvate kinase	*2.7.1.40*
8	Pyruvate carboxylase	*6.4.1.1*
9	PEP carboxykinase (GTP)	*4.1.1.32*

10a	6-Phosphofructo-2-kinase	*2.7.1.105*
10b	Fructose-2,6-bisphosphatase	*3.1.3.46*

B. Fructose 2,6-bisphosphate

Diabetes Mellitus

Diabetes mellitus is a very common metabolic disease that is due to an absolute or a relative **deficiency of insulin**. The lack of this peptide hormone (see p. 342) predominantly affects the metabolism of carbohydrates and lipids.

Diabetes mellitus occurs in two forms. In the case of diabetes **type 1** (insulin-dependent diabetes mellitus, IDDM), the insulin-forming cells of the pancreas are destroyed already at a very young age by an autoimmune reaction. The less severe **type 2** diabetes (non-insulin-dependent form, NIDDM) usually first arises at an older age. It is due to decreased insulin secretion, or caused by a defect in receptor function.

A. Insulin biosynthesis ○

Insulin is produced by the B cells of the **islets of Langerhans** in the pancreas. As is usual with secretory proteins, the precursor of the hormone (preproinsulin) carries a signal peptide that directs the peptide chain to the inside of the endoplasmatic reticulum (see p. 210). There, *proinsulin* is produced by cleavage of the signal peptide and the formation of the disulfide bonds (see p. 58). The propeptide arrives in the Golgi apparatus where mature insulin is synthesized by cleavage of the *C peptide*. It is then packed in so-called *ß-granules* and stored in the form of zinc-containing hexamers until secretion.

B. Effects of insulin deficiency ◐

The effects of insulin on carbohydrate metabolism are discussed on p. 162. In a simplified fashion, they can be described as stimulation of glucose utilization and inhibition of *de novo* synthesis of glucose. In addition, the transport of glucose from the blood into most tissues is also insulin-dependent (exceptions are, for example, the liver and the erythrocytes).

The lipid metabolism of adipose tissue is also influenced by the hormone. In these cells, insulin stimulates the synthesis of fatty acids from glucose, e.g., by activation of *acetyl-CoA carboxylase* and increased availability of NADPH + H$^+$, and inhibits fat degradation (lipolysis). The breakdown of muscle protein is also inhibited by insulin. In the case of insulin deficiency, these effects are impaired, and major disturbances are the result.

A characteristic symptom is the elevation of the glucose concentration in the blood from 5 mM to 9 mM and above (**hyperglycemia**, elevated "blood glucose level"). In muscle and adipose tissue, the two most important glucose consumers, both glucose uptake and glucose utilization are impaired. In the liver, glucose utilization is also reduced. At the same time, gluconeogenesis is stimulated, because of the increased proteolysis in the muscles, among other things. This increases the blood sugar level still further. When the capacity of the kidneys for the reabsorption of glucose is exceeded (at levels of 9 mM or more), then glucose is excreted in the urine (**glucosuria**).

The enhanced degradation of fats as a result of insulin deficiency also has serious consequences. The fatty acids that accumulate in large quantities are taken up by the liver and used for lipoprotein synthesis (leading to **hyperlipidemia**) and, in part, are degraded to acetyl-CoA. Since the citric acid cycle cannot take up all of the acetyl-CoA, most of it is converted into *ketone bodies* (see p. 284), which are released into the blood, where they dissociate into the respective anions and protons. Because of the high ketone body production, diabetics often develop a severe **metabolic acidosis** (diabetic ketoacidosis, see p. 256).

Inadequately treated diabetes mellitus can, in the long run, lead to a variety of secondary complications. These include changes in the walls of the blood vessels (diabetic angiopathy) and damage to the kidneys (nephropathy) as well as to the nervous system and the eyes (e.g., cataracts).

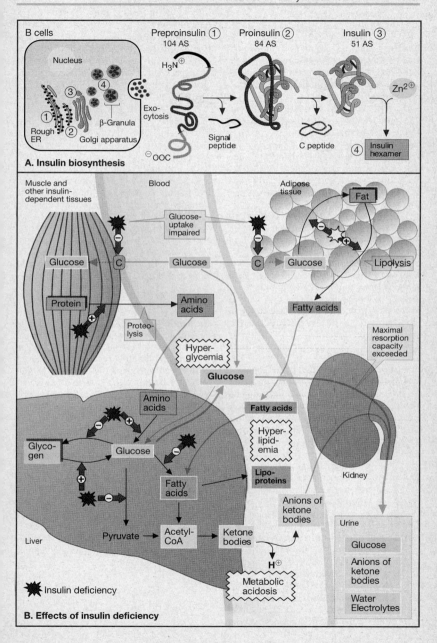

B cells

Preproinsulin ① 104 AS

Proinsulin ② 84 AS

Insulin ③ 51 AS

Nucleus

$H_3N^{\oplus}$

$Zn^{2\oplus}$

③

④

Exo-cytosis

① ②

Rough ER

β-Granula

Golgi apparatus

$^{\ominus}OOC$

Signal peptide

C peptide

④ Insulin hexamer

A. Insulin biosynthesis

Muscle and other insulin-dependent tissues

Blood

Adipose tissue

Fat

Glucose-uptake impaired

⊖

⊖

⊖ ⊕

Glucose

C

Glucose

C

Glucose

Lipolysis

Protein

⊕

Amino acids

Proteo-lysis

Fatty acids

Maximal resorption capacity exceeded

Hyper-glycemia

Glucose

Fatty acids

Amino acids

Hyper-lipid-emia

Glyco-gen

⊖ ⊕

Glucose

⊖

Lipo-proteins

Kidney

⊕

⊖

Fatty acids

Anions of ketone bodies

Urine

Liver

Pyruvate → Acetyl-CoA → Ketone bodies

$H^{\oplus}$

Glucose

Anions of ketone bodies

Water Electrolytes

✦ Insulin deficiency

Metabolic acidosis

B. Effects of insulin deficiency

Fatty Acid Synthesis

The biosynthesis of fatty acids is catalyzed by *fatty acid synthase,* a multifunctional enzyme that is located in the cytoplasm. In a cyclic reaction, a starter molecule, **acetyl-CoA,** is elongated by one C_2 unit at a time. After seven such cycles, a C_{16} acid, **palmitate,** arises as the end product of the reaction. **Malonyl-CoA,** the actual substrate of the elongation step, releases its carboxyl group as CO_2 during condensation with the growing chain. The reducing agent for fatty acid synthesis is **NADPH + H+**. In total, 1 acetyl-CoA, 7 malonyl-CoA and 14 NADPH + H+ are converted to 1 palmitate, 7 CO_2, 6 H_2O, 8 CoA and 14 NADP+.

Acetyl-CoA, the principal substrate of fatty acid synthesis, is produced in the *pyruvate dehydrogenase* reaction. It is transported from the mitochondria into the cytoplasm via the citrate shuttle (see p. 128). Most of the reductant, NADPH +H+, is supplied by the hexose monophosphate pathway (see p. 142). In addition, *"malic enzyme"* (see p. 128) and the *NADP+-dependent isocitrate dehydrogenase* also contribute to the generation of NADPH.

A. Fatty acid synthase ◑

The fatty acid synthase of vertebrate animals is composed of two identical peptide chains, i.e., it is a *homodimer.* Each of the two peptide chains, which are shown here as hemispheres, catalyzes 7 different partial reactions that are required for palmitate synthesis (reactions [1]-[7]). The catalysis of multiple sequential reactions by one protein has fundamental advantages as compared to separate enzymes. Competing reactions are prevented, the reactions occur in a coordinated way, as on a production line, and they are especially efficient, because of high local substrate concentrations (low diffusion losses).

Each half of the fatty acid synthase complex can covalently bind substrates (acyl or acetyl residues) as thioesters at two particular SH groups: a **cysteine** residue (Cys-SH) and a **4′-phosphopantetheine** group (Pan-SH). Pan-SH, which is very similar to coenzyme A, is bound to a domain of the enzyme that is referred to as the **acyl-carrier protein (ACP).** This part of the enzyme works like a long arm, which passes on the substrate from one reaction center to the next. Interestingly, the reaction depends on cooperation between the two halves of the enzyme. Fatty acid synthase is, therefore, only functional as a dimer.

The various activities involved in fatty acid biosynthesis are located within three different domains of the protein. **Domain 1** catalyzes the entry of the substrates acetyl-CoA (or acyl-CoA) and malonyl-CoA by *[ACP]S-acetyl* and *[ACP]S-malonyl transferase* [1, 2] and, subsequently, the condensation of the two partners by *3-oxoacyl-[ACP] synthase* [3]. **Domain 2** reduces the growing fatty acid chain with the help of *3-oxoacyl-[ACP] reductase* [4], *3-hydroxyacyl-ACP dehydratase* [5], and *enoyl-[ACP] reductase* [6]. Finally, *acyl-[ACP] hydrolase* [7] in **domain 3** catalyzes the release of the finished product after 7 chain-elongation steps.

B. Reactions of fatty acid synthesis ◑

The biosynthesis of palmitate begins with the transfer of an **acetyl residue** to the functional cysteine residue of the enzyme [1]. At the same time, a **malonyl residue** is transferred to the 4-phosphopantetheine group of ACP [2]. The chain elongation step occurs by transferring the acetyl (or acyl) residue to C-2 of the malonyl group (blue arrow), during which process the carboxyl group is cleaved off as CO_2 [3]. The three subsequent reactions, reduction of the 3-oxo group [4], the elimination of water [5] and another reduction step [6], lead to the production of a fatty acid with 4 carbons. By means of the acyltransferase [1], this intermediate is relocated from ACP back to the functional cysteine, so that the next cycle can begin by loading of ACP with another malonyl residue [2]. After 7 cycles, acyl-[ACP] hydrolase [7] "recognizes" that the product has reached the correct length, and **palmitate** is released.

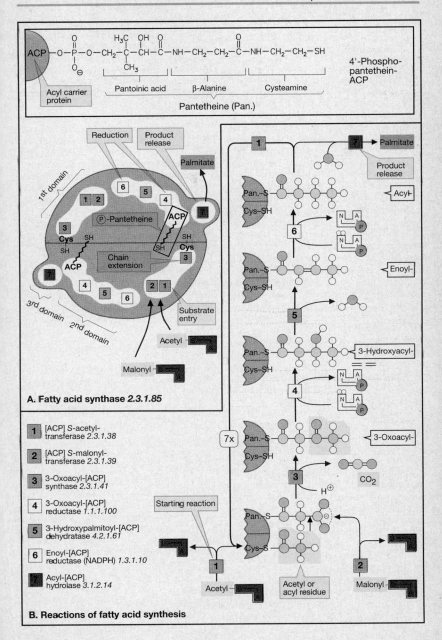

A. Fatty acid synthase 2.3.1.85

1 [ACP] S-acetyl-
   transferase 2.3.1.38

2 [ACP] S-malonyl-
   transferase 2.3.1.39

3 3-Oxoacyl-[ACP]
   synthase 2.3.1.41

4 3-Oxoacyl-[ACP]
   reductase 1.1.1.100

5 3-Hydroxypalmitoyl-[ACP]
   dehydratase 4.2.1.61

6 Enoyl-[ACP]
   reductase (NADPH) 1.3.1.10

7 Acyl-[ACP]
   hydrolase 3.1.2.14

B. Reactions of fatty acid synthesis

Biosynthesis of Complex Lipids

A. Biosynthesis of fats and phospholipids ◑

Complex lipids, such as neutral fats (triacylglycerols), phospholipids and glycolipids, are synthesized via common reaction pathways. Their synthesis begins with **glycerol 3-phosphate**, which is either derived by reduction from the glycolytic intermediate **dihydroxyacetone 3-phosphate** [1], or arises from the ATP-dependent phosphorylation of free **glycerol** [2]. From glycerol 3-phosphate, a **lysophosphatidate** is synthesized by esterification with a long-chain fatty acid at C-1 [3]. The lysophosphatidate is then converted to a **phosphatidate** by another esterification [4]. In this step, unsaturated fatty acids, especially oleic acid, are preferred substrates. Phosphatidates are key molecules in the biosynthesis of fats, phospholipids and glycolipids (see also p. 368).

For the synthesis of fats (above center) the phosphate group of phosphatidates is first removed by hydrolysis [5]. The **diacylglycerol** formed in the process is then converted to a **triacylglycerol** by transfer of a further fatty acid from acyl-CoA [6]. Most neutral fats are stored within the adipose tissue in the form of fat droplets.

Alternatively, **phosphatidylcholine** (lecithin) can arise from diacylglycerols by transfer of phosphocholine ([7], upper right). The phosphocholine group is supplied by CDPcholine, i.e., choline activated by condensation with cytidine diphosphate.

In a similar reaction, **phosphatidylethanolamine** is synthesized from a diacylglycerol and CDPethanolamine. **Phosphatidylserine** is derived from phosphatidylethanolamine by an exchange of the amino alcohols. Further reactions serve to interconvert the phospholipids. For instance, phosphatidylserine can be converted to phosphatidylethanolamine by decarboxylation, and this then to phosphatidylcholine by methylation with S-adenosyl methionine (not shown).

Dietary lipids can also be used for the biosynthesis of neutral fats, phospholipids and glycolipids. In the digestive tract, triacylglycerols are hydrolyzed by lipases to yield fatty acids and 2-monoacylglycerols (see p. 246). The intestinal mucosa reabsorbs these metabolites and once again synthesizes lipids from them. This occurs by two different acyltransferases in the order 2-monoacylglycerol → diacylglycerol → triacylglycerol (neutral fat).

The biosynthesis of **phosphatidylinositol** also begins with phosphatidates. CDPdiacylglycerol is produced by transfer of CMP, and this then reacts with inositol to form phosphatidylinositol [10]. This phospholipid is especially abundant in plasma membranes. There it can be further phosphorylated by ATP to yield phosphatidylinositol 4-phosphate (PInsP) and phosphatidylinositol 4,5-bisphosphate (PInsP$_2$). PInsP$_2$ is a substrate of *phospholipase C*, which catalyzes its hydrolysis to *2,3-diacylglycerol* (DAG) and *inositol-1,4,5-triphosphate* (InsP$_3$, see p. 370).

The following enzymes, among others, are involved in the biosynthesis of complex lipids. Most of them are associated with membranes of the smooth endoplasmatic reticulum, where the pathways described here take place.

[1] Glycerol-3-phosphate dehydrogenase (NAD+) *1.1.1.8*
[2] Glycerol kinase *2.7.1.30*
[3] Glycerol-3-phosphate acyltransferase *2.3.1.15*
[4] 1-Acylglycerol-3-phosphate acyltransferase *2.3.1.51*
[5] Phosphatidate phosphatase *3.1.3.4*
[6] Diacylglycerol acyltransferase *2.3.1.20*
[7] Diacylglycerol cholinephosphotransferase *2.7.8.2*
[8] 2-Acylglycerol acyltransferase *2.3.1.22*
[9] Phosphatidate cytidyltransferase *2.7.7.41*
[10] CDPdiacylglycerol-inositol 3-phosphatidyltransferase *2.7.8.11*
[11] Phosphatidylinositol kinase *2.7.1.67*
[12] Phosphatidylinositol-4-phosphate kinase *2.7.1.68*

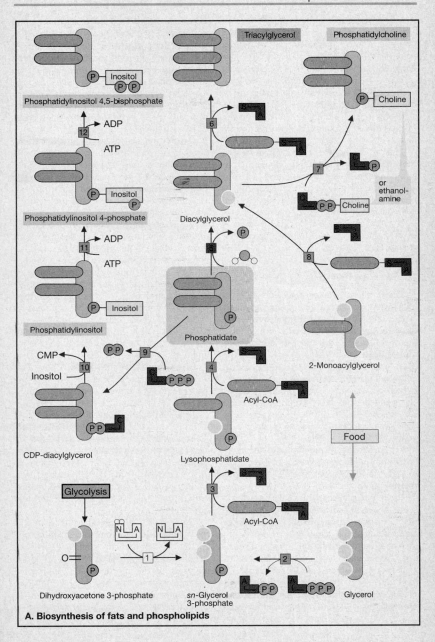

A. Biosynthesis of fats and phospholipids

Biosynthesis of Cholesterol

Cholesterol is a major constituent of the *cell membrane* of animal cells (see p. 202). The daily requirement for cholesterol (1 g) can, in principle, be met by the body itself. However, in the case of a mixed diet, only about half of the cholesterol requirement is synthesized in the intestine, the skin, and, above all, in the liver (about 50%). The rest is derived from the diet.

Most of the cholesterol is incorporated into the lipid layer of plasma membranes. In addition, large amounts are required for conversion to **bile acids** (see pp. 278 and 284). A certain amount of cholesterol is also excreted with the *bile*. In total, about 1 g of cholesterol is excreted per day. A very small fraction is used for the biosynthesis of the **steroid hormones** (see p. 338).

A. Cholesterol biosynthesis ○

The biosynthetic pathway can be divided into four phases. In the first one (**1**), **mevalonate**, a C_6 compound, is produced from three molecules of **acetyl-CoA**. In the second part (**2**), mevalonate is converted to **isopentenyl diphosphate,** the "active isoprene." In the third phase (**3**), six of these C_5 fragments polymerize to give **squalene,** a C_{30} compound. In the last section, three carbons are removed and squalene undergoes cyclization to yield cholesterol (**4**). The illustration shows only the most important intermediates in the pathway (see also p. 369).

(1) Formation of mevalonate. The reactions involved in the conversion of acetyl-CoA to acetoacetyl-CoA and then to **3-hydroxy-3-methyl-glutaryl-CoA** (3-HMG-CoA) are similar to those of the biosynthetic pathway leading to the *ketone bodies* (for details, see p. 284). However mevalonate synthesis occurs in the endoplasmatic reticulum (ER), whereas ketone bodies are synthesized in the mitochondria. The conversion of 3-HMG-CoA to mevalonate involves a reduction reaction and the removal of CoA. *3-HMG-CoA reductase* is the **key regulatory enzyme** in cholesterol biosynthesis. It is subject to various forms of regulation. Firstly, synthesis of the enzyme is *repressed* by oxysterols, the final products of the pathway. Secondly, the enzyme is subject to protein **phosphorylation** (*interconversion*) in response to various different hormones. Here, the phosphorylated form of the enzyme is the inactive one. Insulin and thyroxine stimulate expression of the reductase gene, whereas glucagon inhibits it. 3-HMG-CoA reductase synthesis is also repressed by cholesterol supplied with the diet.

(2) Formation of isopentenyl diphosphate. The conversion of mevalonate to *isopentenyl diphosphate* involves two phosphorylation reactions followed by a decarboxylation. Three ATP molecules are consumed in the process. Isopentenyl diphosphate is the precursor of all isoprenoids (see p. 50).

(3) Formation of sqalene. Isopentenyl diphosphate undergoes isomerization to form dimethylallyl diphosphate. These two C_5 compounds then condense with one another to yield geranyl diphosphate. A further isopentenyl diphosphate is added to give farnesyl diphosphate. This intermediate then undergoes dimerization, in a *head-to-head reaction,* to yield squalene. Farnesyl diphosphate is also the precursor for the synthesis of polyisoprenoids, such as dolichol and ubiquinone (see p. 50).

(4) Formation of cholesterol. Cholesterol is a C_{27} compound that is formed from the linear C_{30} isoprenoid squalene via a complicated series of reactions. The first intermediate containing the gonane ring system (see p. 53) is lanosterol, which is synthesized in a two-step process catalyzed by a CytP450-containing *monooxygenase* (see p. 292) and a *cyclase,* respectively. Later steps, some of which also depend on CytP450 enzymes, lead to the oxidative removal of three methyl groups from the gonane backbone to yield cholesterol.

The pathway described here is located in the *smooth ER*. It is driven by the hydrolysis of energy-rich CoA derivatives and ATP. As usual, the reducing agent is NADPH + H^+. The intermediates of the reaction pathway involve CoA-activated compounds, diphosphates and very lipophilic intermediates (squalen to cholesterol) bound to *sterol carrier proteins*.

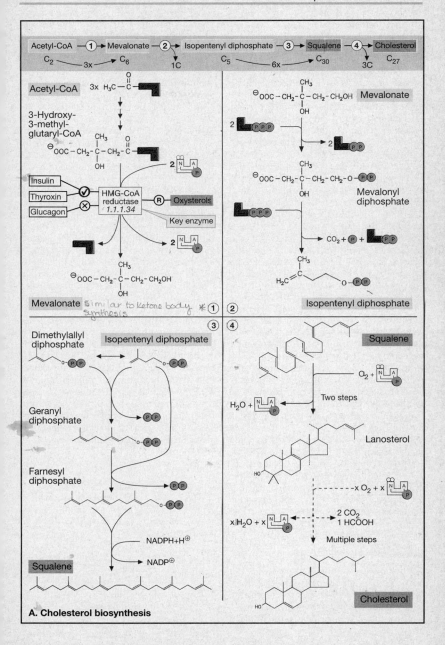

Acetyl-CoA —①→ Mevalonate —②→ Isopentenyl diphosphate —③→ Squalene —④→ Cholesterol

C_2 — 3x → C_6 1C C_5 — 6x → C_{30} 3C C_{27}

Acetyl-CoA

3-Hydroxy-3-methyl-glutaryl-CoA

HMG-CoA reductase *1.1.1.34*

Insulin
Thyroxin
Glucagon

Oxysterols

Key enzyme

Mevalonate   Similar to ketone body synthesis ✳①   ②

Mevalonate

Mevalonyl diphosphate

$CO_2 + P +$

Isopentenyl diphosphate

③ ④

Dimethylallyl diphosphate   Isopentenyl diphosphate

Geranyl diphosphate

Farnesyl diphosphate

NADPH+H⊕
NADP⊕

Squalene

Squalene

$O_2 +$

Two steps

$H_2O +$

Lanosterol

$x\ O_2 + x$

$x\ H_2O + x$ 2 CO_2
1 HCOOH

Multiple steps

Cholesterol

A. Cholesterol biosynthesis

Amino Acid Biosynthesis

A. Symbiotic nitrogen fixation ○

In the atmosphere, elemental nitrogen (N_2) is present in practically unlimited quantities. However, before it can enter the nitrogen cycle of nature, it must be reduced to NH_3 and incorporated ("fixed") into amino acids. Only a few bacteria and blue-green algae are capable of fixing atmospheric nitrogen. Of particular economic importance is the symbiosis between bacteria of the genus *Rhizobium* and legumes (*Fabales*), such as clover, beans, and peas. These plants are nutritionally valuable, because of their high protein content.

In the symbiosis with *Fabales*, the bacteroids live in **root nodules** as so-called *bacteroids* within plant cells. The plant supplies the bacteroids with nutrients, but it also benefits from the fixed nitrogen, which the symbionts make available. The N_2-fixing enzyme of the bacteria is *nitrogenase*. It consists of two components: a *Fe protein which* contains an $[Fe_4S_4]$ center (see p. 132) as a redox system, accepts electrons from *ferredoxin*, and donates them to the second component, the *Fe-Mo protein*. This molybdenum-containing protein transfers the electrons to N_2 and thus, via various intermediate steps, produces **ammonia (NH_3)**. A certain proportion of the reducing equivalents is transferred in a side-reaction to H^+. Thus, besides NH_3, hydrogen is also produced. Nitrogen fixation requires 16 ATP per mol of N_2. The free energy of the ATP hydrolysis is used by the Fe protein to decrease the reduction potential far enough to allow N_2 reduction. Since nitrogenase is extremely sensitive to oxygen, the plant synthesizes an O_2-binding heme protein (*leghemoglobin*), which helps to keep the O_2 partial pressure in the nodules at a low level.

B. Amino acid biosynthesis: overview ◑

The proteinogenic amino acids (see p. 58) can be divided into *five families*, based on their biogenetic origin. The members of each family are derived from the same precursors, which are all produced either in the citric acid cycle or in catabolic carbohydrate metabolism. Here, the biosynthetic pathways are shown in an overview. Further details can be found on pp. 372 and 373.

Whilst plants and micro-organisms can synthesize all amino acids from small precursors, mammals, during the course of evolution, have lost the ability to synthesize approximately half of the 20 proteinogenic amino acids. These **essential amino acids** must therefore be supplied in the diet. So, for instance, the animal metabolism is no longer capable of *de novo* synthesis of the *aromatic amino acids* (tyrosine is non-essential only, because it can be formed from phenylalanine). The *branched-chain amino acids* (valine, leucine, isoleucine, and threonine) as well as *methionine* and *lysine*, also belong to the essential amino acids. *Histidine* and *arginine* are essential for rats; whether this is also true for humans is still under debate. A supply of these amino acids in the diet appears to be essential at least during growth.

The amino acids, which can arise from oxoacids by transamination (*alanine, aspartate*, and *glutamate*) are **non-essential**. *Proline* can be synthesized in sufficient quantities from glutamate, and the members of the serine family (*serine, glycine*, and *cysteine*) can also be synthesized by animal metabolism itself. The nutritional value of proteins critically depends on their content of essential amino acids (see p. 326). Many plant proteins are poor in lysine or methionine, whereas in animal proteins all amino acids are present in a balanced relationship.

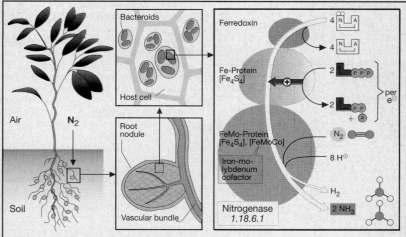

A. Symbiotic nitrogen fixation

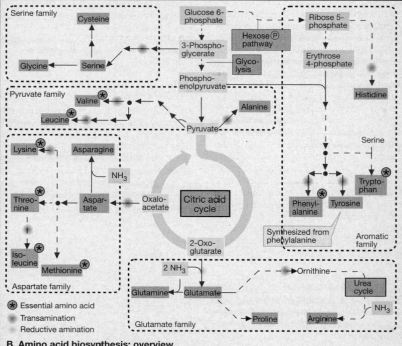

B. Amino acid biosynthesis: overview

Purine and Pyrimidine Synthesis

The bases contained in nucleic acids are derivatives of the heterocyclics purine and pyrimidine (see p. 78). The biosynthesis of purine and pyrimidine bases is complicated, but, as these compounds are indispensable, most cells have the corresponding pathways. The illustration shows a simplified representation of the initial steps in purine and pyrimidine biosynthesis. More details are provided on pp. 374 and 375.

A. Formation of pyrimidine and purine rings ○

The carbon and nitrogen atoms of the pyrimidine ring are contributed by three different precursors. Nitrogen N-1 and carbons C-4 and C-6 are derived from *aspartate*, carbon C-2 from HCO_3^-, and the second nitrogen (N-3) from the amide group of *glutamine*.

The synthesis of the purine ring is more complex. The most important precursor is *glycine*, which donates C-4, C-5, and N-7. Carbon C-6 is derived from HCO_3^-, while nitrogens N-3 and N-9 originate from the amide group of *glutamine*. Nitrogen N-1 is contributed by *aspartate*. In the reaction, aspartate donates an amino group, i.e., it is converted to fumarate, as it is in the urea cycle (see p. 286). Finally, carbons C-2 and C-8 are derived from the formyl group of N^{10}-formyl tetrahydrofolate (see p. 375).

B. Pyrimidine and purine synthesis ○

The most important intermediates in the synthesis of nucleic acid precursors are **uridine monophosphate** (**UMP**) in the case of the pyrimidines, and **inosine monophosphate** (**IMP**, base: hypoxanthine) in the case of the purines. There is a fundamental difference between the pathways for the synthesis of pyrimidines and purines. The synthesis of the pyrimidine ring is complete before it is linked to ribose 5'-phosphate to form a nucleotide, whereas the purine ring is assembled step by step, with ribose phosphate acting as a carrier molecule.

The immediate precursors for the synthesis of the pyrimidine ring (**1**) are **carbamoyl phosphate** (see p. 286) and the amino acid **aspartate**. They are joined by *aspartate carbamoyltransferase* to form *N*-**carbamoyl aspartate** (**1b**). The next step involves ring closure to yield **dihydroorotate**, followed by reduction to give the pyrimidine base **orotate**. Orotate then reacts with **5'-phosphoribosyl 1'-diphosphate (PRPP)** to form the nucleotide **orotate 5'-monophosphate (OMP)**. Finally, OMP is decarboxylated to yield UMP.

As noted above, purine synthesis (**2**) starts with PRPP. Formation of the five-membered ring begins with the addition of an amino group, which is the precursor of N-9 (**2a**). The remaining atoms of the five-membered ring are supplied by glycine and the formyl group of N^{10}-formyl-THF (**2b, 2c**). In the following step, atoms N-3 and N-6 of the later 6-membered ring are attached to the existing structure (**2d, 2e**). The synthesis of the ring then continues with N-1 and C-2. A final cyclization step yields inosine 5'-monophosphate. However, IMP does not accumulate, but is rapidly converted to AMP and GMP via reactions discussed on the following page.

Further information

The control of pyrimidine synthesis by the allosteric regulation of aspartate carbamoyl transferase is discussed on p. 112. The synthesis of IMP is also regulated by feedback inhibition, i.e., ADP and GDP inhibit the synthesis of PRPP from ribose 5'-phosphate. Similarly, the synthesis of 5'-phosphoribosylamine (step **2a**) is inhibited by the end products AMP and GMP.

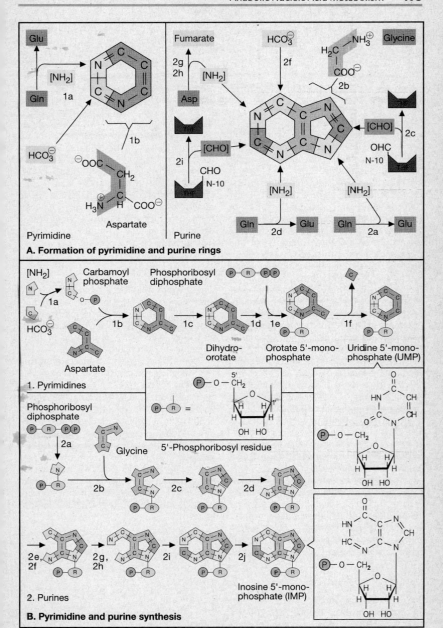

A. Formation of pyrimidine and purine rings

1. Pyrimidines

2. Purines

B. Pyrimidine and purine synthesis

Nucleotide Biosynthesis

A. Overview ◑

The *de-novo* synthesis of purines and pyrimidines yields the monophosphates IMP and UMP, respectively (see p. 174). All other nucleotides and deoxynucleotides are synthesized from these two precursors. Here we present an overview of the pathways involved. More details can be found on pp. 374 and 375. Nucleotide synthesis by recycling of bases is discussed on p. 156.

The synthesis of the **purine nucleotides** (1) begins with **IMP.** Its base, hypoxanthine, is converted in two steps to either adenine or guanine. Thereafter, *phosphotransferases* catalyze the phosphorylation of **AMP** and **GMP** to yield the *diphosphates* **ADP** and **GDP** and the *triphosphates* **ATP** and **GTP**, respectively. The nucleoside triphosphates serve as building blocks for RNA (see p. 218) or as coenzymes (see p. 102). The conversion of the ribonucleotides to deoxyribonucleotides occurs at the level of the diphosphates. The enzyme responsible for this, *nucleoside diphosphate reductase*, is discussed in section **B**.

The biosynthetic pathways for the **pyrimidine nucleotides** are somewhat more complicated. First of all, **UMP** is phosphorylated twice to yield the diphosphate, **UDP**, and the triphosphate, **UTP**. Then *CTP synthase* catalyzes the conversion of UTP to **CTP**. As the reduction of pyrimidine nucleotides to deoxyribonucleotides takes place at the diphosphate level, CTP must first be hydrolyzed by a *phosphatase* to yield **CDP** before **dCDP** and **dCTP** can be produced. The DNA building block deoxythymidine triphosphate (**dTTP**) is synthesized from UDP in several steps. In the process, the base thymine is formed by methylation of **dUMP** at the monophosphate level. *Thymidylate synthase* and its helper enzyme *dihydrofolate reductase* are the targets of cytostatic agents (**C**).

B. Synthesis of deoxyribonucleotides ◑

The reduction of ribonucleotides is a complex process that involves several proteins. The necessary reducing equivalents come from

NADPH + H+, but they are not donated directly to the substrate. Rather, they are transferred via an *electron transport chain* (**1**). In the first step, *thioredoxin reductase* catalyzes electron transfer from NADPH via enzyme-bound FAD to **thioredoxin**, a small redox protein that undergoes reductive cleavage of a disulfide bond. Thioredoxin then reduces a catalytically active disulfide bond in *nucleoside diphosphate reductase* ("ribonucleotide reductase"). The resulting SH groups are the actual electron donors for the reduction of ribonucleotide diphosphates. A **tyrosine radical** in the enzyme also participates in the reaction (**2**). Initially, a substrate radical is produced (a) that loses a water molecule to become a radical cation (b). The deoxyribose residue is ultimately produced by reduction (c).

C. Cytostatic agents ○

Drugs, which inhibit cell proliferation (**cytostatic agents**) are used in cancer chemotherapy. Many of these substances interfere with the biosynthesis of deoxyribonucleotides, and thus inhibit DNA replication in the S phase of the cell cycle (see p. 356).

The adenine analog **6-mercaptopurine** (**1**), after conversion to 6-mercaptopurine mononucleotide, inhibits the synthesis of AMP and GMP from IMP. **Hydroxyurea** inhibits *nucleoside diphosphate reductase* by intercepting the tyrosine radical (see above). 5-Fluoro-dUMP, synthesized from **fluorouracil** (**2**), is a competitive inhibitor of *thymidilate synthase* [3]. It binds just like dUMP, but cannot be further metabolized, because the fluorine atom cannot be cleaved off. The folic acid analogue **methotrexate** is an extremely effective inhibitor of *dihydrofolate reductase* [4].

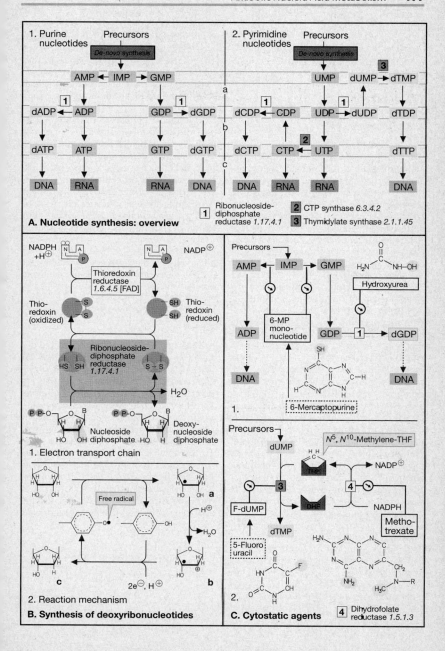

1. Purine nucleotides

Precursors

De-novo synthesis

AMP ← IMP → GMP

dADP ← ADP   GDP → dGDP

dATP   ATP   GTP   dGTP

DNA   RNA   RNA   DNA

2. Pyrimidine nucleotides

Precursors

De-novo synthesis

UMP   dUMP → dTMP   **3**

dCDP ← CDP   UDP → dUDP   dTDP

dCTP   CTP ← UTP   dTTP   **2**

DNA   RNA   RNA   DNA

a
b
c

1 Ribonucleoside-diphosphate reductase *1.17.4.1*

2 CTP synthase *6.3.4.2*

3 Thymidylate synthase *2.1.1.45*

A. Nucleotide synthesis: overview

NADPH +H⊕

Thioredoxin reductase *1.6.4.5 [FAD]*

NADP⊕

Thio-redoxin (oxidized)   Thio-redoxin (reduced)

Ribonucleoside-diphosphate reductase *1.17.4.1*

H₂O

Nucleoside diphosphate   Deoxy-nucleoside diphosphate

1. Electron transport chain

Free radical

2e⊖, H⊕

a
b
c

2. Reaction mechanism

B. Synthesis of deoxyribonucleotides

Precursors

AMP   IMP   GMP

H_2N—C(=O)—NH—OH

Hydroxyurea

6-MP mono-nucleotide

ADP   GDP   **1**   dGDP

DNA   DNA

6-Mercaptopurine

1.

Precursors

dUMP N^5, N^{10}-Methylene-THF

THF   NADP⊕

F-dUMP   **3**   DHF   **4**

dTMP   NADPH

5-Fluoro uracil   Metho-trexate

2.

C. Cytostatic agents

4 Dihydrofolate reductase *1.5.1.3*

Heme Biosynthesis

Almost 85% of heme biosynthesis occurs in the bone marrow, and a much smaller percentage in the liver. Both the mitochondria and the cytoplasm are involved in this pathway.

A. Heme biosynthesis ○

The biosynthesis of the heme group starts in the mitochondria. **Succinyl-CoA** (upper left), an intermediate of the citric acid cycle, condenses with **glycine** to yield an intermediate that is then decarboxylated to give **5-aminolevulinate** (ALA). The reaction is catalyzed by *ALA synthase* [1], the key regulatory enzyme of the pathway. The expression of the ALA synthase gene is repressed, and existing enzyme is inhibited by heme, the end product of the pathway. This is another typical case of "feedback" inhibition.

In the next step, 5-aminolevulinate is transported from the mitochondria into the cytoplasm, where two molecules condense with one another to form **porphobilinogen**, a compound that already contains the pyrrole ring of the heme molecule. The enzyme involved, *porphobilinogen synthase* [2], is inhibited by lead ions. This is why acute lead poisoning is associated with increased concentrations of ALA in the blood and urine.

The next steps in the pathway bring about formation of the **tetrapyrrole structure** characteristic of the porphyrins. *Hydroxymethylbilane synthase* catalyzes the linkage of four porphobilinogen molecules and the removal of an NH_2 group to yield **uroporphyrinogen III**. This reaction also requires *uroporphyrinogen III synthase* [4]. The absence of the synthase leads to the formation of a "wrong" isomer, uroporphyrinogen I.

The tetrapyrrole structure of uroporphyrinogen III is still very different from that of heme, i.e., the central iron atom is missing, and the ring contains only 8 instead of the 11 double bonds found in heme. In addition, the ring carries only charged and thus highly polar side chains (4 acetate and 4 propionate residues). The typically apolar location of heme molecules in the interior of proteins re-quires that these side chains first become less polar. Initially, the four acetate residues (R_1) are decarboxylated to yield methyl groups. The resultant **coproporphyrinogen III** then returns to the mitochondria. All subsequent steps are catalyzed by enzymes located either on or within the inner mitochondrial membrane.

Two of the propionate residues (R_2) are now converted to vinyl residues by an *oxidase*. The formation of **protoporphorinogen IX** brings the modification of the side chains to an end. The next step, also an oxidative one, removes six reducing equivalents to generate the resonance-stabilized π electron system, which gives heme its characteristic red color. Finally, a divalent iron ion is incorporated into the ring. This step also requires a specific enzyme, *ferrochelatase*. The **heme** or *Fe-protoporphyrin IX* thus formed can be incorporated into hemoglobin and myoglobin (see p. 258) where it is bound non-covalently, or into cytochrome c, to which it is attached by covalent bonds (see p. 100).

Further information

A whole series of hereditary or acquired diseases, the so-called **porphyrias**, are associated with disturbances in porphyrin biosynthesis. Some of them can be quite severe. Several of these disorders result in the excretion of heme precursors in the stool or urine, giving them a dark red color. Porphyrins may also accumulate in the skin. In this case, exposure to light causes disfiguring, poorly-healing blisters. Neurological disturbances are also common. It is possible that the medieval legends describing human vampires (Dracula) had their origin in the behavior of porphyria sufferers (light-shyness, bizarre appearance, drinking of blood to compensate for the heme deficiency).

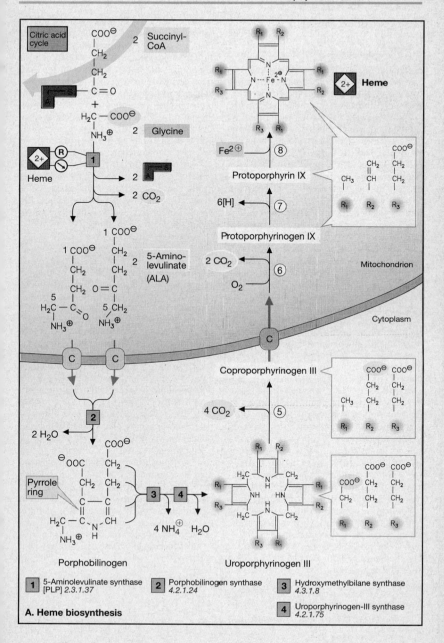

A. Heme biosynthesis

| 1 | 5-Aminolevulinate synthase [PLP] 2.3.1.37 | 2 | Porphobilinogen synthase 4.2.1.24 | 3 | Hydroxymethylbilane synthase 4.3.1.8 |

| 4 | Uroporphyrinogen-III synthase 4.2.1.75 |

Degradation of Porphyrins

A. Hemoglobin degradation ⓘ

In the human body, approximately 100–200 million erythrocytes are broken down every hour. Degradation begins in the endoplasmatic reticulum of reticuloendothelial cells of the liver, spleen, and bone marrow (RES = reticuloendothelial system). Once the protein moiety (globin) has been removed, a NADPH-dependent *heme oxygenase* catalyzes the cleavage of the red **heme** to yield CO (carbon monoxide), an iron atom, which is recycled, and the green pigment **biliverdin**.

Biliverdin reductase then catalyzes the further degradation of biliverdin to the orange-colored **bilirubin**. This color change is readily observed in vivo in a hematome. The intense *color* of heme and other porphyrins (see p. 178) is a result of the numerous conjugated double bonds, which form an extended *resonance-stabilized π system*.

Further degradation of bilirubin takes place in the liver. Bilirubin is only poorly soluble in the plasma, and therefore during transport it is bound to **albumin**. Certain drugs bind to albumin at the same site as bilirubin, and thus may displace it. Bilirubin is eventually taken up from the blood by the parenchymal cells of the liver.

In the hepatocytes, the solubility of bilirubin is increased by conjugation with two molecules of glucuronic acid (substrate: UDP-GlcUA; see p. 290). The formation of this conjugate is catalyzed by *UDPglucuronate glucuronosyltransferase*, an enzyme found in the ER of the liver and, at lower activites, also in the kidneys and intestinal mucosa. Glucuronic acid is attached to the propionate side chains of bilirubin via ester bonds. The resulting **bilirubin diglucuronide** is then transported into the bile by an active process (i.e., against a concentration gradient). This transport is the rate-limiting step in hepatic bilirubin metabolism. Both conjugate formation and transport can be induced by drugs, such as *phenobarbital*.

In the intestine, bacterial β-*glucuronidases* partially hydrolyze bilirubin diglucuronide. The bilirubin that is released is reduced in a stepwise fashion to **urobilinogen** and **stercobilinogen**. In the presence of oxygen, these colorless compounds are oxidized again. The resulting pigments, **urobilin** and **stercobilin**, are orange and yellow. They are, for the most part, excreted in the stool (hence its brown color), or reabsorbed and returned to the liver (*enterohepatic circulation*; see p. 288). Urobilinogen also appears in the urine where oxidative processes partially convert it to urobilin. This contributes to the normal color of urine, and to its darkening on standing.

Not only hemoglobin, but other *hemoproteins*, such as myoglobin, cytochromes, catalase, and peroxidases, also supply heme groups that are degraded by the same pathway. Their contribution to daily bile pigment formation (250 mg) is 10–15%.

Further information

Hyperbilirubinemia. This term is used to describe elevated bilirubin levels in the blood (> 10 mg · l^{-1}). From there bilirubin may diffuse into peripheral tissues, giving them a yellow color. Such a condition is referred to as **jaundice**. It is especially obvious when the white conjunctiva of the eyes is inspected. Jaundice can be caused by

a) excessive formation of bilirubin by increased degradation of erythrocytes ("*hemolytic anemia*," due to hereditary enzyme defects or poisoning);

b) by impaired excretion as a result of hereditary or acquired liver defects ("*hepatocellular anemia*"); or

c) by impaired excretion of gall (e.g., "*obstructive jaundice*" due to gallstones). Unconjugated bilirubin can even cross the blood-brain barrier, leading to brain damage. The exact diagnosis of the cause of hyperbilirubinemia requires the analysis of bilirubin in the blood plasma. Conjugated ("*direct*") and unconjugated ("*indirect*") bilirubin can be distinguished from one another by certain color reactions.

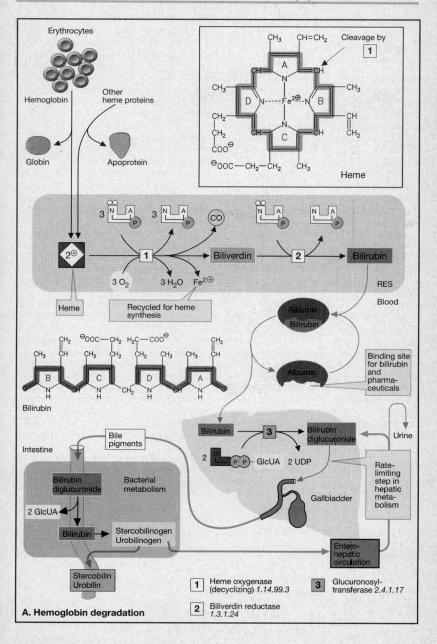

A. Hemoglobin degradation

1 Heme oxygenase (decyclizing) 1.14.99.3
2 Biliverdin reductase 1.3.1.24
3 Glucuronosyltransferase 2.4.1.17

Structure of Cells

A. Comparison of prokaryotes and eukaryotes ●

Present-day organisms are divided into two large groups, the prokaryotes and the eukaryotes. The **prokaryotes** (*eubacteria* and *archaebacteria*) are small unicellular organisms. The **eukaryotes** include *fungi*, *plants*, and *animals*, most of which are multicellular, and only some unicellular. Multicellular eukaryotes are made up of a variety of cell types specialized for different tasks. Eukaryotic cells are much larger than prokaryotic ones (volume ratio approximately 2000:1). Their most important distinguishing feature is the fact that they have a nucleus (Greek: *karyon*, hence the name) and other membrane-enclosed organelles.

With regard to structure and functions, eukaryotic cells are more complex than those of prokaryotes, and also show a greater degree of specialization. Eukaryotic DNA consists of very long, linear molecules (10^7 to more than 10^{10} base pairs). They are located in a nucleus, associated with histones, and contain non-coding regions (*introns*). In contrast, prokaryotic DNA is circular, has no introns, is shorter (up to $5 \cdot 10^6$ base pairs), and is found in the cytoplasm. Eukaryotic cells are *compartmentalized* (see below). The synthesis and maturation of RNA and proteins occur in different compartments, whereas in prokaryotes these processes are much simpler, and proceed in close proximity to one another.

B. Structure of an animal cell ●

Eukaryotic cells show much greater variety of size and structure than prokaryotic cells. In the human body alone, there are at least 200 different cell types. Therefore, the illustration showing the structure of an idealized animal cell is highly simplified.

The eukaryotic cell is organized by a system of membranes. On the outside, it is confined by a **plasma membrane**. Within the cell, there is a large space, the **cytoplasm**, which contains numerous soluble components. The cytoplasm is subdivided by membranes into distinct *compartments* (confined reaction spaces). Well-characterized compartments surrounded by intracellular membranes are known as **cell organelles**.

The largest organelle is the **nucleus** (see p. 194). It is readily visible with the light microscope. The **endoplasmic reticulum,** a closed system of flattened cavities and tubes, merges with the outer membrane of the nucleus. Another membrane-bound organelle is the **Golgi apparatus**, or Golgi complex, which resembles a stack of layered sheets. The **endosomes** and **exosomes** are bubble-shaped compartments (*vesicles*) involved in the exchange of materials between the cell and its surroundings. Probably the most important organelles in cellular metabolism are the **mitochondria**, which are about the size of bacteria. The **lysosomes** and **peroxisomes** are small, globular organelles, which perform specific metabolic tasks. The entire cell is criss-crossed by a scaffolding-like structure made up of proteins, which is referred to as the **cytoskeleton**.

In addition to these organelles, plant cells (see p. 129) have **chloroplasts**, which are the site of photosynthesis (see p. 118), **vacuoles**, which perform both structural and storage functions, and a rigid **cell wall** made of cellulose and other polysaccharides (see p. 38).

The approximate volume of each compartment (as a percentage of the total cell volume, yellow) and the approximate number of each organelle per cell (highlighted in blue) are those for mammalian *hepatocytes* (liver cells). The values markedly differ from one cell type to the next.

The cell organelles and other cellular elements are described in more detail in the following pages.

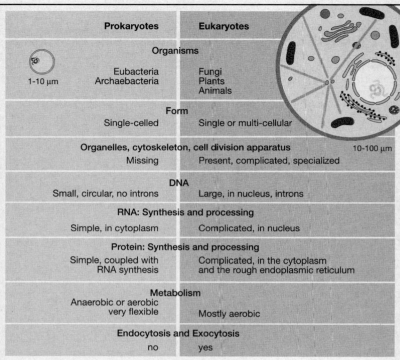

	Prokaryotes	**Eukaryotes**
Organisms		
1-10 μm	Eubacteria Archaebacteria	Fungi Plants Animals
Form		
	Single-celled	Single or multi-cellular
Organelles, cytoskeleton, cell division apparatus		10-100 μm
	Missing	Present, complicated, specialized
DNA		
	Small, circular, no introns	Large, in nucleus, introns
RNA: Synthesis and processing		
	Simple, in cytoplasm	Complicated, in nucleus
Protein: Synthesis and processing		
	Simple, coupled with RNA synthesis	Complicated, in the cytoplasm and the rough endoplasmic reticulum
Metabolism		
	Anaerobic or aerobic very flexible	Mostly aerobic
Endocytosis and Exocytosis		
	no	yes

A. Comparison of prokaryotes and eukaryotes

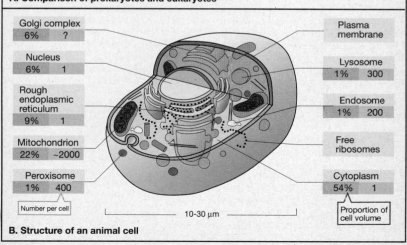

Golgi complex
6%   ?

Nucleus
6%   1

Rough
endoplasmic
reticulum
9%   1

Mitochondrion
22%   ~2000

Peroxisome
1%   400

Number per cell

Plasma
membrane

Lysosome
1%   300

Endosome
1%   200

Free
ribosomes

Cytoplasm
54%   1

Proportion of
cell volume

10-30 μm

B. Structure of an animal cell

Cell Fractionation

A. Isolation of cell organelles ◐

A variety of procedures have been developed for the investigation of individual compartments of the cell. The isolation of the different cell organelles begins with the disruption (**homogenization**) of the tissue of interest. Animal tissues are commonly homogenized in a buffered medium using a *Potter-Elvehjem homogenizer*, i.e., a rotating Teflon pestle in a glass cylinder. This is a relatively gentle method, which is particularly suitable for the isolation of fragile molecules and structures. Other cell disruption procedures include enzymatic lysis with cell wall-degrading enzymes, mechanical disruption by grinding of (frozen) tissue, mincing or chopping with rotating knives, large pressure changes, osmotic shock, and repeated freezing and thawing.

For the isolation of intact organelles, it is important that the homogenization medium be *isotonic*, i.e., it must have the same osmolarity as the interior of the cell. If the solution were hypotonic, the organelles would take up large quantities of water and burst. In hypertonic solutions, on the other hand, they would shrink.

Homogenization is followed by crude filtration through gauze to remove intact cells and connective tissue. The actual fractionation of the cellular components is then achieved by **differential centrifugation**, i.e., centrifugation at different gravitational forces (multiples of the earth's gravitational force). Increasing the gravitational force in a stepwise manner results in the sequential sedimentation of the different organelles, thus separating them according to their density and shape.

Nuclei already sediment at gravitational forces obtainable with bench-top centrifuges. Decantation of the supernatant and careful resuspension of the sediment (pellet) yields a fraction enriched in nuclei. However, this fraction still contains other cellular components as contaminants, e.g., fragments of the cytoskeleton with attached organelles.

Particles smaller and less dense than nuclei are obtained by subjecting the supernatant to stepwise increases in gravitational force. However, this requires more powerful centrifuges, such as refrigerated high-speed centrifuges and ultracentrifuges. The order in which the fractions are obtained is: **mitochondria, membrane vesicles,** and **ribosomes**. Finally, the supernatant of the last centrifugation contains the "**cytosol**", i.e., the soluble components of the cell dissolved in buffer.

The isolation of cell organelles is usually carried out at low temperatures (0–5 °C) to slow down degradative reactions catalyzed by enzymes released in the process of tissue disruption. The addition of thiols and chelating agents serves to protect functional SH groups.

B. Marker molecules ○

During cell fractionation, it is important to analyze the various fractions for purity. Whether the desired organelle is present within a particular fraction, and whether this fraction contains other contaminants, is determined using **marker molecules**. These are molecules that occur exclusively or predominantly in a particular type of organelle. These molecules are often organelle-specific enzymes (**marker enzymes**) whose activity can be measured as an indicator of their presence. The distribution of marker enzymes in the cell reflects the compartmentation of the processes they catalyze. These reactions are discussed in more detail in relation to the particular organelles in which they are found.

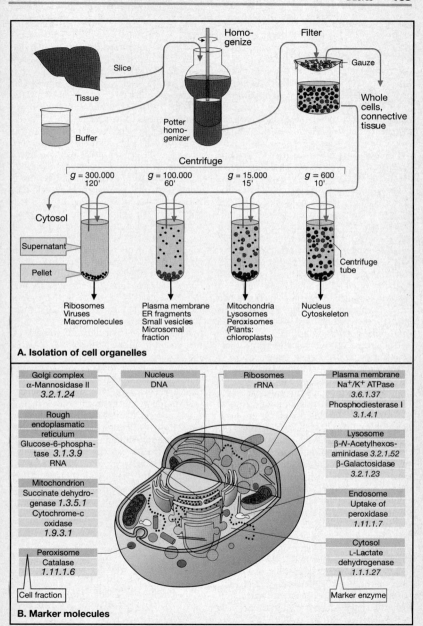

A. Isolation of cell organelles

Homogenize

Filter

Slice

Tissue

Buffer

Potter homogenizer

Gauze

Whole cells, connective tissue

Centrifuge

g = 300.000
120'

g = 100.000
60'

g = 15.000
15'

g = 600
10'

Cytosol

Supernatant

Pellet

Centrifuge tube

Ribosomes
Viruses
Macromolecules

Plasma membrane
ER fragments
Small vesicles
Microsomal
fraction

Mitochondria
Lysosomes
Peroxisomes
(Plants:
chloroplasts)

Nucleus
Cytoskeleton

B. Marker molecules

Golgi complex
α-Mannosidase II
3.2.1.24

Nucleus
DNA

Ribosomes
rRNA

Plasma membrane
Na⁺/K⁺ ATPase
3.6.1.37
Phosphodiesterase I
3.1.4.1

Rough
endoplasmatic
reticulum
Glucose-6-phospha-
tase *3.1.3.9*
RNA

Lysosome
β-N-Acetylhexos-
aminidase *3.2.1.52*
β-Galactosidase
3.2.1.23

Mitochondrion
Succinate dehydro-
genase *1.3.5.1*
Cytochrome-c
oxidase
1.9.3.1

Endosome
Uptake of
peroxidase
1.11.1.7

Peroxisome
Catalase
1.11.1.6

Cytosol
L-Lactate
dehydrogenase
1.1.1.27

Cell fraction

Marker enzyme

Centrifugation

A. Principles of centrifugation ◑

Particles in solution sink ("*sediment*") when their density is higher, or float when their density is lower, than that of their surroundings. The greater the difference in density, the faster they move. When there is no difference in density (*isopycnic conditions*), they remain where they are. When the differences in density are small, particles can be separated in a **centrifuge**, which creates a centrifugal force many times greater than the earth's gravity.

Rotors. The "centrifugal force" in a centrifuge (which, in a strict sense, is not a force but an acceleration) is usually expressed as multiples of the earth's gravitational acceleration ($g = 9.81$ m · s^{-2}). Values of up to 10,000 g are obtained with simple bench-top centrifuges, whereas high-speed refrigerated centrifuges reach 50,000 g, and ultracentrifuges, which operate with refrigeration and under a vacuum, can reach 500,000 g. There are two types of rotors, *fixed-angle* and *swing-out bucket*, which can be used with high-speed centrifuges and ultracentrifuges.

The **sedimentation velocity** (v) of a particle during centrifugation depends on the angular velocity (ω) and effective radius of the rotor (r_{eff}, distance from the axis of rotation) and the sedimentation properties of the particle. These properties are expressed as the **sedimentation coefficient, s,** in Svedberg (1 S = 10^{-13} s). The illustration shows the relationship between density and sedimentation coefficient for various particles in cesium chloride (CsCl) solutions. Note the broad range of s values particles can exhibit. To allow comparison of sedimentation coefficients determined in different media, they are usually corrected to the density and viscosity of water at 20 °C ($s_{20,w}$).

Various different properties of particles determine their sedimentation behavior, i.e., their relative molecular mass (M$_r$), shape (expressed as the frictional coefficient f), and their partial specific volume (v̄, reciprocal particle density). It is apparent from the illustration that proteins, RNA, and DNA have very different densities.

B. Density gradient centrifugation ○

Macromolecules or organelles that differ only slightly in size or density can be separated using density gradient centrifugation.

In **zonal centrifugation**, the sample containing the substances to be separated (e.g., proteins, cells, or organelles) is layered as a narrow band on top of the centrifugation solution. During centrifugation, the particles move through the solution, because their density is greater than that of the solution. The rate of movement depends on their mass and shape (see **A**, formulae). The run is stopped before the particles have reached the bottom of the centrifuge tube. A hole is then punched in the bottom of the tube, and the contents are allowed to drip out. The drops are collected into separate fractions, which contain the different particles. During centrifugation, the solution in the centrifuge tube is stabilized by a density gradient. These are solutions of carbohydrates or colloidal silica gel, the concentration of which increases from the surface of the tube to the bottom. Density gradients prevent the formation of convection currents, which could be detrimental to the separation of the particles.

Isopycnic centrifugation begins with the sample (e.g., DNA, RNA, or viruses) homogeneously distributed throughout the solution (usually cesium chloride, CsCl). Separation, therefore, takes much longer than in zonal centrifugation. The density gradient actually forms during the run by sedimentation and diffusion. With time, each particle moves to the region corresponding to its own *buoyant density*. Centrifugation is stopped when equilibrium is reached. Ultimately, the different fractions from the gradient are analyzed using appropriate measuring techniques.

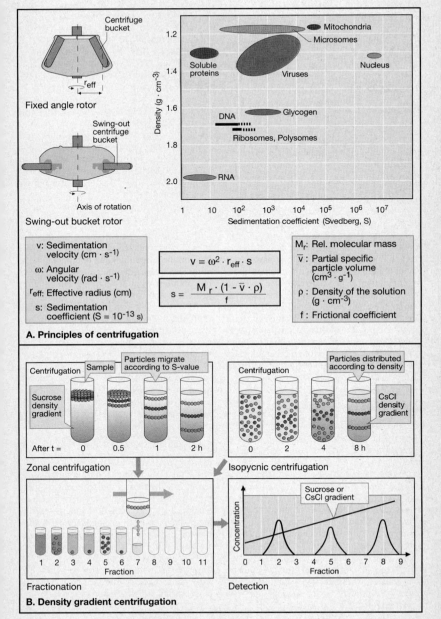

v: Sedimentation
velocity (cm · s⁻¹)

ω: Angular
velocity (rad · s⁻¹)

r_{eff}: Effective radius (cm)

s: Sedimentation
coefficient (S = 10⁻¹³ s)

$$v = \omega^2 \cdot r_{eff} \cdot s$$

$$s = \frac{M_r \cdot (1 - \bar{v} \cdot \rho)}{f}$$

M_r: Rel. molecular mass

$\bar{v}$: Partial specific
particle volume
(cm³ · g⁻¹)

ρ : Density of the solution
(g · cm⁻³)

f : Frictional coefficient

A. Principles of centrifugation

B. Density gradient centrifugation

Cell Components, Cytoplasm

The bacterium *Escherichia coli* (*E. coli*) lives as a symbiont in the intestine of mammals. Its structure and composition have been especially well characterized.

A. Components of a bacterial cell ○

The major component of all cells is **water** (70%). The others are **macromolecules** (proteins, nucleic acids, polysaccharides), **small organic molecules,** and **inorganic ions**. The most common of the macromolecules are the proteins, which account for 55% of the dry weight of the cell.

A single *E. coli* cell has a volume of about 0.88 μm³. One-sixth of this volume is accounted for by membranes, and another sixth by DNA (the "nucleoid"). The remaining internal space is referred to as the **cytoplasm** (not "cytosol", see p. 184). When a number of assumptions are made about the size (average 40 kDa) and distribution of the proteins within the cell, it follows that the cytoplasm of an *E. coli* cell contains approximately 250,000 protein molecules. In eukaryotic cells, which are approximately one thousand times larger, the number of protein molecules has been estimated to be in the order of several billion.

B. View into a bacterial cell ○

The illustration shows, in a schematic way, a view of the **cytoplasm** of an *E. coli* cell. The magnification is approximately one million-fold.

At this magnification, a single carbon atom will be the size of a salt grain, and an ATP molecule would be as big as a grain of rice. The cross section shown has a length of about 100 nm, i.e., a cube with these sides corresponds to 1/800th of the volume of an *E. coli* cell. For the sake of simplicity, small molecules, such as water, cofactors, and metabolites have been omitted from the main illustration (they are shown magnified in the lower right corner). Such a section of the cytoplasm contains:

- Hundreds of **macromolecules** required for the synthesis of proteins, i.e., 30 ribosomes, more than 100 protein factors, 30 aminoacyl-tRNA synthases, 340 tRNA molecules, 2–3 mRNAs (each of which is 10 times the length of the section shown), and 6 molecules of RNA polymerase.
- About 300 other **enzyme molecules** including 130 glycolytic enzymes and 100 citric acid cycle enzymes.
- 30,000 **small organic molecules** with masses ranging from 100 to 1000 Da, e.g., metabolites in intermediary metabolism and coenzymes. These are shown at 10-fold higher magnification in the bottom right hand corner.
- And finally, 50,000 **inorganic ions**. The remainder is water.

The illustration emphasizes the fact that the cytoplasm of cells is densely packed with macromolecules and small organic constituents. The macromolecules are very close to one another; most of them are separated by just a few water molecules.

All of these molecules are in constant motion. However, repeated collisions prevent them from proceeding in any particular direction. Instead, they follow a zigzag path. Due to their large mass, proteins are especially slow. Nevertheless, they still move at an average speed of 5 nm per millisecond, i.e., a distance equal to their own length. From a statistical point of view, a protein can reach any given point in a bacterial cell in less than a second.

C. Biochemical functions of the cytoplasm ○

In eukaryotes, the cytoplasm accounts for 50% of the total cell volume. Thus, in quantitative terms, it it the most important compartment of the cell. The cytoplasm is the *central reaction space* of the cell. It is here that most of the metabolic pathways for nutrient degradation and the synthesis of cellular building blocks are located. The cytoplasm contains much of the *intermediary metabolism* e.g., glycolysis, the hexose monophosphate pathway, gluconeogenesis, the synthesis of fatty acids, the biosynthesis of proteins, and so on.

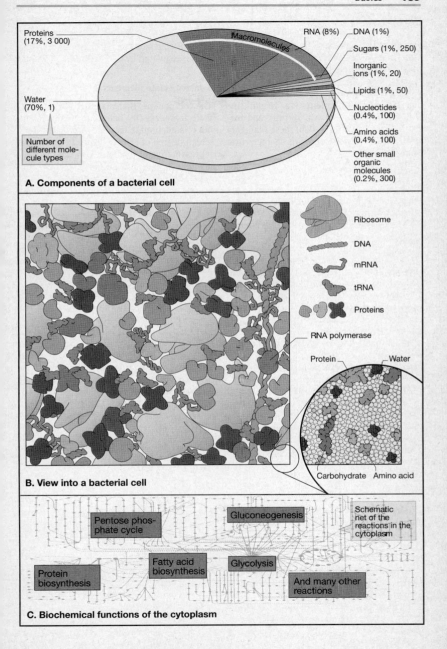

A. Components of a bacterial cell

Proteins (17%, 3 000)
Macromolecules
RNA (8%)
DNA (1%)
Sugars (1%, 250)
Inorganic ions (1%, 20)
Lipids (1%, 50)
Nucleotides (0.4%, 100)
Amino acids (0.4%, 100)
Other small organic molecules (0.2%, 300)
Water (70%, 1)
Number of different molecule types

B. View into a bacterial cell

Ribosome
DNA
mRNA
tRNA
Proteins
RNA polymerase
Protein
Water
Carbohydrate
Amino acid

C. Biochemical functions of the cytoplasm

Pentose phosphate cycle
Gluconeogenesis
Schematic net of the reactions in the cytoplasm
Fatty acid biosynthesis
Glycolysis
Protein biosynthesis
And many other reactions

Composition

The cytoplasm of eukaryotic cells is traversed by a three-dimensional scaffolding structure called the **cytoskeleton**. It is made up of filaments (long fibers). These filaments are divided into three groups, based on their diameters: **microfilaments** (ca. 10 nm), **intermediate filaments** (ca. 10 nm), and **microtubules** (ca. 25 nm). All of these filaments are polymers assembled from characteristic protein subunits.

A. Actin ◑

Microfilaments (actin filaments) are composed of actin, the most abundant protein in eukaryotic cells. Actin can exist either as a monomer (**G actin,** "globular actin") or as a polymer (**F actin,** "filamentous actin"). G actin is an asymmetrical, globular protein (42 kDa) composed of two domains. As the ionic strength increases, G actin aggregates to form helically-wound filaments of F actin.

When G actin associates to F actin, the orientation of the monomers is always the same, and therefore F actin has *polarity*. F actin filaments hav e two different ends, (+) and (-), which show different rates of polymerization. These ends must be stabilized by special proteins (as in muscle cells), or otherwise, at a critical concentration of G actin, the (+) end will continue to extend and the (-)-end will continue to decay. In this case, the actin monomers move slowly through the polymer (*"treadmill effect"*). Under experimental conditions, this process can be blocked by fungal toxins. For example, *phalloidin* binds to the (-) end and inhibits dissociation, whereas *cytochalasins* bind to the (+) end to block polymerization.

Actin-associated proteins: In the cytoplasm of animal cells, there are more than 50 different types of proteins that bind specifically to either G actin or F actin. This binding can have various different functions. It can regulate the size of the G actin pool (example: *profilin*), influence the polymerization rate of G actin (*villin*), stabilize the ends of F actin filament (*fragin*, β-*actinin*), attach filaments to one another or to other cellular components (*villin, α-actinin, spectrin, MARCKS*) or disrupt the helical structure of F actin (*gelsolin*). The activity of these actin-binding proteins is regulated by Ca^{2+} and protein kinases.

B. Intermediate filaments ◑

The components that make up the intermediate filaments (IF) belong to five different, but related, protein families. IF proteins show a high degree of cell-type specificity. Typical representatives include the *cytokeratins, desmin, vimentin, acidic fibrillar gliaprotein,* and *neurofilament.* All of these proteins consist of a central rod-shaped α-helical superhelix ("coiled coil", see keratin on p. 68). Dimers associate with one another in an antiparallel fashion to form tetramers. A staggered head-to-head arrangement leads to stable **protofilaments** that do not have polarity. One IF is composed of eight such protofilaments.

C. Microtubules ◑

Microtubules are made up of two globular proteins α- and β-**tubulin** (53 and 55 kDa). These can associate with one another to α,β-heterodimers. Thirteen of these heterodimers form a ring, which then gives rise to a long tube by polymerization. Like microfilaments, microtubules are polar, dynamic structures with (+) and (-) ends. The (-) end is stabilized by capping proteins, whereas the (+) end shows *dynamic instability*. It can either grow slowly or shrink rapidly. The tubulin monomers of microtubules bind GTP, which is slowly hydrolyzed into GDP. Other proteins can also be associated with the microtubules.

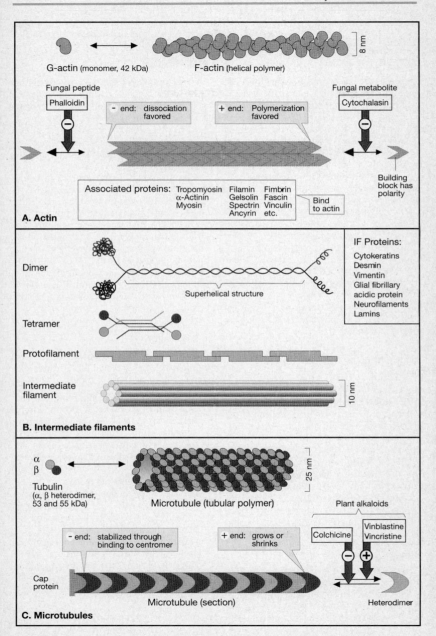

A. Actin

G-actin (monomer, 42 kDa)   F-actin (helical polymer)

8 nm

Fungal peptide

Phalloidin

– end: dissociation favored

+ end: Polymerization favored

Fungal metabolite

Cytochalasin

Building block has polarity

Associated proteins: Tropomyosin   Filamin   Fimbrin
α-Actinin   Gelsolin   Fascin
Myosin   Spectrin   Vinculin
Ancyrin   etc.

Bind to actin

B. Intermediate filaments

Dimer

Superhelical structure

Tetramer

Protofilament

Intermediate filament

10 nm

IF Proteins:
Cytokeratins
Desmin
Vimentin
Glial fibrillary acidic protein
Neurofilaments
Lamins

C. Microtubules

α
β

Tubulin
(α, β heterodimer, 53 and 55 kDa)

Microtubule (tubular polymer)

25 nm

Plant alkaloids

– end: stabilized through binding to centromer

+ end: grows or shrinks

Colchicine

Vinblastine
Vincristine

Cap protein

Microtubule (section)

Heterodimer

Structure and Functions

The cytoskeleton of eukaryotic cells performs three major functions:

1. It acts as a **mechanical scaffolding**. This scaffolding gives the cell its typical shape, and provides connections between membranes and organelles. The scaffolding is a dynamic structure that is being constantly synthesized and degraded in response to changes in the conditions experienced by the cell.
2. It acts as the **motor for movement** of cells. Not only muscle cells (see p. 306), but also cells of non-contractile tissues, contain a variety of different motor proteins. These proteins coordinate the movement of cells and determine the direction in which they move. Organelle transport, cell movement, shape changes during growth, cytoplasmic streaming, and cell division are all made possible by particular components of the cytoskeleton.
3. It serves as a **rail for transport** within the cell. Organelles and other large complexes can move along this rail.

A. Microfilaments and intermediate filaments ◑

The illustration shows a cross-section of the **microvilli** of an intestinal epithelial cell as an example of the structure and function of the components of the cytoskeleton (see **C1**). **Microfilaments** made of *F actin* traverse the microvilli in ordered bundles. These microfilaments are held together by actin-associated proteins, most importantly *fimbrin* and *villin*. *Calmodulin* and a myosin-like *ATPase* connect the microfilaments laterally to the plasma membrane. A further actin-associated protein, *fodrin*, connects the actin fibers at the base. It also anchors them to the cytoplasmic membrane and to a net made of **intermediate filaments**.

Actin is also involved in dynamic processes, such as muscle contraction (see p. 306), cell movement, phagocytosis, the formation of microspikes and lamellipodia (cellular extensions), and acrosomal processes during the fusion of sperm with egg cells.

B. Microtubules ◑

The illustration only shows the **microtubules** of the cell. They radiate out in all directions from a center near the nucleus, the **centrosome**. The (-) end of the centrosome is blocked by associated proteins. The (+) end may also be stabilized by associated proteins, e.g., when the microtubules reach the cytoplasmic membrane.

The microtubules are involved in determining cell shape, and they also serve as guiding rails for the transport of organelles. Together with associated motor proteins (*dynein, kinesin*), microtubules are able to perform mechanical work e.g., during the transport of mitochondria, the movement of cilia (hair-like protrusion of cells in the lungs, intestinal epithelium and oviduct) and the beating of the flagella of sperm. Microtubules also have a special role to play during cell division.

C. Architecture ○

The complex and dense net-like structure of the cytoskeleton is illustrated by three examples:

1. At the border of an intestinal epithelial cell (see B) **microfilaments** (**a**) protrude from the center out into the microvilli. The filaments are firmly held together by an associated protein called *spectrin* (**b**); in addition, they are anchored to **intermediate filaments** (**c**).
2. Only **microtubules** can be seen in this fibroblast. They originate from the microtubule organizing center (centrosome) and radiate out as far as the plasma membrane.
3. In this epithelial cell *keratin filaments* have been highlighted by a specific stain. Keratin fibers belong to the **intermediate filaments** (see pp. 68 and 190; d = nucleus).

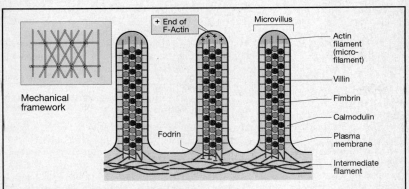

Mechanical framework

+ End of F-Actin

Microvillus

Actin filament (micro-filament)

Villin

Fimbrin

Calmodulin

Plasma membrane

Fodrin

Intermediate filament

A. Microfilaments and intermediate filaments

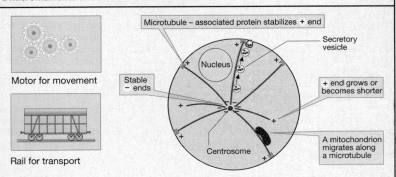

Motor for movement

Rail for transport

Microtubule – associated protein stabilizes + end

Secretory vesicle

Nucleus

Stable – ends

+ end grows or becomes shorter

A mitochondrion migrates along a microtubule

Centrosome

B. Microtubules

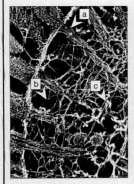

1. Microfilaments

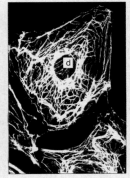

2. Microtubules

3. Intermediate filaments

C. Architecture

Nucleus

A. Nucleus ●

The nucleus is the largest organelle in the eukaryotic cell. With a diameter of about 10 μm, it is easily visible with the light microscope. The nucleus is separated from the rest of the cell by an **envelope** consisting of an outer and inner nuclear membrane.

The region between the two nuclear membranes is called **perinuclear space**. It is continuous with the interior of endoplasmatic reticulum (ER). The outer nuclear membrane resembles the membrane of rough ER, and is often studded with ribosomes. The inner nuclear membrane is lined with special proteins, which act as anchors for various nuclear structures (*nuclear lamina*).

Almost all of the DNA of the cell is located in the nucleus. It is the most important compartment for the *storage, replication*, and *expression of genetic information*. During *interphase* (the phase between cell divisions), much of the DNA in the nucleus is present as **heterochromatin**, i.e., densely packed DNA associated with proteins. More loosely packed DNA is referred to as **euchromatin**; it is the site where *transcription* of DNA into RNA. Nuclei often contain a **nucleolus**, or sometimes several nucleoli, which are the sites of ribosome formation. During cell division, the nuclear structures disintegrate and the chromatin becomes organized into *chromosomes*, i.e., extremely condensed forms of DNA molecules visible under the light microscope.

B. Import of large nuclear proteins ◐

The exchange of macromolecules, such as proteins and RNA between the nucleus and the cytoplasm occurs by **nuclear pores** (diameter ca. 7 nm), which consist of a pore complex made up of proteins. These pores regulate the transport across the nuclear membrane. Peptides and small proteins, such as some histones, are readily able to enter the nucleus. Larger proteins (over 40 kDa), however, can only pass through the nuclear pores when they carry a specific signal sequence. Such **nuclear-targeting sequences** consist of four basic amino acids located within the peptide chain of the protein. Unlike other signal sequences, they are not removed after the protein has been imported (see p. 216).

C. Interactions between nucleus and cytoplasm ◐

Almost all of the RNA in the cell is synthesized in the nucleus. This process, which is referred to as **transcription** utilizes the information stored in DNA (see p. 222). The synthesis of *ribosomal RNA* (rRNA) takes place in the nucleolus, while *messenger RNAs* and *transfer RNAs* (mRNA and tRNA) are synthesized within the euchromatin. **Replication**, the enzyme catalyzed duplication of DNA, also takes place in the nucleus (see p. 220).

The nucleotide building blocks required by the nucleus for replication and transcription have to be imported from the cytoplasm. Their incorporation into RNA results in the production of primary products that are subsequently altered by cleavage, deletion of parts of the molecule, and addition of extra nucleotides (**maturation**). Finally, the mRNA and tRNA formed in the nucleus are transported into the cytoplasm to be used in protein synthesis (**translation**, see p. 218).

Proteins cannot be synthesized in the nucleus, and therefore all nuclear proteins must be imported from the cytoplasm. Examples include the *histones* and the *non-histone proteins*, which are associated with the DNA in chromatin, the polymerases, various hormone receptors, transcription factors, and the ribosomal proteins. The ribosomal proteins begin to associate with rRNA to form ribosomal subunits, while they are still in the nucleolus.

One very special function of the nucleus is the **biosynthesis of NAD⁺**. The precursor of this coenzyme, *nicotinamide mononucleotide* (NMN), is synthesized in the cytoplasm and then transported into the nucleolus for conversion to the dinucleotide NAD⁺. NAD⁺ then returns to the cytoplasm.

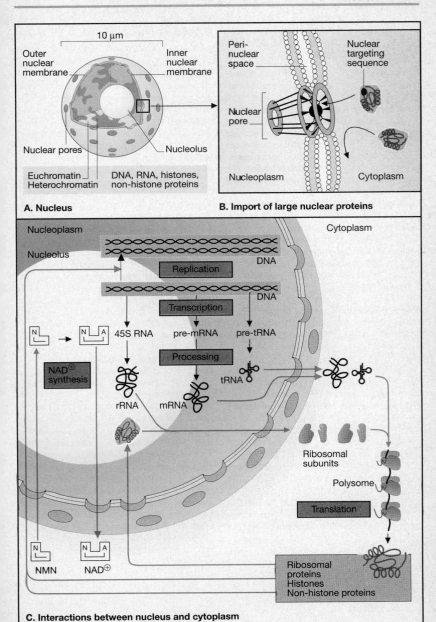

10 µm

Outer nuclear membrane

Inner nuclear membrane

Nuclear pores

Nucleolus

Euchromatin
Heterochromatin

DNA, RNA, histones, non-histone proteins

A. Nucleus

Peri-nuclear space

Nuclear targeting sequence

Nuclear pore

Nucleoplasm

Cytoplasm

B. Import of large nuclear proteins

Nucleoplasm

Cytoplasm

Nucleolus

DNA

Replication

DNA

Transcription

45S RNA   pre-mRNA   pre-tRNA

NAD⊕ synthesis

Processing

tRNA

rRNA   mRNA

N   →   N A

N   N A

NMN   NAD⊕

Ribosomal subunits

Polysome

Translation

Ribosomal proteins
Histones
Non-histone proteins

C. Interactions between nucleus and cytoplasm

Structure and Functions

A. Mitochondrial structure ○)

Mitochondria are organelles the size of bacteria (about 1×2 μm). They are found in almost all eukaryotic cells in very large numbers. Typically, there are about 2000 mitochondria per cell. Their total volume accounts for up to 25% of the cell volume. Mitochondria are delimited by two membranes, a smooth **outer membrane** and a folded **inner membrane** that encloses the **matrix space**. The folding of the inner membrane results in a very large surface area. The folds are referred to as **cristae**, while the space between the inner and the outer membranes is usually called the **intermembrane space**.

Different cell types differ in the number and shape of their mitochondria. There is also a great deal of variation in the number of cristae. The mitochondria of tissues with intensive oxidative metabolism, e.g., heart muscle, have especially large numbers of cristae. Even within one type of tissue, there can be variation in the shape of the mitochondria, depending on their functional status. Mitochondria are variable and plastic organelles.

Mitochondrial membranes contain integral membrane proteins. The outer membrane contains **porins**, which form pores and make the membrane permeable to molecules smaller than 10 kDa (see p. 208). The inner mitochondrial membrane, on the other hand, is impermeable to most molecules (the few exceptions include O_2, CO_2 and H_2O). It contains *cardiolipin*, an atypical phospholipid characteristic of mitochondria (see p. 48). The inner mitochondrial membrane is almost 75% protein — an unusually high percentage. The proteins of the inner mitochondrial membrane include **specific transport proteins** ("carriers") for the controlled exchange of metabolites with the cytoplasm (see p. 198). In addition, it contains enzymes and other components of the **respiratory chain** as well as **ATP synthase**. The matrix space is also very rich in proteins, especially enzymes of the citric acid cycle.

B. Metabolic functions ●

Mitochondria are the *biochemical powerhouse of the cell*, i.e., by oxidative degradation of foods, they produce the majority of the ATP required by the cell. The following processes and pathways are located in the mitochondria: **conversion of pyruvate to acetyl-CoA**, catalyzed by the *pyruvate dehydrogenase complex*, the **citric acid cycle**, the **respiratory chain** coupled to the **synthesis of ATP** (together these are referred to as *oxidative phosphorylation*), degradation of fatty acids by **β-oxidation**, and parts of the **urea cycle**. Mitochondria also supply metabolic intermediates to the rest of the cell, and act as a **calcium store**, which maintains the cytoplasmic Ca^{2+} concentration at a constantly low level (about 1 μM).

The major function of the mitochondria is the *uptake of energy-yielding substrates* (fatty acids, pyruvate, carbon skeletons of amino acids) from the cytoplasm and their oxidative degradation to yield CO_2 and H_2O, coupled to the synthesis of ATP. The reactions of the **citric acid cycle** (see p. 126) result in the complete oxidation of carbon compounds to CO_2, and they also supply reducing equivalents temporarily bound to coenzymes. Most of these processes take place in the matrix space. The **respiratory chain** (see p. 130), which reoxidizes the reduced coenzymes, is located in the inner mitochondrial membrane. It employs NADH and ubiquinol as reducing agents (electron donors) for the reduction of oxygen and the formation of water. This strongly exergonic reaction is coupled to the transport of protons (H^+) from the matrix space to the intramembrane space. This results in the formation of an **electrochemical gradient** across the inner mitochondrial membrane (see p. 116). With catalysis by *ATP synthase*, the mitochondria can exploit this gradient for the synthesis of ATP from ADP and P_i. Certain transport processes also are driven by this gradient (see p. 198).

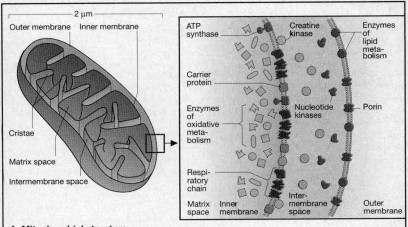

A. Mitochondrial structure

Outer membrane — Inner membrane

├─ 2 μm ─┤

Cristae

Matrix space

Intermembrane space

ATP synthase

Creatine kinase

Enzymes of lipid metabolism

Carrier protein

Enzymes of oxidative metabolism

Nucleotide kinases

Porin

Respiratory chain

Matrix space

Inner membrane

Intermembrane space

Outer membrane

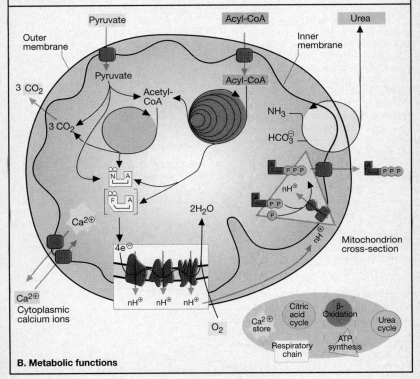

B. Metabolic functions

Pyruvate

Acyl-CoA

Urea

Outer membrane

Inner membrane

Pyruvate

Acetyl-CoA

Acyl-CoA

NH_3

$3 \: CO_2$

$3 \: CO_2$

$HCO_3^{\ominus}$

$Ca^{2\oplus}$

$2H_2O$

$4e^{\ominus}$

$nH^{\oplus}$

$nH^{\oplus}$ $nH^{\oplus}$ $nH^{\oplus}$

O_2

$Ca^{2\oplus}$
Cytoplasmic calcium ions

Mitochondrion cross-section

$Ca^{2\oplus}$ store

Citric acid cycle

β-Oxidation

Urea cycle

Respiratory chain

ATP synthesis

Transport Systems

Mitochondria are surrounded by an inner and an outer membrane (see p. 196). The inner mitochondrial membrane is impermeable to most small molecules. This not only holds for metabolic intermediates, but also for ions, such as H^+ and Na^+. There are *separate pools* of these metabolites in the cytoplasm and mitochondria. In contrast, the outer mitochondrial membrane contains pores (see p. 208), and thus does not constitute a barrier to small molecules.

A. Transport systems ❶

The exchange of molecules between the cytoplasm and the mitochondrial matrix is facilitated by various **transporters** located in the inner membrane. These — either by symport (**S**), by antiport (**A**) or by uniport (**U**, see p. 206) — transfer substances, such as pyruvate, phosphate, ATP, ADP, glutamate, aspartate, malate, 2-oxoglutarate, citrate, and fatty acids across the membrane. There are also carrier systems that regulate the influx and efflux of calcium, and thereby control the cytoplasmic calcium concentration.

Most of the ATP produced by the mitochondria is utilized in the cytoplasm. ATP^{4-} generated in the matrix is exported to the cytoplasm by an *ADP/ATP translocator* in exchange for ADP^{3-}. The *import of phosphate* into the mitochondria is independent of ADP/ATP transport, and occurs as a cotransport with a proton. *Pyruvate uptake* across the inner membrane is also mediated by a pyruvate-specific transporter.

B. Fatty acid transport ❶

Fatty acids have a group-specific transport system. Before they can be taken up, the acyl groups of activated fatty acids (acyl-CoA) must be esterified with *carnitine* in the cytoplasm. The **acyl carnitine** formed is then transported into the matrix space by a **carnitine transporter** in exchange for free carnitine. Once in the matrix, the acyl residues are transferred back to CoA.

C. Malate shuttle ❶

Mitochondria have several "shuttle" systems for the **import** from the cytoplasm **of reducing equivalents** (mostly $NADH + H^+$ produced by glycolysis). In mammals, the main system is the so-called malate shuttle. It is not as complicated as it appears. Basically, it involves the transport of reducing equivalents bound to malate. The oxaloacetate formed by reoxidation of malate in the matrix space does not immediately return to the cytoplasm, but is first transaminated to aspartate. The cycle further involves glutamate and 2-oxoglutarate, which are required for the formation by transamination of sufficient quantities of aspartate

Transport processes in the inner mitochondrial membrane can be driven by *concentration gradients* of metabolites, or by the *electrochemical potential* across the inner mitochondrial membrane (see p. 130). For example, the carnitine system for the transport of fatty acids is driven by the higher concentration of acyl-CoA in the cytoplasm. Phosphate and pyruvate import are driven by the **proton gradient**, whereas the ATP/ATP exchange and the Ca^{2+} influx are dependent on the **membrane potential** of the inner mitochondrial membrane.

Further information

Mitochondria are the major consumers of oxygen in the body. Oxygen deficiency (hypoxia) as a result of insufficient oxygen supply by the blood (ischemia), as in the case of *arteriosclerosis* or a *heart attack*, has especially severe effects on the mitochondria of the affected tissue.

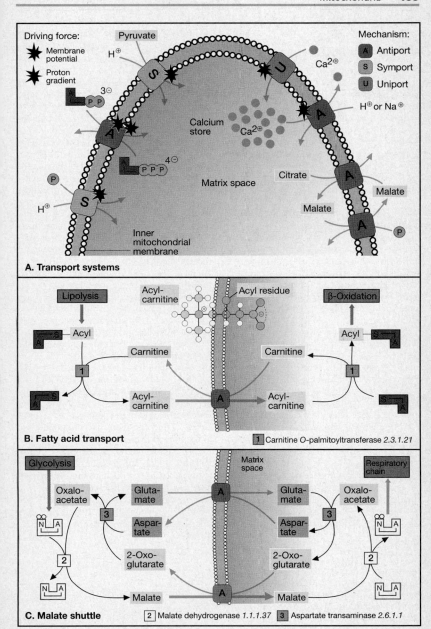

A. Transport systems

Driving force:
- ✦ Membrane potential
- ✦ Proton gradient

Mechanism:
- **A** Antiport
- **S** Symport
- **U** Uniport

Pyruvate
H⊕
Ca²⊕
3⊖
4⊖
Calcium store
Ca²⊕
H⊕ or Na⊕
Citrate
Matrix space
Malate
Malate
P
P
H⊕
Inner mitochondrial membrane

B. Fatty acid transport

Lipolysis
Acyl-carnitine
Acyl residue
β-Oxidation
Acyl
Acyl
Carnitine
Carnitine
Acyl-carnitine
Acyl-carnitine

1 Carnitine O-palmitoyltransferase 2.3.1.21

C. Malate shuttle

Glycolysis
Matrix space
Respiratory chain
Oxalo-acetate
Gluta-mate
Gluta-mate
Oxalo-acetate
Aspar-tate
Aspar-tate
2-Oxo-glutarate
2-Oxo-glutarate
Malate
Malate

2 Malate dehydrogenase 1.1.1.37   **3** Aspartate transaminase 2.6.1.1

Mitochondrial DNA, Peroxisomes

A. Mitochondrial DNA ◑

Mitochondria probably have evolved from aerobic prokaryotic bacteria living in symbiosis with anaerobic host cells (*endosymbiotic theory*).

Due to this history, mitochondria have their own DNA (*mtDNA*), which they use to synthesize a small proportion of their own proteins. However, most mitochondrial proteins are encoded by nuclear DNA. These proteins are synthesized in the cytoplasm on free ribosomes before being imported into the mitochondria (see p. 216). Specialized *translocators* in the inner mitochondrial membrane mediate the import of proteins.

The complete sequence of human **mtDNA** (16,569 base pairs) has ben determined. mtDNA is circular, and not associated with histones. Both strands, the G-rich H strand and the C-rich L strand, contain information. The mtDNA encodes 2 *rRNAs* (green), *tRNAs* for 22 amino acids (blue) and 13 different *polypeptides* (brown).

The **genetic code** of mitochondria (see p. 226) is somewhat different from the nuclear code, and there are also differences among species. In other words, mitochondria "speak a dialect of the nuclear language." The enzymes responsible for transcription and translation in the mitochondria are more like those of prokaryotes than eukaryotes. Prokaryotic and mitochondrial enzymes are similar in size, act in a similar fashion, and show the same pattern of inhibition by antibiotics (see p. 232).

Mitochondria reproduce by simple fission. In sexually reproducing higher organisms (including humans), mitochondria are usually transmitted to the offspring only by the mother.

B. Peroxisome ◑

All animal and plant cells contain peroxisomes. They are similar to lysosomes in size and abundance. In liver cells, for instance, there are about 400 peroxisomes, which account for about 1% of the volume of the cell.

Peroxisomes are surrounded by a single membrane and are free of DNA and ribosomes. They contain enzymes, which use oxygen as the oxdizing agent and produce hydrogen peroxide (H_2O_2) as a byproduct. Some other peroxisomal enzymes use H_2O_2 as an oxidant, and still others destroy it. Some of the most common enzymes in the peroxisomes are *catalase*, *urate oxidase* and D-*amino acid oxidase*.

The H_2O_2 produced by peroxisomal enzymes is toxic. It contributes to the formation of free radicals, which destroy membrane lipids and attack proteins and nucleic acids. Peroxisomal *catalase* is a key enzyme in the detoxification functions of the liver and kidneys. Catalase can oxidize substrates with H_2O_2 as the oxidant (type **1**: *peroxidase reaction*) e.g., ethanol, phenols, formic acid, formaldehyde. This is one of the pathways for the oxidation of ethanol to acetaldehyde in the liver (see p. 294). Catalase can also detoxify H_2O_2 (type **2**: *catalase reaction*). In the reaction O_2 and H_2O are formed from two molecules of H_2O_2 by "dismutation"($2\ O^{-1} \rightarrow 1\ O^o + 1\ O^{-2}$). This latter reaction appears to have less physiological relevance.

Peroxisomes are also very important in the degradation of long-chain fatty acids ($C_n > 18$). Degradation occurs via β-oxidation (see p. 148) with the production of acetyl-CoA and H_2O_2. The latter is formed in the flavin-dependent dehydrogenase step. However, fatty acid degradation in the peroxisomes stops at C_8 fatty acids. These are further metabolized in the mitochondria.

Like mitochondria, peroxisomes reproduce by growth and division. Their proteins are encoded in the nucleus and synthesized in the cytoplasm. Their import is guided by a peroxisome-specific signal sequence (see p. 216). Due to their high protein content, peroxisomes often contain crystalline protein aggregates.

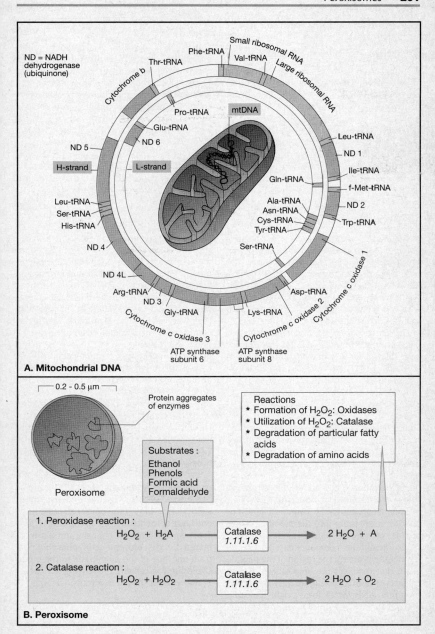

ND = NADH
dehydrogenase
(ubiquinone)

Phe-tRNA
Thr-tRNA
Cytochrome b
Small ribosomal RNA
Val-tRNA
Large ribosomal RNA
Pro-tRNA
mtDNA
Glu-tRNA
ND 6
ND 5
H-strand
L-strand
Leu-tRNA
ND 1
Ile-tRNA
Gln-tRNA
f-Met-tRNA
Ala-tRNA
Asn-tRNA
ND 2
Cys-tRNA
Trp-tRNA
Tyr-tRNA
Leu-tRNA
Ser-tRNA
His-tRNA
ND 4
Ser-tRNA
ND 4L
Asp-tRNA
Arg-tRNA
ND 3
Gly-tRNA
Lys-tRNA
Cytochrome c oxidase 2
Cytochrome c oxidase 1
Cytochrome c oxidase 3
ATP synthase
subunit 6
ATP synthase
subunit 8

A. Mitochondrial DNA

0.2 - 0.5 µm

Protein aggregates
of enzymes

Peroxisome

Substrates :

Ethanol
Phenols
Formic acid
Formaldehyde

Reactions
* Formation of H_2O_2: Oxidases
* Utilization of H_2O_2: Catalase
* Degradation of particular fatty
  acids
* Degradation of amino acids

1. Peroxidase reaction :

$$H_2O_2 + H_2A \longrightarrow \boxed{\begin{array}{c}\text{Catalase}\\1.11.1.6\end{array}} \longrightarrow 2\,H_2O + A$$

2. Catalase reaction :

$$H_2O_2 + H_2O_2 \longrightarrow \boxed{\begin{array}{c}\text{Catalase}\\1.11.1.6\end{array}} \longrightarrow 2\,H_2O + O_2$$

B. Peroxisome

Membrane and Components

A. Structure of the plasma membrane ●

All membranes have the same basic structure. They consist of a continuous *bilayer* of **lipid molecules** of about 5 nm in thickness, in which **proteins** are embedded. Some membranes also contain **carbohydrates** bound to the lipids and proteins. The relative proportions of lipid, protein, and carbohydrate are characteristic of a particular type of membrane and can differ markedly among different types of membranes and cells (see p. 204).

B. Lipids ●

Three classes of lipids are found in membranes — **phospholipids, cholesterol** and **glycolipids**. The phospholipids (structure see p. 48) include phosphatidylcholine (lecithin), phosphatidylethanolamine, phosphatidylserine, phosphatidylinositol, and sphingomyelin. Glycolipids have a carbohydrate group attached to the lipid part of the molecule.

Membrane lipids are *amphipathic* molecules (see p. 22) with a polar, hydrophilic *head* (blue) and an apolar, lipophilic *tail* (yellow). The illustration shows phosphatidylcholine as an example.

C. Proteins ●

Some membrane proteins span the lipid bilayer one or several times. These are referred to as transmembrane or **integral membrane proteins**. Other membrane proteins are loosely associated with the surface of the membrane, and are therefore referred to as **peripheral membrane proteins**. These peripheral proteins may contain a lipid anchor to insert them into the membrane (see p. 214), or may be loosely associated with other membrane components. For integral membrane proteins, the region of the peptide chain that traverses the lipid bilayer usually consists of 21 to 25 predominantly hydrophobic amino acids, which form a right-handed α helix with 6 or 7 turns (a **transmembrane helix**, see p. 122).

D. Carbohydrates ◗

Cell surface proteins and some lipid molecules carry covalently-bound carbohydrate groups on the outside of the membrane. These **glycoproteins** and **glycolipids**, together with additional, loosely attached glycoproteins and polysaccharides, form the cell coat (*glycocalyx*). The oligosaccharides are linked to the proteins via the hydroxyl group of a serine residue (*O-linked glycoproteins*) or the amido group of an asparagine residue (*N-linked glycoproteins*). The oligosaccharides of *N*-linked glycoproteins can be divided into *mannose-rich* and *complex* types.

E. Membrane fluidity ◗

The lipid bilayer of biological membranes shows a high degree of *fluidity*. The lipid molecules within the membrane are highly mobile. They can diffuse back and forth in the plane of the membrane (lateral displacement) and rotate about their own axes. In addition, they are very flexible in the region of their fatty acid chains. *Membrane fluidity* depends on the lipid composition of the membrane and the temperature. An increase in the concentration of unsaturated fatty acid increases the fluidity of the membrane. This is, because the *cis* double bonds disrupt the semi-crystalline structure of the membrane (**B**). The membrane proteins are mobile as well. Unless they are held in place by particular mechanisms, they are free to "float" in the lipid bilayer as in a two-dimensional fluid. Taken together, this system is referred to as a "*fluid mosaic*."

Although the membrane components readily move within the plane of the membrane, most of them, e.g., the proteins, cannot switch sides. Only some lipids can cross from one layer of the membrane to the other ("*flip-flop*"). Special proteins (translocators) are required for the movement of phospholipids from one side of the membrane to the other. Cholesterol is capable of crossing the membrane without the aid of a translocator.

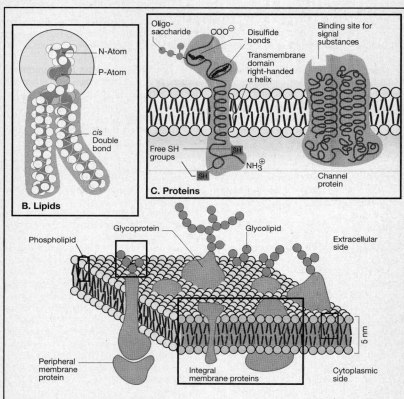

B. Lipids

N-Atom
P-Atom
cis Double bond

C. Proteins

Oligo-saccharide
COO⊖
Disulfide bonds
Transmembrane domain right-handed α helix
Binding site for signal substances
Free SH groups
SH
SH
NH₃⊕
Channel protein

Phospholipid
Glycoprotein
Glycolipid
Extracellular side
Peripheral membrane protein
Integral membrane proteins
Cytoplasmic side
5 nm

A. Structure of the plasma membrane

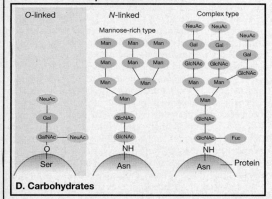

O-linked

N-linked
Mannose-rich type

Complex type

NeuAc
Gal
GalNAc — NeuAc
O
Ser

Man Man Man
Man Man Man
Man Man
Man
GlcNAc
GlcNAc
NH
Asn

NeuAc NeuAc NeuAc
Gal Gal Gal
GlcNAc GlcNAc GlcNAc
Man Man
Man
GlcNAc
GlcNAc — Fuc
NH
Asn — Protein

D. Carbohydrates

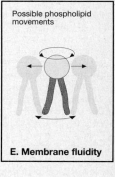

Possible phospholipid movements

E. Membrane fluidity

Functions and Composition

The most important membranes in animal cells are the *plasma membrane*, the inner and outer *nuclear membranes*, the membranes of the *endoplasmic reticulum* (ER) and the *Golgi apparatus*, and the inner and outer *mitochondrial membranes*. *Lysosomes, peroxisomes*, and various *vesicles* are also separated from the cytoplasm by membranes. Plant cells contain additional membranes, i.e., chloroplast, leucoplast, and vacuolar membranes. All membranes show *polarity*, i.e., there is a difference in the composition of their inner layer (facing toward the cytoplasm) and outer layer (facing away from the cytoplasm).

A. Functions of membranes ●

Membranes and their constituents have the following functions:

1. Delimitation and insulation of cells and organelles. The *insulation* provided by the plasma membrane protects cells both mechanically and chemically from their environment. The plasma membrane is essential for the maintenance of concentration differences of metabolites and inorganic ions between the intracellular and extracellular compartments.

2. Regulated metabolite and ion transport. This determines the *internal milieu* and is essential for *homeostasis*, i.e., the maintenance of constant metabolite and inorganic ion concentrations and other physiological conditions. Regulated and selective transport of metabolites and inorganic ions through pores and transporters (see p. 206) is essential, because of the delimitation of the cells and organelles by membrane systems.

3. Reception of extracellular signals and their transfer to the inside of the cell (see p. 110 and 350) as well as the production of signals.

4. Enzymatic catalysis of reactions. There are important *enzymes* located in membranes at the interface between the lipid and aqueous phases. This is where reactions involving apolar substrates occur. Examples include the *biosynthesis of lipids* and the *metab-*

olism of apolar xenobiotics (see p. 286). The most important reactions in energy metabolism, i.e., *oxidative phosphorylation* (see p. 130–133 and 206) and *photosynthesis* (see p. 110) also occur in membranes.

5. Communication with the extracellular matrix, as well as *interactions* with other cells for the purposes of cell fusion and tissue formation.

6. Anchoring of the cytoskeleton (see p. 190) to maintain cell shape and to allow movement.

B. Composition of membranes ◑

Membranes consist of **lipids, proteins** and **carbohydrates** (see p. 202). The amounts of these different components vary from one membrane to another (left). Proteins are usually the most important constituent, accounting for about half of the mass of membranes. Carbohydrates are only found on the side facing away from the cytoplasm; they account for only a few percent of the membrane mass. An example of a very unusual composition is *myelin*, the insulating material of the nerve cells, which is three-quarters lipid. In contrast, the inner *mitochondrial membrane* is characterized by a very low percentage of lipid and an especially high content of protein.

Closer observation of the contributions of the various lipids to the composition of membranes reveals typical patterns for particular cells and tissues. The illustration shows the diversity of the membrane lipids and an approximation of their relative abundance. The most important membrane lipids are the *phospholipids* (indicated by red stars) followed by the *glycolipids* and *cholesterol*. Triacylglycerols (neutral fats) are not found in membranes. One special feature is that *cholesterol* is found almost exclusively in eukaryotic cells. Animal membranes contain much more cholesterol than plant membranes, in which cholesterol is usually replaced by other sterols. In prokaryotes (with few exceptions), there is no cholesterol at all. The inner mitochondrial membrane of eukaryotes is also almost free of cholesterol (*endosymbiotic theory*, see p. 198).

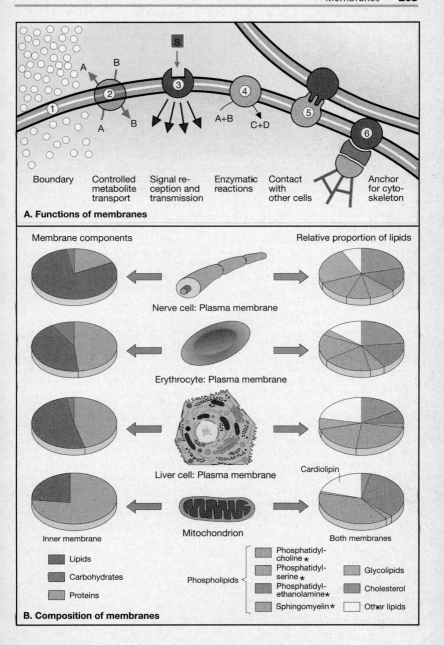

A. Functions of membranes

Boundary | Controlled metabolite transport | Signal reception and transmission | Enzymatic reactions | Contact with other cells | Anchor for cytoskeleton

B. Composition of membranes

Membrane components

Relative proportion of lipids

Nerve cell: Plasma membrane

Erythrocyte: Plasma membrane

Liver cell: Plasma membrane

Cardiolipin

Mitochondrion

Inner membrane

Both membranes

Lipids

Carbohydrates

Proteins

Phospholipids
- Phosphatidyl-choline ★
- Phosphatidyl-serine ★
- Phosphatidyl-ethanolamine ★
- Sphingomyelin ★

Glycolipids

Cholesterol

Other lipids

Transport Processes

A. Permeability of membranes●

Small, uncharged molecules, such as gases, water, ammonia, glycerol, and urea can diffuse freely through membranes. However, as their **size** increases, molecules are no longer able to permeate membranes. For instance, membranes are impermeable to glucose and other sugars.

The **polarity** of a molecule is also important. Apolar substances, such as benzene, ethanol, diethyl ether, and many narcotics are readily able to cross membranes by diffusion. In contrast, biological membranes are impermeable to hydrophilic, especially electrically charged compounds. In order to be able to take up such substances, cells have specialized *transport proteins* (see below and p. 208).

B. Passive and active transport ◑

Free diffusion is the simplest form of transport across a membrane. It is frequently facilitated by particular membrane proteins (*facilitated diffusion*), which can be divided into two groups:

Channel proteins form water-filled pores in the membrane that are permeable to particular ions. For example, there are specific ion channels for Na^+, K^+, Ca^{2+} and Cl^- ions (see, p. 318)

Transporters, on the other hand, actually bind the molecules they transport. A conformational change in the transporter then allows the molecule to cross the membrane. In this respect, transporters (carrier proteins, permeases) are like enzymes. The only difference is that they "catalyze" vectorial transport rather than an enzymatic reaction. They show *specificity* — sometimes group specificity — for the molecule(s) they transport. In addition, they exhibit a particular *affinity* (expressed as the dissociation constant, K_d) and a *maximal transport capacity* (V, compare with p. 90).

Free diffusion and transport processes facilitated by ion channels and transporters are driven by *concentration gradients* or *charge gradients* (together referred to as *elec-trochemical gradients*). These forms of transport are classified as **passive transport**, i.e., they run "downhill" along a concentration gradient. In this way, cells take up metabolites from the blood where their concentrations are much higher.

In contrast, **active transport** processes run "uphill," i.e., against a concentration or charge gradient. They therefore require an *input of energy*, which is usually supplied by the hydrolysis of ATP. Some transport processes are driven by the hydrolysis of other energy-rich compounds, such as phosphoenolpyruvate, or by light energy. In other cases, an endergonic transport is coupled to another, spontaneously occurring transport process (so-called *secondary active transport*). This, for instance, occurs in the epithelial cells of the intestine and the kidneys, where glucose is taken up against a concentration gradient in a cotransport with sodium ions (see p. 248). Here, the driving force for the transport of glucose is provided by the concentration gradient of Na^+.

Passive and active transporters in various membranes help cells regulate their volume, internal pH, and ionic composition. They concentrate metabolites, which are important for energy metabolism and biosyntheses, concomitantly exluding toxic substances. Specific transporters also serve in the formation of *ion gradients*, which are essential for oxidative phosphorylation and the stimulation of muscle and nerve cells (see pp. 130, 318)

C. Transport processes ◑

Transport can occur as a **uniport**, in which case *individual* molecules cross the membrane with the help of a channel or transport protein (example: transport of glucose into liver cells). The coupling of the transport of two different substances can occur as a **symport** (example: transport of glucose together with Na^+ ions into intestinal epithelial cells), or as an **antiport** (example: exchange of HCO_3^- for Cl^- at the erythrocyte membrane).

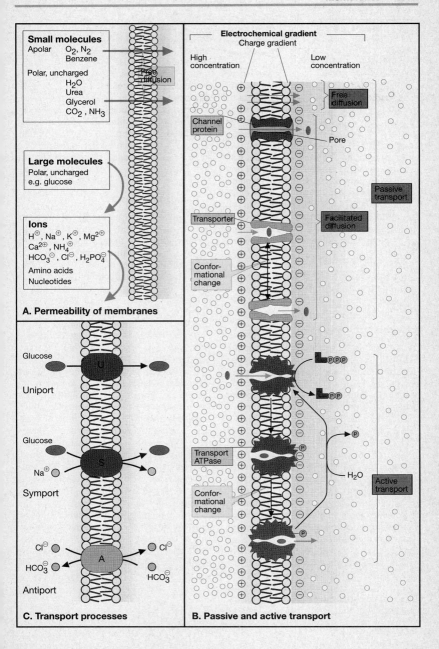

A. Permeability of membranes

Small molecules
Apolar O_2, N_2
         Benzene

Polar, uncharged
    H_2O
    Urea
    Glycerol
    CO_2, NH_3

Free diffusion

Large molecules
Polar, uncharged
e.g. glucose

Ions
$H^{\oplus}$, $Na^{\oplus}$, $K^{\oplus}$, $Mg^{2\oplus}$
$Ca^{2\oplus}$, $NH_4^{\oplus}$
$HCO_3^{\ominus}$, $Cl^{\ominus}$, $H_2PO_4^{\ominus}$
Amino acids
Nucleotides

C. Transport processes

Glucose

Uniport

Glucose

$Na^{\oplus}$

Symport

$Cl^{\ominus}$

$HCO_3^{\ominus}$

$Cl^{\ominus}$

$HCO_3^{\ominus}$

Antiport

B. Passive and active transport

Electrochemical gradient
Charge gradient

High
concentration

Low
concentration

Free
diffusion

Channel
protein

Pore

Transporter

Facilitated
diffusion

Passive
transport

Conformational
change

Transport
ATPase

H_2O

Conformational
change

Active
transport

Transport Proteins

A. Porins ○

Gram-negative bacteria have a complicated cell wall. Their plasma membrane is protected from the surrounding environment by a network of peptidoglycans (*murein*) and an additional *outer membrane*. Metabolites that the bacterial cell wants to absorb or release must be able to cross the outer membrane without hindrance. To achieve this, the bacteria have transmembraneous *channnel proteins* called **porins**. These trimeric proteins contain water-filled pores in the center, which allow the passage of molecules up to a molecular mass of 600 Da (*facilitated diffusion*). In higher organisms, porin-like proteins are found in the membranes of mitochondria and chloroplasts (see p. 196).

The van der Waals model on the right shows the structure of a porin trimer occurring in the outer membrane of the bacterium *Rhodopseudomonas blastica*.

B. Glucose transporter ○

As described on the previous page, the uptake of glucose by cells is usually a passive process. First, glucose is bound to a **glucose transporter** (GLUT) located in the cell membrane. A conformational change in the transporter then allows glucose to cross the membrane.

As membrane proteins are difficult to crystallize, the three-dimensional structures of only a few of them are known to date (see, for instance, A and p. 123) However, in many cases it is possible to predict the number and positions of the membrane-spanning parts from the amino acid sequence. The GLUT molecule shown here schematically contains 12 transmembrane helices and a carbohydrate chain which is located on the outer surface of the membrane.

The glucose transporters are a family of structurally-related proteins with different functions. GLUT1 and GLUT3 have a high affinity for glucose (K_d about 1 mM). They are found in almost all cells, where they ensure the continuous uptake of glucose. GLUT2 occurs on liver and pancreas cells. This transporter has a much lower affinity (K_d 15–20 mM). The uptake of glucose by GLUT2, therefore, is proportional to the fluctuating concentration of glucose in the blood. GLUT4 with a K_d of about 5 mM is found in the plasma membrane of muscle and fat cells. The hormone *insulin* causes an increase in the number of GLUT4 molecules on the cell surface, and thereby stimulates glucose uptake by these tissues (see p. 164). GLUT5 is synthesized by cells of the intestinal epithelium. It is involved in the symport of glucose with Na^+ ions (see p. 248).

C. Transport ATPases ◑

Transport ATPases catalyze the ATP-dependent active transport of molecules across membranes, i.e., they translocate them against a concentration gradient. There are many different types of these ATPases. They can transport such different molecules as cations ("ion pumps"), peptides ("peptide pumps") and apolar compounds (e.g., the so-called multidrug resistance proteins). Many cells use ion pumps for the cations H^+, Na^+, K^+, and Ca^{2+} to generate an *electrochemical gradient* (see p. 116) as a result of these transport processes.

The illustration shows the **Na^+/K^+-exchanging ATPase**, which is found in the plasma membrane of all animal cells. On the inside of the membrane, the enzyme undergoes a reaction cycle in which it becomes alternately phosphorylated and dephosphorylated. This results in an alternation between two different conformational states, which is responsible for the ion transport. For every ATP consumed, 3 Na^+ ions are transported out of the cell and 2 K^+ ions in. The constant operation of the enzyme leads to an imbalance in the number of Na^+ and K^+ ions inside and outside the cell, which is typical for animal cells (see p. 312). Na^+/K^+-ATPase is a membrane-spanning glycoprotein made up of four subunits ($\alpha_2\beta_2$).

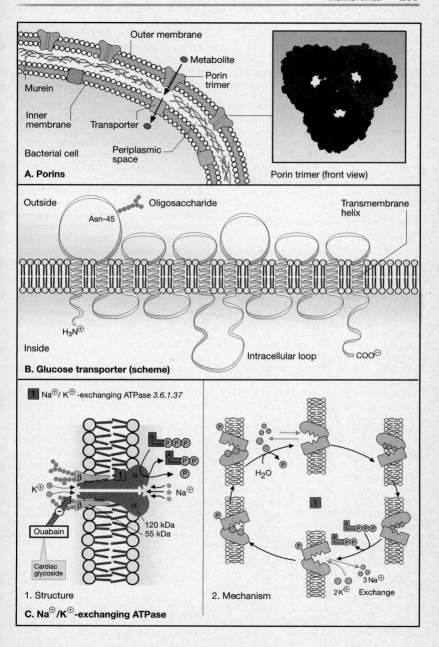

A. Porins

Outer membrane
Metabolite
Porin trimer
Murein
Inner membrane
Transporter
Bacterial cell
Periplasmic space

Porin trimer (front view)

B. Glucose transporter (scheme)

Outside
Asn-45
Oligosaccharide
Transmembrane helix
$H_3N^\oplus$
Inside
Intracellular loop
$COO^\ominus$

C. $Na^\oplus$/$K^\oplus$-exchanging ATPase

1 $Na^\oplus$/$K^\oplus$-exchanging ATPase *3.6.1.37*

$K^\oplus$
$Na^\oplus$
Ouabain
Cardiac glycoside
120 kDa
55 kDa

1. Structure

H_2O
$3 Na^\oplus$
$2 K^\oplus$ Exchange

2. Mechanism

Endoplasmic Reticulum and Golgi Apparatus

The endoplasmic reticulum (ER) is a large intracellular compartment enclosed by a network of membranes and extending throughout the cytoplasm. It is made up of tubular and sack-shaped structures, which are continuous with one another and with the outer membrane of the nucleus. A morphological distinction is made between the **rough ER** (rER) and the **smooth ER** (sER). Membranes of the rER are distinguished by many attached ribosomes, whereas none are associated with the sER.

A. Rough endoplasmic reticulum and Golgi apparatus ◑

The **rER** (**1**) is the site of active *protein biosynthesis*. This is where proteins destined for membranes, lysosomes, or for export from the cell are synthesized. All other proteins are produced in the cytoplasm on ribosomes not bound to ER membranes.

Most proteins synthesized at the rER (**1**) undergo post-translational modifications in its interior (see below). Either they remain within the rER as membrane proteins, or they are transported by **transport vesicles** (**2**) to the Golgi apparatus (**3**). Transport vesicles are formed from existing membranes by budding, and they disappear again by fusing with them (see p. 214).

Like the ER, the **Golgi apparatus** (**3**) is a complex network of membrane-enclosed spaces. It consists of stacks of disk-shaped structures, which are the site of *protein maturation* and *sorting*. There are *cis*, **medial**, and *trans* Golgi regions, as well as a *trans* **Golgi network** (tGN). Post-translational protein modification takes place in the various regions of the Golgi apparatus. The modified proteins are then transported by vesicles to various targets within the cell, i.e., the lysosomes (**4**), parts of the plasma membrane (**6**), or to secretory vesicles (**5**) that release their contents into the extracellular space by fusion with the plasma membrane (**exocytosis**). These transport processes can be *constitutive* (continuously occurring) or they are *regulated* by chemical signals.

B. Protein synthesis in the rER ○

An *N*-terminal sequence of 16–60 amino acids determines which proteins are synthesized in the rER and which in the cytoplasm. This part of the protein, the so-called **signal peptide**, serves as the "address" for the ER (see p. 216).

Translation of the mRNA always begins in the cytoplasm. When the peptide chain being synthesized on a ribosome begins with a signal peptide, a **signal-recognition particle** binds to it and interrupts translation. The ribosome then attaches itself to the rER via a receptor anchored in the rER membrane. The signal peptide forms a loop, which passes through the membrane and penetrates into the lumen of the rER with the help of a special translocator protein. Once this has occurred, protein synthesis can continue. As a first step in the post-translational modification of the protein, a *signal peptidase* within the rER cleaves off the signal peptide (see p. 216).

C. Protein modification: *N*-glycosylation ◑

N-glycosylation involves the *en-bloc* transfer of an oligosaccharide from the long-chain isoprenoid **dolichol** to specific asparagine residues of the protein.

Further protein modifications takes place in the Golgi apparatus (**3**). For example, proteins may become phosphorylated, and *glycosidases* or *glycosyltransferases* may either remove sugars from protein-bound oligosaccharides (so-called *trimming*) or attach further sugars to them. Finally, peptide fragments may be removed from proteins in the secretory vesicles (**5**) prior to the release of their contents by exocytosis. These sequence-specific, protease-catalyzed cleavage processes often serve to convert the protein to its active form (see, for example, p. 164).

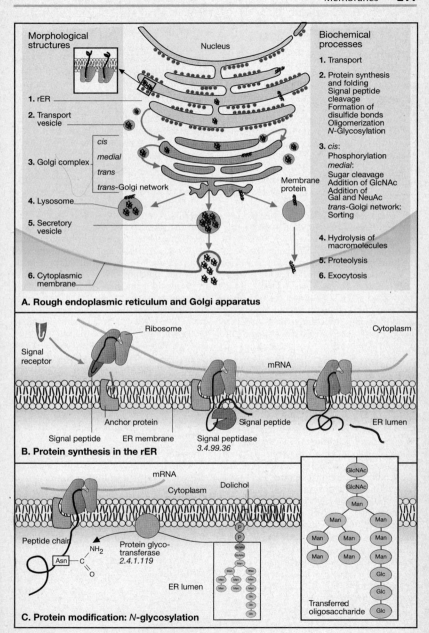

A. Rough endoplasmic reticulum and Golgi apparatus

Morphological structures

1. rER
2. Transport vesicle
3. Golgi complex
   - *cis*
   - *medial*
   - *trans*
   - *trans*-Golgi network
4. Lysosome
5. Secretory vesicle
6. Cytoplasmic membrane

Nucleus

Membrane protein

Biochemical processes

1. Transport
2. Protein synthesis and folding
   Signal peptide cleavage
   Formation of disulfide bonds
   Oligomerization
   N-Glycosylation
3. *cis*:
   Phosphorylation
   medial:
   Sugar cleavage
   Addition of GlcNAc
   Addition of Gal and NeuAc
   trans-Golgi network:
   Sorting
4. Hydrolysis of macromolecules
5. Proteolysis
6. Exocytosis

B. Protein synthesis in the rER

Signal receptor

Ribosome

Cytoplasm

mRNA

Anchor protein

Signal peptide   ER membrane

Signal peptide

Signal peptidase
3.4.99.36

ER lumen

C. Protein modification: *N*-glycosylation

mRNA

Cytoplasm

Dolichol

Peptide chain

Asn — C
       NH₂
       O

Protein glyco-transferase
2.4.1.119

ER lumen

GlcNAc
GlcNAc
Man
Man   Man
Man   Man   Man
Man   Man   Man
              Glc
              Glc
              Glc

Transferred oligosaccharide

Lysosomes

A. Structure and contents ◑

Lysosomes are organelles that are 0.2 to 2.0 μm in diameter and surrounded by a single membrane. They can take on very different shapes. Usually, there are several hundred lysosomes per cell. They are the "gut" of the cell, i.e., their main function is the degradation of various cell components. To this end, lysosomes are filled with about 40 different types of degradative enzymes. All of these enzymes are *hydrolases* with acidic pH optima. The marker enzyme for lysosomes is *acid phosphatase*.

The lysosomal matrix is moderately acidic, with a pH of 4.5–5.0. A *V-type* ATP-driven *proton pump* (see p. 206) is responsible for the maintenance of the proton gradient between the lysosomal matrix and the cytoplasm (pH 7–7.3). All of the lysosomal enzymes have pH optima near pH 5 corresponding to the pH of the lysosomes. At the near-neutral pH values found in the cytoplasm, they exhibit low levels of activity. Apparently, this is a mechanism to protect the cells from digesting themselves.

B. Functions ◑

The main function of lysosomes is the enzymatic degradation of unwanted cellular components e.g., aged mitochondria. First of all, they engulf the organelles and macromolecules to be disposed of (1). **Primary lysosomes** derived from the Golgi apparatus then become **secondary lysosomes**, in which the actual digestion takes place (2). Finally, **residual bodies** are formed, containing the indigestible residues of the lysosomal degradation process.

Lysosomes are also responsible for the degradation of macromolecules and particles taken up by cells via *endocytosis* and *phagocytosis*, e.g., lipoproteins, proteohormones, and bacteria. In the process, lysosomes fuse with the **endosomes** (3) that deliver the substances to be degraded.

C. Biosynthesis of lysosomal proteins ○

Lysosomal proteins are synthesized in the rER, where they become glycosylated with oligosaccharide residues (1; see p. 210). In the next step, which is specific for lysosomal proteins, terminal mannose residues (Man) are phosphorylated at the C-6 position (right-hand side of diagram). The reaction proceeds in two steps. First, GlcNAcphosphate is incorporated into the protein, and then GlcNAc is released. Thus, lysosomal proteins carry a terminal **mannose 6-phosphate** residue (Man-6-P; 2) that directs them into the lysosomes during the sorting process.

The membranes of the Golgi apparatus contain **receptor** molecules, which are specific for Man-6-P residues and thus recognize and selectively bind lysosomal proteins (3). Local accumulation of these receptors occurs with the help of *clathrin*. This protein allows the pinching-off of the appropriate membrane fragments, followed by their transport as transport vesicles to the **endolysosomes** (4). These then mature to give **primary lysosomes** (5). Finally, the phosphate group is removed from the Man-6-P residues (6). The *Man-6-P receptors* are recycled. The drop in the pH in the endolysosomes results in dissociation of the proteins from the receptors (7). The latter are then transported back to the Golgi apparatus by transport vesicles, (8).

Further information

A number of rare hereditary disorders are due to genetic defects of certain lysosomal enzymes. Most frequently, the breakdown of glycogen, lipids, and proteoglycans is affected, leading to *glycogenoses*, *lipidoses*, and *mucopolysaccharidoses*, respectively. Products that cannot be further metabolized due to defective or missing enzymes, accumulate in *residual bodies*, eventually leading to irreversible cell damage. As a result, the affected organs gradually become non-functional.

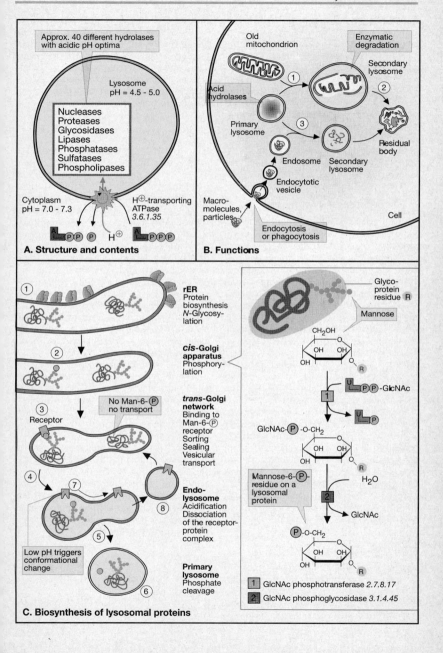

A. Structure and contents

Approx. 40 different hydrolases with acidic pH optima

Lysosome pH = 4.5 - 5.0

Nucleases
Proteases
Glycosidases
Lipases
Phosphatases
Sulfatases
Phospholipases

Cytoplasm pH = 7.0 - 7.3

$H^{\oplus}$-transporting ATPase 3.6.1.35

B. Functions

Old mitochondrion

Enzymatic degradation

Secondary lysosome

Acid hydrolases

Primary lysosome

Residual body

Endosome

Secondary lysosome

Endocytotic vesicle

Macro-molecules, particles

Endocytosis or phagocytosis

Cell

C. Biosynthesis of lysosomal proteins

rER
Protein biosynthesis
N-Glycosylation

cis-Golgi apparatus
Phosphorylation

trans-Golgi network
Binding to Man-6-(P) receptor
Sorting
Sealing
Vesicular transport

No Man-6-(P) no transport

Receptor

Endo-lysosome
Acidification
Dissociation of the receptor-protein complex

Low pH triggers conformational change

Primary lysosome
Phosphate cleavage

Glyco-protein residue R

Mannose

GlcNAc-(P)-O-CH₂

Mannose-6-(P)-residue on a lysosomal protein

H_2O

GlcNAc

(P)-O-CH₂

1 GlcNAc phosphotransferase 2.7.8.17
2 GlcNAc phosphoglycosidase 3.1.4.45

Protein Translocation

Like their structure, the **localization** of intracellular and extracellular proteins is genetically predetermined. Many of these proteins carry the "address" for their final destination as a **structural signal**. A series of signal-dependent decisions during and after translation then determines where the protein ends up once its synthesis is complete. A cellular guidance system directs them to their final destination. The mechanisms and signals involved are briefly described on this and the following page.

A. Transport across membranes ⟳

Proteins that have to be transported ("translocated") across membranes carry a **signal sequence** (see p. 216) at their N-terminus or elsewhere in the sequence, which is recognized by a **protein translocator** in the respective membrane. The protein must be in the unfolded state during passage through the membrane. As all proteins are unfolded at the beginning of protein synthesis, a growing peptide chain is readily able to cross the ER membrane (see p. 210), provided it carries an N-terminal signal peptide. In contrast, mature proteins, which are already folded, must first be unfolded before they can cross membranes. The ATP-dependent unfolding process involves proteins known as "*chaperones*". These molecules, which include the so-called *heat-shock proteins* (**hsp**), help to stabilize folding intermediates. The actual transport across the membrane is also energy-dependent, i.e., it is driven by ATP hydrolysis, a proton gradient, or the membrane potential. Chaperones are also involved in the correct re-folding. Incorrectly folded protein molecules are recognized and degraded by proteolysis.

B. Vesicular transport ⟳

Vesicles detach themselves from membranes by the process of "budding-off". Such vesicles carry proteins either in their **lumen** or bound to **membrane receptors**. Once the vesicle has fused with its target compartment, the receptors release their ligands, and are transported by the vesicles back to the original membrane. The weakening of the receptor-ligand binding is caused by a decrease in the pH value. Various different proteins, e.g., *clathrin* and small *G proteins* with GTPase activity (see p. 350) are involved in the processes of sorting and transport.

C. Lipid anchors ○

Proteins that are synthesized at the rER and carry a "stop-transfer-signal" (see p. 216) remain in the ER membrane and become fixed there by hydrophobic interactions (*integral membrane proteins*). Alternatively, fixation of a protein in a membrane can be brought about by the attachment of a **lipophilic anchor**. These so-called *peripheral membrane proteins* become linked enzymatically to lipids during or after translation. The necessary signal is usually contained within the protein in the form of a peptide sequence.

Both fatty acids and isoprenoids can serve as membrane anchors. Proteins can become *acylated* by reaction with derivatives of **palmitic acid** (C_{16}) or **myristic acid,** and "*prenylated*" by linkage with either **farnesol** (C_{15}) or **geranyl geranol** (C_{20}).

Many viral proteins, most membrane-localized hormone receptors (e.g., the receptors for insulin and acetylcholine), the products of cellular oncogenes (e.g., some *ras* proteins) and structural membrane proteins (e.g., ankyrin, vinculin, glycophorin, lamins) have lipid anchors.

Other membrane proteins are linked to **glycosylated phosphatidylinositol**. This "GPI anchor" is bound to the *C*-terminal amino acid. Some adhesion molecules (e.g., N-CAM, proteoglycans), and many membrane enzymes, such as alkaline phosphatase, acetylcholine esterase, and various proteinases carry GPI anchors.

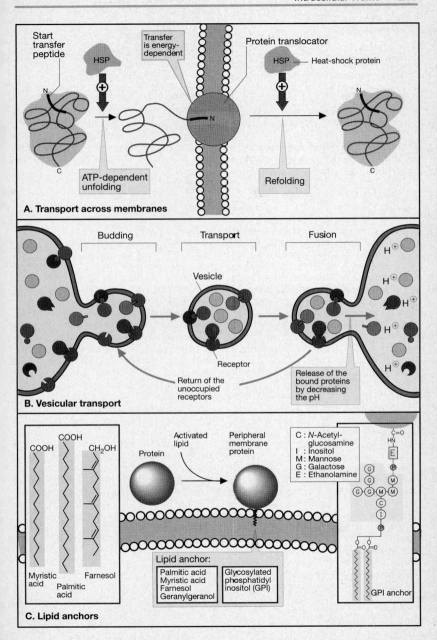

A. Transport across membranes

Start transfer peptide

HSP

Transfer is energy-dependent

Protein translocator

HSP — Heat-shock protein

N

C

ATP-dependent unfolding

Refolding

N

C

B. Vesicular transport

Budding

Transport

Fusion

Vesicle

H⁺

Receptor

Return of the unoccupied receptors

Release of the bound proteins by decreasing the pH

C. Lipid anchors

COOH

COOH

CH₂OH

Activated lipid

Protein

Peripheral membrane protein

C : N-Acetyl-glucosamine
I : Inositol
M : Mannose
G : Galactose
E : Ethanolamine

Myristic acid

Palmitic acid

Farnesol

Lipid anchor:

Palmitic acid
Myristic acid
Farnesol
Geranylgeranol

Glycosylated phosphatidyl inositol (GPI)

GPI anchor

Protein Sorting

A. Protein targeting ○

The biosynthesis of all proteins begins on free ribosomes (above, center). Proteins that carry a signal peptide for the ER (**1**) follow the *secretory pathway* (right), whereas all other proteins follow the *cytoplasmic pathway* (left).

Secretory pathway: A **signal peptide** for the secretory pathway directs the growing polypeptide chain and the associated ribosome to attach to the rough ER. With the help of a specific transport protein, the chain then passes through the membrane and into the lumen of the rER (see p. 210). The ultimate destination of such proteins is determined either by the absence of any additional signal sequences (indicated by *) or by the presence of further signal peptides or signal regions.

When the growing peptide chain of a protein has a special "**stop-transfer sequence**" (**4**), the protein remains in the membrane of the ER and becomes an *integral membrane protein.* Many of these molecules are further transferred from the ER membrane to other membranes by vesicular transport (see p. 214).

For proteins, which enter the lumen of the rER, the next destination is usually the Golgi apparatus. Proteins destined to remain in the rER, e.g., enzymes involved in protein modification, return to the ER only after they have been in the Golgi complex. This also requires a special signal sequence (**2**). From the Golgi apparatus, the remaining proteins take the pathway leading to lysosomes (**3**) or to the cell membrane (*; integral membrane proteins or constitutive exocytosis) or they are released into the extracellular space (**9**; inducible exocytosis) by secretory vesicles (**8**).

Cytoplasmic pathway: Proteins lacking a signal peptide for the ER are synthesized in the cytoplasm on free ribosomes, and remain in that compartment unless they contain special signals for targeting to mitochondria (**5**), the nucleus (**6**), or peroxisomes (**7**).

B. Signals for protein sorting ○

Signal sequences are short stretches of amino acids located at the *N*-terminus of a protein, the *C*-terminus, or even in the middle of the peptide chain. It is not so much the amino acid sequence that constitutes these signals, but rather the physical properties of these sequences e.g., their polarity or positive or negative charge. **Signal regions** are three-dimensional domains on the surface of a protein made up of different fragments of the same peptide chain or by different chains altogether.

The illustration summarizes some of the known signal sequences and signal regions. The sequences are shown using the single letter code for amino acids (see p. 58). For example, the sequence KDEL-COO⁻ (**2**) corresponds to the *C*-terminal amino acid sequence -Lys-Asp-Glu-Leu. This sequence targets the protein to the ER.

There are two different ways in which structural signals can be read. Usually, they are recognized and bound by *receptors* (**A: R**) located on the membranes of organelles. The receptors then maneuver the proteins — often in an energy-dependent manner — into the respective organelles. This ensures that protein import is selective. However, signal sequences can also serve as *recognition sites for enzymes*, which modify the proteins, altering their properties and bringing about a decided change in their fate (see p. 210).

Once they have fulfilled their function, some signal sequences are removed by sequence-specific hydrolases. Here, these enzymes are illustrated as a pair of scissors. In the case of proteins with multiple signal sequences, these must be removed sequentially. In contrast, sequences that are to be read more than once (e.g., those for the nucleus) are not removed.

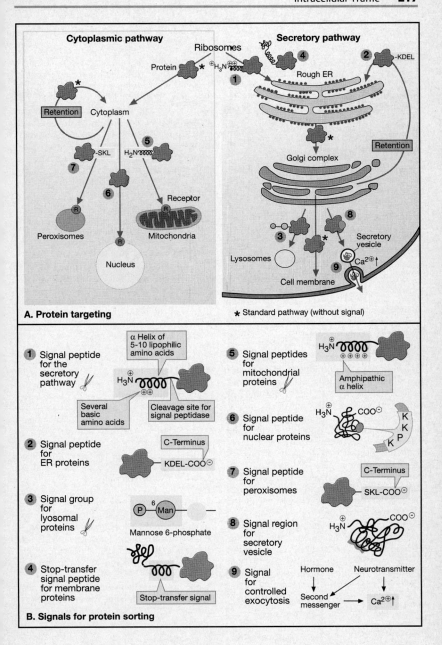

A. Protein targeting

★ Standard pathway (without signal)

Cytoplasmic pathway

Ribosomes
Protein ★ +H₃N₀₀₀₀
Retention   Cytoplasm
★
-SKL   H₂N₀₀₀₀
5
7
6
Peroxisomes   Receptor
Mitochondria
Nucleus

Secretory pathway

Ribosomes
4
1
2   -KDEL
Rough ER
★
Golgi complex   Retention
3
8
Lysosomes   Secretory vesicle
★   Ca²⁺↑
9
Cell membrane

B. Signals for protein sorting

1 Signal peptide for the secretory pathway
α Helix of 5–10 lipophilic amino acids
+H₃N₀₀₀₀
Several basic amino acids
Cleavage site for signal peptidase

2 Signal peptide for ER proteins
C-Terminus
KDEL-COO⁻

3 Signal group for lyosomal proteins
P⁶Man
Mannose 6-phosphate

4 Stop-transfer signal peptide for membrane proteins
Stop-transfer signal

5 Signal peptides for mitochondrial proteins
+H₃N₀₀₀₀
Amphipathic α helix

6 Signal peptide for nuclear proteins
+H₃N   COO⁻
K K P K

7 Signal peptide for peroxisomes
C-Terminus
SKL-COO⁻

8 Signal region for secretory vesicle
+H₃N   COO⁻

9 Signal for controlled exocytosis
Hormone   Neurotransmitter
Second messenger → Ca²⁺↑

The Genome

On this and the following pages, we discuss the processes responsible for the inheritance of genetic information (**replication**) and its conversion into functional biomolecules (**expression**). To begin with, we summarize these processes in an overview.

A. Expression and transmission of genetic information ●

Genetic information is encoded in the base sequence of DNA molecules (see p. 82). DNA sequences whose information is converted into functional molecules are referred to as **genes**. The sequence of one of the two strands of DNA (the coding strand) is converted into a complementary RNA sequence by a process known as **transcription** (see p. 222). One type of RNA formed by this process, so-called **mRNA**, serves as a template for the synthesis of proteins, while other RNA molecules, i.e., **rRNA**, **tRNA and snRNA**, take over special functions during gene expression. Eukaryotic genes are usually made up of multiple **exons** and **introns**. Only the exons code for proteins. The non-coding **introns**, which lie interspersed between the exons, are removed from the initially synthesized **hnRNA** after transcription. This takes place during **RNA maturation** (see p. 224), which also involves modification of the two ends of the mRNA. Mature mRNAs leave the nucleus and bind to **ribosomes** in the cytoplasm. There, the mRNA sequence is converted into a corresponding protein sequence by a process called **translation** (see p. 228–230). It involves tRNAs that recognize particular RNA sequences. These tRNAs are pre-loaded with the appropriate amino acid in a process called **amino acid activation**.

During cell division, the genetic information has to be passed on to the daughter cells. To achieve this, the total DNA in the cell is copied by **replication** (see p. 220). In this way, the daughter cells obtain identical sets of DNA molecules.

B. Chromatin ◑

In the nuclei of eukaryotes, the DNA is associated with proteins and RNA molecules (see p. 194). The resulting nucleoprotein complexes are referred to as **chromatin**. Details of chromatin are only visible with the light microscope when it condenses to form **chromosomes**, but not in the uncondensed form. During interphase, most of the chromatin is uncondensed. In this state, a morphological distinction can be made between tightly packed **heterochromatin** and more loosely packed **euchromatin**. Euchromatin is the site of active transcription.

The proteins found in chromatin can be classified as either *histone* or *non-histone* proteins. The histones are small, basic proteins that show a high degree of sequence similarity from one eukaryote to the next. They are directly associated with the DNA, and contribute to the structural organization of chromatin. In addition, their basic amino acids neutralize the negatively charged phosphate groups of the DNA, thus allowing it to pack more tightly in the nucleus. This is why it is possible for the 46 DNA molecules of the diploid human genome ($5 \cdot 10^9$ base pairs, with a total length of about 2 m) to be accommodated in a nucleus with a diameter of only 10 μm.

In chromatin histone molecules of the types **H2A, H2B, H3, and H4** (2 of each) associate with one another to form an octameric complex (**histone octamer**) that is wrapped in 146 base pairs of DNA constituting $1^3/_4$ turns. These particles have a diameter of 7 nm, and are referred to as **nucleosomes**. Another histone (**H1**) binds to DNA (linker DNA) not directly associated with the octamers. It supports the formation of fibrous superstructures with diameters of 30 nm, so-called **solenoids**. These, in turn, form **loops** that are about 200 nm long. The end of each loop is bound to a supporting structure made of proteins and referred to as the **nuclear scaffolding**.

The **non-histone proteins** are highly heterogeneous. They include *structural proteins, enzymes,* and *transcription factors* (see p. 224).

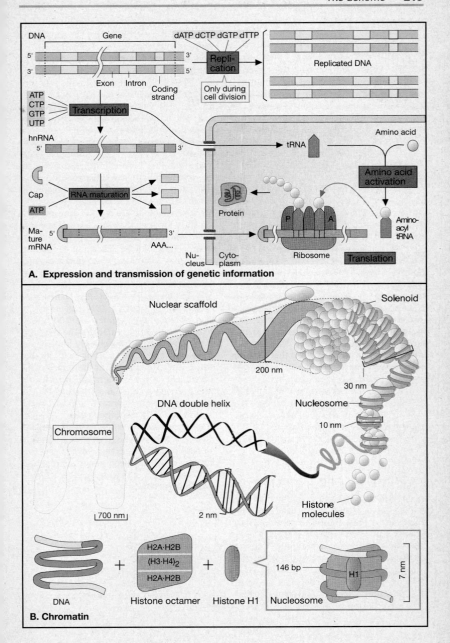

A. Expression and transmission of genetic information

B. Chromatin

Replication

A. Mechanism of DNA polymerases ○

Before genetic information can be passed on to daughter cells during cell division, it first has to be be copied. The most important enzymes catalyzing DNA **replication** are *DNA-dependent DNA polymerases*. They use one of the strands of double-helical DNA, the so-called template strand, to synthesize a second complementary strand (see p. 82). The result is a DNA double helix identical to the original one. The substrates of DNA polymerase are the four deoxyribonucleotide triphosphates *dATP, dGTP, dCTP, and dTTP.* During replication, the incoming nucleotide of the new strand binds to the current nucleotide of the template strand by specific base pairing. The α-phosphate residue of the newly bound nucleotide is then subject to nucleophilic attack by the $3'$- OH group of the previously incorporated nucleotide. This is followed by the elimination of diphosphate and the formation of a new *phosphoric acid diester bond*. These steps are repeated again and again as the DNA polymerase moves from one base to the next along the template strand. This mechanism means that the template DNA strand is read in the $3' \rightarrow 5'$ direction (see p. 78).

Most cells contain several DNA polymerases. Besides those that actually catalyze replication, there are also polymerases, which are involved in DNA repair processes (see p. 234) and (in eukaryotes) in the replication of mitochondrial DNA (see p. 200).

B. Replication in *E. coli* ◑

We now have a good understanding of replication in prokaryotes, while many aspects of eukaryotic replication remain unclear. However, it is true to say that the process is basically the same in most cells. Here we show a simplified scheme of replication in the bacterium *Escherichia coli.*

In bacteria, replication begins at the **origin of replication**, a specific point in the circular DNA molecule, and continues in both directions. As a result, two **replication forks** are

formed that move in opposite directions, i.e., both strands are replicated at the same time. In the figure, the original DNA is shown in blue, and newly synthesized strands in red and orange.

Each fork (**2**) has at least two molecules of *DNA polymerase III*, along with various helper proteins. The latter include *DNA topoisomerases*, which unwind the tightly coiled double-stranded DNA, *helicases*, which separate it into two strands, and *single-strand binding proteins*. Since the template strand is always read in the $3' \rightarrow 5'$ direction (see above), only one of the strands can undergo continuous replication (dark blue/red; **2**). The direction in which the other strand (light blue) is read is the opposite of the direction of movement of the fork. The result is short pieces of DNA called **Okazaki fragments**, after their discoverer (green/orange). Each fragment begins with a short starter sequence (**primer**) made of RNA (green), which is essential for the functioning of the DNA polymerase and is synthesized by a special *RNA polymerase* ("*primase*", not shown). *DNA polymerase III* then extends this primer by 1000–2000 deoxyribonucleotides (orange). The synthesis of the fragment is subsequently interrupted, and a new one begun starting with another RNA primer synthesized in the interim. The individual Okazaki fragments are initially not bound to one another and still have RNA at the 5' end (**3**). Some distance away from the replication fork, *DNA polymerase I* now begins to replace the RNA primer with DNA. The gaps still remaining are closed by a *DNA ligase.* In the DNA double helices thus formed, only one of the strands has been newly synthesized. This is referred to by saying that replication is **semiconservative**.

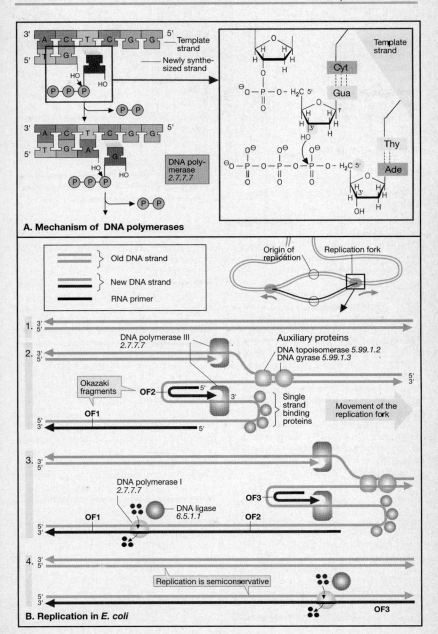

A. Mechanism of DNA polymerases

B. Replication in *E. coli*

Transcription

Before the information stored within DNA can be made use of, it must first be rewritten (*transcribed*) into RNA. During transcription, DNA only serves as a template, i.e., it is not altered in any way. Transcribable DNA segments, which code for defined products are called **genes**. It has been estimated, that the mammalian genome contains at least 50,000 individual genes, which together account for less than 20% of the DNA. The function of the remaining DNA sequences is not yet clear.

A. Transcription and maturation of RNA: overview ◑

Transcription is catalyzed by *DNA-dependent RNA polymerases*. They act in a similar way to DNA polymerases (see p. 220), except that they incorporate ribonucleotides instead of deoxyribonucleotides into the newly synthesized strand. Eukaryotic cells usually contain at least three different types of RNA polymerase. *RNA polymerase I* catalyzes the synthesis of an RNA with a sedimentation coefficient (see p. 186) of 45S, which serves as precursor for three different ribosomal RNAs (see p. 228). The products of *RNA polymerase II* activity are so-called **hnRNAs**, which serve as precursors for mRNA and snRNAs (see p. 228). Finally, *RNA polymerase III* transcribes genes that code for tRNAs, 5S rRNA, and certain snRNAs. These are the precursors that ultimately give rise to functional RNA molecules by a process called **RNA maturation** (see p. 224). Polymerases II and III are inhibited by α-*amanitin*, a poison produced by the mushroom *Amanita phalloides*.

B. Organization of the β-globin gene ○

As an example of the organization of a small eukaryotic gene, the illustration shows the gene that encodes the β chain of hemoglobin (146 amino acids, see p. 258). This gene consists of just over 2000 base pairs (bp). However, only 438 of them provide information for the amino acid sequence of the protein. Three coding regions (dark blue) are separated by two non-coding regions (**introns**, I1

and I2). At the 5′ end of the gene is the **promoter region** (pink) a stretch of about 200 bp that has regulatory functions (see p. 224). Transcription starts at the 3′ end of the promoter region and continues until the so-called **polyadenylation sequence** is reached (see below). The resulting primary transcript (**hnRNA**) of the β-globin gene is about 1600 bp long. During RNA maturation, the noncoding sequences are removed from the hnRNA, and both ends of the molecule undergo modification. The mature β-globin mRNA still contains about 40% of the hnRNA, as well as an additional 3′ terminal sequence made up of approximately 200 adenylate residues — a poly(A) sequence (see p. 224).

C. Process of transcription ◑

RNA polymerase II binds to the 3′ end of the promoter region. One sequence that is very important for this binding is the so-called **TATA box**, a short A- and T-rich region that varies somewhat from one gene to the next. A typical base sequence ("*consensus sequence*") is ..TATAAA.. Several proteins, the so-called *basal transcription factors*, are essential for the interaction of the polymerase with this region. Additional factors can either enhance or inhibit this process (transcriptional control, see p. 108). Following **initiation** (2), the RNA polymerase moves in the 3′→5′ direction of the template strand. The enzyme separates a short stretch of the DNA double helix into two single strands. In this region, complementary nucleoside triphosphates are bound by base pairing, and become linked to the growing RNA molecule (3). Shortly after the beginning of **elongation**, the 5′ end of the transcript is protected by a "**cap**" (see p. 224). Once the polyadenylation sequence has been reached (typical base sequence: ..AATAAA..), the transcript is released (4), and shortly thereafter, the polymerase stops transcribing and dissociates from the DNA.

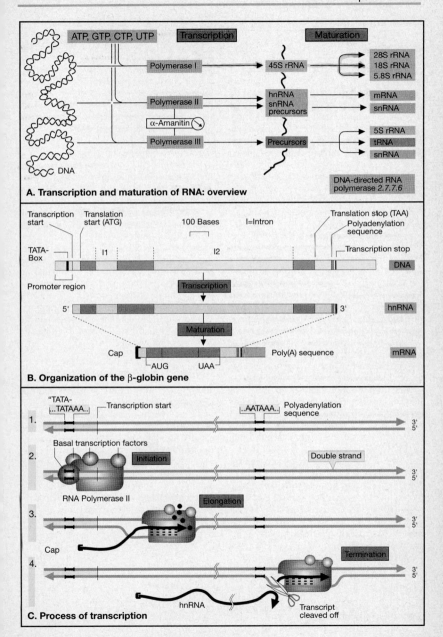

A. Transcription and maturation of RNA: overview

ATP, GTP, CTP, UTP

Transcription — Maturation

Polymerase I → 45S rRNA → 28S rRNA / 18S rRNA / 5.8S rRNA

Polymerase II → hnRNA → mRNA / snRNA precursors → snRNA

α-Amanitin

Polymerase III → Precursors → 5S rRNA / tRNA / snRNA

DNA

DNA-directed RNA polymerase 2.7.7.6

B. Organization of the β-globin gene

Transcription start
Translation start (ATG)
100 Bases
I=Intron
Translation stop (TAA)
Polyadenylation sequence
Transcription stop

TATA-Box
I1
I2
Promoter region

DNA

Transcription

5'   3'   hnRNA

Maturation

Cap   Poly(A) sequence   mRNA
AUG   UAA

C. Process of transcription

"TATA-
..TATAAA.."
Transcription start
..AATAAA..
Polyadenylation sequence

1.   3'/5'

Basal transcription factors
Initiation
Double strand

2.   3'/5'

RNA Polymerase II

Elongation

3.   3'/5'

Cap

Termination

4.   3'/5'

Cap
hnRNA
Transcript cleaved off

RNA Maturation

A. Transcriptional control ○

Every cell of an organism contains the complete genome of that organism, but only a small proportion of this information is ever used. Only those genes that code for structural proteins and enzymes of intermediary metabolism are constantly being transcribed. Besides these *"house-keeping genes"*, there are many other genes that are only active in certain cell types, under certain metabolic conditions, or during differentiation. Which genes are transcribed in a special situation, and which are not, is determined by a regulatory system that is only just beginning to be understood.

Transcriptional control involves an interaction between two different types of structures. Most genes contain within their *promoter region* (see p. 222) several short DNA segments (**control elements**) to which **transcription factors** can bind and thus regulate transcription.

Control elements only affect the gene with which they are associated, and are therefore, called *cis-acting elements*. The transcription factors, on the other hand, are proteins, i.e., the products of other, independent genes. They are therefore referred to as *trans-acting factors*.

Gene transcription requires not only RNA polymerase, but also other proteins called **basal transcription factors**. The most important of these are designated A-H. In eukaryotes, the first basal transcription factor to bind to the TATA box (see p. 222) is TF-IID. This is followed by binding of other basal transcription factors and RNA polymerase. Additional factors can affect the initiation of transcription by binding to neighboring control elements, so-called enhancers. From there, they interact with the basal complex to either activate or inhibit it. The activating factors include, for instance, the complexes of steroid hormones with their receptors (see p. 346).

B. Splicing ◑

Following transcription, the non-coding sequences of the hnRNA (derived from DNA introns, see p. 218) are removed. This process, **RNA splicing**, is catalyzed by complexes of proteins with RNA known as *"small ribonucleoprotein particles"* (**snRNP**). The intervening introns in the hnRNA have characteristic sequences at either end (**a**). The first step in splicing involves the OH group of an adenosine residue located within the intron, which attacks and cleaves the phosphate diester bond at the introns's 5′ end (**b**). At the same time, a new bond is formed within the intron, which gives it the shape of a lariat (**c**). In the second step, the terminal OH group of the 5′ exon attacks the bond at the 3′ end of the intron. As a result, the two exons become linked and the intron is released (**d**).

Five different snRNPs (**U1, U2, U4, U5, and U6**) are involved in the reaction. Each consists of multiple protein molecules and one molecule of **snRNA** (see p. 82). During splicing, complexes made up of hnRNA and snRNPs form **spliceosomes**. It is thought that the snRNAs in the spliceosome form base pairs with one another and with the hnRNPs, and thus fix and orient the reacting groups. Catalysis itself is due to the RNA portion of the spliceosome. Catalytic RNAs of this type are known as **"ribozymes"**.

C. 5′- and 3′-modification of mRNA ◑

Shortly after the onset of transcription in eukaryotes, the 5′ end of the growing RNA molecule becomes blocked by a structure referred to as a **"cap"**. In the case of mRNAs, the cap consists of a 7′-methyl-GTP residue. The cap protects the RNA from digestion by 5′ exonucleases. At the end of transcription, a **polyadenylate "tail"** made up of approximately 200 adenylate (AMP) residues is added to the 3′ end. Then, and only then, does the mature mRNA leave the nucleus.

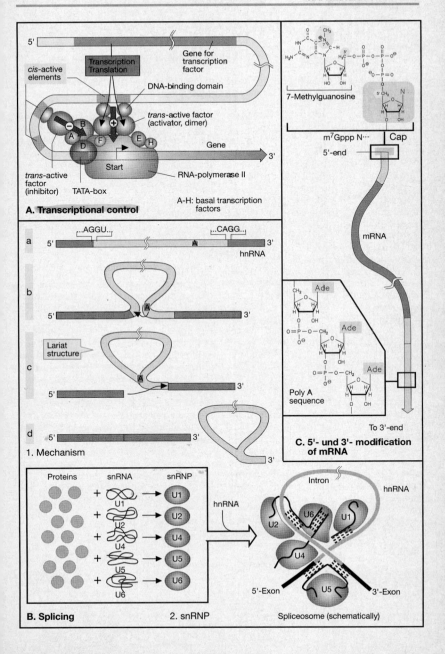

A. Transcriptional control

5'

Gene for transcription factor

Transcription Translation

DNA-binding domain

cis-active elements

trans-active factor (activator, dimer)

trans-active factor (inhibitor)

TATA-box

Start

Gene

RNA-polymerase II

3'

A–H: basal transcription factors

B. Splicing

a   5'   ..AGGU..   ..CAGG..   A   3'
                                              hnRNA

b   5'   A   3'

Lariat structure

c   5'   A   3'

d   5'   3'

1. Mechanism

Proteins   snRNA   snRNP

+   U1   →   U1
+   U2   →   U2
+   U4   →   U4
+   U5   →   U5
+   U6   →   U6

2. snRNP

hnRNA

Intron   hnRNA

U2   U6   U1

U4

5'-Exon   U5   3'-Exon

Spliceosome (schematically)

C. 5'- und 3'- modification of mRNA

7-Methylguanosine

m⁷Gppp N···   Cap

5'-end

mRNA

Poly A sequence

Ade

Ade

Ade

To 3'-end

Genetic Code, Amino Acid Activation

A. The genetic code ●

A large fraction of the genetic information contained in the DNA encodes the amino acid sequences of proteins. Expression of the genetic information involves the translation of a "text" written in the "nucleic acid language" into one written in the "protein language." This is the origin of the term **translation**, used to describe protein biosynthesis. The process of translation is governed by the **genetic code**.

There are 20 different proteinogenic amino acids, and therefore the nucleic acid language must contain at least 20 words (**codons**). However, there are only four letters (A, G, C, and U or T) in the nucleic acid alphabet, so that each word must be at least three letters long to obtain 20 different words. The codons do, in fact, consist of three sequential bases (**triplets**). In Fig. **1**, the standard DNA code (sequence of triplets in the non-coding strand) is represented as a wheel that is read from the inside to the outside. The DNA codons are identical to those of mRNA, except that, in mRNA, U replaces T. As an example of how the DNA code is read, Fig. **2** shows short stretches of normal and a mutated β-globin gene, along with the corresponding mRNA and amino acid sequences.

There are a potential $4^3 = 64$ codons in the DNA code, but only 20 amino acids. Thus, most amino acids have more than one synonymous codon. Besides the triplets that encode the amino acids, there are also three triplets that signal the end of translation (**stop codons**). A further special codon, the **start codon**, marks the beginning of translation. The genetic code shown in the illustration is virtually universal. Only mitochondria (see p. 200) and certain micro-organisms somewhat deviate from this pattern.

B. Transfer RNA (tRNA) ◑

The intermediaries between the nucleic acids and proteins are the **transfer RNAs (tRNAs)**, small RNA molecules of 70–90 bases. They recognize mRNA codons by base pairing. Attached to the 3' end of the tRNA molecule (sequence: ..CCA) is the amino acid encoded by the mRNA codon that is recognized by the tRNA. The actual process of translation is performed by enzymes that load the tRNAs with the appropriate amino acids (**C**).

The base sequence (**1**) and the tertiary structure (**2**) of Phe-tRNA from yeast are typical of all tRNAs (see p. 85). The molecule contains a high proportion of unusual and modified bases (dark green; ψ: pseudouridine, D: dihydrouridine; T: thymidine, otherwise only found in DNA; Y: a highly modified base called wybutosine; *: methylated nucleotides). The conformation of the molecule is stabilized by multiple base pairings, some of which do not follow the usual pattern (see p. 84).

C. Amino acid activation ◑

For each of the 20 amino acids, there is a corresponding *amino acid tRNA ligase* that attaches it to the appropriate tRNA. This is referred to as **amino acid activation**. It is a two-step process. First of all, the amino acid reacts with ATP to form an "energy-rich" mixed acid anhydride. Then the aminoacyl residue is transferred to the terminal 3'-OH group of the tRNA. The carboxyl group of the amino acid in the **aminoacyl-tRNA** is esterified with the ribose residue of the terminal adenosine.

The accuracy of translation depends, first and foremost, on the substrate specificity of the amino acid-tRNA ligases. Erroneously incorporated amino acid residues fail to be recognized by the ribosome. A "proof-reading mechanism" in the active center of the ligase ensures that incorrectly incorporated amino acid residues are immediately removed.

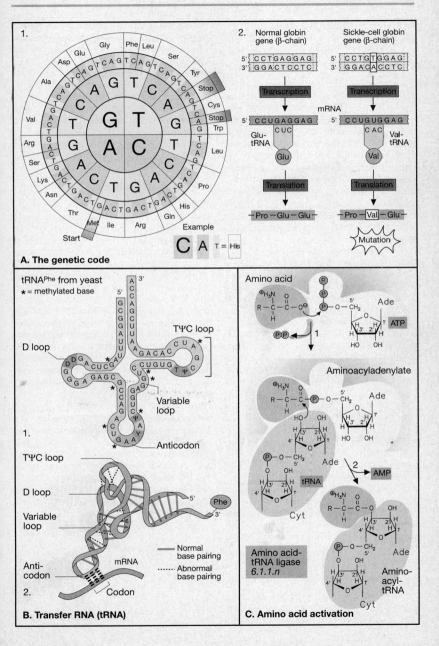

A. The genetic code

B. Transfer RNA (tRNA)

C. Amino acid activation

Ribosomes, Initiation of Translation

A. Structure of eukaryotic ribosomes ●

Protein biosynthesis, like amino acid activation (see p. 226), takes place in the cytoplasm. It is catalyzed by complex nucleoprotein particles called **ribosomes**. Ribosomes consist of two different subunits, each made up of **ribosomal RNA (rRNA)** and a large number of **proteins**. The ribosomes and their subunits are usually classified according to their sedimentation coefficients (see p. 186) rather than their molecular mass. Thus, the sedimentation coefficient for the complete eukaryotic ribosome is about 80 Svedberg units (80S) and that of its subunits 40S and 60S (S values are not additive).

The smaller 40S subunit consists of one molecule of 18S rRNA and 30–40 protein molecules. The larger 60S subunit contains three types of rRNA, with sedimentation coefficients of 5S, 5.8S, and 28S, as well as 40–50 proteins. When mRNA is present, the subunits associate with one another to form a complete ribosome with a mass about 650 times that of a hemoglobin molecule. Ribosomes have a diameter of 20–25 nm, and are therefore readily visible with the electron microscope. The structural organization of ribosomes is not yet entirely clear. However, it is known that the filamentous mRNA passes through a cleft near a characteristic "horn" on the small subunit located between the two subunits. The tRNAs also bind near this site. The illustration shows a tRNA molecule for size comparison.

Prokaryotic ribosomes are similar in structure to, but somewhat smaller than, eukaryotic ribosomes (sedimentation coefficient: 70S for the complete ribosome, 30S and 50S for the subunits). Mitochondrial and chloroplast ribosomes closely resemble prokaryotic ribosomes.

B. Polysome ◑

Cells where active protein synthesis is taking place often contain ribosomes lined up next to one another like a string of pearls, so-called

polysomes. This is due to the fact that a single mRNA molecule can be translated by several different ribosomes at the same time. The initial step in translation is the binding of the ribosome to the **start codon** (AUG, see p. 226) near the 5′ end of the mRNA (above). As translation proceeds, the ribosome moves toward the 3′ end until it encounters a **stop codon** (UAA, UAG, or UGA). Once such a codon is reached, the newly synthesized protein is released, and the ribosome dissociates into its two individual subunits.

C. Initiation of translation in *E. coli* ◑

Protein synthesis in prokaryotes is basically similar to that in eukaryotes, except that it is simpler and better understood. Here and on the following page, we use the bacterium *Escherichia coli* as an example to illustrate the details of translation.

The first phase in translation is **initiation**, which can be divided into several steps. In the first step, two proteins, the so-called **initiation factors IF-1** and **IF-3**, bind to the 30S subunit (**1**). A further factor, **IF-2**, then binds as a complex with GTP. This facilitates the association of the 30S subunit with the mRNA and the binding of a special tRNA at the start codon (**3**). In prokaryotes, this starter tRNA carries the substituted amino acid *N*-formyl-methionine (fMet). In eukaryotes, it carries an unsubstituted methionine. Finally, the 50 S subunit binds to the above complex (**4**). During steps 3 and 4, the initiation factors are released again, and the GTP bound to IF-2 is hydrolyzed to GDP and P_i. The **70S initiation complex** thus formed contains fMet-tRNA^{fMet} at a tRNA binding site that is referred to as the **peptidyl site (P)**. A second binding site, the **acceptor site (A)**, is not yet occupied during this phase of translation.

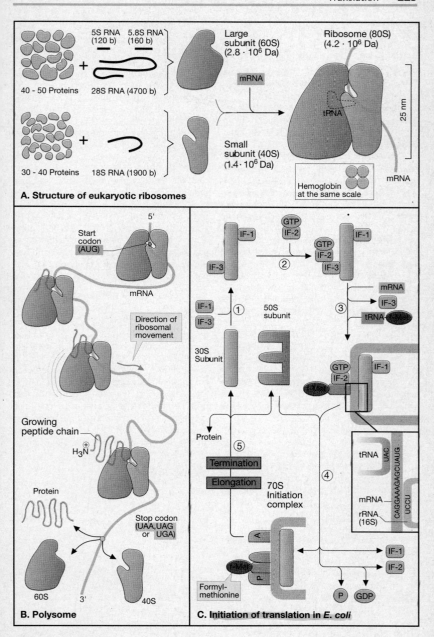

A. Structure of eukaryotic ribosomes

5S RNA (120 b)   5.8S RNA (160 b)
28S RNA (4700 b)
40 – 50 Proteins
Large subunit (60S) (2.8 · 10⁶ Da)
Ribosome (80S) (4.2 · 10⁶ Da)
mRNA
tRNA
25 nm
30 – 40 Proteins
18S RNA (1900 b)
Small subunit (40S) (1.4 · 10⁶ Da)
mRNA
Hemoglobin at the same scale

B. Polysome

5'
Start codon (AUG)
mRNA
Direction of ribosomal movement
Growing peptide chain
H_3N^+
Protein
Stop codon (UAA, UAG or UGA)
60S   3'   40S

C. Initiation of translation in *E. coli*

GTP IF-2
IF-1
IF-3
IF-1
②
GTP IF-2 IF-3
IF-1
mRNA
IF-3
③
tRNA f-Met
IF-1 IF-3
①
50S subunit
GTP IF-2 IF-1
f-Met
30S Subunit
Protein
⑤
Termination
Elongation
70S Initiation complex
④
tRNA   UAC
mRNA   CAGGAAAGAGCUAUG
       UCCU
rRNA (16S)
A
f-Met
P
IF-1
IF-2
P   GDP
Formyl-methionine

Elongation, Termination

A. Elongation and termination of protein biosynthesis in E. coli ◑

Following the initiation of translation (see p. 228), further amino acid residues are added to the growing peptide chain (**elongation**) until the ribosome reaches a stop codon on the mRNA and the process is interrupted (**termination**).

Elongation can be divided into three phases. First, the *peptidyl site (P)* of the ribosome becomes occupied by a tRNA that carries the complete peptide chain formed until then at its 3′ end (above left). Then, a second tRNA, charged with the appropriate amino acid (Val-tRNA^{Val} in the example shown), binds via its complementary anticodon (see p. 226) to the mRNA codon exposed at the *acceptor site* (in this case GUG).

The tRNA binds as a complex with a GTP-containing protein, the *elongation factor Tu* (EF-Tu) (**1a**). Dissociation of EF-Tu only occurs after the bound GTP has been hydrolyzed to GDP and phosphate (**1b**). Prior to GTP hydrolysis, the binding of the tRNA to the mRNA is relatively weak. Thus GTP hydrolysis by EF-Tu serves as a delaying factor that allows time to check whether the correct tRNA is bound. A further protein, the elongation factor Ts (EF-Ts), then catalyzes the exchange of GDP for GTP, and thus regenerates the EF-Tu·GTP complex.

It is during the next step (**2**) that the actual synthesis of the peptide bond occurs. The ribosomal *peptidyltransferase* (probably a ribozyme) catalyzes the transfer of the growing peptide chain from the tRNA at the P site to the amino group of the valine residue attached to the Val-tRNA at the A site. After the transfer, which does not require ATP, the free tRNA dissociates from the P site (**3**) and a further GTP-containing elongation factor (EF-G·GTP) binds to the ribosome. The hydrolysis of the GTP by this factior provides the energy for the **translocation** of the ribosome (**3**). During this process, the ribosome moves 3 bases along the mRNA in the direction of the 3′ end. As the tRNA carrying the polypeptide chain remains stationary relative to the

mRNA, it arrives at the P site of the ribosome, while the subsequent mRNA codon (in this case GAG) appears at the A site. The ribosome is then ready for the next elongation cycle (**4**).

When a stop codon (UAA, UAG, or UGA) appears at the A site, **termination** begins (**5**). There are no complementary tRNAs for the stop codons. Instead, two *releasing factors* bind. One of these factors (RF-1) catalyzes the hydrolytic cleavage of the ester bond between the tRNA and the C-terminus of the peptide, thus releasing the protein. The GTP-containing factor RF-3 supplies the energy for the dissociation of the complex into its component parts (**6**).

The energy requirement for protein synthesis is high. Three energy-rich phosphoric anhydride bonds are hydrolyzed per amino acid. In addition, amino acid activation uses one ATP per amino acid (see p. 226), and two GTPs are consumed per elongation cycle. Finally, initiation and termination each require one GTP.

Further information

In eukaryotic cells, the number of initiation factors is higher, and thus the structure of the initiation complex is more intricate, than in prokaryotes. The cap structure at the 5′ end of eukaryotic mRNA (see p. 224) plays a central role in initiation. By contrast, elongation and termination are rather similar in both types of organisms. The structural differences between prokaryotic and eukaryotic ribosomes are the basis for the selective action of many antibiotics (see p. 232).

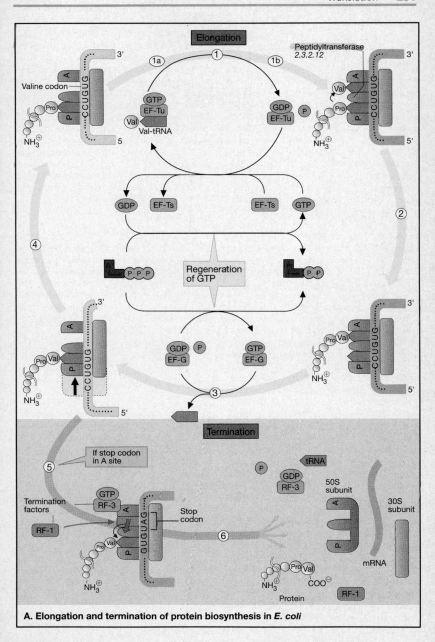

A. Elongation and termination of protein biosynthesis in *E. coli*

Antibiotics

A. Antibiotics: overview ○

Antibiotics can be defined as substances, which, at low concentrations, inhibit the reproduction of bacteria and fungi. Nowadays, it is impossible to think about the treatment of infectious diseases without considering antibiotics. Many antibiotics have their effect on their target cells by interfering with either the replication or transcription of DNA or the synthesis of proteins. Substances that only restrict the reproduction of bacteria are termed *bacteriostatic* (in the case of fungi, fungistatic), whereas those that actually kill bacteria are called *bactericidal* (or fungicidal, for fungi). Most antibiotics are produced by micro-organisms. Especially important are bacteria of the genus *Streptomyces* and certain fungi. However, there are also synthetic antibacterial substances, such as the sulfonamides and the gyrase inhibitors.

The illustration shows some of the therapeutically important antibiotics and their site of action in bacterial metabolism. So-called **intercalators**, such as rifamycin and actinomycin D (**1**) insert themselves into the DNA double helix and thereby interfere with replication and transcription (see **B**). Since the DNA is the same in all cells, intercalators are also toxic to eukaryotes, and are therefore only used as cytostatic agents (see p. 174) in particular cases. Synthetic inhibitors of DNA-topoisomerase (see p. 218), the so-called **gyrase inhibitors** (**2**), inhibit replication and thus bacterial reproduction. A large group of antibiotics have their effect on the ribosomes, i.e., they are **translation inhibitors** (**3**). These include the *tetracyclines*, which inhibit a wide spectrum of bacteria by interfering with tRNA binding to the 30S subunit. The *aminoglycosides*, the best-known being streptomycin, affect all phases of translation. *Erythromycin* impairs the normal function of the large ribosomal subunit, whereas *chloramphenicol* inhibits the peptidyltransferase. Chloramphenicol is one of the few biomolecules that contain a nitro group. Finally, *puromycin* mimics an aminoacyl-tRNA, and thereby causes premature termination of elongation.

Other frequently used agents are the β-lactam antibiotics (**4**). The members of this group contain a reactive β-lactam ring. The *penicillins* and *cephalosporins* are synthesized by fungi. They are mostly used against Gram-negative bacteria, where they inhibit cell wall synthesis (see **C**). The first fully synthetic antibiotics (1935) were the **sulfonamides** (**5**). These are analogues of p-aminobenzoic acid, and as such have their effect on the synthesis of *folic acid*, an essential precursor of the coenzyme THF (see p. 102). The **transport antibiotics** (6) act like ion channels. Their incorporation into the plasma membrane leads to a loss of ions that damages the bacterial cells.

B. Intercalators ○

The mode of action of intercalators is illustrated using the **daunomycin-DNA complex** as an example. Two daunomycin molecules (red) lying close together are inserted into the DNA double helix (blue). The ring system of the antibiotic inserts itself between G/C base pairs (below), while the sugar moiety occupies the minor groove (above). This leads to a localized alteration in the structure of the DNA that inhibits replication and transcription.

C. Penicillin as "suicide substrate" ○

The site of action of the β-lactam antibiotics is the enzyme *muramoylpentapeptide carboxypeptidase (3.4.17.8),* which is essential for the cross-linking of bacterial cell walls. The antibiotic has a structure similar to that of the substrate of this enzyme (the dipeptide D-Ala-D-Ala). It is first bound reversibly in a way that brings the β-lactam ring into close proximity with an essential serine residue of the enzyme. Nucleophilic substitition then results in the formation of a stable covalent bond between the enzyme and the inhibitor, blocking the active site. The loss of activity of the enzyme leads to the formation of unstable cell walls and eventually to the death of dividing bacteria.

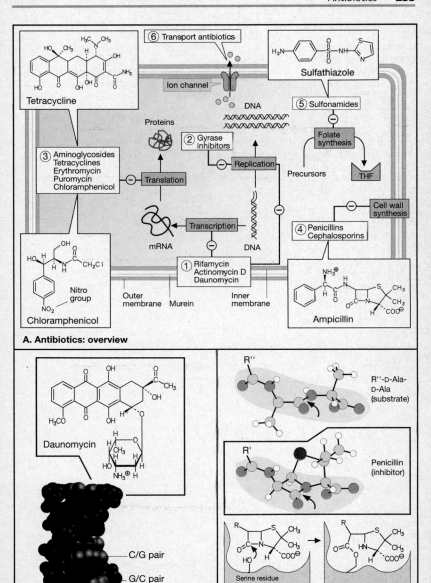

A. Antibiotics: overview

Tetracycline

Chloramphenicol
Nitro group

③ Aminoglycosides
Tetracyclines
Erythromycin
Puromycin
Chloramphenicol

Proteins

Translation

Transcription

mRNA

② Gyrase inhibitors

Replication

DNA

DNA

① Rifamycin
Actinomycin D
Daunomycin

⑥ Transport antibiotics

Ion channel

Sulfathiazole

⑤ Sulfonamides

Folate synthesis

Precursors

THF

Cell wall synthesis

④ Penicillins
Cephalosporins

Ampicillin

Outer membrane
Murein
Inner membrane

B. Intercalators

Daunomycin

Daunomycin-DNA complex

C/G pair

G/C pair

C. Penicillin as "suicide substrate"

R''-D-Ala-D-Ala (substrate)

Penicillin (inhibitor)

Serine residue

Enzyme-inhibitor complex

Covalent acyl enzyme

Mutation and Repair

Genetic information is set down in the base sequence of the DNA, and this is why changes in the DNA bases or their sequence are *mutagenic*. Many mutagenic substances also damage growth regulation in cells, and are therefore *carcinogenic*. Gene alterations (**mutations**) are an important factor in biological evolution. On the other hand, when mutation rates are too high, they can threaten the survival of individual organisms or entire species. This is why cells possess **repair mechanisms** that correct most of the DNA alterations caused by mutations.

A. Mutagenic agents ◑

Mutations can arise as a result of physical or chemical damage, or they can be due to errors during DNA replication and recombination.

The most important physical mutagen is **ionizing radiation**. It results in the production of *free radicals* (molecules with unpaired electrons) in the cell, which are extremely reactive and can damage DNA. Short-wavelength *ultraviolet light* (UV) also has a mutagenic effect, especially on skin cells. The most common chemical change due to UV exposure is the formation of **thymine dimers** in which two neighboring thymine bases become covalently linked to one another (**1**). This results in errors when the DNA is read during replication and transcription.

Since there are so many **chemical mutagens**, only a few examples can be shown here. *Nitrous acid* (HNO_2) and *hydroxylamine* (NH_2OH) both deaminate bases, e.g., they convert cytosine to uracil and adenine to inosine. *Alkylating compounds* carry reactive groups that can form covalent bonds with DNA bases. *Methyl nitrosamines* (**2**) release the reactive methyl cation (CH_3^+), which methylates OH and NH_2 groups in DNA. The well-known carcinogen, *benzo(a)pyrene*, is an aromatic hydrocarbon, which is harmless as such, but is converted to a mutagenic derivative in the liver (see p. 292). A reactive epoxide that arises as a result of hydroxylations of one of the rings (**3**) reacts with the amino group of guanine and other targets in DNA.

B. Effects ◑

Nitrous acid causes **point mutations (1)**. For example, C is converted to U, which, in the next round of replication, pairs with A instead of G. The alteration thus becomes fixed. Mutations involving the incorporation or removal of a number of nucleotides that is not a multiple of three lead to reading errors in whole segments of DNA, because they move the reading frame (**frame shift mutations, 2**). Starting at the additionally inserted C, the resulting mRNA will be interpreted in a different way during translation, giving rise to an entirely new protein sequence.

C. Repair mechanisms ○

An important mechanism for the removal of DNA damage is **excision repair (1)**. In this process, a *nuclease* removes a complete segment of DNA on both sides of the error site (**b**). The segment is then replaced by a *DNA polymerase*, using the the opposite strand as a template (**d**). Finally, *DNA ligase* closes the gaps (**e**). Thymine dimers can be removed by **photoreactivation (2)**. A *photolyase* binds at the site of the defect (b) and, upon illumination, cleaves the dimer to yield two single bases (**c**). A third mechanism is **recombination repair (3**, shown in simplified form). In this process, the region containing the defect is omitted during replication (**b**). The resulting gap is closed by shifting the corresponding segment from the correctly replicated strand (**c**). The new gap thus formed is then filled by polymerases and ligases. Finally, the original defect is corrected by excision, as shown in **1**.

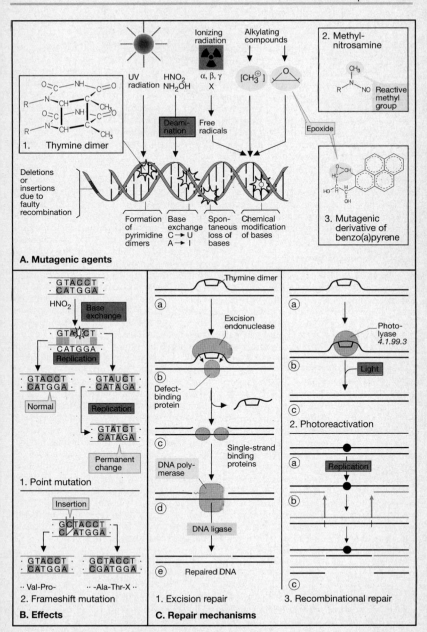

A. Mutagenic agents

Ionizing radiation
Alkylating compounds

2. Methyl-nitrosamine

UV radiation
HNO₂ NH₂OH

α, β, γ
X

[CH₃⁺]
O

Reactive methyl group

1. Thymine dimer

Deamination
Free radicals

Epoxide

Deletions or insertions due to faulty recombination

Formation of pyrimidine dimers

Base exchange
C → U
A → I

Spontaneous loss of bases

Chemical modification of bases

3. Mutagenic derivative of benzo(a)pyrene

B. Effects

GTACCT
CATGGA

HNO₂   Base exchange

GTAUCT
CATGGA

Replication

GTACCT
CATGGA

Normal

GTAUCT
CATAGA

Replication

GTATCT
CATAGA

Permanent change

1. Point mutation

Insertion

GCTACCT
CL ATGGA

GTACCT
CATGGA

GCTACCT
CGATGGA

·· Val-Pro-   ·· Ala-Thr-X ··

2. Frameshift mutation

C. Repair mechanisms

Thymine dimer

ⓐ

Excision endonuclease

ⓑ Defect-binding protein

ⓒ
Single-strand binding proteins

DNA polymerase

ⓓ

DNA ligase

ⓔ   Repaired DNA

1. Excision repair

Thymine dimer

ⓐ
Photolyase 4.1.99.3

ⓑ   Light

ⓒ

2. Photoreactivation

ⓐ   Replication

ⓑ

ⓒ

3. Recombinational repair

DNA Cloning

The main reason for the rise of molecular genetics since the 1970s has been the development and refinement of methods for the analysis and manipulation of DNA. **"Gene technology,"** as it is known today, has many practical applications. For example, it has developed new methods for the diagnosis and therapy of diseases, and it has made feasible targeted changes to particular characteristics of organisms. It is impossible to totally exclude biological risks when dealing with such applications, and therefore it is imperative to take a responsible approach when using gene technology. On this and the following pages, we present a brief overview of procedures used in gene manipulation.

A. Restriction endonucleases ◑

Most procedures in gene technology involve the isolation of defined DNA fragments, followed by joining them to other DNA fragments to produce new combinations. To do this, it is necessary to use enzymes that cut and rejoin DNA in a specific fashion. The most important of these enzymes are the *restriction endonucleases*, a group of bacterial enzymes that catalyze sequence-specific cleavage of double-stranded DNA. Numerous restriction enzymes are known. Their names are usually abbreviated forms of the names of their organism of origin. In the example here, we use *Eco*RI, a nuclease isolated from *Escherichia coli*. *Eco*RI, like most other restriction endonucleases, cleaves DNA at the site of a palindromic sequence, i.e., a short segment of DNA in which both strands have the same sequence when read in the $5' \rightarrow 3'$ direction. The sequence recognized by *Eco*RI is 5'-GAATTC-3'. The enzyme, a homodimer, then cleaves both strands between G and A. This results in the formation of complementary *overhanging ends* (AATT), which are held together by base pairing. These can be readily separated, however, e.g., by mild heating. When the fragments are cooled, the overhanging ends re-hybridize in the correct orientation. The cleavage sites can then be sealed again by a *DNA ligase*.

B. DNA cloning ◑

In most cells, the absolute amount of any given DNA segment, e.g., a gene, is very low. In order to experiment with such DNA sequences, it is therefore necessary first to produce many identical copies (**"clones"**) of them. The classical procedure for the cloning of DNA in the laboratory takes advantage of the ability of bacteria to take up and replicate short, circular DNA fragments known as **plasmids**. The DNA fragment to be cloned is first cleaved from the DNA of origin with the help of *restriction endonucleases* (see above). For the sake of simplicity, we show cleavage by *Eco*RI alone. In practice, two different enzymes are usually employed. The vehicle (*"vector"*) for the DNA to be cloned is a plasmid with only one *Eco*RI cleavage site. The circular plasmid is first linearized by cleavage with *Eco*RI, and then mixed with the isolated DNA fragments. Since the fragment and the vector have the same overhanging ends, some of them will hybridize with one another, rather than rehybridizing with themselves. *DNA ligase* is then used to reseal the site of cleavage. The result is the integration of the DNA to be cloned into the vector and the formation of a new (*"recombinant"*) plasmid. Pretreatment of a large number of host cells in an appropriate manner results in the uptake, by at least some of them, of the recombinant plasmid (**transformation**). The cells then replicate the plasmid along with their own genome. In order to ensure that only the bacteria containing the plasmid replicate, plasmids are used that confer *resistance* to a particular antibiotic on the host. When the bacteria are incubated in the presence of this antibiotic, only the cells containing the plasmid will replicate. The DNA can then be isolated from these cells, and multiple copies of the cloned DNA fragment can be obtained by cleavage with *Eco*RI.

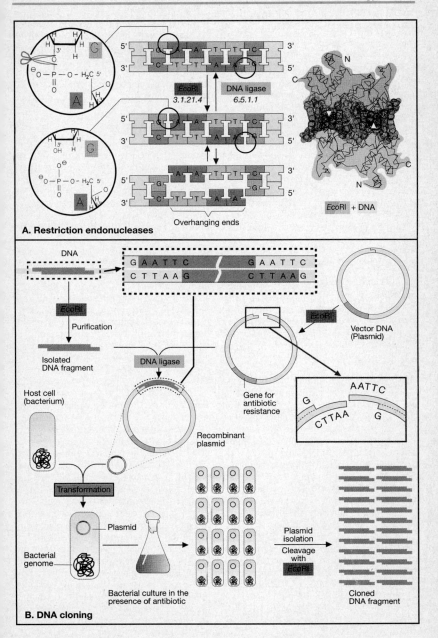

A. Restriction endonucleases

*Eco*RI
3.1.21.4

DNA ligase
6.5.1.1

Overhanging ends

*Eco*RI + DNA

B. DNA cloning

DNA

*Eco*RI
Purification

Isolated
DNA fragment

Host cell
(bacterium)

DNA ligase

Vector DNA
(Plasmid)

*Eco*RI

Gene for
antibiotic
resistance

Recombinant
plasmid

Transformation

Plasmid

Bacterial
genome

Bacterial culture in the
presence of antibiotic

Plasmid
isolation
Cleavage
with
*Eco*RI

Cloned
DNA fragment

DNA Libraries, DNA Sequencing

A. Gene libraries ○

In gene technology, it is often necessary to isolate a DNA fragment that has not yet been fully characterized. This may occur, for example, when the aim is to determine the complete nucleotide sequence of that fragment.

One method for the isolation of a particular DNA sequence is based on the construction of a so-called DNA library. A DNA library consists of a large number of vector DNA molecules containing different fragments of foreign DNA. For example, it is possible to take all of the mRNA molecules in a cell and convert them into DNA fragments (these are called copy-DNA or cDNA), which are then randomly incorporated into vector molecules. A genomic DNA library can be made by isolating DNA from a cell, cleaving it into small fragments with restriction endonucleases (see p. 236), and then incorporating these into vector DNA. The best vectors for DNA libraries are derived from *bacteriophages* (abbreviated: phages). Phages are viruses that only infect bacteria, where they are replicated together with the bacterial genome (see p. 362).

The first step in isolating the DNA fragment of interest from the library is to dilute a small portion of the library (10^5 to 10^6 phages), mix the dilution with host bacteria, and then plate out the mixture onto nutrient medium. The bacteria subsequently grow and form a continuous, turbid layer of cells. Initially, the bacteria infected with the phage grow more slowly than uninfected cells. This results in the formation of clear circular zones or *plaques.* The cells in the plaque contain the progeny of one and only one of the phages from the original library. The next step is to make an imprint of the plate on a synthetic filter. The filter is then heated to make the phage DNA adhere to it. Incubation of the filter with a DNA fragment that hybridizes to the piece of DNA of interest (a *DNA probe*) will result in binding of the probe at sites on the imprint at which the DNA being sought is located. Binding of the DNA probe can be detected by pre-labeling it with radioactivity, or by another method. The final step is to go back to the original plate, pick out the positive plaques, and replicate the phage within them. Large quantities of the desired DNA fragment can then be obtained by restriction digestion.

B. Sequencing of DNA ○

The current method of choice for the determination of the nucleotide sequence of DNA molecules is the **chain termination method**. First of all, the DNA fragment (**a**) is cloned into M13, a phage that allows isolation of the coding strand of the DNA fragment as single-stranded DNA. This is then allowed to hybridize to a short piece of DNA called a *primer*, which binds to the 3′ end of the single-stranded DNA fragment (**b**). Deoxyribonucleoside triphosphates (dNTP, i.e., dGTP, dATP, dTTP, dCTP) are then added, and a *DNA polymerase* is allowed to catalyze the synthesis of a second strand, using the coding strand as a template (**c**). In addition to dNTPs, small quantities of *dideoxynucleoside triphosphates* (ddNTP) are also added. Incorporation of a ddNTP leads to the *termination of second-strand synthesis.* This can occur whenever the corresponding dNTP ought to be incorporated. The illustration depicts the principle of the method using ddGTP as an example. In this case, fragments are produced that include the primer plus 3, 6, 8, 13, or 18 other nucleotides. To obtain a DNA sequence, *four separate reactions*, each with a different ddNTP, have to be performed (**c**). The products are then loaded side by side onto a synthetic gel slab, and separated by *electrophoresis* (**d**, see p. 252). The distance moved is inversely related to the length of the chain. Once the DNA fragments have been *visualized* (**e**), the nucleotide sequence can be read directly from the gel simply by noting the order in which the fragments appear in the individual lanes, starting from the bottom and proceeding toward the top of the gel (**f**). The right-hand side of the illustration shows part of an actual sequencing gel. Under optimal conditions, it is possible to determine the sequence of hundreds of nucleotides in a single experiment.

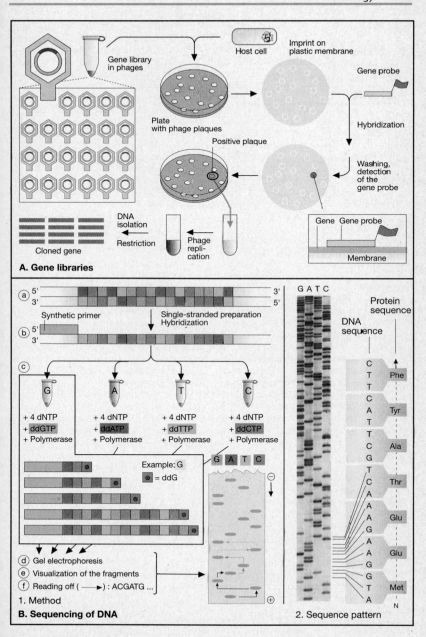

A. Gene libraries

Gene library in phages

Host cell

Imprint on plastic membrane

Plate with phage plaques

Gene probe

Hybridization

Positive plaque

Washing, detection of the gene probe

DNA isolation

Restriction

Phage replication

Cloned gene

Gene   Gene probe

Membrane

B. Sequencing of DNA

(a) 5' — 3'
   3' — 5'

Synthetic primer

Single-stranded preparation
Hybridization

(b) 5' 3'

(c) G   A   T   C

+ 4 dNTP +ddGTP + Polymerase

+ 4 dNTP +ddATP + Polymerase

+ 4 dNTP +ddTTP + Polymerase

+ 4 dNTP +ddCTP + Polymerase

Example: G
= ddG

G   A   T   C

(d) Gel electrophoresis
(e) Visualization of the fragments
(f) Reading off (—▶): ACGATG ...

1. Method

G A T C

DNA sequence

Protein sequence

C
T   Phe
T
C
A   Tyr
T
T
C   Ala
G
T
C   Thr
A
A
A   Glu
G
A
A   Glu
G
G
T   Met
A
N

2. Sequence pattern

PCR, RFLP

Here we describe applications of gene technology that are of particular relevance to medicine.

A. Polymerase chain reaction (PCR) ◐

The polymerase chain reaction allows the **amplification** of a chosen DNA fragment without the need for restriction enzymes, vectors, or host cells (see p. 236). All that is needed is 1. two oligonucleotides (*primers*), one binding to each strand on either side of the DNA fragment to be amplified; 2. sufficient quantities of the four deoxyribonucleoside triphosphates; and 3. a special heat-tolerant *DNA polymerase.*

In the first step, the double-stranded DNA is heated to denature into single-stranded DNA (**a**). The mixture is then cooled to allow hybridization of the primers (**b**). Complementary DNA strands are synthesized in both directions, starting from the primers (**c**). This process (cycle 1) is *repeated 20–30 times* with the same reaction mixture (cycle 2 and subsequent cycles). After the third cycle, double-stranded DNA fragments equal in length to the distance between the two primers are beginning to accumulate. Their relative abundance doubles after each cycle, until almost all of the newly synthesized fragments correspond to the original fragment delimited by the primers. The cyclical heating and cooling is controlled by a computerized thermostat.

B. Analysis of RFLP ○

Restriction endonucleases cleave DNA at specific sites, which are usually palindromic sequences (see p. 236). When a mutation alters one of the bases making up this sequence, the restriction enzyme no longer cleaves the DNA at this site. Mutations can also give rise to new restriction sites. As a result, DNA obtained from different individuals differing in their genetic makeup may produce restriction fragments of different lengths. This is referred to as *restriction fragment length polymorphism*, or **RFLP.** Fig. **1** shows how RFLPs are detected. First of all, the

DNA fragment containing the sequence of interest is *amplified* by PCR (**A**). The amplified fragment is then digested with a suitable *restriction enzyme.* The fragments are subsequently separated by *gel electrophoresis* and detected using a specific *gene probe* (see p. 238). Fig. **2** illustrates how the method can be applied in forensic science. The restriction pattern of a DNA sample isolated from tissue found at the scene of a crime clearly corresponds to the pattern obtained for suspect 2, but not for suspect 1. Even very small quantities of DNA-containing material are sufficient for the production of such a "*genetic fingerprint.*" A second important application of RFLP analysis is the identification of genetic mutations leading to hereditary disorders.

C. Overexpression of proteins ○

The treatment of diseases sometimes requires proteins, e.g., proteohormones, which are in very low abundance in animals, and therefore are extremely difficult to isolate in sufficient quantities. It is now possible to obtain large amounts of such proteins by *overexpression* in bacterial or eukaryotic host cells. To do this, it is necessary first to isolate the appropriate *gene* and incorporate it into an **expression plasmid**. Besides the gene of interest, the plasmid must also contain DNA sequences that direct the host cell to replicate and transcribe the plasmid DNA. Once the appropriate host cells have been *transformed* and *replication* has taken place, *transcription* of the gene is initiated by the process of *induction.* The desired protein is then produced in large quantities by *translation* of the mRNA formed.

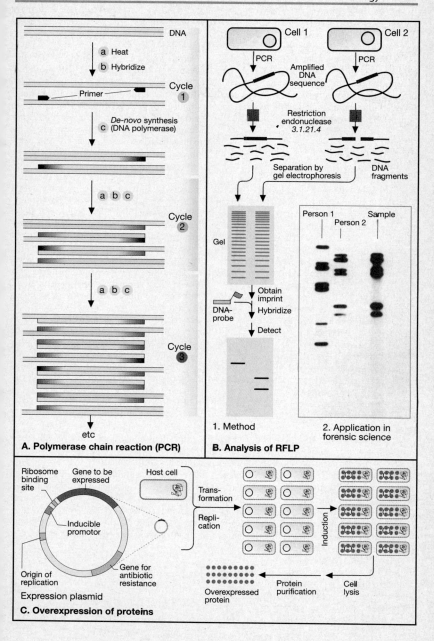

A. Polymerase chain reaction (PCR)

DNA

a Heat
b Hybridize

Primer — Cycle 1

De-novo synthesis
c (DNA polymerase)

a b c — Cycle 2

a b c — Cycle 3

etc

B. Analysis of RFLP

Cell 1   Cell 2

PCR   PCR

Amplified DNA sequence

Restriction endonuclease 3.1.21.4

Separation by gel electrophoresis   DNA fragments

Gel

Obtain imprint
DNA-probe   Hybridize
Detect

1. Method

Person 1   Person 2   Sample

2. Application in forensic science

C. Overexpression of proteins

Ribosome binding site   Gene to be expressed   Host cell

Inducible promotor

Transformation
Replication

Induction

Origin of replication   Gene for antibiotic resistance

Expression plasmid

Overexpressed protein   Protein purification   Cell lysis

Digestion: Overview

Most nutrients in the diet (see p. 326) have first to be broken down to smaller molecules before they can be taken up and utilized by the body. The mechanical and enzymatic breakdown of foods and the absorption of the products of these processes are referred to collectively as **digestion**.

A. Hydrolysis and resorption of dietary constituents ●

The first step in digestion is the mechanical fragmentation of foods by chewing. This is followed by enzymatic degradation, catalyzed by *digestive enzymes* found either in the digestive secretions or on the surface of the intestinal epithelium (see p. 244). Almost all digestive enzymes are *hydrolases* (class 3 enzymes, see p. 68). Hydrolases catalyze degradative reactions involving the incorporation of water into the reaction products.

Dietary **proteins** are first denatured by hydrochloric acid in the stomach (see p. 246), making them more susceptible to attack by *endopeptidases* present in gastric and pancreatic secretions. The peptides thus released are subsequently degraded to **amino acids** by *exopeptidases*, also found in the intestine. The amino acids are then absorbed by intestinal mucosa in a cotransport with Na^+ ions (*secondary active transport,* see p. 248). There are several groups of amino acids that share common transport systems.

Dietary **carbohydrates**, such as starch and glycogen, are degraded in a stepwise fashion by various *glycosidases* secreted by the pancreas. The products are oligosaccharides, which are further degraded by intestinal glycosidases to yield **monosaccharides**. The uptake of glucose and galactose into the cells of the intestinal epithelium is also mediated by secondary-active cotransport with Na^+ ions. However, passive transport systems for monosaccharides also exist.

Nucleic acids are degraded by *nucleases* produced in the pancreas and small intestine. The products are **nucleobases** (purine and pyrimidine derivatives), **pentoses** (riboses and deoxyriboses), **phosphate,** and **nu-**cleosides (pentose + base). These cleavage products are resorbed by the intestinal mucosa of the jejunum.

Lipids pose a special problem for the digestive process, because of their insolubility in water. The digestion of lipids begins with their *emulsification* by bile salts and bile *phospholipids* (see p. 244). The hydrolysis of neutral fats is catalyzed in the small intestine by pancreatic *lipase* with the assistance of an auxiliary protein known as *colipase*. This reaction, and the hydrolysis of other lipids, occur at the water/lipid interface on the surface of *micelles* formed by emulsification. The main products of lipid hydrolysis are **fatty acids, 2-monoacylglycerols, glycerol, cholesterol,** and **phosphate**. Once these have been taken up by cells of the intestinal epithelium, fatty acids, glycerol and the 2-monoacylglycerols are resynthesized into fats and deposited in the lymph system (see p. 246). Lipids in milk are already present in emulsified form, and therefore they are much more easily digested. Their degradation yields mostly short chain fatty acids.

Inorganic dietary **constituents**, such as water, salts as well as the **vitamins** are absorbed directly by the intestine. High molecular weight, indigestable constituents, e.g., **dietary fiber** from plant cell walls (predominantly cellulose and lignin) pass through the intestine unchanged. Before being excreted with the feces, dietary fiber promotes digestion by binding of water and by stimulating peristaltic movement of the intestine.

Most digestion products that are absorbed by the epithelial cells of the jejunum and ileum are transported to the liver via the **portal vessel.** Exceptions to this rule are fats, cholesterol, and lipid-soluble vitamins (see p. 330). These are released by the epithelial cells into the **lymph system**. From there, they enter the blood via the thoracic duct in the form of *chylomicrons* (see p. 254).

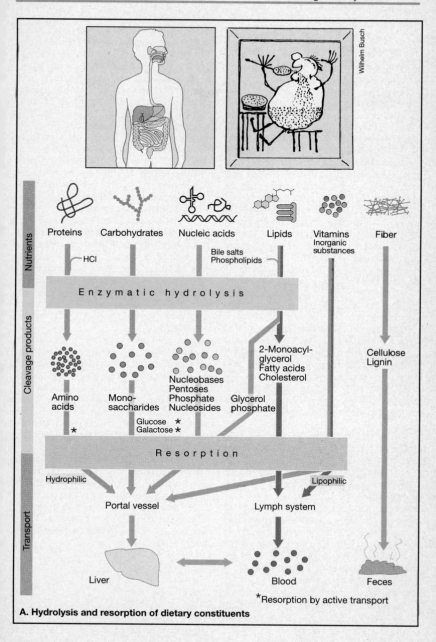

Wilhelm Busch

Nutrients

Proteins   Carbohydrates   Nucleic acids   Lipids   Vitamins Inorganic substances   Fiber

— HCl   Bile salts Phospholipids —

E n z y m a t i c   h y d r o l y s i s

Cleavage products

Amino acids   Mono-saccharides   Nucleobases Pentoses Phosphate Nucleosides   2-Monoacyl-glycerol Fatty acids Cholesterol   Glycerol phosphate   Cellulose Lignin

*   Glucose *
Galactose *

R e s o r p t i o n

Transport

Hydrophilic   Lipophilic

Portal vessel   Lymph system

Liver   Blood   Feces

*Resorption by active transport

A. Hydrolysis and resorption of dietary constituents

Digestive Secretions

A. Digestive juices ◑

Saliva. The salivary glands produce secretions that contain water, inorganic salts, and *glycoproteins* (mucins) that act as lubricants. In addition, *antibodies* and *enzymes* are also present. One of the salivary enzymes, α-*amylase*, catalyzes the partial hydrolysis of starch and glycogen whilst they are still in the mouth.

Gastric juice. In the stomach, the chyme is mixed with a strongly acidic secretion of the gastric mucosa containing *hydrochloric acid* (see p. 246), *mucins*, *inorganic salts*, and the precursors of enzymes, so-called *proenzymes* ("zymogens"). The most important digestive enzymes of the stomach are the *pepsins* (a group of enzymes with somewhat varying specificities), which belong to the *aspartate proteinases* (see p. 150). They undergo autocatalytic activation at low pH (see p. 246). The stomach also secretes a glycoprotein referred to as "*intrinsic factor*", which binds vitamin B_{12} (extrinsic factor) and thus protects it from degradation.

The intense enzymatic digestion that begins in the stomach takes 1 to 3 hours, and is mainly directed against dietary proteins. The acidic contents of the stomach are subsequently released, batchwise, into the *duodenum*, where they are mixed with (and thus neutralized by) the alkaline secretion of the pancreas. In addition, cystic bile is added at this point.

Pancreatic secretions. In the acinus cells of the pancreas, a secretion is produced that has an HCO_3^--based buffer capacity sufficiently high to neutralize the hydrochloric acid coming from the stomach. The pancreatic secretion also contains a number of enzymes that catalyze the hydrolysis of various high-molecular weight dietary constituents (see p. 242). All of these enzymes are *hydrolases* with pH optima in the neutral or weakly alkaline range. Most of them are produced and secreted as proenzymes.

Trypsin, chymotrypsin, and elastase are *endopeptidases*, i.e., they cleave peptide bonds located within the peptide chain. They all belong to the *serine proteinases* (see p. 150). *Trypsin* catalyzes the hydrolysis of peptide bond next to the basic amino acids Arg and Lys, whereas *chymotrypsin* is specific for peptide bonds of apolar amino acids like Tyr, Trp, Phe and Leu. *Elastase* usually cleaves bonds adjacent to the aliphatic amino acids Gly, Ala, Val, and Ile. In each case, the point of attack is located on the *carboxyl side* of the respective amino acid. Smaller peptides are attacked by *carboxypeptidases*, which are exopeptidases i.e., they remove amino acids one by one from the *C*-terminal end of the peptide (see p. 150). Pancreatic α-*amylase* acts as an *endoglycosidase*. It degrades α1→4-linked polymeric carbohydrates, such as starch and glycogen, to maltose, maltotriose, and a mixture of other oligosaccharides.

Several pancreatic enzymes are involved in the hydrolysis of lipids. These include pancreatic *lipase* with its auxiliary protein *colipase*, *phospholipase* A_2, and *cholesterol esterase*. These enzymes are dependent on the presence of bile acids (see below).

The most important hydrolases for the breakdown of dietary nucleic acids are *ribonuclease* (RNAase) and *deoxyribonuclease* (DNAase).

Bile. Hepatic bile is a watery secretion of the liver, which, after being concentrated and stored in the gallbladder, is released into the duodenum as cystic bile. The most important constituents, besides water and inorganic salts, are *bile salts*, *phospholipids*, *bile pigments*, and *cholesterol*. Bile salts (see p. 288), together with phospholipids, facilitate the *emulsification* of insoluble dietary lipids (see p. 250).

Secretions of the small intestine. Further digestive enzymes are secreted into the intestine by the Lieberkühn and Brunner glands of the small intestine. Together with membrane-bound enzymes on the surface of the intestinal epithelium, these enzymes complete the hydrolysis of the various dietary constituents.

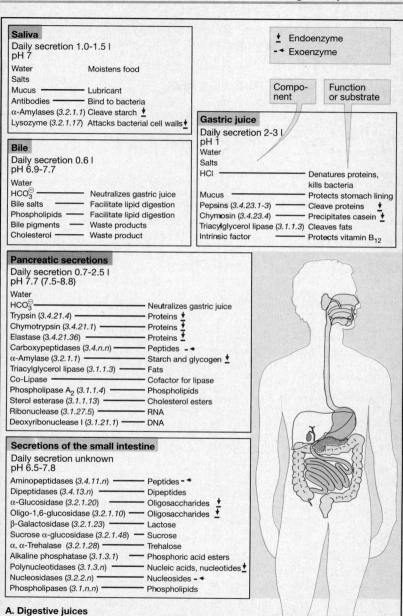

Saliva

Daily secretion 1.0-1.5 l
pH 7

Water	Moistens food
Salts	
Mucus	Lubricant
Antibodies	Bind to bacteria
α-Amylases (3.2.1.1)	Cleave starch ↨
Lysozyme (3.2.1.17)	Attacks bacterial cell walls ↨

Bile

Daily secretion 0.6 l
pH 6.9-7.7

Water	
$HCO_3^{\ominus}$	Neutralizes gastric juice
Bile salts	Facilitate lipid digestion
Phospholipids	Facilitate lipid digestion
Bile pigments	Waste products
Cholesterol	Waste product

Pancreatic secretions

Daily secretion 0.7-2.5 l
pH 7.7 (7.5-8.8)

Water	
$HCO_3^{\ominus}$	Neutralizes gastric juice
Trypsin (3.4.21.4)	Proteins ↨
Chymotrypsin (3.4.21.1)	Proteins ↨
Elastase (3.4.21.36)	Proteins ↨
Carboxypeptidases (3.4.n.n)	Peptides -◂
α-Amylase (3.2.1.1)	Starch and glycogen ↨
Triacylglycerol lipase (3.1.1.3)	Fats
Co-Lipase	Cofactor for lipase
Phospholipase A_2 (3.1.1.4)	Phospholipids
Sterol esterase (3.1.1.13)	Cholesterol esters
Ribonuclease (3.1.27.5)	RNA
Deoxyribonuclease I (3.1.21.1)	DNA

Secretions of the small intestine

Daily secretion unknown
pH 6.5-7.8

Aminopeptidases (3.4.11.n)	Peptides -◂
Dipeptidases (3.4.13.n)	Dipeptides
α-Glucosidase (3.2.1.20)	Oligosaccharides ↨
Oligo-1,6-glucosidase (3.2.1.10)	Oligosaccharides ↨
β-Galactosidase (3.2.1.23)	Lactose
Sucrose α-glucosidase (3.2.1.48)	Sucrose
α, α-Trehalase (3.2.1.28)	Trehalose
Alkaline phosphatase (3.1.3.1)	Phosphoric acid esters
Polynucleotidases (3.1.3.n)	Nucleic acids, nucleotides ↨
Nucleosidases (3.2.2.n)	Nucleosides -◂
Phospholipases (3.1.n.n)	Phospholipids

↨ Endoenzyme
-◂ Exoenzyme

Compo-nent	Function or substrate

Gastric juice

Daily secretion 2-3 l
pH 1

Water	
Salts	
HCl	Denatures proteins, kills bacteria
Mucus	Protects stomach lining
Pepsins (3.4.23.1-3)	Cleave proteins ↨
Chymosin (3.4.23.4)	Precipitates casein ↨
Triacylglycerol lipase (3.1.1.3)	Cleaves fats
Intrinsic factor	Protects vitamin B_{12}

A. Digestive juices

Digestive Processes

During a meal, foodstuffs are chewed, mixed with saliva, and then swallowed. In the stomach, gastric juice is added, and this is where enzymatic digestion really begins.

Gastric juice is the product of several different cell types. The *parietal cells* produce hydrochloric acid (HCl), so-called *chief cells* release pepsinogens (see below), and *accessory cells* and other epithelial cells secrete *mucus*, a complex mixture of glycoproteins and other components that protects the epithelium from digestion by pepsins and damage inflicted by HCl. When acid secretion and mucus formation are unbalanced, *gastric ulcers* may arise.

A. Formation of hydrochloric acid ◑

The HCl present in gastric juice is an approximately equimolar mixture of H^+ and Cl^- ions. It activates pepsinogens, kills micro-organisms, and denatures dietary proteins so that they are more readily degraded by proteinases.

To generate H^+ ions, the cells take up CO_2 from the blood and hydrate it, with catalysis by *carbonate dehydratase* ("carbonic anhydrase"), to yield carbonic acid (H_2CO_3). This spontaneously dissociates into hydrogen carbonate (HCO_3^-) and protons H^+. The protons are secreted into the intestinal lumen in exchange for K^+ ions by an H^+/K^+-exchanging ATPase located in the luminal membrane. This energy-dependent "uphill" transport results in a 1 million-fold higher H^+ concentration in the intestinal lumen than in the parietal cells.

The chloride ions (Cl^-) are also derived from the blood. They enter the parietal cells via an antiporter (see p. 206), which exchanges them for the HCO_3^- ions derived from CO_2 hydration. Cl^- ions are therefore taken up at about the same rate as protons are released. HCl secretion is regulated by hormones (e.g., the *gastrins*, a family of related peptides), by *histamine* (see p. 342), and other factors.

Pepsin activation. The hydrolysis of dietary proteins is initiated by pepsins in the stomach. These are proteinases with somewhat varying specificities that arise as precursors called *pepsinogens*. When they come into contact with the acidic environment inside the stomach, the pepsinogens are converted into their active form, the *pepsins*, by autocatalytic cleavage of a blocking peptide. The pepsins are *endopeptidases* (see p. 150) with an unusually acidic pH optimum (pH 2). With the help of two aspartyl residues in the active site (*aspartate proteinases*), they cleave peptide bonds of proteins preferentially next to the amino acids Phe and Leu.

B. Activation of digestive enzymes from the pancreas ◑

The pancreatic secretion contains further inactive precursors of digestive enzymes, so-called **proenzymes** or *zymogens*. Among the different proenzymes of pancreatic origin, the most important is *trypsinogen*. In the intestine, small amounts are converted into trypsin by the action of *enteropeptidase*, an endopeptidase located on the inner surface of the duodenum. It removes a short peptide from trypsinogen so that functional groups can rearrange themselves to form the active site of the resulting *trypsin* (see p. 151).

The active trypsin then catalyzes further peptide cleavage reactions, which result in the activation of additional trypsinogen (*autocatalytic activation*) and of the other pancreatic zymogens. The pancreatic enzymes activated in this way then continue the digestion of the foodstuffs.

Self-digestion of the pancreas, as seen in acute pancreatitis, is usually prevented by several different methods. 1. As already mentioned, most hydrolytic enzymes are synthesized in the pancreas as precursors (zymogens) and then only converted to active hydrolases in the intestine by the action of trypsin. 2. Premature activation of trypsin is inhibited by a pancreatic *trypsin inhibitor* that forms a highly stable complex with trypsin. 3. The inner surfaces of the stomach and the intestine are coated with *mucins*, which provide protection from enzymatic damage.

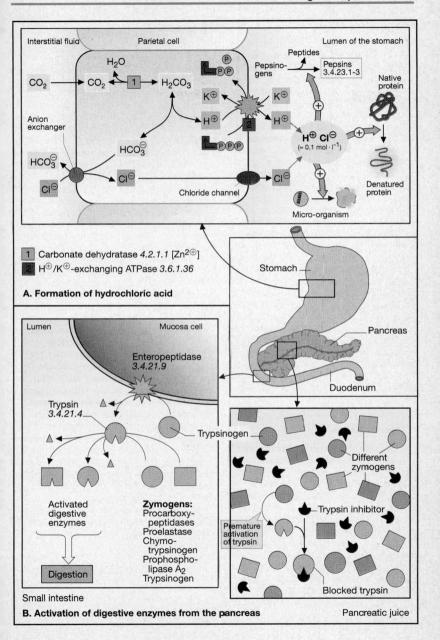

A. Formation of hydrochloric acid

1 Carbonate dehydratase *4.2.1.1* [Zn^{2+}]

2 H^{+}/K^{+}-exchanging ATPase *3.6.1.36*

B. Activation of digestive enzymes from the pancreas

Resorption

Enzymatic hydrolysis in the digestive tract ensures that foodstuffs are broken down to fragments that can be absorbed by epithelial cells. *Resorption* of the products of digestion occurs primarily in the small intestine. Only ethanol and short-chain fatty acids can be taken up by the stomach.

The process of absorption is facilitated by the large inner surface of the intestine, with its brush-border cells. Lipophilic molecules can cross the plasma membrane of the cells by simple *diffusion*, whereas polar molecules require transporters (*facilitated diffusion*). Many nutrients are absorbed by carrier-mediated cotransport with Na^+ ions. In this instance, the concentration difference of sodium ions (high in the intestinal lumen and low in the mucosa cells) drives the import of nutrients against a concentration gradient (*secondary-active transport*).

The failure of these intestinal transport systems can cause a variety of diseases.

Amino acids. Proteins are degraded into amino acids by a variety of *peptidases* (see p. 242). For the uptake of amino acids, there are *group-specific amino acid transport systems*, some of which employ cotransport with Na^+ ions (secondary active transport), while others facilitate diffusion in an Na^+-independent manner. Small peptides can also be taken up.

A. Monosaccharides ◑

The cleavage of polymeric carbohydrates by α-*amylase* [1] leads to oligosaccharides, which are further cleaved by *glycosidases* (oligosaccharidases, disaccharidases [2]) most of which are located on the outer surface of brush-border cells. The resulting monosaccharides enter the intestinal epithelium through various *sugar-specific transporters*. **Glucose** and **galactose** are taken up by *secondary—active transport* (cf. p. 208). Another transporter on the opposite side of the cell then releases them into the blood. **Fructose** is taken up by facilitated diffusion.

B. Lipids ◑

As fats and many other lipids are insoluble in water, they enter the small intestine as lipid droplets. The lipids in such a droplet are largely inaccessible to attack from digestive enzymes. In order to increase their accessible surface, the droplets habe to be broken down into small micelles (i.e., *emulsified*) by the action of bile salts and bile phospholipids (see p. 294). Only milk fats are well emulsified from the beginning. Their digestion by lipases from the saliva and gastric juice already begins in the stomach.

The hydrolysis of the bulk of fats (triacylglycerols) is catalyzed by pancreatic *triacylglycerol lipase* [3]. Even in the presence of bile salts, this enzyme is inactive unless *colipase* is also present. This is a small protein (10 kDa), also of pancreatic origin, that anchors lipase to the micelles. Cleavage by lipases usually occurs at positions 1 and 3 of the glycerol residue, resulting in the removal of two fatty acid residues. Therefore, in quantitative terms, **fatty acids** and **2-monoacylglycerol** are the main products of fat hydrolysis. However, some **glycerol** is also produced when hydrolysis goes to completion.

The products of hydrolysis enter the intestinal epithelium by passive diffusion. Once inside the cells, long-chain fatty acids undergo activation by linkage with coenzyme A [4], and are then recycled for the *resynthesis of triacylglycerols*. The resulting fats are deposited into the lymph in the form of *chylomicrons* (see p. 254), which bypass the liver and enter the bloodstream via the thoracic duct. Short-chain fatty acids with less than 12 carbons are directly released into the blood, and reach the liver via the portal vein. Glycerol can also take this route.

Cholesterol is also absorbed by passive diffusion. From the many plant sterols, only *ergosterol* (provitamin D_2) is taken up by animals.

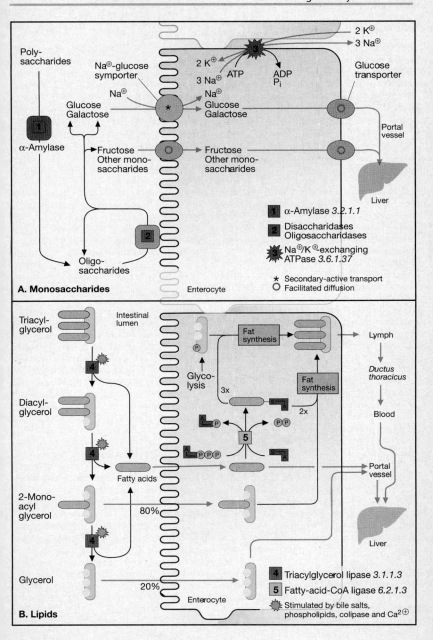

A. Monosaccharides

Poly-saccharides

α-Amylase

1

Glucose
Galactose

Fructose
Other mono-saccharides

Oligo-saccharides

2

Na⊕-glucose symporter

Na⊕

*

3 Na⊕
Na⊕
Glucose
Galactose

Fructose
Other mono-saccharides

2 K⊕
3 Na⊕

2 K⊕
3 Na⊕
ATP
ADP
Pi

3

Glucose transporter

Portal vessel

Liver

Enterocyte

1 α-Amylase 3.2.1.1

2 Disaccharidases
Oligosaccharidases

3 Na⊕/K⊕-exchanging
ATPase 3.6.1.37

★ Secondary-active transport
○ Facilitated diffusion

B. Lipids

Triacyl-glycerol

Diacyl-glycerol

2-Mono-acyl glycerol

Glycerol

Intestinal lumen

4

4

Fatty acids

80%

4

20%

Glyco-lysis

P

3x

5

P

P P

A P P P

Fat synthesis

Fat synthesis

2x

Enterocyte

Lymph

Ductus thoracicus

Blood

Portal vessel

Liver

4 Triacylglycerol lipase 3.1.1.3

5 Fatty-acid-CoA ligase 6.2.1.3

Stimulated by bile salts,
phospholipids, colipase and Ca$^{2\oplus}$

Composition and Functions

The blood accounts for about 8% of the mass of the human body. It is composed of whole **cells, cell fragments,** and an aqueous solution of proteins and low molecular- weight compounds called the **blood plasma**. The various cellular elements make up approximately 45% of the total volume of the blood. This percentage is referred to as the *hematocrit*.

A. Functions of the blood ●

The blood performs a variety of different functions in the body. It is the major means of transport, it serves to maintain a constant "internal milieu" in the body (homeostasis), and it plays a major role in the defense of the body against foreign substances.

Transport. The blood transports the *gases* oxygen and carbon dioxide. It also carries *nutrients* absorbed in the intestine to the liver and other organs. Yet another function of the blood is to relieve the various tissues of their *metabolic wastes* and to deliver these waste materials to the lungs, liver, and kidneys for disposal. The blood also takes care of the distribution of *hormones* throughout the organism (see p. 338).

Homeostasis. The blood maintains the *water balance* between the blood vessels, the cells (intracellular space), and the extracellular space. In cooperation with the lungs, the liver, and the kidneys it helps to maintain a *constant intracellular and extracellular pH* (see p. 256). The maintenance of a constant *body temperature* also depends on the controlled transport of heat by the blood.

Defense. The body employs both specific and non-specific means to defend itself against invading molecules and foreign cells. This specific defense mechanism is made up of the cells of the immune system, soluble antibodies and other components (see pp. 264 f.).

Blood clotting. Within the blood, there are factors, that prevent the loss of blood following damage to a blood vessel. The process is referred to as blood clotting or hemostasis (see p. 262). At the same time, the blood pro- vides a system for the breakdown of blood clots (fibrinolysis).

B. Cellular elements ◑

The solid elements in the blood are the *erythrocytes* (red blood cells), the *leukocytes* (white blood cells), and the *thrombocytes* (blood platelets). The **erythrocytes** are especially important, and therefore their biochemical characteristics are discussed separately on pages 258 to 262.

The **leukocytes** include various types of *granulocyte, monocyte,* and *lymphocyte* (see p. 264). These cells differ from one another in their shape, function, and site of synthesis. **Thrombocytes** are cell fragments that bud off from large precursor cells, the *megakaryocytes*, in the bone marrow. Their most important function is the promotion of blood clotting.

C. Blood plasma: composition ◑

Blood plasma is an aqueous solution containing electrolytes, nutrients, metabolites, proteins, vitamins, trace elements, and hormones. The determination of the **electrolyte composition** of the blood plasma is very important in clinical diagnosis. As compared to the cytoplasm of cells, blood plasma contains relatively high concentrations of sodium, calcium, and chloride ions. In contrast, the concentrations of potassium, magnesium, and phosphate ions in the blood are comparatively low. The concentration of proteins (proteinaceous anions) is also higher in the cells than in the plasma. The electrolyte composition of blood plasma is similar to that of sea water. This fact probably reflects the widely accepted hypothesis that life once evolved in the sea.

The table lists the most important **metabolites** in the blood plasma. **Plasma proteins** are discussed on the next page. The liquid phase remaining after blood has clotted is referred to as **serum**. Unlike plasma, serum does not contain fibrinogen or other proteins consumed during blood clotting (see p. 262).

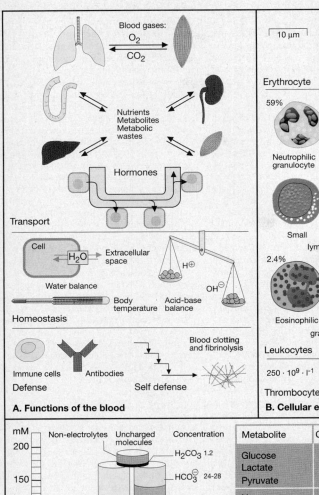

A. Functions of the blood

Blood gases:
O_2
CO_2

Nutrients
Metabolites
Metabolic
wastes

Hormones

Transport

Cell — H_2O → Extracellular space

Water balance

$H^\oplus$
$OH^\ominus$

Body temperature   Acid-base balance

Homeostasis

Immune cells   Antibodies

Blood clotting and fibrinolysis

Defense   Self defense

B. Cellular elements

10 µm

Erythrocyte $5000 \cdot 10^9 \cdot l^{-1}$

59%   6.5%

Neutrophilic granulocyte   Monocyte

31%

Small   Large

lymphocyte

2.4%   0.6%

Eosinophilic   Basophilic
granulocyte

Leukocytes $7 \cdot 10^9 \cdot l^{-1}$

$250 \cdot 10^9 \cdot l^{-1}$

Thrombocytes

C. Blood plasma: composition

mM   Non-electrolytes   Uncharged molecules   Concentration

200

150

H_2CO_3 1.2
$HCO_3^\ominus$ 24-28

100   136-145 $Na^\oplus$ $Cl^\ominus$ 100-110

50

$HPO_4^{2\ominus}$ 1.1-1.5

3.5-5.0 $K^\oplus$
$SO_4^{2\ominus}$ 0.3-0.6

2.1-2.6 $Ca^{2\oplus}$ Organic acids

0   0.6-1.0 $Mg^{2\oplus}$ Cations   Anions   Proteins

Metabolite	Concentration (mM)
Glucose	3.6 - 6.1
Lactate	0.4 - 1.8
Pyruvate	0.07 - 0.11
Urea	3.5 - 9.0
Uric acid	0.18 - 0.54
Creatine	0.06 - 0.13
Amino acids	2.3 - 4.0
Ammonia	0.02 - 0.06
Lipids (total)	5.5 - 6.0 g · l^{-1}
Triacylglycerols	1.0 - 1.3 g · l^{-1}
Cholesterol	1.7 - 2.1 g · l^{-1}

Plasma Proteins

Proteins are the most important solutes in the blood plasma. The protein concentration in the plasma is 60–80 g · l^{-1}. Thus, the plasma proteins account for 4% of total body protein.

A. Plasma proteins ◗

Blood plasma contains almost 100 different proteins. Based on their behavior during electrophoresis (see below), they are commonly divided into five different groups. These are the **albumins** and the α_1-, α_2-, β- and γ-**globulins**. The distinction between the albumins and the globulins was originally based on differences in their solubility, i.e., the albumins are soluble in pure water, whereas the globulins only dissolve in the presence of salts.

The most important protein in the blood, on a quantitative basis, is **serum albumin**. Albumin plays a crucial role in the maintenance of the osmotic pressure of the blood. It also functions as a carrier protein for lipophilic substances, such as fatty acids, bilirubin, some steroid hormones, vitamins, and calcium ions. The albumin fraction also includes *transthyretin* (pre-albumin). This protein acts in concert with albumin and a thyroxin-binding protein to transport the hormone thyroxine and its metabolites (iodothyronins, see p. 340).

The table further summarizes the names, sizes, and functions of some important **globulins** of the blood plasma. These proteins are involved in the transport of lipids (see p. 254), hormones, vitamins, and metal ions, and they constitute an important part of the blood clotting system (see p. 262). In addition, they include the antibodies (γ-globulins) that constitute a major part of the immune system (see p. 266).

B. Electrophoresis ◗

Electrophoresis is a procedure for the separation of proteins and other electrically charged molecules, based on their migration in an electric field. It can be either analytical or preparative. The velocity of migration of a molecule is determined by its net charge, its size and shape, and the magnitude of the voltage applied. The net charge, in turn, is dependent on the number of negatively and positively charged residues and the pH of the electrophoresis buffer. At physiological pH, most proteins are anions, i.e., the number of negative charges exceeds the number of positive charges. The most simple type of electrophoresis employs *cellulose acetate sheets* as support. Using this procedure, the serum proteins can be separated into the five fractions already mentioned. After the run, the separated protein bands are made visible by staining, and then analyzed semi-quantitatively by densitometry. Some diseases result in an alteration in the concentration of individual proteins (so-called "*dysproteinemias*").

More efficient electrophoresis procedures employ polyacrylamide (PAGE = **polyacrylamide gel electrophoresis**) or agarose gels as supports. The electrophoretic conditions and the detection system can be modified to obtain very efficient and specific separation procedures. For example, gels with a gradient in pore size or pH can be made, or detergents, such as *sodium dodecylsulfate* (SDS) can be added to improve separation. In the analysis of complex mixtures, antibodies can serve to detect individual proteins. Detailed descriptions of standard procedures can be found in relevant laboratory manuals.

Further information

Most plasma proteins are synthesized in the liver. Important exceptions are the immunoglobulins and proteohormones. Except for albumin, all of the plasma proteins are *glycoproteins*. They contain *N*- and *O*-linked oligosaccharides attached to amino acid residues (see p. 40). Their half-life, which is usually several days, is determined by the number of sialic acid molecules on their surface. Plasma proteins often exist as *isoforms* that differ from one another by a single amino acid exchange (genetic polymorphism).

Group	Protein	M_r in kDa	Function
Albumins:	Transthyretin	50–66	Transport of thyroxin and triiodothyronin
	Albumin: 45 g · l⁻¹	67	Maintenance of osmotic pressure; transport of fatty acids, bilirubin, bile acids, steroid hormones, pharmaceuticals and inorganic ions.
α_1-Globulins:	Antitrypsin	51	Inhibition of trypsin and other proteases
	Antichymotrypsin	58–68	Inhibition of chymotrypsin
	Lipoprotein (HDL)	200–400	Transport of lipids
	Prothrombin	72	Coagulation factor II, thrombin precursor (3.4.21.5)
	Transcortin	51	Transport of cortisol, corticosterone and progesterone
	Acid glycoprotein	44	Transport of progesterone
	Thyroxin-binding globulin	54	Transport of iodothyronins
α_2-Globulins:	Ceruloplasmin	135	Transport of copper ions
	Antithrombin III	58	Inhibition of blood clotting
	Haptoglobin	100	Binding of hemoglobin
	Cholinesterase (3.1.1.8)	ca. 350	Cleavage of choline esters
	Plasminogen	90	Precursor of plasmin (3.4.21.7), breakdown of blood clots
	Macroglobulin	725	Binding of proteases, transport of zinc ions
	Retinol-binding protein	21	Transport of vitamin A
	Vitamin D-binding protein	52	Transport of calciols
β-Globulins:	Lipoprotein (LDL)	2.000–4.500	Transport of lipids
	Transferrin	80	Transport of iron ions
	Fibrinogen	340	Coagulation factor I
	Sex hormone-binding globulin	65	Transport of testosterone and estradiol
	Transcobalamin	38	Transport of vitamin B_{12}
	C-reactive protein	110	Complement activation
γ-Globulins:	IgG	150	Late antibodies
	IgA	162	Mucosa-protecting antibodies
	IgM	900	Early antibodies
	IgD	172	B-lymphocyte receptors
	IgE	196	Reagins

A. Plasma proteins

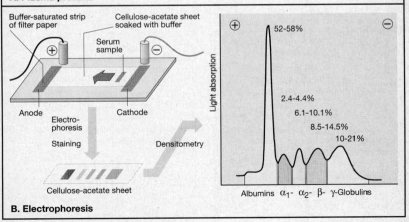

B. Electrophoresis

Lipoproteins

The lipoproteins can be divided into two separate groups. One group of proteins contains covalently-bound lipids, whereas in the other group lipids are associated with proteins via non-covalent interactions. Covalently bound lipids form **lipid anchors**, which fix the protein to membranes (see p. 214). The composition of lipoproteins of the second group is not as well defined. They are actually aggregates of proteins with lipids (**lipoprotein complexes**). The size and composition of such complexes is variable. Several types are found in the blood plasma, where they serve in the transport of water-insoluble lipids.

A. Composition of lipoprotein complexes ○

Basically, lipoprotein complexes are spherical aggregates of lipids and proteins. They consist of a core of *apolar lipids* (triacylglycerols and acyl esters of cholesterol) and a shell (about 2 nm thick) made up of *apoproteins* and *amphipathic lipids* (phospholipids and cholesterol). The shell is polar on its outside, and thus keeps the lipids dissolved in the plasma. The larger the lipid core (i.e., the higher the proportion of lipids), the lower the density of the lipoprotein complex.

Lipoprotein complexes are subdivided into five different groups. In the order of decreasing size and increasing density, these are the **chylomicrons** and **chylomicron remnants, VLDL** (very low density lipoproteins), **IDL** (intermediate density lipoproteins), **LDL** (low density lipoproteins), and **HDL** (high density lipoproteins). These lipoprotein complexes have characteristic apoproteins on the outside, which either "float" in the shell, or are only loosely associated with it (here we show LDL as an example). The apoproteins play a major role in determining the fate of the lipoproteins. They serve as *recognition molecules* for membrane receptors (see below), and are essential partners for enzymes and proteins that are involved in the metabolism and exchange of lipids.

B. Transport of triacylglycerols and cholesterol ○

The vehicles for the transport of *dietary lipids* from the intestine to the liver are the **chylomicrons**. They are assembled in the intestinal mucosa and transported to the blood via the lymph system (see p. 248). Once they reach the muscles and adipose tissues, their apoprotein C-II activates a *lipoprotein lipase* present on the inner surface of the blood vessels. Due to the action of this enzyme, the chylomicrons rapidly lose most of their triacylglycerols. The **chylomicron remnants** formed in this way are taken up by the liver.

VLDL, IDL, and **LDL** are closely related to one another. They transport triacylglycerols, cholesterol, and phospholipids from the liver to other tissues. VLDLs are formed in the liver (see p. 278) and can be converted to IDL or LDL by degradation of triacylglycerols, as described for chylomicrons. The LDLs thus produced supply the tissues with cholesterol. In contrast, **HDLs** transport excess cholesterol formed in other tissues back to the liver. During transport, cholesterol becomes acylated by transfer of fatty acids from lecithin. The enzyme involved is *phosphatidylcholine sterol acyltransferase* ("LCAT," 2.3.1.43). An exchange of lipids and proteins between HDL and VLDL also occurs.

Receptor-mediated endocytosis. In preparation for cholesterol uptake, cells bind LDLs with the aid of membrane receptors that recognize apo-B100 and apo-E. The whole complex is then taken up by **endocytosis**. Uptake occurs in the so-called "*coated pits*" i.e., regions of the membrane that are lined on the inside with the protein *clathrin*. Clathrin facilitates the invagination and pinching-off of vesicles ("*coated vesicles*"). Inside the cell, clathrin is separated from the vesicles and recycled, while the vesicles associate with lysosomes, where their contents are degraded (see p. 212).

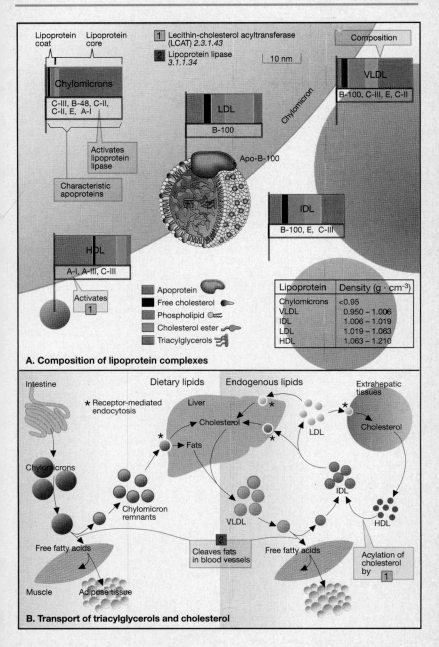

A. Composition of lipoprotein complexes

Lipoprotein coat   Lipoprotein core

1 Lecithin-cholesterol acyltransferase (LCAT) 2.3.1.43
2 Lipoprotein lipase 3.1.1.34

10 nm

Chylomicrons
C-III, B-48, C-II, C-II, E, A-I

Activates lipoprotein lipase

Characteristic apoproteins

LDL
B-100

Chylomicron

Apo-B-100

Composition

VLDL
B-100, C-III, E, C-II

IDL
B-100, E, C-III

HDL
A-I, A-III, C-III

Activates 1

Apoprotein
Free cholesterol
Phospholipid
Cholesterol ester
Triacylglycerols

Lipoprotein	Density (g · cm⁻³)
Chylomicrons	<0.95
VLDL	0.950 – 1.006
IDL	1.006 – 1.019
LDL	1.019 – 1.063
HDL	1.063 – 1.210

B. Transport of triacylglycerols and cholesterol

Intestine

Dietary lipids   Endogenous lipids   Extrahepatic tissues

★ Receptor-mediated endocytosis

Liver

Cholesterol

LDL

Cholesterol

Fats

Chylomicrons

Chylomicron remnants

VLDL

IDL

HDL

Free fatty acids

Free fatty acids

2 Cleaves fats in blood vessels

Acylation of cholesterol by 1

Muscle   Adipose tissue

Acid-Base Balance

A. Hydrogen ion concentration of the plasma

The concentration of hydrogen ions in the plasma and the extracellular space of mammals is about 40 nM. This corresponds to a **pH of 7.40**. Living organisms have to maintain a relatively constant pH value, since large changes in the proton concentration can be life-threatening.

Maintenance of a constant pH is, in part, achieved by **buffer systems** in the blood (**C**), which mitigate short-term changes in the acid-base metabolism. In the long term, however, proton loss and proton gain have to be balanced. The pH value of the plasma may be affected by a malfunctioning of the buffer system, or when the acid-base balance is disturbed, e.g., due to kidney diseases or altered breathing frequency (*hypoventilation* or *hyperventilation*). A decrease by more than 0.03 units below pH 7.4, is referred to as **acidosis**, and an increase of the same magnitude is called **alkalosis**.

B. Acid-base balance ❶

There are two major sources of protons – the free acids in the diet and the sulfur-containing amino acids in proteins. The *acids* obtained from the diet, for example citric acid, dissociate – i.e., release their protons – at the alkaline pH of the intestinal tract. In terms of proton balance, however, this effect is less important than that of the amino acids **methionine** and **cysteine**, which are derived from protein degradation. In the liver, the sulfur atom of these amino acids is oxidized to sulfuric acid, which immediately dissociates into sulfate and protons.

During anaerobic glycolysis in the muscles and erythrocytes, substantial amounts of glucose are converted to **lactic acid** (see p. 310), which dissociates to give lactate and a proton. The ketone bodies **acetoacetic acid** and **3-hydroxybutyric acid**, formed in the liver (see p. 284), also release protons at physiological pH. Under normal conditions, these acids are completely metabolized to CO_2 and H_2O, thus having no effect on the overall proton balance. When large amounts of these acids are formed, (e.g., during *starvation*, or in the case of *diabetes mellitus*, see p. 164) this may result in overloading of the buffer systems of the plasma and a consequent decrease in pH (**metabolic acidosis**).

Excretion of protons. The site of excretion of protons is the kidney, where it occurs as active transport, or as exchange with Na^+ ions (see p. 300).

C. Buffer systems of the plasma ❶

The most important buffer system of the plasma is the **carbon dioxide-bicarbonate buffer** system. It consists of the weak acid **carbonic acid** ($pK_1 = 6.1$) and its conjugate base **hydrogen carbonate** ("bicarbonate"). In aqueous solution carbonic acid exists in equilibrium with its anhydride, CO_2. The rate at which this equilibrium is attained is increased by the enzyme *carbonate dehydratase* ("carbonic anhydrase"), which is found in large amounts in the erythrocytes. At the pH of the plasma, HCO_3^- and CO_2 are present in a ratio of about 20 : 1. The CO_2 dissolved in the blood is constantly exchanging with the CO_2 in the gas phase of the alveoli of the lungs, and its concentration thus remains essentially constant. This is why the HCO_3^-/CO_2 system is such an effective *open buffer system*. As the breathing frequency changes, so does the CO_2 concentration leading to a shift in the pH of the plasma (*respiratory acidosis* or *alkalosis*). Thus, the lungs exert a very rapid and pronounced effect on the pH of the plasma without actually being involved in proton excretion as such.

Amino acid side chains of blood proteins – especially **hemoglobin** in the erythrocytes – are a second major factor in maintaining constant plasma pH (see p. 26). A minor role is played by **phosphates**.

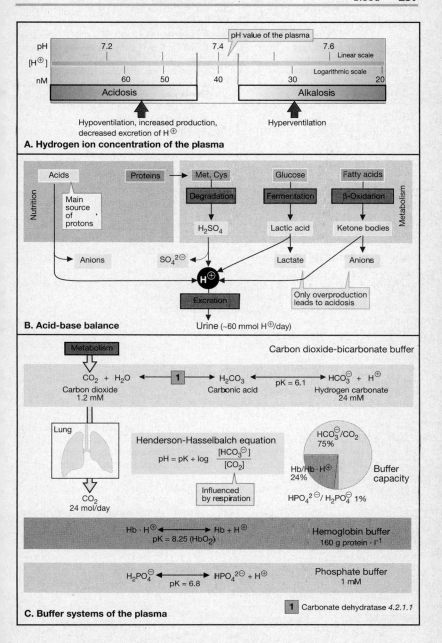

A. Hydrogen ion concentration of the plasma

B. Acid-base balance

C. Buffer systems of the plasma

Gas Transport

The most important function of the erythrocytes is the *transport of O_2 and CO_2* between the lungs and the various tissues of the body. **Hemoglobin**, a protein only found in erythrocytes, plays a key role in this process.

A. Solubility of oxygen ○

Oxygen (O_2) is only poorly soluble in water, and higher organisms therefore require a special system to transport it throughout the body. One liter of blood plasma dissolves only about 3.2 ml of O_2, whereas the hemoglobin present in human blood (about 160 g · l^{-1}) can bind 220 ml O_2 per l of blood, i.e., a 70-fold higher amount.

B. Structure of hemoglobin ◑

The hemoglobin of adults (**Hb A**) is a tetramer made up of two α and two β subunits, each with a mass of about 16 kDa (**1**). Although the amino acid sequences of the α and β chains differ from one another, they fold in a similar way. About 80% of the amino acids of the globin moiety are involved in α helices, which are designated A–H. Each subunit contains one heme group with a central ferrous ion (Fe^{2+}). During O_2 binding (**oxygenation**), the oxidation level of the heme iron (+2) does *not* change. *Methemoglobin*, formed in small quantities by oxidation of the iron to Fe^{3+}, is unable to transport O_2.

Four of the six coordination sites of the iron are occupied by heme nitrogens, and a fifth by a histidine residue of the globin moiety. The sixth is occupied by oxygen in *oxyhemoglobin* and water in *deoxyhemoglobin*.

C. Saturation curve ◑

The amount of O_2 transported in the blood depends on the Hb concentration and the O_2 concentration in the capillaries of the lungs and the peripheral tissues, respectively. This is best illustrated by a diagram showing the fraction of the heme groups occupied by oxygen (the *saturation S* of the system) as a function of O_2 concentration (traditionally given as *partial pressure, pO_2*, in mmHg). Allosteric effects (see p. 260) make the O_2 saturation curve of hemoglobin S-shaped (*sigmoidal*). From the figure, it is obvious that under normal circumstances, only half of the O_2 bound in the lungs is subsequently released into the tissues ($\Delta S < 0.5$). Since one liter of blood contains about 10 mmol of heme, the amount of O_2 transported is obtained by multiplying this value by ΔS.

D. Hemoglobin and CO_2 transport ◑

Hb is also involved in the transport of carbon dioxide (CO_2). The direction of this process, from the tissues to the lungs, is the opposite of that for oxygen. About 90% of the CO_2 is not transported as such, but rather as the more soluble **hydrogen carbonate** ion (**HCO_3^-**). In the lungs, HCO_3^- has to be converted back to CO_2 before it can be exhaled. Both processes—HCO_3^- formation in the tissues and CO_2 release in the lungs — are coupled to the oxygenation and deoxygenation of hemoglobin. Deoxy-Hb is a stronger base than oxy-Hb. As a result, it has a tendency to bind additional protons (about 0.7 H^+ per tetramer), an effect that promotes the formation of HCO_3^- from CO_2 in the tissues (below). HCO_3^- formation is also accelerated by *carbonate dehydratase* ("carbonic anhydrase"), an enzyme present in the erythrocytes in high concentrations. Most of the HCO_3^- formed is taken up by the blood in exchange for Cl^- ions and transported to the lungs. There the reactions described above occur in reverse order, i.e., deoxyhemoglobin becomes oxygenated and releases protons. These protons then displace the HCO_3^-/H_2CO_3 equilibrium and promote the release of CO_2. A small part of the CO_2 (about 5%) is transported covalently bound to Hb as *carbamino-Hb* (not shown). Together, the mechanisms described above are responsible for about 60% of CO_2 transport.

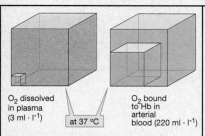

A. Solubility of oxygen

O₂ dissolved in plasma (3 ml · l⁻¹)   at 37 °C   O₂ bound to Hb in arterial blood (220 ml · l⁻¹)

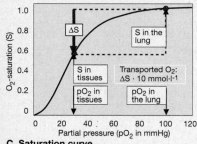

C. Saturation curve

Hemoglobin A ($\alpha_2 \beta_2$) M_r 65 kDa

B. Structure of hemoglobin

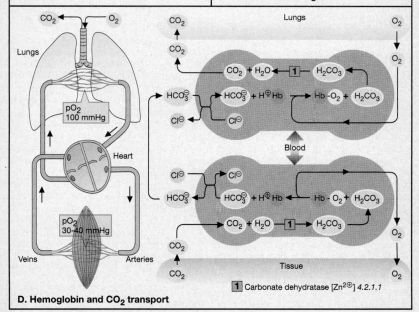

D. Hemoglobin and CO₂ transport

1 Carbonate dehydratase [Zn^{2+}] *4.2.1.1*

Hemoglobin, Erythrocyte Metabolism

A. Hemoglobin: allosteric effects ○

The first observation of allosteric behavior by a protein (see p. 112) was made with hemoglobin as early as 1911. The molecular basis of this behavior was later elucidated by *M. F. Perutz* using X-ray crystallography. Fig. **A** shows a simplified version of the way in which Hb, like aspartate carbamoyl-transferase, can exist in two different forms, the T form and the R form (see also p. 268A). The **T form** ("tense", left) has a much *lower O_2 affinity* than the R form (right). The subunits are held together by electrostatic interactions (here schematically represented by clamps). The binding of O_2 to one of the subunits of the T form leads to a local conformational change that weakens the association between subunits. Increasing O_2 partial pressure thus results in more and more molecules being converted to the high-affinity **R form** ("relaxed"). This *cooperative interaction* between the subunits leads to a marked increase in the O_2 affinity of Hb with increasing O_2 concentration, i.e., the O_2 saturation curve becomes *sigmoidal* (see p. 112). The concentrations of various **allosteric effectors** (see **B**) determine the equilibrium between the T and the R forms (yellow arrows).

B. Regulation of O_2 transport ◑

Besides oxygen itself, the allosteric effectors of hemoglobin include CO_2, H^+ ions, and **2,3-bisphosphoglycerate (BPG)**, a metabolite formed only in erythrocytes. BPG is synthesized from 1,3-bisphosphoglycerate, a normal intermediate of glycolysis. It can reenter glycolysis by conversion to 2-phosphoglycerate (**1**; note that this bypass does not yield ATP). BPG only binds to deoxy-Hb, thus shifting the equilibrium in favor of this form. The result is increased O_2 release at constant pO_2. In the diagram, this corresponds to displacement of the saturation curve toward the *right* (**2**). H^+ ions (i.e., a lower pH) have the same effect as BPG. As CO_2 hydration releases protons (see preceeding page), high CO_2 levels in the plasma also result in a *right shift* of the oxygen saturation curve. This has long been known as the *Bohr effect*. The effects of CO_2/H^+ and BPG are *additive*. In the presence of both effectors, the saturation curve of isolated Hb is similar to that of whole blood. The figure also shows the saturation curve of myoglobin (see p. 310) for comparison. It is non-sigmoidal and not affected by either BPG or CO_2.

C. Erythrocyte metabolism ○

Erythrocytes lack mitochondria and other organelles, and thus their cytoplasmic metabolism is much reduced. However, they degrade externally supplied glucose into lactate via **glycolysis**. The main function of the ATP thus synthesized is to supply energy the Na^+/K^+-ATPase (see p. 208), which maintains the ionic status of the erythrocytes. BPG (see **B**) is another important product of glucose degradation. A small portion of the glucose is metabolized via the **hexose-monophosphate pathway**. The NADPH thus formed is important for processes involved in protecting the erythrocytes against oxidative damage.

Oxygen and its derivatives may convert hemoglobin into inactive methemoglobin. In addition, in the presence of O_2, highly reactive peroxides are formed from membrane lipids. It is essential that these be disposed of enzymatically. A selenium-containing *glutathione peroxidase* [3] converts peroxide groups into harmless hydroxyl functions, with the help of **glutathione (GSH)** as the reductant. Glutathione is an atypical tripeptide (γ-Glu-Cys-Gly). It is the *thiol group* in the cysteine moiety, which is oxidized to the corresponding *disulfide (GSSG)* during reduction of methemoglobin and peroxides. The regeneration of reduced GSH is catalyzed by *glutathione reductase* [1], which, in turn, uses NADPH + H^+ as the reducing agent.

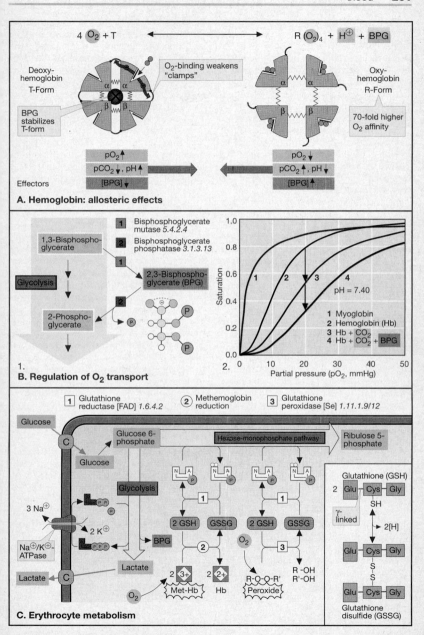

A. Hemoglobin: allosteric effects

B. Regulation of O_2 transport

C. Erythrocyte metabolism

Hemostasis

Hemostasis is the process that prevents excessive loss of blood following injury to blood vessels. There are actually two parts to this process — the prevention of bleeding, and the promotion of blood clotting (coagulation). Major contributors to hemostasis are the *thrombocytes* and various components *(coagulation factors)* derived from the blood plasma and the walls of the blood vessels. Here, our attention will be restricted to the enzymatic reactions associated with **blood clotting** and **fibrinolysis**, the breaking down of blood clots.

The nomenclature used to describe the coagulation factors is somewhat confusing. They are usually designated by roman numerals, with the addition of an **a** to indicate the activated form (e.g., factor Xa). Most of the coagulation factors are proteinases. Their *inactive precursors* are represented as complete circles here, whereas the *active enzymes* are shown as circle sectors.

A. Blood clotting ○)

Blood clotting is the enzymatic conversion of the soluble plasma protein **fibrinogen** (factor I) into a fibrous network of insoluble polymers. The enzyme involved is *thrombin* (factor IIa), which catalyzes the proteolytic removal of two peptides from the fibrinogen molecule. As a result of this, two binding sites are exposed, which allow the fibrin molecules to aggregate into an insoluble **fibrin polymer.** The amino acid side chains of fibrin subsequently form a covalent network. This crosslinking is catalyzed by a special *transglutaminase* (factor XIIIa), which creates isopeptide bonds between the molecules. Ultimately, a solid molecular plug (thrombus) is formed.

Blood clotting can be initiated in two different ways. The first is triggered by an injury to tissues (*extravascular pathway*, right), and the second is initiated by contact with the damaged inner surface of a blood vessel (*intravascular pathway*, left). In both cases, a *cascade of proteolytic cleavages* takes place. In the process, active *serine proteinases* (symbol: circle sectors) arise from inactive precursors (zymogens, symbol: full circle). The active

enzymes then proceed to attack other inactive proteins. Both reaction pathways require Ca²⁺ and phospholipids, and both culminate in the conversion of inactive *prothrombin* (factor II) to active *thrombin* (IIa). This last step is catalyzed by *factor Xa* (thrombokinase).

The **intravascular reaction pathway** is mainly triggered by collagen, which is not normally exposed on the surface of blood vessels. This contact results in the activation of *factor XII.* The **extravascular reaction pathway** begins with the release of *factor III* from injured cells. Within a few seconds of the release of this factor, clotting begins in the region of the wound.

B. Control of blood clotting ○)

Blood clotting is regulated by activators and inhibitors in dynamic equilibrium. For instance, the extent of clotting is controlled by highly effective proteinase inhibitors found in the plasma. In addition, thrombin activates *protein c,* an enzyme catalyzing the proteolytic degradation of factors V and VIII. This is one of several mechanisms of feedback inhibition of blood clotting.

The coagulation factors II, VII, IX, and X all require Ca²⁺ ions for activity. These ions bind to γ-**carboxyglutamyl residues** (Gla) present in these factors. The synthesis of Gla by carboxylation of glutamate residue in the liver depends on **vitamin K** (see p. 330)

C. Fibrinolysis ○)

Fibrin networks are degraded by the proteinase *plasmin* (precursor: *plasminogen*). Still other proteinases from various tissues catalyze the activation of plasminogen. For example, there is a *plasminogen activator* formed in the the kidneys (urokinase) and *tissue plasminogen activator* (t-PA) produced by the vascular endothelium.

A proteinase inhibitor in the plasma, α2-*antiplasmin,* is involved in the negative regulation of plasmin activity. Fibrinolytic proteinases such as urokinase, t-PA, and *streptokinase* (a bacterial enzyme) are administered to heart attack victims to promote thrombus degradation.

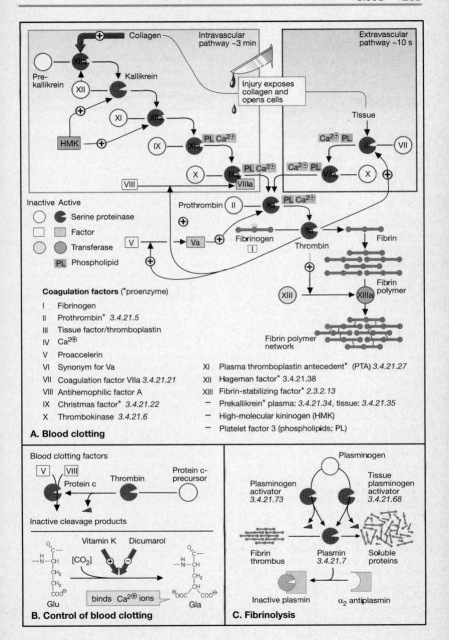

A. Blood clotting

Coagulation factors (*proenzyme)

I   Fibrinogen
II   Prothrombin* 3.4.21.5
III   Tissue factor/thromboplastin
IV   Ca²⊕
V   Proaccelerin
VI   Synonym for Va
VII   Coagulation factor VIIa 3.4.21.21
VIII Antihemophilic factor A
IX   Christmas factor* 3.4.21.22
X   Thrombokinase 3.4.21.6

XI   Plasma thromboplastin antecedent* (PTA) 3.4.21.27
XII   Hageman factor* 3.4.21.38
XIII Fibrin-stabilizing factor* 2.3.2.13
—   Prekallikrein* plasma: 3.4.21.34, tissue: 3.4.21.35
—   High-molecular kininogen (HMK)
—   Platelet factor 3 (phospholipids; PL)

B. Control of blood clotting

C. Fibrinolysis

Immune Response

Viruses, bacteria, fungi, and parasites that enter the body of vertebrates are recognized as foreign and destroyed by the **immune system**. Some of the body's own cells that have undergone alterations, e.g., tumor cells, are also recognized and eliminated by this system. The immune system mounts a response to stimuli from anything identified as being foreign. The response is specific, and the event causing it is stored in the immune system's memory.

The response to foreign substances, so-called **antigens**, is mediated by cells specific to the immune system. Among these, the **lymphocytes**, of which there are several different types, are especially important (see p. 250). *Cellular immunity* is mediated by **T lymphocytes** (T cells). These are leukocytes that undergo key steps of their differentiation in the thymus, hence the name. T cells act on virus-infected body cells, for instance, and are also involved in defense against fungi and parasites. Moreover, T cells play a role in the rejection of foreign tissues (grafts). Based on their functions, they are classified into three groups: **cytotoxic T cells** (green), **helper T cells** (blue), and **T suppressor cells** (not shown).

Humoral immunity is mainly due to the activity of **B lymphocytes** (B cells, light brown), which, in contrast to the T cells, are formed in the bone marrow rather than in the thymus. B cells carry **antibodies** (see p. 266) on their surface or, after differentiation, secrete them in large amounts into the plasma. These antibodies are capable of specific binding to antigens. The binding of antibodies to antigens labels the foreign substances for the defense system, and thus helps to eliminate extracellular viruses and bacteria.

The "memory" of the immune system is provided by so-called **memory cells**. These are especially long-lived cells, which can derived from any of the immune cells listed above.

A. Simplified scheme of the immune response ◐

A virus that gains entry into the body (above) is endocytosed by **macrophages**, and then partially degraded in the ER (**1**). This gives rise to peptide fragments (recognizable as foreign), which are displayed on the surface of the macrophage (**2**). Presentation of the fragments on macrophage surfaces is facilitated by a special group of membrane-bound proteins (**MHC proteins**, see p. 272). Complexes of MHC proteins with virus fragments are detected and bound by T cells with the help of specific receptors. Although there are many different T cells, only a few of them have the appropriate receptors. (**3**). Binding leads to activation of these T cells and selective replication of them (**4**, *clonal selection*). This process is promoted by various hormone-like signaling proteins, the **interleukins** (IL). These are secreted by the cells of the immune system that have been activated by the binding of T cells. For instance, macrophages release **IL-I** (**5**), while T cells excrete **IL-II** (**6**), which stimulates their own clonal replication as well as that of T helper cells.

T cells replicated and activated by clonal selection take on different functions, depending on their type. **Cytotoxic T cells** (green) are able to recognize and bind body cells that have been infected by viruses, and therefore carry virus fragments on their MHC receptors (**7**). The cytotoxic T cell then releases a protein that perforates the membrane of the infected cell and kills it (**8**). In contrast to cytotoxic T cells, **helper T cells** (blue) bind to the **B cells**, which present MHC protein-bound virus fragments on their surface (**9**). This results in clonal selection of individual B cells and massive proliferation of them. Following stimulation by interleukins (**10**), the B cells mature to become **plasma cells** (**11**), which synthesize and secrete large amounts of antibodies (**12**).

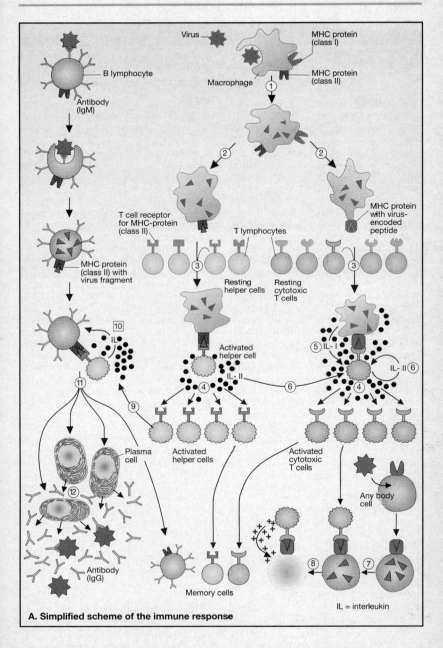

A. Simplified scheme of the immune response

IL = interleukin

Antibodies

A. Domain structure of immunoglobulin G ◑

The antibodies found in the blood plasma of mammals are, in the simplest case, Y-shaped tetrameric glycoproteins (see p. 263). Shown here is the most important of these, immunoglobulin G (IgG). Like all immunoglobulins, it consists of two identical **light chains** (L, yellow) and two identical **heavy chains** (H, orange). The light chains are composed of two *globular domains*, C_L and V_L, whereas the heavy chains consist of four such domains, called V_H, C_H1, C_H2 and C_H3 (the letters C and V refer to *constant* and *variable* regions of the antibody). Disulfide bonds connect the two heavy chains with each other, as well as the heavy with the light chains. Further disulfide bridges are found within the domains, where they stabilize the tertiary structure of the molecule. The domains that are homologous to one another have lengths of about 110 amino acids (aa). Apparently, the present structure of antibodies arose as a result of gene duplication.

Antibodies can be cleaved by proteinases. Cleavage by *papain* yields two identical **F$_{ab}$ fragments** ("antigen-binding") and an **F$_c$ fragment** ("crystallizable"). Isolated F$_{ab}$ fragments are still able to bind antigens. Studies of the F$_c$ fragment have shown that this part of the IgG molecule is not involved in antigen binding. Instead, the F$_c$ fragment mediates binding of the molecule to cell surfaces, e.g., in phagocytosis, interaction with the complement system, and transport of antibodies throughout the cells.

The antibody molecules are most flexible in the *"hinge"* region that connects the F$_{ab}$ "arms" with the F$_c$ part.

B. Classes of immunoglobulins ◑

Human immunoglobulins are divided into five different classes: **IgA** (two subgroups), **IgD**, **IgE**, **IgG** (four subgroups), and **IgM**. The differences between the classes lie in their heavy chains (**H chains**), which are designated by the Greek letters α, δ, ε, γ and μ, while the light chains (**L chains**), ϰ and λ, are

the same. IgD, IgE, and IgG are tetramers of the type (LH)$_2$. IgM may exist in various different forms. Secreted IgMs consist of five such tetramers. IgA, on the other hand, is made up of only two or three tetramers. Oligomers of IgM and IgA are held together by a "joining" or **J protein**.

The IgGs of all five classes are secretory proteins. They are released into the blood by mature B cells (plasma cells). Early varieties of IgM and IgD are also found as integral membrane proteins on the surface of B cells (see p. 272).

The different antibodies also have different functions. **IgMs** are the first immunoglobulins to be produced following contact with an antigen. They are especially active against micro-organisms. The most important immunoglobulins, on a quantitative basis, are the **IgGs**. These are found in the blood and in the interstitial fluid. In the presence of appropriate receptors, they are able to cross the placenta, i.e., they can be transferred from mother to fetus. Most **IgAs** occur in the intestinal tract and in body secretions (saliva, sweat, tears, milk, respiratory and intestinal secretions). Their release into these fluids is facilitated by binding to special peptides. In the plasma of healthy organisms, the levels of **IgE** are usually very low. They increase, however, during allergic reactions and after infection with parasites. The amounts of **IgD**, the function of which is not yet known, are also very low.

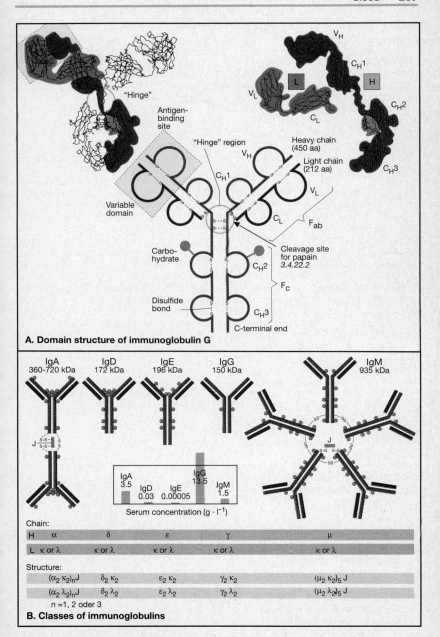

A. Domain structure of immunoglobulin G

"Hinge"

Antigen-binding site

"Hinge" region

V_H

C_H1

Variable domain

Carbohydrate

Disulfide bond

C_H2

C_H3

C-terminal end

Heavy chain (450 aa)

Light chain (212 aa)

V_L

C_L

F_{ab}

Cleavage site for papain
3.4.22.2

F_C

V_H

C_H1

L

H

V_L

C_L

C_H2

C_H3

B. Classes of immunoglobulins

IgA	IgD	IgE	IgG	IgM
360–720 kDa	172 kDa	196 kDa	150 kDa	935 kDa

J

Serum concentration (g · l^{-1})

IgA 3.5   IgD 0.03   IgE 0.00005   IgG 13.5   IgM 1.5

Chain:					
H	α	δ	ε	γ	μ
L	κ or λ	κ or λ	κ or λ	κ or λ	κ or λ

Structure:					
	$(\alpha_2 \kappa_2)_n J$	$\delta_2 \kappa_2$	$\varepsilon_2 \kappa_2$	$\gamma_2 \kappa_2$	$(\mu_2 \kappa_2)_5 J$
	$(\alpha_2 \lambda_2)_n J$	$\delta_2 \lambda_2$	$\varepsilon_2 \lambda_2$	$\gamma_2 \lambda_2$	$(\mu_2 \lambda_2)_5 J$

n = 1, 2 oder 3

Molecular Models: Hemoglobin, Immunoglobulin G

The facing page shows space-filling models of two important blood proteins: **hemoglobin** (Hb), which makes up most of the dry mass of erythrocytes (see p. 260), and a soluble **antibody** of the type **IgG**, which circulates in the blood plasma (see p. 266). The models are based on structural analyses of protein crystals using X-ray crystallography.

A. Deoxyhemoglobin and oxyhemoglobin ○

The illustration shows a comparison between the **deoxygenated** (left) and the **oxygenated** (right) forms of human **hemoglobin**. As discussed on p. 260, the hemoglobin of adults (HbA) is composed of 4 subunits, two α *chains* (141 amino acid residues, shown in yellow) and two β *chains* (146 residues, blue). Each chain contains a *heme group* (red).

The models illustrate the marked structural difference between the *T conformation* (left) and the *R conformation* (right) of tetrameric Hb. As discussed in more detail on p. 260, binding of one or two oxygen molecules to heme groups in the T state induces a conformational change giving rise to the R ("relaxed") state, which has a much higher affinity for oxygen. During the interconversion, the subunits move relative to one another, while the folding of the individual chains remains almost the same. However, as shown by the models, the heme groups become more accessible to oxygen in the R state. Since the concentration ratio of the T and R forms, and thus the effective oxygen affinity of Hb, vary with oxygen concentration, the saturation curve of Hb is distinctly S-shaped (sigmoidal, see p. 258).

B. Immunoglobulin G (IgG) ◑

As discussed on p. 266, immunoglobulins are soluble **antibodies** that circulate in the blood plasma. The illustration shows the structure of immunoglobulin G (**IgG**), the most important of the immunoglobulins. The Y-shaped molecule is a tetramer made up of two identical heavy chains (**H chains**, shown in orange) and two identical light chains (**L chains**, yellow). Each of the H chains bears an oligosaccharide consisting of 9 sugar residues (violet, see also pp. 40 and 72).

The illustration on the left shows an entire IgG molecule. Because of its flexibility, the so-called "hinge" region (see p. 266) is not clearly visible in X-ray structures. To generate the model, therefore, computer-based simulation (*molecular modeling*) has been used to predict possible conformations of the hinge region from its amino acid sequence. One of these structures was then combined with experimentally determined data for F_{ab} and F_c fragments to obtain a complete molecule.

The structure shown on the right was prepared from experimental data. It shows the complex of an isolated F_{ab} fragment (yellow/orange) with an antigen, the enzyme *lysozyme* (green). Note the excellent fit between the complementary surfaces of antibody and antigen, respectively. Residues highlighted in red belong to the so-called **hypervariable region** of the H and L chains. These regions show an especially high sequence variation between one IgG molecule and the next (see p. 270). From the figure, it is obvious, that most of the hypervariable residues are in direct contact with the antigen.

β chain α chain

Heme group

Deoxy-Hb (T form) Oxy-Hb (R form)

A. Deoxy- and oxyhemoglobin

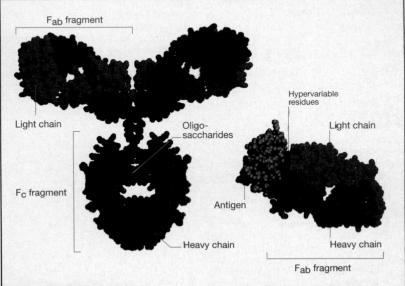

F_ab fragment

Light chain

F_c fragment

Oligo-
saccharides

Heavy chain

Hypervariable
residues

Light chain

Antigen

Heavy chain

F_ab fragment

B. Immunoglobulin G   (IgG)

Antibody Biosynthesis

A. Variability of immunoglobulins ◐

Immunoglobulins (Ig) all share the same basic structure (see p. 266). Despite this fact, they are extremely diverse in terms of specificity. It has been been estimated that 10^8 different antibody variants occur in every human being. Both the heavy and the light chains of the immunoglobulins show marked **variability**. The variable regions are highlighted in red in the illustration.

There are two different types of light (L) chains ($\varkappa$, λ) and five different types of heavy (H) chains (α, δ, ε, γ, and μ, see p. 266). These differences are referred to as **isotypic variation**. During the biosynthesis of the immunoglobulins, plasma cells can switch between one isotype and another ("gene switch"). The term **allotypic variation** refers to the allelic variability within a species, i.e., the genetically-fixed variations from one individual to the next. **Idiotypic variation** is a result of variations in the binding site of the antigen. This involves the *variable domains* of the light and heavy chains. Some sites, the *hypervariable regions*, show greater than average variation (see p. 268).

B. Origins of antibody variety ◐

The extreme diversity shown by antibodies is made possible by three different factors:

1. **Multiple genes**. There are multiple genes that encode the variable protein domains, but only one of these is selected and expressed.
2. **Somatic mutations**. During differentiation of B cells to plasma cells, mutations occur in the coding genes. Thus, the *"primordial"* germ-line genes can become different *somatic genes* in the individual B cell clones.
3. **Somatic recombination**. The genes are divided into fragments, of which there are multiple versions. These different fragments combine in a variety of ways during B cell maturation. New gene combinations, referred to as *mosaic genes*, arise as a result of this.

C. Biosynthesis of a light chain ○

Taking the biosynthesis of a mouse $\varkappa$ chain as an example, let us consider the basic features of immunoglobulin gene organization and immunoglobulin expression. The various regions of the gene encoding light chains are designated L, V, J, and C. In germ-line DNA of mice, they are located on chromosome 6 (in humans, on chromosome 2) and they are separated from one another by introns (see p. 222) of different lengths.

Approximately 150 identical **L gene segments** encode the signal peptide (leader sequence, 17–20 residues) for the secretion of the product (see p. 216). The greater part of the variable domains (95 of the 108 residues) is encoded by about 150 different **V segments,** located next to the L segments. L and V segments always occur together i.e., in tandem. In contrast, there are only five variants of the **J segments** (joining segments) at most. These encode a peptide of 13 amino acids that links the variable to the constant portion of the $\varkappa$ chain. The constant part of the light chain (84 residues) is encoded by a single **C-segment.**

During differentiation of the B lymphocytes, unique **V/J combinations** arise in each B cell. One of the 150 L/V tandem segments is selected, and linked to one of the five J segments. This gives rise to a **somatic gene** that is much smaller than the germ-line gene. Transcription of this genes leads to the formation of an hnRNA for the $\varkappa$ chain, from which introns and extra J segments are removed by splicing (see p. 224). The mature RNA contains the segments L, V, J, and C . Once transported into the cytoplasm, it is ready for translation. The subsequent steps in Ig biosynthesis are, in general, the same as for other membrane-bound or secretory proteins (see p. 214).

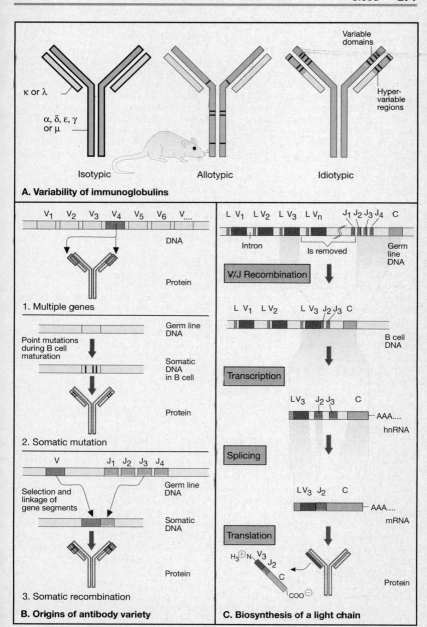

A. Variability of immunoglobulins

κ or λ

α, δ, ε, γ or μ

Isotypic Allotypic Idiotypic

Variable domains

Hyper-variable regions

B. Origins of antibody variety

V₁ V₂ V₃ V₄ V₅ V₆ V....

DNA

Protein

1. Multiple genes

Germ line DNA

Point mutations during B cell maturation

Somatic DNA in B cell

Protein

2. Somatic mutation

V J₁ J₂ J₃ J₄

Selection and linkage of gene segments

Germ line DNA

Somatic DNA

Protein

3. Somatic recombination

C. Biosynthesis of a light chain

L V₁ L V₂ L V₃ L Vₙ J₁ J₂ J₃ J₄ C

Intron

Is removed

Germ line DNA

V/J Recombination

L V₁ L V₂ L V₃ J₂ J₃ C

B cell DNA

Transcription

L V₃ J₂ J₃ C

AAA....

hnRNA

Splicing

L V₃ J₂ C

AAA....

mRNA

Translation

H₃⁺N V₃ J₂ C

COO⁻

Protein

MHC Proteins

A. Proteins of the immunoglobulin gene superfamily ◑

The illustration shows the most important members of the immunoglobulin superfamily. All of these proteins have a common function, i.e., the ability to *specifically recognize* and differentiate between various molecules by binding to them. In some cases, this binding can lead to *signal transduction* into the cell.

All of the proteins in this family are *homologous* to one another, i.e., they are similar in structure. Not only the immunoglobulins, but also all other members of the family, are made up of characteristic domains (blue box). The constant regions are shown in brown and green, and the variable ones in orange. Homologous domains are shown in the same color. Most of the proteins of the superfamily carry transmembrane regions near their C termini, i.e., they are *integral membrane proteins* (the exceptions to this are the secreted forms of the immunoglobulins IgA, IgD, IgE, IgG and IgM). Disulfide bonds located at specific positions may be either intramolecular or intermolecular.

T cell receptors are located on the surfaces of T cells (see p. 264). They are glycosylated heterodimers consisting of an α chain and a β chain. Their constant and variable domains are closely related to those of the immunoglobulins, but they have only *one* specific binding site. As with the immunoglobulins, multiple coding regions (exons) encode this binding site, which is ultimately generated by somatic recombination (see p. 270). This ensures the structural diversity that is the prerequisite for specific binding of target molecules by the receptor.

T cells receptors bind to MHC proteins (**B**) of other cells (bottom left). The illustration shows a T cell binding to a B cell that is presenting a foreign peptide. This interaction results in the activation of the T cells (see p. 264).

The **immunoglobulins** (Ig) are discussed on the preceding pages. The **IgM dimer** shown here makes its first appearance on the surface of B cells, and is later secreted by plasma cells.

The **MHC proteins** are so called, because they are encoded by a stretch of DNA known as the *major histocompatability complex*. They are glycosylated membrane proteins that are expressed in all vertebrate animals (see p. 264). In humans, the MHC proteins are also referred to as *HLA antigens* (human leukocyte-associated antigens). MHC proteins are extremely diverse. Their polymorphism is so great that it is very unlikely that any two individuals have the same set of MHC proteins, except for identical twins.

MHC proteins can be divided into two large groups. **Class I** MHC proteins are found on the surfaces of almost all nucleated cells. They are the reason why one individual rejects tissue transplants from another (this is the origin of the older term "transplantation antigens"). Class I MHC proteins consist of only one α chain associated with β_2-*microglobulin*, a small invariant protein.

Class II MHC proteins are similar to class I proteins, except that they consist of two membrane-spanning chains, an α chain and a β chain. These are restricted to the surfaces of the cells of the immune system. Their presence allows the body to differentiate between the cells of the immune system and other cells.

B. Structure of a human class I MHC protein ○

Structural analyses of MHC proteins have shown that on the apex of the molecule, there is a cleft between two domains of the α chain. An oligopeptide of 10–20 amino acids (magenta) is frequently found within this cleft. This is the way in which fragments of intracellularly degraded foreign proteins are presented to the *T cell receptors*. However, the T cells only bind when, in addition to the exposed antigen, further helper molecules are present (not shown here).

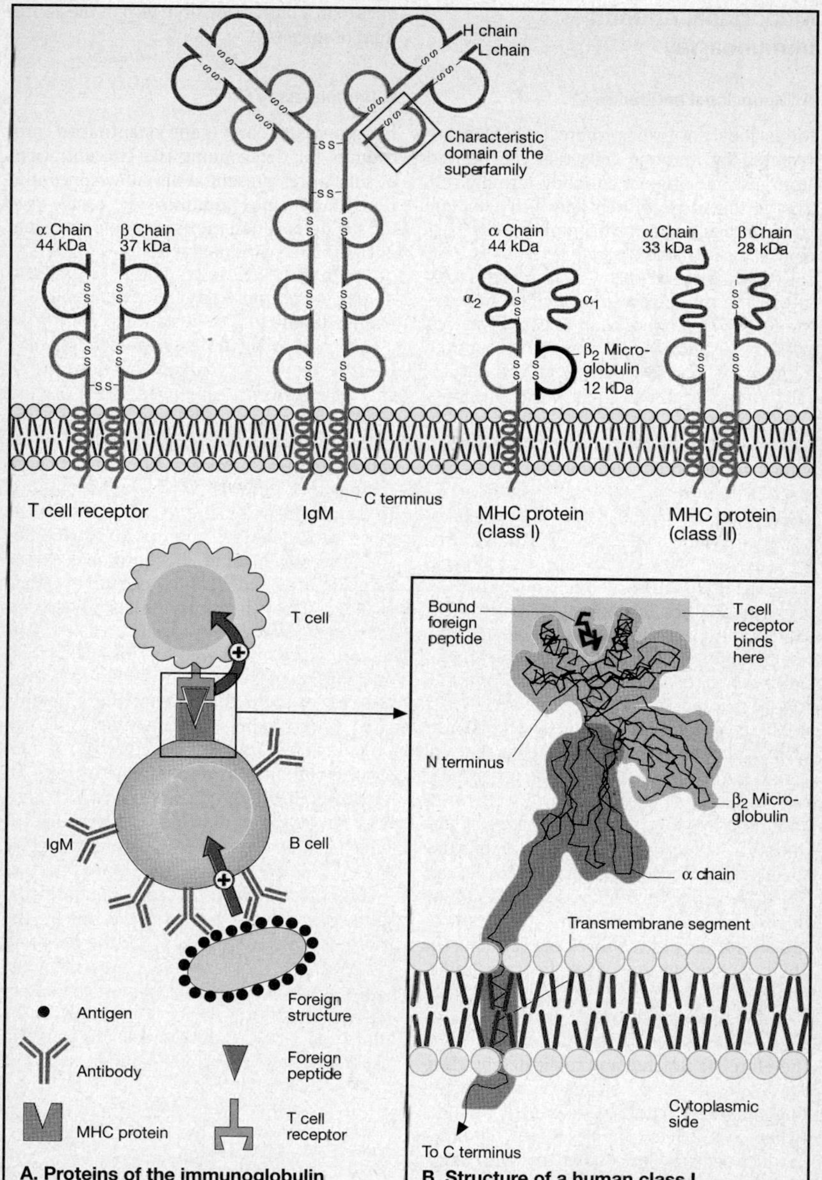

Characteristic domain of the superfamily

α Chain 44 kDa β Chain 37 kDa

α Chain 44 kDa

α Chain 33 kDa β Chain 28 kDa

H chain
L chain

α_2 α_1

β_2 Micro-globulin 12 kDa

C terminus

T cell receptor   IgM   MHC protein (class I)   MHC protein (class II)

T cell

B cell

IgM

Antigen

Antibody

MHC protein

Foreign structure

Foreign peptide

T cell receptor

Bound foreign peptide

T cell receptor binds here

N terminus

β_2 Micro-globulin

α chain

Transmembrane segment

Cytoplasmic side

To C terminus

A. Proteins of the immunoglobulin gene superfamily

B. Structure of a human class I MHC protein (type HLA-A2)

Monoclonal Antibodies, Immunoassay

A. Monoclonal antibodies ○

Monoclonal antibodies (abbreviated Mab) are secreted by immune cells that are derived from just one original antibody-forming cell. This is the basis of their specificity for one particular epitope of an immunogenic substance, a so-called *antigenic determinant*. For the production of Mabs, **lymphocytes** isolated from the spleen of immunized mice (**1**) are fused with mouse tumor cells (**myeloma cells, 2**). This fusion is necessary, because the lifespan of antibody-secreting lymphocytes in culture is limited to a few weeks. Fusion of the lymphocytes with tumor cells results in the formation of cell hybrids, or **hybridomas**, which are potentially immortal.

Successful fusions (**2**) are quite rare, but their frequency can be increased by the addition of polyethylene glycol (PEG). Selection for successfully fused cells involves a long period of **primary culture** in HAT medium (**3**), a nutrient solution containing **h**ypoxanthine, **a**minopterin, and **t**hymidine. *Aminopterin*, an analog of dihydrofolic acid, competitively inhibits *dihydrofolate reductase*, and thus inhibits the synthesis of dTMP (see p. 176). As dTMP is essential for DNA synthesis, myeloma cells cannot survive in the presence of aminopterin. Spleen cells, on the other hand, can circumvent the inhibitory effect of aminopterin by using hypoxanthine and thymidine as DNA precursors. However, they fail to survive, because of their limited lifespan. This is why only hybridomas survive culture in HAT medium, because they possess both the immortality of the myeloma cells and the ability of the spleen cells to tolerate aminopterin.

Only a few hybridoma cells actually produce antibodies. Those, which do must, therefore, be isolated and replicated by **cloning (4)**. By testing the clones for antibody formation, positive cultures are selected, and enriched by further cloning (**5**). The ultimate products of this procedure are hybridomas synthesizing monoclonal antibodies. Production of Mabs by these cells can be carried out *in vitro* in a bioreactor, or *in vivo* in the ascites fluid of mice (**6**).

B. Immunoassay ○

Immunoassays are (semi)quantitative procedures for determining the concentrations of substances present at very low concentrations. In principle, immunoassays can be used for the determination of any compound that elicits the formation of antibodies.

The basis of this procedure is the *antigen-antibody "reaction,"* i.e., the specific binding of an antibody to the substance to be assayed (see p. 268). Many different procedures have been developed e.g., *radioimmunoassay* (RIA), *chemoluminescence immunoassay* (CIA) etc. Here we present only one example, the **enzyme-linked immunoassay** (EIA). The sample to be assayed, e.g., blood serum containing the hormone thyroxine, is pipetted into a microtiter plate (**1**). The "wells" of the plate have been precoated with **antibodies** that specifically bind the hormone. At the same time, a small quantity of pure hormone is added, to which an enzyme is covalently attached, the so-called **tracer, (1)**. The number of antibodies in the well to which the tracer and the hormone can bind is limited. Thus, the two of them compete for these binding sites. Once binding has taken place (**2**), unbound molecules are washed away. A substrate solution (**chromogenic solution**) for the tracer enzyme is then added, which initiates an indicator reaction, the products of which can be assayed by spectrophotometry (**4**).

The more enzyme/tracer is bound to the antibodies on the walls of the vessel, the more dye will be produced by the indicator reaction. The more hormone present in the sample, on the other hand, the less binding of the tracer will take place. The assay can be made quantitative with the help of a standard calibration curve run in parallel with the assay.

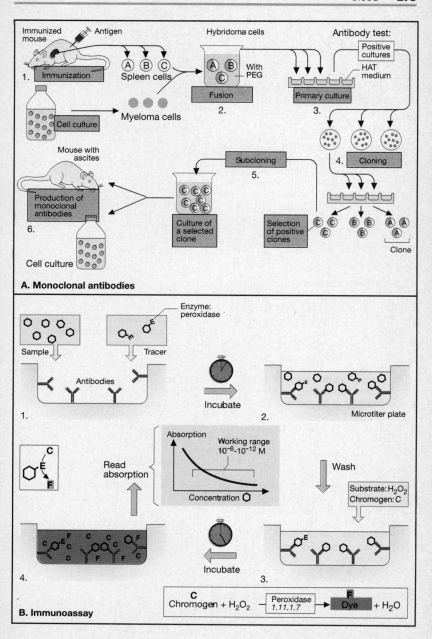

A. Monoclonal antibodies

B. Immunoassay

Functions

The liver weighs about 1.5 kg, and is thus one of the largest organs in the human body. Despite the fact that it constitutes only 2–3% of the body mass, it accounts for 20–30% of the total oxygen consumption.

A. Diagram of a hepatocyte ◑

There are approximately 300 billion cells in the liver, and 80% of them are hepatocytes. The liver cells, especially the hepatocytes, are central to the *intermediary metabolism* of the body. Therefore, the hepatocytes have come to be considered the prototype cell in biochemistry (see p. 182).

B. Functions of the liver ●

The most important functions of the liver are:

1. The **uptake** of nutrients delivered from the digestive tract via the portal vein.
2. The synthesis, storage, interconversion, and degradation of metabolites (**metabolism**).
3. The regulated **supply** of energy-rich intermediates and building blocks for biosynthetic reactions.
4. The **detoxification** of harmful compounds by biotransformation.
5. The **excretion** of substances with the bile, as well as the synthesis and degradation of many blood plasma constituents (not shown).

C. Liver metabolism ●

The liver is involved in the metabolism of most metabolites. Its primary function is the maintenance of sufficient plasma levels of these metabolites (*homeostasis*).

Carbohydrate metabolism. The liver takes up glucose and other monosaccharides from the blood plasma. These sugars are then converted to glucose 6-phosphate and other intermediates of glycolysis (see p. 282). Subsequently, they are either stored as the reserve carbohydrate glycogen or degraded. Another large part is converted into fatty acids, and only a small fraction is used for the generation of ATP. When there is a major decline in the blood glucose level, the glucose flux is reversed, and the liver secretes rather than takes up glucose. This glucose is derived from the *glycogen store*. If the glycogen store is already exhausted, glucose can also be synthesized by *gluconeogenesis* from lactate, glycerol, or the backbones of amino acids.

Lipid metabolism. The liver synthesizes fatty acids from acetate units (see p. 144). The fatty acids formed are then used for the synthesis of fats and phospholipids. These are subsequently released into the blood as complexes with proteins, i.e., *lipoproteins* (see p. 254). On the other hand, the liver can take up fatty acids from the plasma. Of particular importance is the ability of the hepatocytes to convert fatty acids into ketone bodies, which are also excreted (see p. 284).

The liver also uses acetate units to synthesize cholesterol, which is then transported to other organs as a constituent of plasma lipoproteins. Excess cholesterol is converted into bile acids or excreted with the bile.

Amino acid and protein metabolism. The liver controls the amino acid level of the plasma, i.e., it breaks down excess amino acids. The nitrogen thus released is converted to urea and transported to the kidneys. The carbon skeletons of the amino acids enter intermediary metabolism, where they are consumed either in the synthesis of glucose or in the production of energy. In addition, the liver is the site of the synthesis and degradation of most proteins and peptides in the blood plasma.

Biotransformations. Steroid hormones and degradation products of the blood pigment hemoglobin are taken up by the liver, where they are inactivated and converted into highly polar metabolites. Drugs, ethanol, and other foreign substances (*xenobiotics*), are metabolized by similar reactions, to prepare them for excretion (see p. 290).

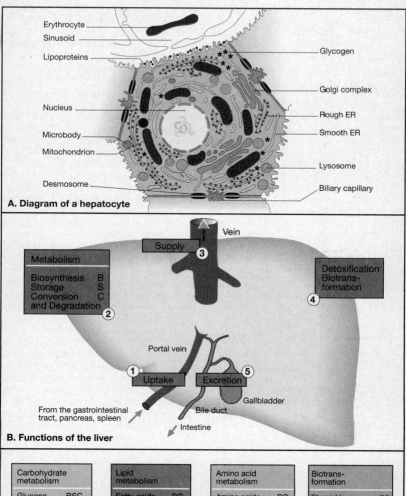

A. Diagram of a hepatocyte

Erythrocyte
Sinusoid
Lipoproteins
Nucleus
Microbody
Mitochondrion
Desmosome

Glycogen
Golgi complex
Rough ER
Smooth ER
Lysosome
Biliary capillary

B. Functions of the liver

Vein
Supply ③
Metabolism
Biosynthesis   B
Storage   S
Conversion   C
and Degradation
②

Detoxification
Biotrans-
formation
④

Portal vein
① Uptake   Excretion ⑤
Gallbladder
From the gastrointestinal
tract, pancreas, spleen
Bile duct
Intestine

C. Liver metabolism

Carbohydrate metabolism		Lipid metabolism	
Glucose	BSC	Fatty acids	BC
Galactose	C	Fats	BC
Fructose	C	Ketone bodies	B
Mannose	C	Cholesterol	BEC
Pentoses	BC	Bile acids	BE
Lactate	C	Vitamins	SC
Glycerol	BC		
Glycogen	BSC		

Amino acid metabolism	
Amino acids	BC
Urea	B

Plasma proteins	
Lipoproteins	BC
Albumin	BC
Coagulation factors	BC
Hormones	BC
Enzymes	BC

Biotrans-formation	
Steroid hormones	EC
Bile pigments	EC
Ethanol	C
Drugs	EC

B	Biosynthesis
C	Conversion and degradation
E	Excretion
S	Storage

Metabolism in the Well-Fed State

The tissues of higher organisms are dependent on a constant supply of "energy-rich" metabolites to provide energy and as precursors for the synthesis of complex macromolecules. The timing of the provision of these fuels in the diet, and the amounts provided, are often quite variable. This variation is compensated for by the liver, which, in conjunction with other tissues, acts as a *buffering* and *storage organ.*

In the biochemistry of nutrition, we differentiate between the **well-fed (absorptive) state** and **starvation (postresorptive state)**. The transition between these two states depends on the levels of energy-rich metabolites in the blood plasma. It is brought about by the joint action of hormones and signals from the nervous system. In the fed state, the energy requirements of the tissues are predominantly met by glucose, whereas in the starved state, they are met mainly by fatty acids, amino acids, and ketone bodies (exceptions: nerve tissue and erythrocytes, see pp. 316 and 260).

A. Absorptive state ○)

The absorptive state sets in directly after a meal, and lasts for about 2 to 4 hours. Immediately following digestion of the meal, there is a transient increase in the levels of **glucose, amino acids,** and **neutral fats** (triacylglycerols) in the blood plasma. The pancreas responds to this by changing the amounts of the hormones it secretes. There is an increase in *insulin* release, and a decrease in the release of *glucagon*. The elevation of the insulin/glucagon ratio, combined with the availability of metabolic fuels, stimulates the tissues to enter an *anabolic phase*. In the liver, metabolites are converted into glycogen and fat, while in the muscles glycogen and proteins are synthesized, and fat in the adipose tissues. These stored substances constitute the organism's energy reserves. There is a continued supply of glucose to the heart and the nervous tissue that proceeds regardless of changes in the insulin level.

B. Liver metabolism in the well-fed state ○)

Veinous blood from the intestine and the pancreas is transported directly to the liver via the portal vein.

Carbohydrate metabolism. Uptake of glucose and other sugars across the plasma membrane of hepatocytes occurs with the help of transporters (a process not affected by insulin). Once inside the cell, glucose is rapidly phosphorylated to glucose 6-phosphate (see p. 140). This intermediate can enter a number of different metabolic pathways. There is an increase in glycogen synthesis (caused by insulin). The *hexose monophosphate pathway* (**1**) also becomes more active, and the increased amounts of NADPH thus synthesized are used for the synthesis of fatty acids. *Glycolysis* (**2**) is also accelerated by activation of the key enzymes *phosphofructokinase* and *pyruvate kinase* (combined effects of insulin and glucagon). The acetyl-CoA formed is used in *fatty acid synthesis* (**3**). The rate of *gluconeogenesis* (**4**), on the other hand, decreases, mainly because of an inhibition of *pyruvate carboxylase* and *fructose 1,6-bisphosphatase* (see p. 162).

Fat metabolism. The liver is the most important site of *de novo* fatty acid synthesis. The substrates for this pathway, acetyl-CoA and NADPH, are supplied by the degradation of glucose. The availability of fatty acids increases as a result of their synthesis in the liver and the degradation of lipids supplied in the blood. This leads to increased synthesis of neutral fats. Glycerol 3-phosphate, which is required for fat synthesis, is derived from glycolysis (not shown). The fats thus formed are packaged into *lipoproteins* of the class VLDL (see p. 254) and released into the blood. Their constituents are absorbed predominantly by adipose tissue and the muscles.

Amino acid metabolism. In the absorptive state, the supply of amino acids in the liver exceeds the consumption of them. Some of the amino acids are partially degraded, i.e., deaminated. The carbon skeletons derived from glucogenic amino acids are fed into intermediary metabolism (see p. 150), while the others are used for lipid synthesis. Protein synthesis in the liver is also increased to some extent.

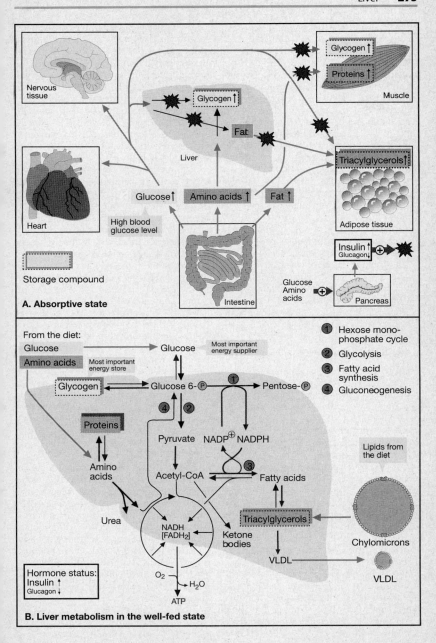

A. Absorptive state

Nervous tissue

Glycogen ↑

Proteins ↑

Muscle

Liver

Glycogen ↑

Fat

Triacylglycerols↑

Heart

Glucose↑   Amino acids ↑   Fat ↑

High blood glucose level

Adipose tissue

Storage compound

Insulin ↑
Glucagon↓

Intestine

Glucose
Amino acids

Pancreas

B. Liver metabolism in the well-fed state

From the diet:

Glucose

Amino acids

① Hexose mono-phosphate cycle

② Glycolysis

③ Fatty acid synthesis

④ Gluconeogenesis

Glucose

Most important energy supplier

Most important energy store

Glycogen

Glucose 6-℗

Pentose-℗

①

④ ②

Proteins

Pyruvate

NADP⊕ NADPH

Lipids from the diet

③

Amino acids

Acetyl-CoA

Fatty acids

Urea

NADH
[FADH₂]

Ketone bodies

Triacylglycerols

Chylomicrons

VLDL

Hormone status:
Insulin ↑
Glucagon ↓

O₂ → H₂O

ATP

VLDL

Metabolism During Starvation

A. Postresorptive state ◑

When animals are deprived of food for a sufficiently long time, there is a switch to a **state of starvation**. This occurs in response to a change in the pattern of hormone release from the pancreas. There is an increase in the amount of *glucagon* being released by the A cells, and a decrease in the amount of *insulin* being secreted by the B cells. The decreased insulin/glucagon ratio in the plasma triggers a switch in intermediary metabolism. Energy-yielding fuels are shunted between the liver, fatty tissues, muscle and brain. This serves to:

- Maintain blood glucose at a level sufficient to ensure a constant supply to the brain, the medulla of the suprarenal gland, and the erythrocytes, all of which are dependent on a constant provision of glucose.
- Supply energy to the remaining tissues through the mobilization of fatty acids from adipose tissues and the formation of ketone bodies in the liver.

B. Liver metabolism during starvation ◑

The liver is the central organ in metabolism, and as such it is responsible, together with the adipose tissue, for the continued provision of energy-yielding fuels during periods of starvation.

Carbohydrates. When energy supplies are low (e.g., at night or during fasting or starvation) the first response of the liver is to mobilize its glycogen reserves. Glycogen is converted to glucose 1-phosphate by *glycogen phosphorylase* and then to glucose 6-phosphate. Following the removal of the phosphate group, glucose is released into the blood (see p. 110). Phosphorylase is activated by the higher level of cAMP resulting from the low insulin/glucagon ratio (see p. 162). The glycogen reserves of the liver only yield about 150 g of glucose, and therefore they are exhausted after 6 to 12 hours of starvation. As mentioned above, it is essential that the plasma glucose level be maintained (e.g., for the brain). Thus, once the glycogen reserves

have been depleted, there is an increase in the rate of *gluconeogenesis* for the synthesis of glucose (**1**, see p. 158). Gluconeogenesis is the most important source of glucose during prolonged fasting. The main substrates for gluconeogenesis are the carbon skeletons of glucogenic amino acids and glycerol derived from fat degradation. Fatty acids cannot be used for the formation of glucose, because the acetyl-CoA arising from the β-oxidation of fatty acids is completely oxidized to CO_2 in the citric acid cycle.

Fat metabolism. The most important energy-yielding fuels in the liver during prolonged periods of starvation are the fatty acids mobilized in the adipose tissue by lipolysis. In the liver, the fatty acids are degraded to acetyl-CoA, and then, for the most part, converted into *ketone bodies* (**3**). Ketone body synthesis is initiated, because under these conditions, the amounts of acetyl-CoA formed exceed the degradative capacity of the citric acid cycle (see p. 146). The principal ketone body is hydroxybutyrate, but small amounts of acetoacetate also arise (see p. 284). A third ketone body, acetone, cannot be used for the provision of energy. The ketone bodies produced by the liver are used to meet the energy requirements of other tissues. The rate of ketone body synthesis increases rapidly during the first few days of fasting, but remains constant thereafter (up to several weeks). All tissues except the liver can utilize ketone bodies for ATP synthesis. The oxidation of ketone bodies saves on glucose, and reduces the necessity for the degradation of proteins.

Energy storage in the body. A human being of normal weight possesses large stores of energy-yielding metabolites, which suffice for many weeks. Fats are the major source of energy, while proteins and glycogen constitute much smaller reserves. The utilization of proteins as fuels is restricted, because of the important roles they play as enzymes, transporters, structural components, etc., Proteins are not normally mobilized for the production of energy before starvation has become very advanced.

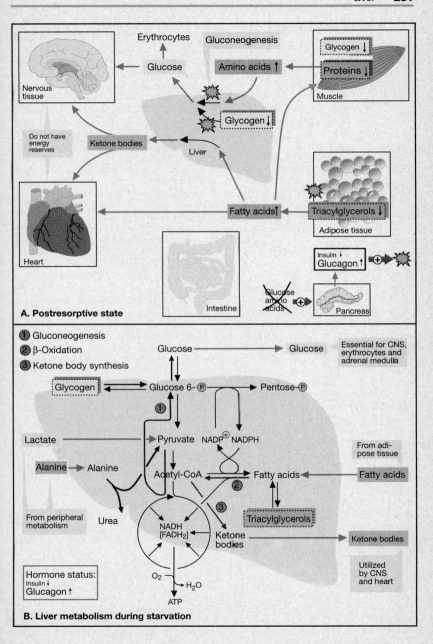

A. Postresorptive state

Erythrocytes

Gluconeogenesis

Glucose ← Amino acids ↑

Nervous tissue

Do not have energy reserves

Ketone bodies ← Liver

Glycogen ↓

Glycogen ↓

Proteins ↓

Muscle

Heart

Fatty acids↑ ← Triacylglycerols ↓

Adipose tissue

Insulin ↓
Glucagon ↑ ⊕

Glucose amino acids ⊕ Pancreas

Intestine

B. Liver metabolism during starvation

① Gluconeogenesis
② β-Oxidation
③ Ketone body synthesis

Glucose → Glucose

Essential for CNS, erythrocytes and adrenal medulla

Glycogen ⇌ Glucose 6-Ⓟ → Pentose-Ⓟ

①

Lactate → Pyruvate

NADP⊕ NADPH

From adipose tissue

Alanine → Alanine

Acetyl-CoA ⇌ Fatty acids ← Fatty acids

②

From peripheral metabolism

Urea

③

Triacylglycerols

NADH [FADH₂] ← Ketone bodies → Ketone bodies

Utilized by CNS and heart

Hormone status:
Insulin ↓
Glucagon ↑

O₂ → H₂O

ATP

Carbohydrate Metabolism

After the fatty acids and the ketone bodies, glucose is the most important energy-yielding metabolite in animals. The concentration of glucose in the blood, the "*blood glucose level*," is maintained constant at 4–6 mM ($0.8–1 \; g \cdot l^{-1}$). This is achieved by the precise regulation of glucose-forming and glucose-utilizing pathways, a process in which the liver plays a major role.

In the liver, glucose is either derived from other sugars (e.g., fructose and galactose) or synthesized from smaller metabolites. Further pathways in the liver, such as the conversion of lactate to glucose via the *Cori cycle* and the conversion of alanine to glucose via the *alanine cycle*, are especially important for peripheral tissues.

A. Cori and alanine cycles ◑

Cells that have no mitochondria (e.g., erythrocytes) or tissues with an oxygen demand that temporarily exceeds the oxygen supply (e.g., muscles during heavy exercise) obtain ATP by fermentation, i.e., the conversion of glucose to lactate (**anaerobic glycolysis**; see p. 136). The lactate thus formed is transported by the blood to the liver, where it is converted back into glucose in **gluconeogenesis** with an input of ATP. The glucose formed in the liver is then sent back to its site of utilization. This is referred to as the *Cori cycle*.

A similar cycle exists for pyruvate and alanine. The *alanine cycle* is fed by the degradation of proteins. The amino acids produced as a result of proteolysis are converted by *transamination* into 2-oxoacids, most of which enter the citric acid cycle (see p. 154). The amino groups gained are transferred to pyruvate, giving rise to alanine. Alanine is then released into the blood and transported to the liver. Thus, the *alanine cycle* is responsible for the simultaneous transfer of glucose precursors and nitrogen to the liver. Once in the liver, the nitrogen is converted into urea.

B. Fructose and galactose metabolism ◑

The main pathway for the degradation of **fructose** in the liver is via glycolysis (left side of the illustration). Before it can enter glycolysis, however, fructose first has to be phosphorylated by *ketohexokinase* [1] to yield fructose 1-phosphate. This is then cleaved by a special *aldolase* [2] to yield dihydroxyacetone phosphate, an intermediate of glycolysis (center), and glyceraldehyde. Glyceraldehyde is subsequently phosphorylated by *triokinase* [3] to glyceraldehyde 3-phosphate, another glycolytic intermediate. Some glyceraldehyde is reduced to glycerol [4] or oxidized to glycerate, both of which can enter glycolysis following phosphorylation (not shown). The reduction of glyceraldehyde [4] requires NADH + H⁺. This is why fructose stimulates ethanol degradation (see p. 294). In tissues other than the liver, the entry of fructose into carbohydrate metabolism occurs by reduction at C-2 to yield sorbitol, and subsequent dehydration at C-1 to yield glucose (*polyol pathway*, not shown).

Galactose is also metabolized in the liver (right side of the illustration). As for all sugars, the metabolism of galactose begins with a phosphorylation, in this case to yield galactose 1-phosphate [5]. The entry of galactose into glycolysis is mediated by epimerization of galactose 1-phosphate at C-4 to yield the glucose derivative. This occurs in a reaction that also involves the uridine diphosphate (UDP) derivatives of the two sugars where UDP is attached at C-1 [6] [7]. Galactose *biosynthesis* also proceeds via this pathway. Except for reaction [5], the process is freely reversible.

Further information

A rare hereditary disease leading to *fructose intolerance* is due to a deficiency in *fructose-bisphosphate aldolase*. The presence of fructose in the diet leads to the accumulation of fructose 1-phosphate, which inhibits glycolysis. Disturbances of glucose metabolism as a result of the accumulation of galactose 1-phosphate are caused by an hereditary disorder known as *galactosemia*. The basis of this disorder is a deficiency of *hexose 1-phosphate-uridyl transferase* [6].

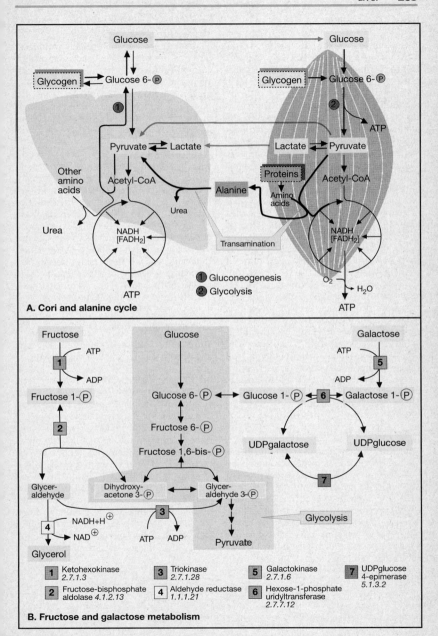

A. Cori and alanine cycle

① Gluconeogenesis
② Glycolysis

B. Fructose and galactose metabolism

1 Ketohexokinase *2.7.1.3*
2 Fructose-bisphosphate aldolase *4.1.2.13*
3 Triokinase *2.7.1.28*
4 Aldehyde reductase *1.1.1.21*
5 Galactokinase *2.7.1.6*
6 Hexose-1-phosphate uridyltransferase *2.7.7.12*
7 UDPglucose 4-epimerase *5.1.3.2*

Lipid Metabolism

The liver is the most important site for the formation of *fatty acids, fats, ketone bodies*, and *cholesterol*. Adipose tissues also synthesize fats, but their main function is the storage of lipids.

The metabolism of lipids in the liver is closely linked to that of carbohydrates and amino acids. In the well-fed, or *absorptive state* (see p. 278), the liver converts glucose via acetyl-CoA into fatty acids. The liver can also retrieve fatty acids from lipids supplied with *chylomicrons* from the intestine. The fatty acids from both sources are converted into neutral fats and phospholipids. Lipoprotein complexes (**VLDL**, see p. 254) are then formed in the smooth ER of the hepatocytes by association with apolipoproteins. The VLDLs are released into the blood to provide extrahepatic tissue with fatty acids, above all the adipose and muscle tissue.

In the *postabsorptive state* (see p. 280), especially during prolonged fasting, starvation, or in the case of *diabetes mellitus* (see p. 164), there is a shift in lipid metabolism. Since glucose and lipids are no longer being supplied in the diet, the organism has to fall back on its own reserves. Under these conditions, the adipose tissue releases fatty acids that are taken up by the liver from the blood, oxidatively degraded to acetyl-CoA, and finally converted to ketone bodies.

A. Biosynthesis of ketone bodies ◑

When the concentration of acetyl-CoA in the liver mitochondria is high, two molecules condense to form **acetoacetyl-CoA** [1]. The addition of a further acetyl group [2] gives rise to 3-hydroxy-3-methylglutaryl-CoA, which, by removal of acetyl-CoA, yields **acetoacetate** (*Lynen cycle*). Acetoacetate can be converted to **3-hydroxybutyrate** by reduction [4], or breaks down to **acetone** by non-enzymatic decarboxylation [5]. Somewhat imprecisely, these three compounds are referred to as "*ketone bodies* ".

The ketone bodies are released by the liver into the blood, in which they are readily soluble. Therefore, the levels of ketone bodies in the blood are elevated during periods of starvation. 3-Hydroxybutyrate and acetoacetate then serve, together with fatty acids, as the key metabolites in energy production. Acetone, which has no metabolic significance, is exhaled via the lungs. After 1–2 weeks of starvation, the nerve tissue also begins to utilize ketone bodies as energy sources. However, it still requires a minimum amount of glucose to replenish the citric acid cycle.

When the production of ketone bodies exceeds their use outside the liver, ketone bodies accumulate in the plasma (*ketonemia*), and are eventually excreted in the urine (*ketosuria*). Both phenomena are observed after prolonged starvation or in *diabetes mellitus* (see p. 164). Since ketone bodies are moderately strong acids (with pK_a values around 4), greatly increased ketone body formation can markedly lower the plasma pH value (*ketoacidosis*). Severe ketoacidosis quickly leads to electrolyte shifts and loss of consciousness, and is therefore life-threatening.

C. Cholesterol metabolism ◑

There are two sources of cholesterol, the diet and *de novo* synthesis within the body. A significant amount of cholesterol is synthesized in the liver. The synthetic pathway (see p. 170) starts with acetyl-CoA. Some cholesterol is required for the synthesis of bile acids (see p. 288), some serves as a building block for cell membranes (see p. 204), and some is stored in the form of lipid droplets, following esterification with fatty acids. The rest of the cholesterol in free or esterified form, is incorporated in lipoprotein complexes of very low density (**VLDLs**), which are then released into the blood to supply other tissues. In addition, the liver takes up from the blood and degrades lipoprotein complexes containing cholesterol and cholesterol esters (HDL, IDL, and LDL, see p. 254).

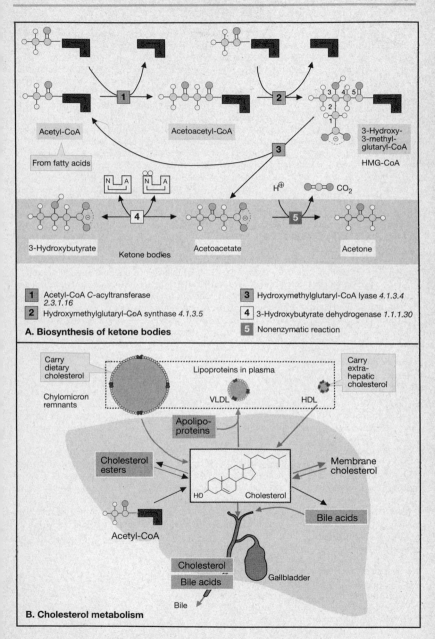

A. Biosynthesis of ketone bodies

Acetyl-CoA

From fatty acids

Acetoacetyl-CoA

3-Hydroxy-3-methyl-glutaryl-CoA
HMG-CoA

$H^{\oplus}$ CO_2

3-Hydroxybutyrate

Ketone bodies

Acetoacetate

Acetone

1 Acetyl-CoA *C*-acyltransferase 2.3.1.16

2 Hydroxymethylglutaryl-CoA synthase 4.1.3.5

3 Hydroxymethylglutaryl-CoA lyase 4.1.3.4

4 3-Hydroxybutyrate dehydrogenase 1.1.1.30

5 Nonenzymatic reaction

B. Cholesterol metabolism

Carry dietary cholesterol

Lipoproteins in plasma

Carry extra-hepatic cholesterol

Chylomicron remnants

VLDL

HDL

Apolipo-proteins

Cholesterol esters

Membrane cholesterol

HO   Cholesterol

Bile acids

Acetyl-CoA

Cholesterol

Bile acids

Gallbladder

Bile

Urea Cycle

The main site of amino acid degradation is the liver. In the process, significant quantities of ammonia are released, either directly or indirectly (see p. 152). The same is true for the degradation of purines and pyrimidines (see p. 156).

Ammonia (above left) — a moderately strong base — is toxic to cells. At higher concentrations, it is especially damaging to the nerve cells. Therefore, ammonia has to be rapidly inactivated and excreted. In humans, this occurs primarily through the formation of **urea.**

In nature, there are various different ways of inactivating and excreting ammonia. Aquatic animals can excrete ammonia directly. For example, fish dispose of it via the gills (*ammonotelic animals*). Terrestrial animals, including humans, excrete only small amounts of ammonia in the urine. Instead, most of the ammonia is converted to urea prior to excretion (*ureotelic animals*). In contrast, birds and reptiles (*uricotelic animals*) form *uric acid*, which is excreted predominantly as a solid in order to save water.

A. Urea cycle ⏵

Urea is the diamide of carbonic acid. In contrast to ammonia, it is *neutral* and *non-toxic*. Being a small, uncharged molecule, it can cross membranes by diffusion. For this reason, and also because it is highly soluble in water, it can be easily transported by the blood and excreted in the urine.

Urea is formed in a cyclic series of reactions. Its nitrogen atoms are derived from **ammonia** and **aspartate,** respectively, and the carbonyl carbon from **hydrogen carbonate**. In the first step [1], hydrogen carbonate (HCO_3^-) and ammonia are condensed to **carbamoylphosphate**, which, because of its acid-anhydride bond, possesses a very high reaction potential. Two ATP molecules are consumed in this reaction. In the next step [2], the carbamoyl residue is transferred to **ornithine,** resulting in the formation of **citrulline**. The second amino group of the urea molecule is provided by the reaction of citrulline with **aspartate** (enzyme [3], lower right). This reaction also requires energy in the form of ATP, which, in this special case, is cleaved to AMP and diphosphate ("pyrophosphate"). In order to shift the equilibrium of the reaction into the favorable direction, the diphosphate formed is subsequently hydrolyzed to inorganic phosphate (not shown). The cleavage of fumarate from arginosuccinate produces **arginine** [4], which is then hydrolyzed. The isourea molecule released in the process [5] immediately rearranges to give **urea**. The other product, ornithine, is ready to reenter the urea cycle.

Fumarate formed in the urea cycle is converted via malate to oxaloacetate, employing two steps of the citric acid cycle [6, 7]. The oxaloacetate thus formed is then transaminated [9] to regenerate aspartate, which can serve as amino group donor in another turn of the cycle.

A substantial amount of energy is required for the synthesis of urea. In total, four acid-anhydride bonds are cleaved for the synthesis of one molecule of urea. Two are used for the synthesis of carbamoyl phosphate and two (!) in the synthesis of argininosuccinate (ATP → AMP + PP_i, PP_i → P_i + P_i).

Urea is synthesized almost exclusively *in the liver*. The urea cycle is divided into two compartments, the mitochondria and the cytoplasm. The intermediates citrulline and ornithine pass the inner mitochondrial membrane with the help of specific transporters (see p. 198). Neither of these two amino acids is found in proteins.

The rate of urea formation is limited by the first reaction of the cycle [1]. Its key enzyme, *carbamoyl phosphate synthase,* needs the presence of *N*-acetylglutamate to be active. The concentration of this allosteric effector varies markedly with the status of liver metabolism (arginine level, energy supply).

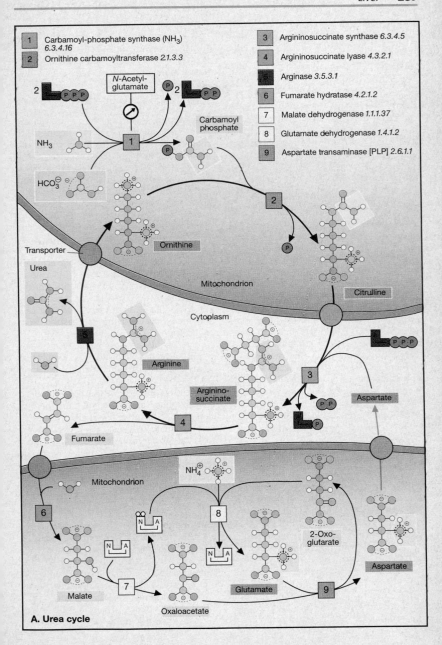

1 Carbamoyl-phosphate synthase (NH₃) 6.3.4.16
2 Ornithine carbamoyltransferase 2.1.3.3
3 Argininosuccinate synthase 6.3.4.5
4 Argininosuccinate lyase 4.3.2.1
5 Arginase 3.5.3.1
6 Fumarate hydratase 4.2.1.2
7 Malate dehydrogenase 1.1.1.37
8 Glutamate dehydrogenase 1.4.1.2
9 Aspartate transaminase [PLP] 2.6.1.1

A. Urea cycle

Bile Acid Metabolism

A. Cholate ◑

Bile acids are are steroids with 24 carbons, which carry from one to three α-hydroxyl groups as well as a side chain with a carboxylate group at its end. They are synthesized in the liver from cholesterol (see p. 55). The most important bile acid in man is **cholic acid**. Under the weakly alkaline conditions found in the bile, it is present as the anion **cholate**. Cholate and the other bile acids are distinctly *amphipathic* (see below).

B. Bile acids and bile salts ◑

Besides cholic acid, there is a second bile acid, *chenodeoxycholic acid*. It differs from cholic acid in that it lacks a hydroxyl group at C-12. Both molecules are referred to as **primary bile acids**. They are quantitatively the most important metabolites of cholesterol.

Two further bile acids, *deoxycholic acid* and *lithocholic acid*, are synthesized in the intestine by dehydroxylation of the primary bile acids at C-7 (see below), and are therefore referred to as **secondary bile acids**.

Before releasing them into the bile, the liver links some of the bile acids with an amino acid (*glycine* or *taurine*) via a peptide bond. The conjugated bile acids formed this way are called **bile salts**. They are more acidic than the corresponding bile acids.

C. Micelles ◑

Because of the α position of their hydroxyl groups (see p. 52), the bile acids and bile salts have a polar and an apolar side. They are *amphipathic*, and thus act as detergents (see p. 22). Their role is to facilitate the solubilization of dietary lipids during the process of digestion by promoting micelle formation. This emulsifying action increases the effectiveness of pancreatic lipases, and promotes the absorption of lipids from the intestine (see p. 246).

The diagram shows in a schematic way how bile acid molecules "float" in the outer layer of a lipid micelle, thus rendering it water-soluble. A molecule of t*riacylglycerol lipase* is associated with the micelle to cleave the fats (triacylglycerols) in its apolar interior. A small auxiliary protein (*colipase*) is also required in the process (not shown, cf. p. 246).

D. Metabolism of bile salts ◑

Bile salts are exclusively synthesized in the liver. Their biosynthesis (**1**) begins with cholesterol, into which hydroxyl groups are introduced. The double bond of cholesterol in ring B (see p. 53) is reduced, and its side chain is shortened by three carbons. The rate-determining step in bile acid formation is the hydroxylation at position 7 catalyzed by a *7α-hydroxylase*. The reaction is inhibited by cholic acid (*end-product inhibition*). The bile acids thus affect the rate of cholesterol degradation.

Prior to leaving the liver, the bile acids are conjugated with the amino acids glycine or taurine (**2**). To facilitate the synthesis of the peptide bond, the bile acids are activated by reaction with CoA. In this way *glycocholic acid* and *taurocholic acid* are synthesized from cholic acid.

The *liver bile* secreted by the liver is concentrated in the gallbladder by removal of water to yield *cystic bile* (**3**), which ultimately reaches the duodenum.

Intestinal bacteria in the colon produce enzymes that can attack and alter the bile salts (**4**). Their peptide bonds are hydrolyzed, and secondary bile acids are formed by removal of the hydroxyl group at C-7 (**5**). Most of the bile acids are reabsorbed from the intestine (**6**) and, following transport to the liver, returned once again into the bile (*enterohepatic circulation*). This is the reason why only about 0.5 g of bile acids appear in the feces per day, while 15 to 30 g are released into the intestine. The excreted portion has to be replaced via the *de-novo* synthesis of cholesterol, which also amounts to about 0.5 g per day.

Unfavorable composition of the bile can lead to crystallization of individual components. The result is *gallstones*, which usually contain cholesterol, calcium salts of the bile salts and, in some cases, bile pigments.

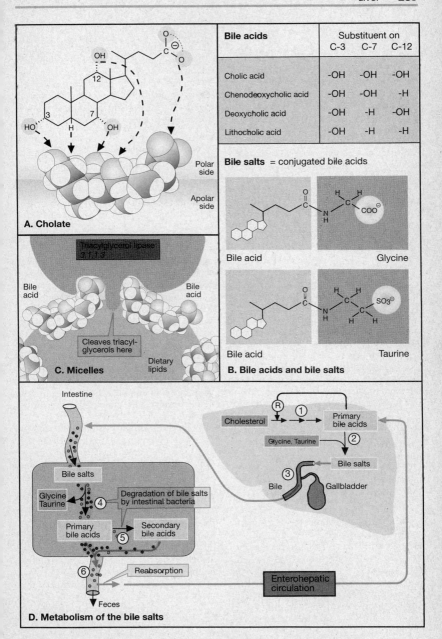

A. Cholate

Polar side

Apolar side

Bile acids	Substituent on		
	C-3	C-7	C-12
Cholic acid	-OH	-OH	-OH
Chenodeoxycholic acid	-OH	-OH	-H
Deoxycholic acid	-OH	-H	-OH
Lithocholic acid	-OH	-H	-H

Bile salts = conjugated bile acids

Bile acid   Glycine

Bile acid   Taurine

B. Bile acids and bile salts

Triacylglycerol lipase
3.1.1.3

Bile acid

Bile acid

Cleaves triacyl-glycerols here

Dietary lipids

C. Micelles

Intestine

Cholesterol → R → ① → Primary bile acids

Glycine, Taurine ②

Bile salts

③

Bile   Gallbladder

Bile salts

Glycine
Taurine ④

Degradation of bile salts by intestinal bacteria

Primary bile acids

Secondary bile acids

⑤

⑥

Reabsorption

Enterohepatic circulation

Feces

D. Metabolism of the bile salts

Biotransformations

A. Biotransformation and detoxification ◑

Foreign substances are constantly being absorbed from the diet and taken up from the environment via the skin and the lungs. These substances can be of natural origin (*xenobiotics*), or may have been synthetically produced by man. Many of them are toxic — especially at high concentrations. Therefore, higher organisms have developed very effective mechanisms for the detoxification and elimination of foreign substances. Collectively, these processes are referred to as **biotransformations**. The chemical reactions involved are similar to those used for the detoxification and elimination of the body's own wastes, such as bile pigments and steroid hormones. Biotransformation occurs predominantly in the liver.

Phase I reactions (interconversions). The so-called phase I reactions either incorporate new functional groups into apolar compounds, or alter groups that are already present in the molecule. The result is usually an increase in the *polarity* and a decrease in the *biological activity* or *toxicity* of the substance. In some cases, foreign substances (e.g., drugs, carcinogens) only become biologically active (i.e., beneficial or toxic) once they have been subjected to phase I reactions.

The most important phase I reactions are:

- *Oxidative reactions*: hydroxylation, epoxide formation, sulfoxide formation, de-alkylation, deamination
- *Reductive reactions*: reduction of carbonyl-, azo- or nitro compounds, dehalogenation
- *Methylation*
- *Desulfuration*

All of these reactions take place in the smooth endoplasmatic reticulum of the hepatocytes. Many of the oxidative reactions are catalyzed by **cytochrome P450** systems (see following page). In general, the enzymes involved have a broad specificity, i.e., most of them accept a wide range of different substrates (exception: enzymes of steroid metabolism). Typically, the enzyme systems involved are often induced by their substrates.

Phase II reactions (conjugate formation). These reactions couple substrates (bilirubin, metabolites of xenobiotics, drugs and steroid hormones) to highly polar, often negatively charged molecules. The enzymes involved are, without exception, transferases, and their products are called **conjugates**.

B. Conjugate formation ◑

The most common type of conjugation reaction is coupling with **glucuronate** (GlcUA), yielding *O*- or *N*-glucuronides. The coenzyme involved is the uridine diphosphate derivative of glucuronic acid, i.e., "active glucuronate" (**1**).

Glucuronate is highly polar, and therefore conjugate formation between glucuronate and apolar molecules makes the target molecule more polar. The example shown here is the conjugate between 2-naphthol and glucuronate (**2**). 2-Naphthol, a compound foreign to animal metabolism, becomes sufficiently water-soluble by conjugate formation with glucuronate and can be excreted with the urine.

The biosynthesis of sulfate esters with the help of **phosphoadenosinephosphosulfate** (PAPS, "*active sulfate*," **3**), also plays a role in conjugate formation. Further conjugates are amides involving the amino acids *glycine* and *glutamine*.

Conjugates with masses exceeding 300 Da are eliminated from the liver either with the bile, or *renally*, i.e., via the blood and kidneys (for conjugates with molecular masses < 300).

Further information

In the liver, there is a cysteine-rich protein called *metallothioneine*, which has a high affinity for divalent metal ions, such as Cd^{2+}, Cu^{2+}, Hg^{2+}, and Zn^{2+}. This protein binds to and detoxifies metals, some of which are required by the organism in trace amounts. The synthesis of metallothioneine is induced by metal ions.

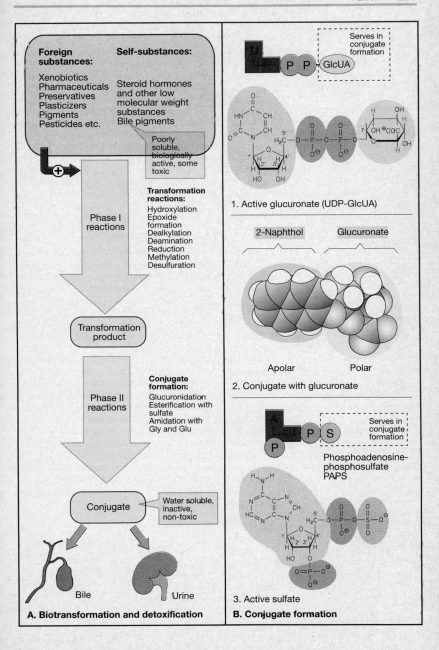

Foreign substances:

Xenobiotics
Pharmaceuticals
Preservatives
Plasticizers
Pigments
Pesticides etc.

Self-substances:

Steroid hormones
and other low
molecular weight
substances
Bile pigments

Poorly soluble, biologically active, some toxic

Phase I reactions

Transformation reactions:

Hydroxylation
Epoxide formation
Dealkylation
Deamination
Reduction
Methylation
Desulfuration

Transformation product

Phase II reactions

Conjugate formation:

Glucuronidation
Esterification with sulfate
Amidation with Gly and Glu

Conjugate

Water soluble, inactive, non-toxic

Bile   Urine

A. Biotransformation and detoxification

Serves in conjugate formation

1. Active glucuronate (UDP-GlcUA)

2-Naphthol   Glucuronate

Apolar   Polar

2. Conjugate with glucuronate

Serves in conjugate formation

Phosphoadenosine-phosphosulfate
PAPS

3. Active sulfate

B. Conjugate formation

Cytochrome P450 Systems

During the biotransformations that take place in the liver (see preceding page), apolar compounds that are only weakly reactive are enzymatically hydroxylated (*phase I reactions*). This makes it easier to conjugate them with polar substances. The hydroxylating enzymes are *monooxygenases* that contain a *heme* group as the redox-active coenzyme (see p. 100). In the reduced form, this heme group can bind carbon monoxide (CO), giving rise to a characteristic light absorption at 450 nm. This is the origin of the name **cytochrome P450** (Cyt P450) for the whole group of enzymes.

Cyt P450 systems are also involved in many other metabolic processes, e.g., the biosynthesis of steroid hormones (see pp. 304 and 338), bile acids (see p. 288), eicosanoids (see p. 350), and unsaturated fatty acids (see p. 365).

A. Cytochrome P450-dependent monooxygenases: reactions ◑

Cyt P450 enzymes catalyze the reductive cleavage of molecular oxygen (O_2). One of the two oxygen atoms is incorporated into the substrate, while the other is released as water. The necessary reducing equivalents are derived from NADPH + H$^+$; they reach the enzyme via an *electron-transferring flavoprotein*.

Many different types of Cyt P450 enzymes are found in the liver, adrenal cortex (synthesis of steroid hormones), and other organs. The specificity of the liver enzymes is generally low. Apolar compounds containing aliphatic or aromatic rings are particular good substrates. These include the body's own substances, such as steroid hormones, as well as foreign substances like medicinal drugs, which are usually inactivated by phase I reactions. This is why Cyt P450 enzyme are so important in pharmacology. The degradation of ethanol in the liver is also catalyzed by Cyt P450 ("microsomal ethanol oxidizing system," see p. 294). Since alcohol and medicinal drugs are degraded by the same enzyme system, the simultaneous intake of both can lead to life-threatening enhancements of their effects.

Only a few of the numerous Cyt P450-dependent reactions are shown here as examples. The hydroxylation of aromatic rings (**a**) plays a central role in the metabolism of medicinal drugs and steroids. Aliphatic methyl groups can be oxidized to hydroxyl functions (**b**). The epoxidation of aromatics (**c**) by Cyt P450 yields products that are very reactive and often toxic. For example, the mutagenic properties of benzo(a)pyrene (see p. 234) are the result of just such an interconversion in the liver. Cyt P450-dependent dealkylations (**d**) result in the release of alkyl groups from O, N, or S atoms.

B. Reaction mechanism ○

The principles of Cyt P450 catalysis are now well understood. The most important function of the heme group is to convert molecular oxygen into an especially reactive, atomic form, which is responsible for all of the reactions described above. In the resting state, the heme iron is trivalent. Initially, the substrate binds near the heme group (**1**). This allows the reduction of the iron to its divalent form and the resultant binding of an O_2 molecule (**2**). A further electron transfer (**3**) and a change in the valence of the iron then reduce the bound oxygen to the peroxide. A hydroxyl ion is subsequently cleaved (**4**) from this intermediate, giving water and the above-mentioned reactive form of oxygen. In this free radical, the iron is formally tetravalent. The activated oxygen can insert itself into a C-H bond of the substrate, and thus form a hydroxyl group (**5**). Following the release of the product (**6**), the enzyme returns to its original conformation.

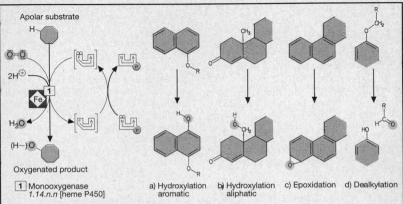

a) Hydroxylation aromatic   b) Hydroxylation aliphatic   c) Epoxidation   d) Dealkylation

1 Monooxygenase
1.14.n.n [heme P450]

A. Cytochrome P450-dependent monooxygenases: reactions

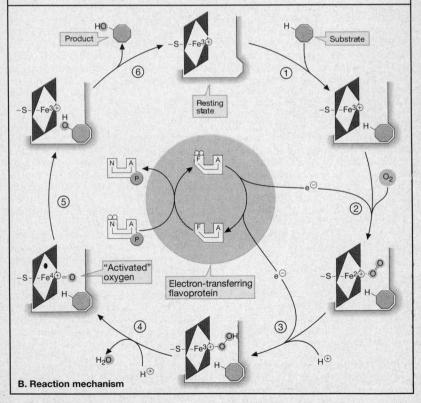

B. Reaction mechanism

Ethanol Metabolism

A. Blood alcohol levels ❶

Trace quantities of **ethanol** (EtOH, "alcohol") are found in fermented fruits, while much higher concentrations occur in alcoholic beverages. The alcohol content of alcoholic beverages is usually given as percent by volume (100-proof = 57.1% v/v). When estimating blood alcohol levels following the consumption of alcoholic beverages, it is helpful to convert this quantity to gram ethanol (density = 0.79 kg · l⁻¹). Thus a bottle of beer (0.5 l with 4% v/v alcohol) contains 20 ml = 16 g of ethanol, whereas a bottle of wine (0.7 l with 12% v/v alcohol) contains 84 ml = 66 g ethanol.

Ethanol is rapidly taken up from the digestive tract by diffusion. The maximum blood alcohol level is reached within as little as 60–90 minutes after the consumption of an alcoholic beverage. However, the *rate of absorption* is influenced by many different factors. An empty stomach, a warm drink, or the presence of sugar and carbonic acid (as in champagne) all serve to promote the absorption of alcohol, whereas a heavy meal has the opposite effect. Alcohol is rapidly distributed throughout the organism. A large amount is taken up by muscles and the brain, but little by adipose tissue or bones. Roughly 70% of the body volume is accessible to alcohol. Thus, the rapid and complete absorption of the alcohol contained in a bottle of beer (16 g) by a 70 kg person leads to a blood alcohol level of 0.033% (distribution in 70 · 0.7 kg = 49 kg gives about 0.33 g /l or 7.2 mM). The lethal alcohol level is 0.3–0.4%.

B. Ethanol metabolism ❶

The major site of ethanol degradation is the liver, where *alcohol dehydrogenase* oxidizes ethanol to ethanal (acetaldehyde), which is further oxidized to acetate by *aldehyde dehydrogenase*. Acetate is then converted to acetyl-CoA by *acetate-CoA ligase* in an ATP-dependent reaction. The production of acetyl-CoA constitutes the link between ethanol degradation and intermediary metabolism.

Besides cytoplasmic alcohol dehydrogenase, other enzymes are also involved in ethanol degradation, i.e., *catalase* (see p. 200) and Cyt P450, formerly referred to as *"microsomal alcohol oxidizing system,"* or MEOS (see preceding page).

The rate-limiting step of ethanol degradation by the liver is the alcohol dehydrogenase reaction. Therefore, the availability of NAD⁺ is a limiting factor. Low concentrations of ethanol are already sufficient to reach the maximum rate of ethanol degradation. Thus, after alcohol intake, there is a slow and constant decline in the amount of ethanol in the body. The *caloric* content of ethanol is 29.4 kJ · g⁻¹. Alcoholic drinks can therefore comprise a major part of the energy intake, especially for alcoholics.

A great deal of attention has been focused on the actions of alcohol. Nevertheless, the exact mechanisms involved are still not fully understood. The acute effects of ethanol are similar to those of narcotics. Many of them can be explained by effects on the membranes of neurons.

C. Liver damage ❶

Elevated consumption of ethanol over many years may lead to liver damage. The limit for a healthy man is about 60 g daily, and for a woman about 50 g, assuming a regular daily intake of ethanol. These values are strongly dependent on body weight, health and the use of medication.

In the liver, the high levels of NADH and acetyl-CoA resulting from the metabolism of ethanol inhibit citric acid cycle activity and ketogenesis, whereas they exert a stimulatory effect on the synthesis of neutral fats and cholesterol. An increase in the storage of lipids leads to a **fatty liver**. This increase in the fat content of the liver (from less than 5% to more than 50% of the dry weight) is usually reversible. It is only with chronic alcoholism that liver cell death becomes a severe problem. Once **cirrhosis** of the liver has set in, the damage to the liver reaches an irreversible stage, which is characterized by progressive loss of liver function.

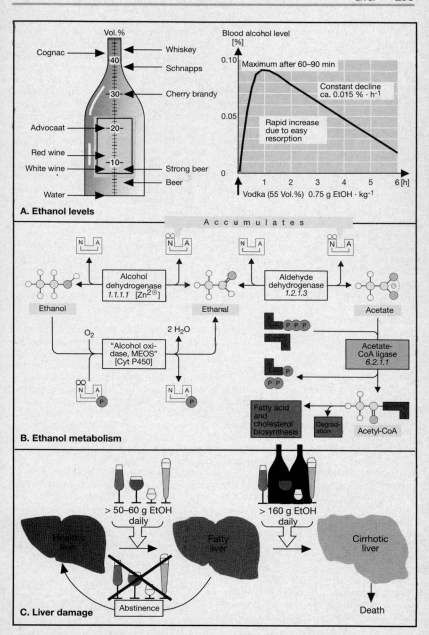

A. Ethanol levels

Vol.%

Cognac — Whiskey
40 — Schnapps
–30– — Cherry brandy
Advocaat
–20–
Red wine
White wine — –10– — Strong beer
— Beer
Water

Blood alcohol level [%]

0.10 Maximum after 60–90 min

Constant decline ca. 0.015 % · h⁻¹

Rapid increase due to easy resorption

0.05

0

1   2   3   4   5   6 [h]

Vodka (55 Vol.%) 0.75 g EtOH · kg⁻¹

B. Ethanol metabolism

Accumulates

Alcohol dehydrogenase 1.1.1.1 [$Zn^{2\oplus}$]

Ethanol

Ethanal

Aldehyde dehydrogenase 1.2.1.3

Acetate

"Alcohol oxidase, MEOS" [Cyt P450]

O_2 $2 H_2O$

Acetate-CoA ligase 6.2.1.1

Fatty acid and cholesterol biosynthesis

Degradation

Acetyl-CoA

C. Liver damage

Healthy liver

> 50–60 g EtOH daily

Fatty liver

> 160 g EtOH daily

Cirrhotic liver

Abstinence

Death

Functions

A. Functions of the kidneys ●

The kidneys are the most important organ involved in regulating the volume and composition of the body fluids. They are extremely well supplied with blood. About 1500 l of blood is filtered through them every day, giving rise to approximately 150 l of ultrafiltrate, which then becomes concentrated by reabsorption of water. As a result, only 0.5 to 2.0 l of **urine** is excreted per day.

The most important function of the kidneys is the **excretion** of water and water-soluble substances (**1**, see p. 298). This is closely associated with their role in the regulation of intracellular pH and electrolyte balance (**homeostasis, 2**; see pp. 300 and 302). Both excretion and homeostasis are under hormonal control. The kidneys are also involved in the synthesis of several **hormones** (**3**; see p. 304). Finally, the kidneys play a role in **intermediary metabolism** (**4**), especially in gluconeogenesis and amino acid degradation (see p. 302).

B. Urine formation ●

The functional unit of the kidney is the **nephron** (shown here schematically). There are about one million of these in the human kidney. The formation of urine in the nephron occurs in three phases.

Glomerular filtration. The ultrafiltration of the plasma in the glomerulus gives rise to primary urine, which is *isotonic* relative to the plasma. The pores that filter the plasma have an average effective diameter of 2.9 nm. Thus, all constituents of the plasma with a molecular mass less than 5 kDa can pass through unhindered. Particles with higher masses are increasingly retarded, but they are not completely excluded from entering the urine unless their mass exceeds 65 kDa. Most blood proteins have masses above 54 kDa (see p. 252), and therefore only very small quantities of them are found in the urine.

Reabsorption. In the proximal and distal tubule, the glomerular filtrate becomes highly concentrated as a result of the removal of water (the final volume is about one-hundredth of original). At the same time, many other low molecular weight constituents, especially **glucose, amino acids,** and **organic** and **inorganic ions,** are reabsorbed by active transport (see p. 302). There are several genetic diseases known that result from disturbances of group-specific transport systems for the recovery of amino acids (e. g *cystinuria, glycinuria,* and *Hartnup's disease*).

Secretion: Some of the substances that have to be excreted from the body are released into the urine in the kidneys by *active transport* processes. These substances include H^+ and K^+ ions, urea, creatinine, and drugs, such as penicillin.

Further information

The mechanisms for concentrating urine and for the transport of various substances require large amounts of energy. This is why the energy consumption of the kidneys is very high. The necessary ATP is derived from the oxidative metabolism of glucose, lactate, pyruvate, fatty acids, glycerol, citrate, and amino acids absorbed from the blood. Not only the liver, but also the kidneys are capable of *gluconeogenesis*. The main substrates for this are the carbon skeletons of amino acids. The ammonia derived from these amino acids serves to buffer the pH of the urine (see p. 300). Enzymes of amino acid metabolism (*amino acid oxidases, amine oxidases, glutaminase*) are found in the kidneys at high activity.

The extent to which a substance is excreted by the kidneys can be determined by measuring its *renal clearance*. This is the volume of plasma that is cleared of a particular substance per unit time. *Inulin*, a polyfructosan (mass about 6 kDa) that is neither actively excreted nor actively reabsorbed, has a clearance of 120 ml min^{-1}. The clearance volume for actively secreted substances is higher, and that for actively absorbed substances lower, than this value.

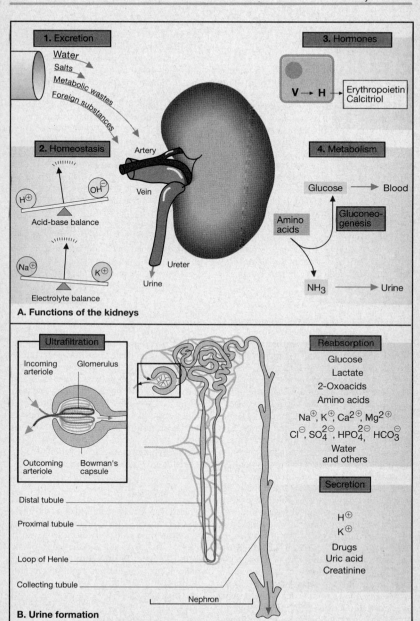

A. Functions of the kidneys

1. Excretion

Water
Salts
Metabolic wastes
Foreign substances

2. Homeostasis

$H^{\oplus}$ $OH^{\ominus}$
Acid-base balance

$Na^{\oplus}$ $K^{\oplus}$
Electrolyte balance

Artery
Vein
Ureter
Urine

3. Hormones

$V \rightarrow H$ → Erythropoietin
Calcitriol

4. Metabolism

Glucose → Blood

Amino acids

Gluconeogenesis

NH_3 → Urine

B. Urine formation

Ultrafiltration

Incoming arteriole
Glomerulus
Outcoming arteriole
Bowman's capsule

Distal tubule
Proximal tubule
Loop of Henle
Collecting tubule

Nephron

Reabsorption

Glucose
Lactate
2-Oxoacids
Amino acids
$Na^{\oplus}$, $K^{\oplus}$, $Ca^{2\oplus}$, $Mg^{2\oplus}$
$Cl^{\ominus}$, $SO_4^{2\ominus}$, $HPO_4^{2\ominus}$, $HCO_3^{\ominus}$
Water
and others

Secretion

$H^{\oplus}$
$K^{\oplus}$
Drugs
Uric acid
Creatinine

Urine

A. Urine ●

Water and water-soluble compounds are excreted with the urine. The volume of urine and its composition vary greatly depending on diet, body weight, age, gender, physical activity, state of health, and environmental conditions, such as temperature and humidity. There is also a pronounced diurnal variation in the excretion of urine. To account for this variation, the quantity and composition of urine excreted are usually averaged over a 24-hour period.

An adult human typically produces 0.5–2.0 l urine per day, 95% of which is water. In general, urine is weakly acidic (about pH 5.8), but this varies markedly with metabolic status. For example, following the digestion of large quantities of plant material, the pH of the urine can rise to a value higher than 7.

B. Organic constituents ◑

The most important organic constituents of the urine are nitrogen-containing compounds. **Urea**, which is synthesized in the liver, is the form in which nitrogen derived from the breakdown of amino acids is disposed of. The amount of urea excreted is therefore directly related to the amount of protein metabolized. The degradation of 70 g protein, for instance, yields 30 g of urea (see p. 150). **Uric acid** is the end product of purine metabolism (see p. 156). **Creatinine** is a product of muscle metabolism, where it is formed spontaneously and irreversibly from **creatine** (see p. 310). The amount of creatinine excreted per day by a given individual is constant, and directly proportional to muscle mass. As a result of this, it can be used as a reference for the quantitation of the other constituents of the urine. The extent of amino acid excretion is strongly dependent on the diet and on the efficiency of liver function. Derivatives of amino acid are also found in the urine (example: **hippurate**). Modified amino acids present in special proteins, e.g., *hydroxyproline* in collagen or *3-methylhistidine* in actin and my-

osin, can serve as indicators of the degradation of these proteins.

Other solutes in the urine are conjugated with sulfuric acid, glucuronic acid, glycine, and other polar compounds. These are synthesized in biotransformation reactions in the liver (see p. 290). Metabolites of hormones (catecholamines, steroids, serotonin) also appear in the urine. Their analysis can provide information about the production of these hormones. *Chorionic gonadotropin* (hCG, mass 36 kDa) is a proteohormone formed at the onset of pregnancy. It circulates in the blood and, because of its small size, also appears in the urine. The immunological detection of hCG in the urine forms the basis of a widely used *pregnancy test*.

The yellow color of urine is due to *urochromes*, which are related to the bile pigments produced by heme degradation (see p. 180). If urine is left to stand long enough, oxidation of the urochromes may lead to a darkening in color.

C. Inorganic constituents ◑

Urine contains substantial amounts of the cations Na^+, K^+, Ca^{2+}, Mg^{2+}, and NH_4^+, the anions Cl^-, SO_4^{2-} and HPO_4^{2-}, and trace quantities of other ions. (The concentrations of calcium and magnesium are higher in the feces than in the urine). The amounts of the various inorganic constituents in the urine are determined largely by the composition of the diet. During *acidosis*, there can be a marked increase in the excretion of ammonia (see p. 300). In many cases, the excretion of ions is regulated by hormones (see p. 304).

Further information

Alterations in the concentrations of the physiological constituents of the urine and the appearance of *pathological constituents* can be used as indicators for the diagnosis of diseases. Examples of such constituents include glucose and ketone bodies in *diabetes mellitus* (see p. 164).

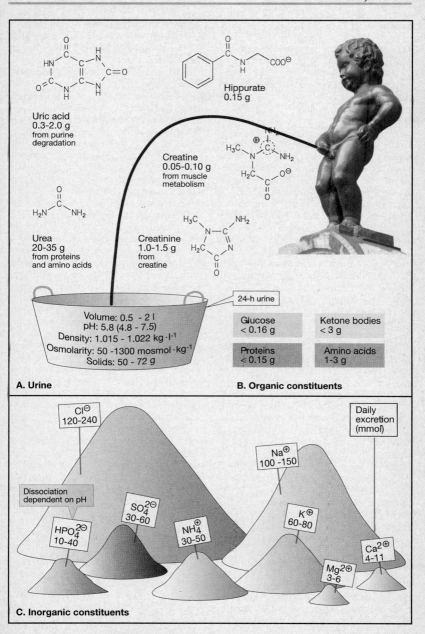

Uric acid
0.3-2.0 g
from purine
degradation

Hippurate
0.15 g

Creatine
0.05-0.10 g
from muscle
metabolism

Urea
20-35 g
from proteins
and amino acids

Creatinine
1.0-1.5 g
from
creatine

24-h urine

Volume: 0.5 - 2 l
pH: 5.8 (4.8 - 7.5)
Density: 1.015 - 1.022 kg · l⁻¹
Osmolarity: 50 -1300 mosmol · kg⁻¹
Solids: 50 - 72 g

Glucose
< 0.16 g

Ketone bodies
< 3 g

Proteins
< 0.15 g

Amino acids
1-3 g

A. Urine

B. Organic constituents

$Cl^\ominus$
120-240

Daily
excretion
(mmol)

$Na^\oplus$
100 -150

Dissociation
dependent on pH

$SO_4^{2\ominus}$
30-60

$NH_4^\oplus$
30-50

$K^\oplus$
60-80

$Ca^{2\oplus}$
4-11

$HPO_4^{2\ominus}$
10-40

$Mg^{2\oplus}$
3-6

C. Inorganic constituents

Proton and Ammonia Excretion

The kidneys and the lungs are the principal organs for maintaining the pH of the extracellular fluid (see p. 256). The kidneys make their contribution by actively excreting protons into the urine.

A. Proton secretion ◑

The tubule cells of the kidneys are capable of secreting **hydrogen ions** ("protons") into the urine despite the fact that the H^+ concentration in the urine is up to 1000-fold higher than in the blood. To achieve this, they absorb carbon dioxide (CO_2) from the blood and then hydrate it to carbonic acid (H_2CO_3) with the help of *carbonate dehydratase* [1]. Carbonic acid then dissociates to hydrogen carbonate ("bicarbonate," HCO_3^-) and a proton. The proton is exported to the urine by an ATP-driven, membrane-localized transport system [2], while hydrogen carbonate returns to the blood. This is the mechanism whereby the kidneys contribute to the CO_2/HCO_3^- buffering of the plasma (see p. 256). In order to maintain the electroneutrality of the tubule cells, for each excreted proton a **Na^+ ion** is taken up from the urine and also returned to the blood, i.e., there is an active excretion of protons in exchange for sodium ions.

About 60 mmol of protons are excreted with the urine per day. However, as the urine contains several systems that buffer these protons, its pH is only weakly acidic (down to about pH 4.8). One important buffer system in the urine is *phosphate* ($HPO_4^{2-}/H_2PO_4^-$). *Ammonia* also makes a contribution. Phosphate excretion somewhat depends on the composition of the diet, whereas the excretion of ammonia can vary markedly, depending on the body's current needs (see below).

B. Ammonia excretion ◑

The detoxification processes occurring in the liver (urea cycle, see p. 286) keep the plasma concentration of ammonia very low. Therefore, the kidneys obtain the ammonia they require from **glutamine**, the most abundant amino acid in the plasma (concentration 0.5–0.7 mM). Glutamine, a vehicle for nitrogen transport in the blood, is derived from the muscles, brain, and liver. Once the kidneys have taken up glutamine, they release ammonia by hydrolysis of the amide bond [4]. The **glutamate** thus formed can liberate a further molecule of ammonia by oxidative deamination to **2-oxoglutarate** (see p. 152). This reaction is catalyzed by *glutamate dehydrogenase* [5], which uses either NAD^+ or $NADP^+$ as coenzyme. Oxoglutarate then enters the citric acid cycle. Several other amino acids, in particular alanine, serine, glycine, and aspartate, can also supply ammonia.

Since ammonia (NH_3) is uncharged, it can diffuse through the cell membrane and into the urine. In the acidic environment of the urine it is converted into an ammonium ion by binding a proton. The charged ammonium ion can no longer cross the tubular membrane. The acidity of the urine, therefore, facilitates ammonia excretion, which occurs at a rate of 30–50 mmol per day.

The excretion of ammonia can be either greatly reduced or strongly increased, depending on the metabolic state of the organism. The deciding factor is the pH value of the plasma, which is usually kept near 7.4 (see p. 256). When the pH value becomes more acidic (*acidosis*), the excretion of protons and ammonia increases. In contrast, a shift of the plasma pH to a more alkaline value (*alkalosis*) results in a lower renal excretion of ammonia.

Acidosis may, for instance, result from extensive ketone body formation, because of starvation or *diabetes mellitus*. The increased ammonia excretion by which the kidneys respond is, in part, mediated by a higher *glutaminase* activity, which in turn results from the enhanced transcription of the glutaminase gene (enzyme induction, see p. 108).

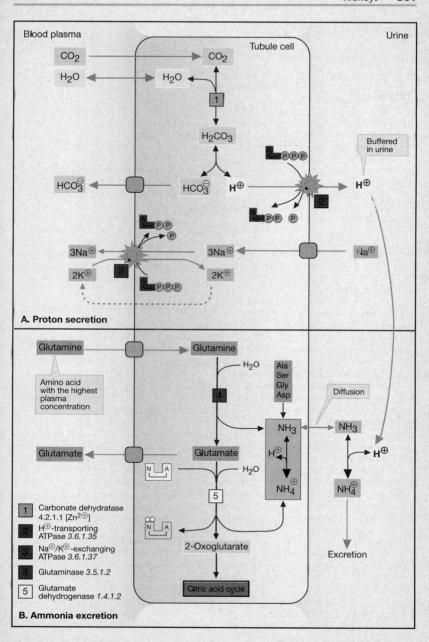

Blood plasma

Tubule cell

Urine

A. Proton secretion

Amino acid with the highest plasma concentration

Buffered in urine

Diffusion

1. Carbonate dehydratase 4.2.1.1 [$Zn^{2\oplus}$]
2. $H^\oplus$-transporting ATPase 3.6.1.35
3. $Na^\oplus/K^\oplus$-exchanging ATPase 3.6.1.37
4. Glutaminase 3.5.1.2
5. Glutamate dehydrogenase 1.4.1.2

B. Ammonia excretion

Electrolyte and Water Recycling

A. Electrolyte and water recycling ◑

All electrolytes and other low molecular weight substances in the blood plasma quantitatively enter the glomerular filtrate (right). They are subsequently salvaged from the urine by selective reabsorption processes, most of which are energy-dependent. Only water, chloride ions, and urea are transported in a passive fashion. The extent of reabsorption determines the amounts of the various substances that remain in the urine and are thus excreted. The illustration is highly simplified and ignores the *zoning* of the various transport processes in the nephron (see physiology textbooks for more details).

Almost all of the **calcium** and **phosphate ions** found in the primary filtrate are reabsorbed. The process requires energy in the form of ATP. The extent of reabsorption of calcium ions is greater than 99%; for phosphate, it is about 80–90%. The reabsorption of these two electrolytes is jointly regulated by three different hormones, parathyrin, calcitonin and calcitriol.

Parathyrin (parathormone, PTH) a peptide hormone produced by the parathyroid gland, stimulates the absorption of calcium in the kidneys and at the same time inhibits the resorption of phosphate. In addition, parathyrin promotes calcium absorption from the intestine and its release from the bones. The overall effect of elevated parathyrin levels is therefore an increase in plasma calcium level and a decline in plasma phosphate level.

Calcitonin, a peptide produced in the C cells of the thyroid gland, inhibits the reabsorption of both calcium and phosphate. The result is an overall decline in the plasma level of either ion. With respect to calcium reabsorption, calcitonin is an antagonist of parathyrin.

The steroid hormone **calcitriol**, which is formed in the kidneys (see p. 304), has a stimulatory effect on the reabsorption of both calcium and phosphate ions.

Sodium ions. Another crucial function of the tubule cells is the reabsorption of Na^+ from the glomerular filtrate. This is a very efficient process, with more than 97% of the filtered amount being reabsorbed. The steroid hormone **aldosterone** (see p. 55) stimulates Na^+ retention, whereas **atrial natriuretic peptide (ANP),** a hormone from the atrium of the heart, inhibits it. Both hormones probably affect Na^+/K^+-ATPase (see p. 208), which, in the distal tubules and collecting ducts of the kidneys, is located on the surface adjacent to the blood plasma. The enzyme "pumps" Na^+ ions from the urine back into the plasma in exchange for K^+ (see p. 300).

Water. The reabsorption of water is a *passive process*. It occurs spontaneously in response to gradients of osmotically active substances, principally that of Na^+ ions. In the distal part of the nephron, water can only enter the cells when the hormone **vasopressin** (antidiuretic hormone, ADH) is present. Vasopressin is a cyclic peptide produced in the neurohypophysis. ANP inhibits water reabsorption, i.e., it stimulates diuresis.

B. Gluconeogenesis and glucose recycling ◑

Only the liver and the kidneys are capable of *de novo* glucose synthesis (*gluconeogenesis*, see p. 158). In the kidneys, the chief substrate for gluconeogenesis is **glutamine**. However, other amino acids and further compounds, such as **lactose**, **glycerol**, and **fructose** can also be used. All of these metabolites are obtained from the blood plasma. As in the liver, gluconeogenesis in the kidneys is induced by **cortisol** (see p. 55). Since the kidneys are also major consumers of glucose, their glucose balance is approximately even.

The reabsorption of glucose from the glomerular filtrate is an energy-dependent process, which is independent of gluconeogenesis. Glucose uptake is not directly coupled to ATP hydrolysis, but occurs as a compulsory cotransport with Na^+ ions. Thus, it is driven by the concentration gradient of Na^+ between the urine and the interior of the cells (*secondary active transport*). The amino acids and ketone bodies are taken up by similar mechanisms.

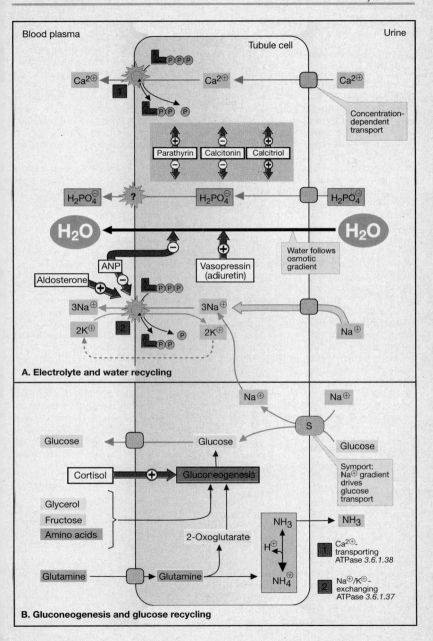

A. Electrolyte and water recycling

Blood plasma

Urine

Tubule cell

Ca²⁺ ← Ca²⁺ ← Ca²⁺

$\boxed{1}$

Concentration-dependent transport

Parathyrin ⊕ ⊖ | Calcitonin ⊖ ⊖ | Calcitriol ⊕ ⊕

$H_2PO_4^{\ominus}$ ← ? ← $H_2PO_4^{\ominus}$ ← $H_2PO_4^{\ominus}$

H_2O ← H_2O

Water follows osmotic gradient

ANP ⊖

Aldosterone ⊕ ⊖

Vasopressin (adiuretin) ⊕

$3Na^{\oplus}$ ← $3Na^{\oplus}$ ← $Na^{\oplus}$

$2K^{\oplus}$ $\boxed{2}$ $2K^{\oplus}$

B. Gluconeogenesis and glucose recycling

$Na^{\oplus}$ $Na^{\oplus}$

S

Glucose ← Glucose ← Glucose

Symport: $Na^{\oplus}$ gradient drives glucose transport

Cortisol ⊕ Gluconeogenesis

Glycerol
Fructose
Amino acids

2-Oxoglutarate

NH_3 → NH_3

$H^{\oplus}$

$NH_4^{\oplus}$

Glutamine → Glutamine

$\boxed{1}$ $Ca^{2\oplus}$-transporting ATPase 3.6.1.38

$\boxed{2}$ $Na^{\oplus}/K^{\oplus}$-exchanging ATPase 3.6.1.37

Hormones of the Kidneys

A. Hormones of the kidneys ◑

Besides their involvement in metabolism and excretion, the kidneys also have important endocrine functions. They produce the hormones **erythropoietin** and **calcitriol**. In addition, they are involved in the production of the hormone **angiotensin**, due to the fact that they excrete the enzyme *renin*.

Calcitriol (1α,25-dihydroxycholecalciferol) is a steroid-related hormone involved in calcium homeostasis. It is formed in the liver from calcidiol by hydroxylation at C-1. The activity of the hydroxylase (*calcidiol-1-monooxygenase*, [1]) is regulated by the hormone *parathyrin*.

Erythropoietin is a polypeptide hormone that is formed predominantly by the kidney, but also by the liver. Together with other factors, so-called *colony stimulating factors*, it controls the differentiation of the bone marrow stem cells. The release of erythropoietin is stimulated by hypoxia (low pO_2). Within hours, the hormone ensures that the precursor cells in the bone marrow are converted to erythrocytes, so that their concentration in the blood increases. Damage to the kidneys leads to a decline in the release of erythropoietin, which in turn results in *anemia*. It is now possible to successfully treat renally caused anemias with erythropoietin overexpressed by genetic engineering (see p. 240).

B. Renin-angiotensin system ◑

Renin [2] is an aspartate proteinase (see p. 150). It is formed by the kidneys as a precursor (prorenin), which, on demand, is processed to yield active renin, which is then released into the blood. In the blood plasma, renin acts on **angiotensinogen**, a glycoprotein belonging to the $α_2$-globulins, which is synthesized in the liver (see p. 252). The decapeptide cleaved off by renin is called **angiotensin I**. It is converted to **angiotensin II** by *peptidyl dipeptidase A* ([3], also known as "angiotensin-converting enzyme," ACE), which is located in the membranes of the blood vessels, especially in the lung. Angi-

otensin II, an octapeptide, acts as a hormone and a neurotransmitter. It is rapidly degraded by other peptidases (so-called angiotensinases [4]), which occur in many different tissues. The half-life of angiotensin in the plasma is only about 1 minute.

The level of plasma angiotensin II is mainly determined by the rate of release of renin. Renin is synthesized and released by the so-called juxtaglomerular cells of the kidneys in response to decreasing levels of Na^+ ions and declining blood pressure.

Effects of angiotensin II. Angiotensin II has its effects on the kidneys, brain stem, pituitary gland, adrenal cortex, blood vessel walls, and heart via membrane-localized receptors. In the kidney, it promotes the *retention of Na^+ and water*. In the brain stem and at the axon terminals of the sympathetic nervous system, angiotensin II promotes increased neurotransmitter release resulting in the development of the *sensation of thirst*. In the pituitary gland, it stimulates the release of vasopressin (antidiuretic hormone, see p. 302) and corticotropin (ACTH). In the adrenal cortex, there is a stimulation of the *biosynthesis and release of aldosterone*, which promotes sodium and water retention in the kidney. *Vasoconstriction* leads to an increase in blood pressure. Thus, all of the effects of angiotensin II lead, either directly or indirectly, to an *increase in blood pressure* coupled to the *retention of sodium and water*. In this complex hormone system, several components are susceptible to the action of *inhibitors*. Such compounds are useful in the treatment of *hypertension* (elevated blood pressure). They include:

– Analogs of angiotensinogen that inhibit renin
- Analogs of angiotensin I that competitively inhibit ACE (enzyme [3]). It should be noted that ACE also cleaves other signal peptides found in the plasma, such as bradykinin
– Peptide hormone antagonists that block the angiotensin receptor.

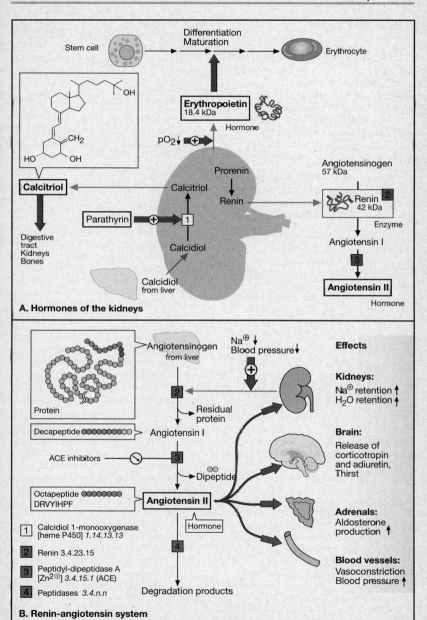

A. Hormones of the kidneys

Stem cell → Differentiation Maturation → Erythrocyte

Erythropoietin 18.4 kDa
Hormone

pO₂↓ ⊕

Calcitriol

Prorenin
↓
Renin →

Angiotensinogen 57 kDa

Renin 42 kDa **2**
Enzyme

Calcitriol

Calcidiol **1**

Parathyrin ⊕ →

Calcidiol

Calcidiol from liver

Digestive tract
Kidneys
Bones

Angiotensin I
3
Angiotensin II
Hormone

B. Renin-angiotensin system

Angiotensinogen from liver

Na⊕ ↓
Blood pressure↓

Effects

Protein

2

⊕

Residual protein

Kidneys:
Na⊕ retention ↑
H₂O retention ↑

Decapeptide → Angiotensin I

ACE inhibitors — ⊘ **3**

Dipeptide

Brain:
Release of corticotropin and adiuretin,
Thirst

Octapeptide DRVYIHPF

Angiotensin II

Hormone

4

Adrenals:
Aldosterone production ↑

1 Calcidiol 1-monooxygenase [heme P450] *1.14.13.13*

2 Renin 3.4.23.15

3 Peptidyl-dipeptidase A [Zn²⊕] *3.4.15.1* (ACE)

4 Peptidases *3.4.n.n*

Degradation products

Blood vessels:
Vasoconstriction
Blood pressure ↑

Contraction

A. Organization of striated muscle ●

Striated muscle consists of parallel bundles of **muscle fibers**. Each fiber is a single large multinucleate cell. Within this cell, **myofibrils** 1 to 2 μm thick extend the full length of the muscle fiber. The *striation* of the fibers is characteristic of the skeletal muscle. It results from the regular arrangement of molecules of differing density (see also physiology textbooks).

The **sarcomers** are made up from two types of overlapping structures, thick **myosin** filaments lying parallel to one another, and thin filaments made up of **F-actin** and also arranged in parallel (see p. 190). The dark outer regions of the A band contain both thick and thin filaments, while in the central H zone there are only myosin filaments. The Z lines correspond to sites where the thin filaments are anchored to a so-called Z disc.

Quantitatively, **myosin** is the most important protein of the myofibrils (65% by weight). It is a hexamer consisting of two identical heavy chains (223 kDa) and four light chains (16 kDa and 20 kDa). A single myosin molecule is shaped like a golf club (below right). Each heavy chain has a globular "head" at its NH_2-terminal end and a long "tail" of about 150 nm length. In the tail region, the two chains intertwine to form a *superhelix*, while the four smaller subunits are attached to the globular heads. In muscle fibers, myosin is arranged in *thick myosin filaments*, which are bundles of many hundreds of molecules stacked adjacent to one another. The head portion of the molecule acts as a Ca^{2+}-dependent *ATPase* (*3.6.1.32*), the activity of which is modulated by the small subunits.

Actin (42 kDa) is the main constituent of the *thin filaments*. It accounts for ca. 20–25% of the muscle proteins. **F-actin** is also an constituent of the cytoskeleton (see p. 190) where the filamentous F-actin polymer is in equilibrium with its monomer, **G-actin**.

The muscle proteins also include tropomyosin and troponin. **Tropomyosin** (64 kDa), a very elongated molecule, binds to F-actin and connects approximately seven actin mono-

mers. **Troponin** (78 kDa), a complex of three different subunits, is bound to one end of the tropomyosin molecule. Other proteins found in muscles at much lower concentration are α- and β-*actinin, desmin, titin,* and *vimentin.*

B. Mechanism of muscle contraction ◑

The mechanism of muscle contraction is explained by the *sliding filament model*, in which the thick and thin filaments slide past one another. Muscle contraction is the result of the following reaction cycle:

(**1**) In the relaxed state, the myosin heads are attached to actin. Upon binding of ATP, the heads detach themselves from actin.

(**2**) The myosin head ATPase hydrolyzes ATP to ADP and P_i, but does not release the two reaction products. ATP cleavage results in an allosteric change that builds up tension in the myosin head.

(**3**) Next, the myosin head forms new bonds with a neighboring actin molecule.

(**4**) Now actin facilitates the release of P_i, and shortly thereafter that of ADP. This results in the conversion of the allosteric tension in the myosin head into a conformational change that acts like a rowing stroke. The cycle continues as long as ATP is available. Thus the thick filaments are constantly moving along the thin filaments in the direction of the Z disc.

Each power stroke of the 500 myosin heads of a thick filament produces a contraction of about 10 nm. During strong contraction, the process is repeated about 5 times per second. ATP cleavage facilitates the interaction between actin and myosin, and thus causes a directed movement of these structural molecules against each other.

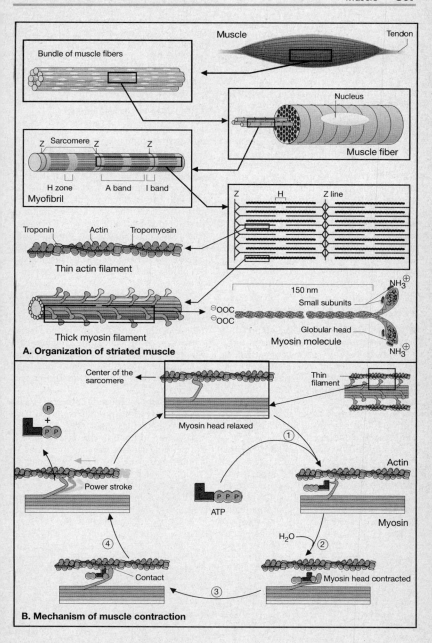

A. Organization of striated muscle

Muscle

Tendon

Bundle of muscle fibers

Nucleus

Muscle fiber

Z Sarcomere Z   Z

H zone   A band   I band
Myofibril

Z   H   Z line

Troponin   Actin   Tropomyosin

Thin actin filament

150 nm

NH₃⊕

Small subunits

⊖OOC
⊖OOC

Globular head

Thick myosin filament

Myosin molecule

NH₃⊕

B. Mechanism of muscle contraction

Center of the sarcomere

Thin filament

Myosin head relaxed

①

Actin

P
+
P P

Power stroke

ATP   P P P

Myosin

H₂O

②

④

Contact

Myosin head contracted

③

Control of Muscle Contraction

A. Electromechanical coupling ◑

Muscle contraction is controlled by motor neurons that release the neurotransmitter **acetylcholine** at **neuromuscular junctions** (see p. 320). Acetylcholine then diffuses across the narrow synaptic cleft, and binds to *acetylcholine receptors* on the membrane of the muscle cell. This results in the opening of ion channels within the receptor molecules, in such a way that a depolarizing, synaptic ion current can flow. This current triggers an all-or-nothing response in the form of an **action potential** across the plasma membrane of the muscle cell (red curve; see pp. 318 and 320). The action potential moves out in all directions from the neuromuscular junction, resulting in stimulation of the entire muscle fiber. Within a few milliseconds, the contractile mechanism responds and the fiber contracts (black curve).

B. Sarcoplasmic reticulum (SR) ◑

Inside the muscle fibers, the action potential is translated into a transient increase in the concentration of Ca^{2+} ions in the cytoplasm. In resting cells, the Ca^{2+} concentration is very low (less than 10^{-7} M). In the so-called **sarcoplasmic reticulum** (SR), however, the Ca^{2+} level is much higher (about 10^{-3} M). The SR is a branched organelle similar to the ER, which surrounds individual myofibrils like a net (the top half of the illustration shows the SR of a heart muscle cell as an example). The high calcium concentration in the SR is maintained by *Ca^{2+}-transporting ATPases*. In addition, there is a special protein called **calsequestrin** (55 kDa), which, because of its many acidic amino acid residues, is able to bind numerous Ca^{2+} ions per molecule.

The transfer of the action potential to the SR of individual myofibrils is facilitated by **transverse tubules** (T tubules), which are open to the extracellular space. The depolarization of the plasma membrane is transmitted via the T tubules to a voltage-gated membrane protein referred to as the "SR foot", which, in turn, triggers opening of **Ca^{2+}** channels in the SR membrane. This results in an influx of calcium ions from the SR, which increases the Ca^{2+} concentration in the cytoplasm of the muscle cell to a level greater than 10^{-7} M. This ultimately stimulates the myofibrils to contract.

C. Regulation by calcium ions ◑

In relaxed skeletal muscle tissue, the complex between **troponin** (subunits = T, C, I) and **tropomyosin** blocks access of the myosin heads to the actin. The rapid increase in the cytoplasmic calcium concentration that follows opening of the calcium channels of the SR results in the binding of Ca^{2+} to the **C subunit** of troponin. The latter is very similar to calmodulin (see p. 352). Ca^{2+} binding triggers a conformational change in troponin, such that the troponin-tropomyosin complex becomes distorted and the myosin binding site is exposed (red). This initiates the contraction cycle (see p. 306).

In the absence of a further stimulus, an ATP-driven calcium pump in the SR membrane ensures that the Ca^{2+} concentration in the cytoplasm rapidly returns to the resting level. As a consequence, the troponin-Ca^{2+} complex dissociates, and troponin returns to its original conformation, which does not allow myosin binding to actin. The muscle then relaxes.

In summary, the following processes occur during **contraction** of striated muscle. The muscle cell becomes depolarized. This signal is passed on to the Ca^{2+} channels in the SR. The Ca^{2+} ion channels open, and the intracelluar Ca^{2+} level rises. Ca^{2+} binds to troponin C, causing a conformational change in troponin that distorts tropomyosin and allows the myosin heads to bind to actin. The actin-myosin cycle begins, and the muscle fibers contract.

At the **end of contraction**, the Ca^{2+} level falls as a result of transport of Ca^{2+} back to the SR, troponin C releases Ca^{2+}, and tropomyosin returns to its original position on the actin molecule, blocking the actin-myosin cycle. The result is muscle relaxation.

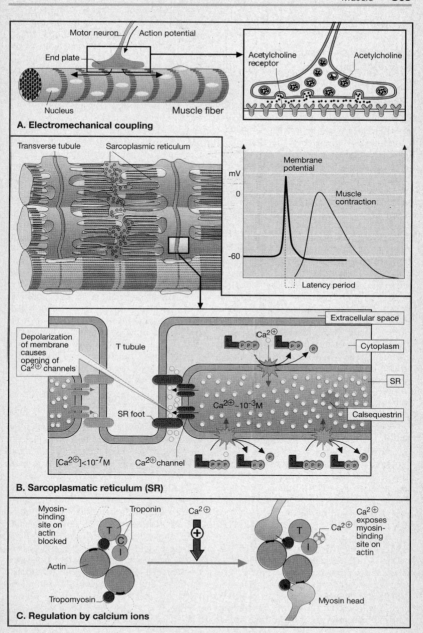

A. Electromechanical coupling

Motor neuron — Action potential
End plate — Acetylcholine receptor — Acetylcholine
Nucleus — Muscle fiber

B. Sarcoplasmatic reticulum (SR)

Transverse tubule — Sarcoplasmic reticulum

mV — Membrane potential — 0 — Muscle contraction — -60 — Latency period

Extracellular space
Depolarization of membrane causes opening of Ca²⁺ channels
T tubule — Ca²⁺ — Cytoplasm
SR
SR foot — Ca²⁺ ~10⁻³M — Calsequestrin
[Ca²⁺] < 10⁻⁷M — Ca²⁺ channel

C. Regulation by calcium ions

Myosin-binding site on actin blocked — Troponin — Ca²⁺ — Ca²⁺ exposes myosin-binding site on actin
Actin — Tropomyosin — Myosin head

Metabolism

A. Energy metabolism ◑

The principal task of the muscles is to perform mechanical work at the expense of ATP. In skeletal muscles, ATP is synthesized from a range of different metabolites, e.g., glucose, fatty acids and ketone bodies. During light exercise, these fuels are completely degraded to CO_2 and H_2O. As this process is O_2-dependent, muscle contraction leads to a strong increase in oxygen consumption. After the onset of contraction, the ATP present in resting muscles would be consumed in less than a second were it nor for highly efficient metabolic pathways that continuously resynthesize ATP.

Rapid replenishment of ATP can be achieved by transferring a phosphate group from **creatine phosphate** to ADP in a reaction catalyzed by *creatine kinase* [1]. Creatine phosphate is an energy reserve typical of muscle tissue. It contains an "energy-rich" bond between the phosphate residue and the nitrogen of the guanido group. However, this store only lasts for a few seconds.

Another mechanism for short-term increase in the level of ATP is conversion of ADP to ATP and AMP, catalyzed by *adenylate kinase* (myokinase [2]). The **AMP** produced in this reaction can subsequently be converted to IMP by *AMP deaminase* [4] in order to "pull" the equilibrium of the reaction in the favorable direction.

The most important long-term reserve in the muscles is **glycogen** (see p. 160). Glycogen is synthesized from glucose in resting muscle, and can account for up to 2% of muscle mass. As in the liver, mobilization of glycogen occurs by phosphorolysis. The glucose 1-phosphate formed is isomerized to glucose 6-phosphate, which yields ATP either via oxidative phosphorylation or — when oxygen is in short supply — through conversion to lactate by anaerobic glycolysis.

Oxidative phosphorylation (see p. 130) is the most efficient pathway for the generation of ATP, and therefore supplies almost all of the ATP required for the continuous contraction and relaxation of cardiac muscle. This is

why the heart is dependent on a constant and adequate supply of oxygen (*heart attacks* occur as a result of interruption of the oxygen supply).

In slowly contracting skeletal muscles, oxygen supply is facilitated by the presence of **myoglobin** (Mb). This hemoglobin-like protein is able to store oxygen. Rapidly contracting muscles do not have red myoglobin, and therefore appear white. They show a preference for **anaerobic glycolysis** to satisfy their high ATP requirements during heavy exercise. Although they are capable of rapid contraction, these muscles can only continue to work for short periods of time, because of the relatively low ATP yield of glycolysis. After a, while, the muscle becomes exhausted as a result of the decline in the pH of the muscle cells. The lactate that accumulates during anaerobic glycolysis is released into the blood and transported to the liver, where it is, in part, utilized for the re-synthesis of glucose (*Cori cycle*, see p. 282).

Glycogen degradation in the muscle is under hormonal control (see p. 110). It is activated by **epinephrine** via the cAMP-dependent activation of *phosphorylase kinase* (see p. 110). The increase in **Ca^{2+}** concentration during muscle contraction also leads to the activation of *glycogen phosphorylase*.

B. Metabolism of proteins and amino acids ◑

Skeletal muscle is actively involved in amino acid metabolism. It is the most important site for the degradation of *branched-chain amino acids* (Val, Leu, Ile). Some other amino acids are also preferentially degraded by muscle cells. *Alanine* and *glutamine* are synthesized in the process and released into the blood. These amino acids are vehicles for the transport of protein nitrogen to the liver (*alanine cycle*, see p. 282) and kidneys (see p. 300). **Muscle proteins** constitute an energy reserve during periods of prolonged starvation. They are degraded to amino acids, which are transported to, and metabolized by, the liver (see p. 280).

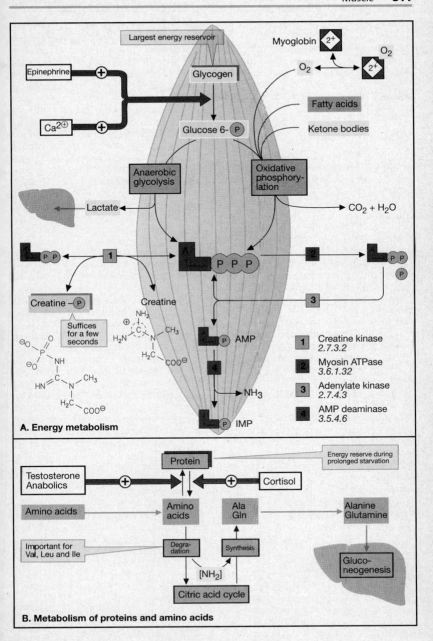

A. Energy metabolism

Largest energy reservoir

Epinephrine +

Ca²⁺ +

Glycogen

Glucose 6-P

Myoglobin 2+ · O₂ · 2+

O₂

Fatty acids

Ketone bodies

Anaerobic glycolysis

Oxidative phosphory-lation

Lactate

CO₂ + H₂O

Creatine – P

Suffices for a few seconds

Creatine

AMP

NH₃

IMP

1 Creatine kinase
2.7.3.2

2 Myosin ATPase
3.6.1.32

3 Adenylate kinase
2.7.4.3

4 AMP deaminase
3.5.4.6

B. Metabolism of proteins and amino acids

Protein

Energy reserve during prolonged starvation

Testosterone Anabolics +

Cortisol +

Amino acids

Amino acids

Ala Gln

Alanine Glutamine

Important for Val, Leu and Ile

Degra-dation

Synthesis

[NH₂]

Citric acid cycle

Gluco-neogenesis

Collagens

The collagens are the most abundant proteins in animals. They account for 25% of all proteins. The collagens form insoluble fibers, which are found in the form of extracellular structures throughout the matrix of the connective tissues.

A. Structure of collagen ◑

Typical collagen molecules consists of a right-handed **triple helix** of three polypeptides (α chains), which, in turn, are made up from multiple repetitions of the characteristic sequence **Gly-X-Y** (see p. 68). Every third amino acid in such sequences is a glycine. *Proline* (**Pro**) is frequently found at positions X or Y, while the Y position is often occupied by *4-hydroxyproline* (**4Hyp**). Sometimes *3-hydroxyproline* (**3Hyp**) or *5-hydroxylysine* (**5Hyl**) also occur. These hydroxylated amino acids are characteristic of collagen. Their synthesis occurs posttranslationally, i.e., proline or lysine residues undergo hydroxylation when already incorporated in the peptide chain.

There are at least 12 variants of collagen. Each has a particular combination of different types of α chains (α1-α3 and additional subtypes). The common type I collagen has the quaternary structure $[α1(I)]_2α2(I)$. It is a long, filamentous molecule with a mass of about 285 kDa (see p. 63).

Collagens tend to spontaneously associate with one another to form larger, more complex structures. Most types create **cylindrical fibrils** (diameter 20–500 nm) with a characteristic staggered banding pattern that repeats itself every 64–67 nm. At either end, the individual collagen molecules are crosslinked via modified lysine side chains. The number of such crosslinks increases as the organism ages.

B. Biosynthesis ◑

The precursor molecule of collagen (pre-pro-protein) is synthesized by ribosomes associated with the rER. Before attaining its mature form, the precursor undergoes extensive *posttranlational modification* in the ER and Golgi apparatus (see p. 210). Cleavage of the signal peptide (**1**) gives rise to **procollagen**. This molecule still carries long **propeptides** at either end. Many proline residues, and some of the lysines of procollagen become hydroxylated (**2**). Some of the hydroxylysine residues additionally undergo glycosylation (**3**). Oxidation of cysteine groups within the propeptides generates intramolecular and intermolecular disulfide bonds (**4**), which ensure the correct assembly of the peptide strands to form a triple helix (**5**). These steps have to be completed correctly before procollagen can be secreted into the extracellular space. This is where the *N*- and *C*-terminal propeptides are removed by hydrolysis (**6**) leading to the staggered assemblage of the collagen molecule to form fibrils (**7**). Finally, the side chains of some of the lysine residues undergo enzyme-catalyzed oxidation to yield aldehyde groups (**8**). Condensation reactions then lead to intramolecular and intermolecular crosslinks (**9**). Only then do the collagen fibrils attain their final structure, which is characterized by high *tensile strength* and a high degree of *resistance to proteinases*.

The **biological roles** of collagen are many and varied (see p. 314). The central importance of collagens is emphasized by the existence of numerous *hereditary genetic defects*, which result from mutations in the collagen molecule itself or in the enzymes involved in its synthesis. These defects can have an impact on the structure and function of the skeleton, the joints, tendons, skin, eyes, blood vessels, hair, and on the overall growth of the body.

The hydroxylations of proline and lysine residues of procollagen are catalyzed by *dioxygenases*, which have iron atoms in their active centers. **Ascorbate** (vitamin C) is essential for the maintenance of the function of these enzymes (see p. 334). Many symptoms of vitamin C deficiency, e.g., loss of teeth, bleeding, skin damage (*scurvy*), can be explained by disturbances in collagen biosynthesis.

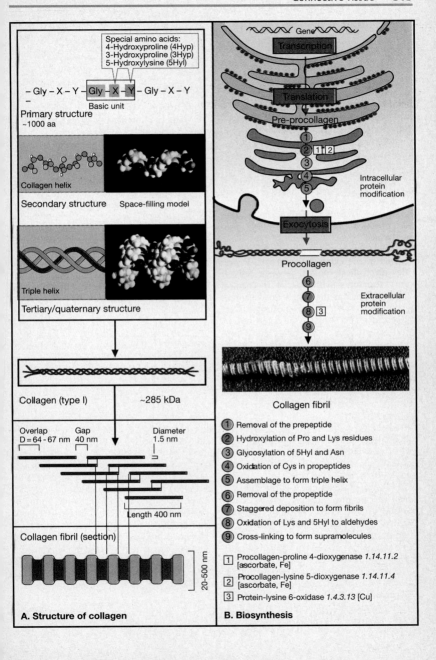

Special amino acids:
4-Hydroxyproline (4Hyp)
3-Hydroxyproline (3Hyp)
5-Hydroxylysine (5Hyl)

– Gly – X – Y – Gly – X – Y – Gly – X – Y –

Basic unit

Primary structure
~1000 aa

Collagen helix

Secondary structure   Space-filling model

Triple helix

Tertiary/quaternary structure

Collagen (type I)   ~285 kDa

Overlap   Gap   Diameter
D = 64 - 67 nm   40 nm   1.5 nm

Length 400 nm

Collagen fibril (section)

20-500 nm

A. Structure of collagen

Gene

Transcription

Translation

Pre-procollagen

1
2   1 2
3
4
5

Intracellular
protein
modification

Exocytosis

Procollagen

6
7
8   3
9

Extracellular
protein
modification

Collagen fibril

① Removal of the prepeptide
② Hydroxylation of Pro and Lys residues
③ Glycosylation of 5Hyl and Asn
④ Oxidation of Cys in propeptides
⑤ Assemblage to form triple helix
⑥ Removal of the propeptide
⑦ Staggered deposition to form fibrils
⑧ Oxidation of Lys and 5Hyl to aldehydes
⑨ Cross-linking to form supramolecules

1 Procollagen-proline 4-dioxygenase *1.14.11.2*
   [ascorbate, Fe]
2 Procollagen-lysine 5-dioxygenase *1.14.11.4*
   [ascorbate, Fe]
3 Protein-lysine 6-oxidase *1.4.3.13* [Cu]

B. Biosynthesis

Composition of the Extracellular Matrix

A. Extracellular matrix ◑

Animal cells are surrounded by a complex intercellular substance known as the **extracellular matrix**. In tissues, such as muscle and liver, the space occupied by the extracellular matrix is insignificant, whereas in other tissues it is very wide. In *connective tissues*, *cartilage*, and *bones*, the extracelluar matrix is the most important part of the tissue. The illustration presents a simplified view of the three main constitutents of the extracellular matrix — **collagens**, which provide tensile strength, network-forming **adhesive proteins**, and the space-filling **proteoglycans**.

A wide variety of functions can be attributed to the extracellular matrix. It provides mechanical contact between cells. It forms structures with special mechanical properties, such as the bones, cartilage, tendons, and joints. It is responsible for the filtering properties of the kidneys. It insulates cells and tissues from one another, e.g., by minimizing friction at joints and facilitating the movement of cells. It guides cells by forming pathways along which they can orient themselves, for example during embryonic development. The extracellular matrix is just as varied in its chemical composition as it is in its functions.

Collagens (see p. 312), of which there are at least 12 different varieties, form fibers, fibrils, nets and banded ligaments. Their characteristic properties are *tensile strength* and *flexibility*.

Adhesive proteins connect the different constituents of the extracellular matrix. Important examples include **laminin, fibronectin**, and **elastin.** These multifunctional proteins are characterized by their ability to bind to several other matrix constituents at the same time. The adhesive proteins mediate the attachment of cells to the intracellular matrix via cell surface receptor proteins or **integrins**.

Proteoglycans act as packaging material. Because of their polar nature and their many negative charges, they bind large numbers of water molecules and cations, thus forming hydrated gels. The properties of the amorphous ground substance of the extracellular matrix are mainly due to the proteoglycans.

B. Fibronectins ○

Typical representatives of the adhesive proteins are the fibronectins. These are filamentous dimers of closely related peptide chains (size 250 kDa) linked to one another via disulfide bonds. The fibronectin molecules are divided into different **domains**. These domains bind to cell-surface receptors, collagens, fibrin and various proteoglycans. This is the basis of the "molecular glue" properties of fibronectins.

Fibronectins have a modular structure. They consist of a small number of different types of **peptide modules**, which are repeated numerous times. Each of the more than 50 modules is encoded by an individual exon in the fibronectin gene. Fibronectins of varying composition can be formed by *alternative splicing* of the RNA transcript of the fibronectin gene. One of the modules contains the characteristic amino acid sequence -*Arg-Gly-Asp-Ser*-. These are the residues that are responsible for the binding of fibronectin to cell-surface receptors ("*integrins*").

C. Proteoglycans ◑

Proteoglycans are extremely large molecular complexes (mass > $2 \cdot 10^6$ Da) made up of carbohydrates (95%) and proteins (5%) and resembling a bottlebrush in shape. Numerous protein molecules are associated with an axis of **hyaluronate** (a polysaccharide, see p. 40), many of them carrying additional polysaccharide chains. The polysaccharides found in proteoglycans typically contain acetylated amino sugars and are, therefore, referred to as **glycosaminoglycans** (see p. 40). The basic units of the various glycosaminoglycans are repeating disaccharide units, which contain a uronic acid and an amino sugar as building blocks.

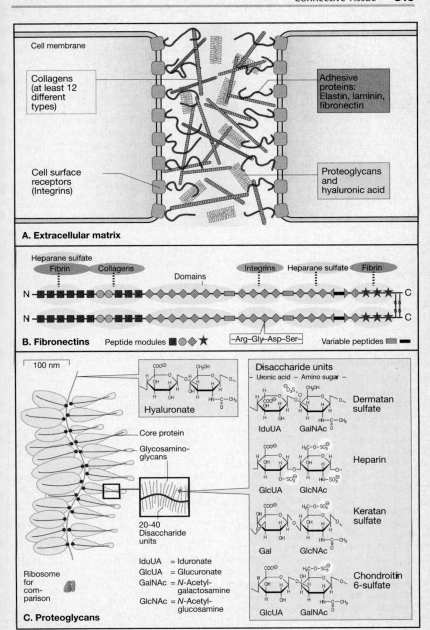

A. Extracellular matrix

Cell membrane

Collagens (at least 12 different types)

Cell surface receptors (Integrins)

Adhesive proteins: Elastin, laminin, fibronectin

Proteoglycans and hyaluronic acid

B. Fibronectins

Heparane sulfate
Fibrin Collagens Domains Integrins Heparane sulfate Fibrin

N C

N C

Peptide modules ■ ● ◆ ★   –Arg–Gly–Asp–Ser–   Variable peptides ▬ ▬

C. Proteoglycans

100 nm

Hyaluronate

Core protein

Glycosamino-glycans

20–40 Disaccharide units

Ribosome for comparison

IduUA = Iduronate
GlcUA = Glucuronate
GalNAc = N-Acetyl-galactosamine
GlcNAc = N-Acetyl-glucosamine

Disaccharide units
– Uronic acid – Amino sugar –

Dermatan sulfate
IduUA GalNAc

Heparin
GlcUA GlcNAc

Keratan sulfate
Gal GlcNAc

Chondroitin 6-sulfate
GlcUA GalNAc

Nerve Tissue

A. Structure of a nerve cell ●

Nerve cells (neurons) have a characteristic structure. They consist of a **cell body** (soma), from which projections called **dendrites** and **axons** branch out. Neurons receive stimuli via the dendrites, and transfer them via the axon. Axons are often surrounded by *Schwann cells*, which form a myelin-containing sheath. This sheath is responsible for the excellent electrical insulation of the axon, and it increases the rate of signal transmission.

The transfer of stimuli occurs at **synapses**, which are located at the ends of **axon terminals**. Synapses mediate contacts between individual neurons, and also between neurons and muscle cells. The synapses store *neurotransmitters* (see p. 322), i.e., chemical signal substances that can be released to excite neighboring neurons or muscle cells.

Nerve cells have a high lipid content, which may account for about 50% of their dry weight. These lipids include a wide variety of phospholipids, glycolipids, and sphingolipids (see p. 204).

B. Energy metabolism of the brain ◑

The brain is very well supplied with blood, and has an active energy metabolism. Although it accounts for only about 2% of the body mass, the brain consumes about 20% of the oxygen and 60% of the glucose, which is fully oxidized to CO_2 and H_2O via glycolysis and the citric acid cycle.

Glucose is the only energy-yielding fuel used by the nervous system. Thus, the brain is dependent on a constant supply of glucose. It is only following prolonged starvation that the brain begins to use **ketone bodies** as an additional energy source (see p. 284). The glycogen reserves of nerve cells are insignificant. Fatty acids, which are transported in the plasma bound to albumin, cannot reach the cells of the brain, because of the *blood-brain barrier*, and thus are unavailable for energy generation. Amino acids cannot be used for ATP production in the brain either, because the neurons are not capable of gluconeogenesis.

The dependence of the brain on glucose means that a decline in the blood glucose level can be life-threatening, e.g., when a diabetic receives an overdose of insulin.

The most important energy-consuming reaction in nerve cells is the hydrolysis of ATP by the *Na^+/K^+-transporting ATPase* in the cell membrane ([1], see p. 206). The active transport of Na^+ and K^+ compensates for the continuous flow of these ions through membrane channels. Of course, many other processes in the brain consume ATP as well.

C. Amino acid metabolism of the brain ◑

The brain is actively involved in amino acid metabolism. The intracellular concentrations of amino acids in the brain are higher than those found in the liver. **Glutamate** (ca. 5–10 mM) and **aspartate** are particularly abundant. These amino acids are formed by transamination of 2-oxoglutarate and oxaloacetate, both intermediates of the citric acid cycle (see p. 152).

In many regions of the brain, γ-**aminobutyrate (GABA**, 4-aminobutyrate) is formed by decarboxylation of glutamate. Biosynthesis and degradation of this neuron-specific amino acid occur in a bypass of the citric acid cycle (**"GABA shunt"**), which, in contrast to the cycle, does not result in the synthesis of GTP. The GABA shunt is typical of the brain, but of little importance in other tissues.

In certain neurons amino acids, including *glycine, aspartate, glutamate*, and *GABA*, serve as **transmitter substances** that are stored in the synapses and released upon stimulation (see p. 322). The released transmitters regulate the electrical behavior of neighboring neurons, either by inducing or inhibiting the generation of an action potential.

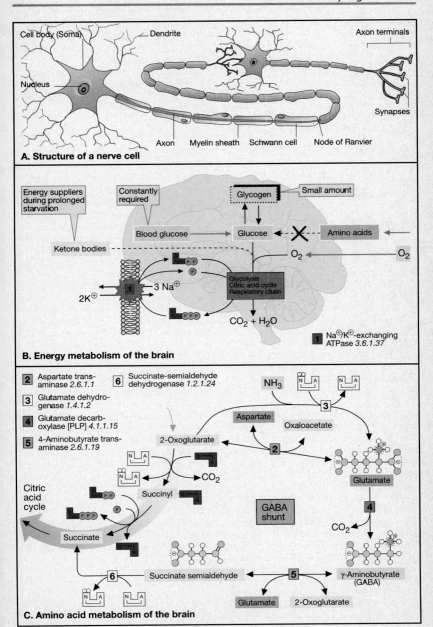

A. Structure of a nerve cell

Cell body (Soma) — Dendrite — Axon terminals

Nucleus

Synapses

Axon   Myelin sheath   Schwann cell   Node of Ranvier

B. Energy metabolism of the brain

Energy suppliers during prolonged starvation

Constantly required

Glycogen — Small amount

Blood glucose → Glucose ⟵✗⟶ Amino acids

Ketone bodies

O_2 ← O_2

$2K^{\oplus}$ — $3\,Na^{\oplus}$

Glycolysis
Citric acid cycle
Respiratory chain

$CO_2 + H_2O$

■ $Na^{\oplus}/K^{\oplus}$-exchanging ATPase 3.6.1.37

C. Amino acid metabolism of the brain

2 Aspartate trans-aminase 2.6.1.1

3 Glutamate dehydro-genase 1.4.1.2

4 Glutamate decarb-oxylase [PLP] 4.1.1.15

5 4-Aminobutyrate trans-aminase 2.6.1.19

6 Succinate-semialdehyde dehydrogenase 1.2.1.24

NH_3

3

Aspartate — Oxaloacetate

2-Oxoglutarate

2

Glutamate

CO_2

Citric acid cycle

Succinyl

Succinate

GABA shunt

4

CO_2

6

Succinate semialdehyde

5

γ-Aminobutyrate (GABA)

Glutamate   2-Oxoglutarate

Resting Potential, Action Potential

Membranes, including the plasma membrane, are impermeable to ionized atoms or molecules unless special transporters facilitate their passage. An important prerequisite for the generation of an electrochemical potential difference across a membrane ("membrane potential," $\Delta\psi$; see p. 116) is the presence of *specific ion channels*. Ion transport via these channels occurs in response to concentration gradients and to the membrane potential itself.

A. Resting potential ◖

Nerve cell membranes contain **channel proteins** for Na^+, K^+, Cl^-, and Ca^{2+} ions. These channels are usually closed, but they may open very briefly to let ions pass through. They can be divided into channels that are regulated by the membrane potential (*voltage-gated*) and those regulated by ligands (*ligand-gated*). The channels are integral membrane proteins made up of several subunits. They are subject to allosteric regulation. Depending on the type of channel involved, interconversions between the different conformational states (open, closed, etc.) are triggered either by alterations in the membrane potential or by ligands, e.g., *neurotransmitters* and *neuromodulators* (see p. 322).

The enzyme *Na+/K+-ATPase* located in the plasma membrane is constantly pumping Na^+ ions out of the cell in exchange for K^+ ions (see p. 206). As a result, there is a large difference in the concentrations of Na^+ and K^+ ions between the cytoplasm and the extracellular space (see table). Three Na^+ ions are exchanged for two K^+ ions, and the transport is therefore *electrogenic*, i.e., it results in an excess negative charge inside the cell relative to the outside. As some of the K^+ channels of nerve cells are open even in the resting state, there is a constant efflux of K^+ into the extracellular space, which results in the generation of a negative **resting potential** of about -60 mV. As indicated by the *permeability coefficients* for the respective ions (see table), the Na^+ and Cl^- channels are usually closed. The membrane is also impermeable to phosphate ions and organic anions, e.g., proteins. Calculations using the **Nernst equation** show that the resting potential of nerve cells is largely determined by the concentration gradient of K^+. This is, because, in the resting state, K^+ ions make by far the greatest contribution to the conductivity of the membrane.

B. Action potential ◖

An action potential is produced in a nerve cell in response to a chemical or electrical stimulus. It consists of a transient increase in the membrane potential from -60 mV to about +30 mV, which starts at a specific point and spreads over the entire membrane. At any given location, the membrane potential returns to its resting value within about 1 ms. The process begins with the opening of Na^+ channels. Positively charged Na^+ ions flow into the cell, and cause a *localized reversal of the membrane potential* (depolarization, **2**). The transient increase of the membrane potential results in the opening of voltage-gated K^+ channels (**2**). These then allow K^+ ions to flow in the opposite direction, so that the membrane potential rapidly returns to a negative value (**3**), which briefly even exceeds the resting potential (**4**). The K^+ channels also close after a few milliseconds. Subsequently, the nerve cell is once again susceptible to stimulation.

All in all, only a very small portion of the unevenly distributed Na^+ and K^+ ions cross the membrane during each action potential. Thus, the process can be repeated again and again whenever the cell receives a new stimulus. The propagation of an action potential along the nerve cell membrane is based on the fact that a local increase of the membrane potential stimulates neighboring voltage-gated ion channels to open. This creates a wave of membrane excitation (*depolarization wave*) that spreads over the whole cell surface.

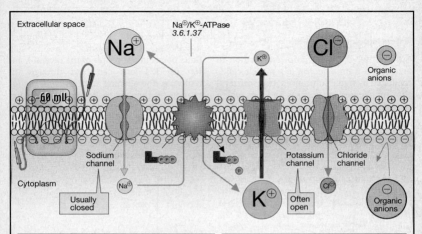

Ion	Concentrations		Permeability coefficient
	Cytoplasm (mM)	Extracellular space (mM)	$(cm \cdot s^{-1} \cdot 10^9)$
$K^\oplus$	139	4	500
$Na^\oplus$	12	145	5
$Cl^\ominus$	4	116	10
Organic anions	138	34	0

Gas constant — Absolute temperature

$$\Delta\Psi = \frac{R \cdot T}{F \cdot z} \ln \frac{c_o}{c_i}$$

Potential-difference — Concentrations outside and inside

Faraday constant — Charge of the ion

e.g. for $K^\oplus$:

$$\Delta\Psi_{K^\oplus} = \frac{0.059}{1} \cdot \log \frac{4}{139} \approx -91 \text{ mV}$$

Nernst equation

A. Resting potential

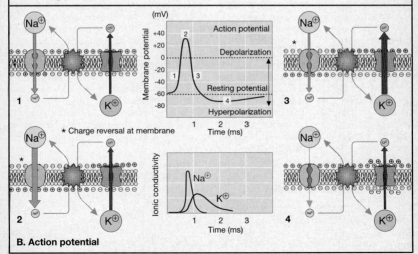

B. Action potential

Synapses

A. Cholinergic synapse ➊

The transfer of signals from one neuron to another, or from a neuron to a muscle cell is mediated by signal substances (**transmitters**) that are released at synapses, i.e., sites where the cells are very close to one another. At a synapse, the *presynaptic membrane* of the signaling cell is separated from the *postsynaptic membrane* of the receiving one by a narrow *synaptic cleft*. The transmitters are released into this cleft by *exocytosis*. From there, they diffuse to **receptors** on the postsynaptic membrane, where they bind and thereby transmit the signal to the neighboring cell. The receptors are either *ligand-gated ion channels* (see p. 318) or membrane proteins that regulate ion channels via G proteins (see p. 350).

The illustration shows a cholinergic synapse between a motor neuron and a muscle cell. Such synapses contain the neurotransmitter **acetylcholine**.

Signal transmission occurs as follows. An **action potential** reaches the presynaptic membrane (**1**), causing **voltage-gated Ca²⁺ channels** to open (**2**). This results in the influx of Ca²⁺ ions from the extracellular space, and a dramatic increase in their concentration in the synapse. The Ca²⁺ ions trigger an exocytotic process in which many **synaptic vesicles** release their contents (acetylcholine) into the synaptic cleft (**3**). Acetylcholine molecules diffuse across the synaptic cleft and bind to, and thus activate, **postsynaptic receptor molecules** (**4**). The acetylcholine receptors are ligand-gated Na⁺ (and K⁺) ion channels. The influx of Na⁺ ions raises the resting potential of the postsynaptic nerve or muscle cell to the point at which **voltage-gated Na⁺ channels** open and an action potential is triggered (**5**, see p. 318).

B. Nicotinic acetylcholine receptor ○

The nicotine-responsive ("nicotinic") acetylcholine receptor is a pentameric membrane protein ($\alpha_2\beta\gamma\delta$, mass 250–270 kDa) that acts as a *ligand-gated ion channel* for Na⁺ and K⁺

ions. The protein complex has binding sites for acetylcholine on the extracellular surfaces of both α subunits. When the ligand binds, a funnel-shaped structure made up of five subunits opens in the center of the receptor, and Na⁺ (or K⁺) ions can pass through the molecule. The opening and closing of the ion channel is thought to be due to a conformational change involving charged peptide sequences inside the molecule.

Various pharmaceutics also bind to the receptor. As mentioned above, nicotine acts as an *agonist* of acetylcholine. In contrast, atropin acts as an *antagonist*, i.e., it displaces acetylcholine from its binding site without triggering an allosteric alteration or opening the ion channel.

C. Metabolism of acetylcholine ➊

Acetylcholine, the acetic acid ester of choline, is synthesized in the cytoplasm of the axon from acetyl-CoA and choline [1]. It is stored in the **synaptic vesicles** each of which contains 1000 to 10,000 acetylcholine molecules. Following its release, acetylcholine diffuses across the synaptic cleft. When not bound to receptors, acetylcholine is hydrolyzed by *acetylcholine esterase* [2]. This enzyme has an extremely high turnover number, ensuring the rapid removal of the signal substance. The cleavage products **choline** and **acetate** are taken up actively by the presynaptic neuron and recycled for the synthesis of acetylcholine [3].

Substances that block the essential serine residue in the active center of acetylcholine esterase prolong the effects of acetylcholine, and thereby act as *neurotoxins*. By contrast, the arrow poison *curare* competitively blocks the acetylcholine binding site on the receptor.

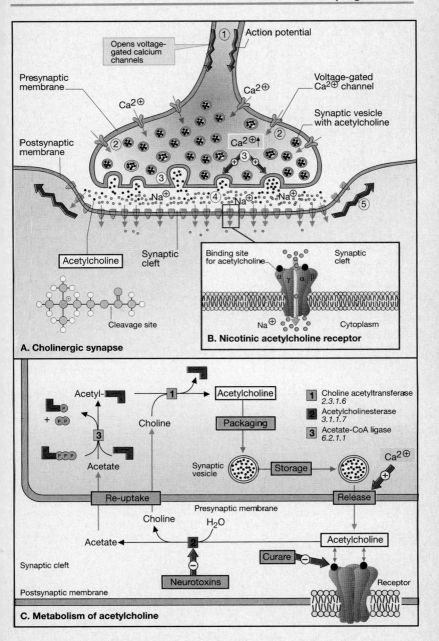

① Action potential

Opens voltage-gated calcium channels

Presynaptic membrane

Ca²⊕

Postsynaptic membrane

Voltage-gated Ca²⊕ channel

Synaptic vesicle with acetylcholine

Ca²⊕↑

Na⊕   Na⊕   Na⊕

Acetylcholine

Synaptic cleft

Cleavage site

A. Cholinergic synapse

Binding site for acetylcholine

Synaptic cleft

α γ α β

Na⊕   Cytoplasm

B. Nicotinic acetylcholine receptor

Acetyl-

P + P·P

P·P·P

Acetate

Choline

Acetylcholine

Packaging

1 Choline acetyltransferase 2.3.1.6

2 Acetylcholinesterase 3.1.1.7

3 Acetate-CoA ligase 6.2.1.1

Ca²⊕

Synaptic vesicle

Storage

Release

Re-uptake

Choline

Presynaptic membrane

Acetate   H₂O

Synaptic cleft

Neurotoxins

Acetylcholine

Curare

Receptor

Postsynaptic membrane

C. Metabolism of acetylcholine

Neurotransmitters

A. Neurotransmitters and neurohormones ●

Nerve cells regulate many functions of the body by releasing chemical substances collectively referred to as *neurosecretions*. These can be divided into two different groups. The first includes the short-lived **neurotransmitters**, which are released into synaptic clefts and only have short-range effects on neighboring cells. The second includes longer-lived **neurohormones**, which are released into the blood and thus can have long-range effects. However, the distinction between the two groups is somewhat blurred, i.e., neurotransmitters can function simultaneously as neurohormones.

A number of different criteria have to be satisfied before a signal substance can be classified as a **neurotransmitter** (or *neuromodulator*). First of all, it must be produced by neurons. Secondly, it must be stored in the synapses. Thirdly, it must be released into the synaptic cleft upon stimulation. Fourthly, it must bind specifically to receptors on the postsynaptic membrane of either another neuron or a muscle cell. Lastly, this binding must trigger an ion flux that regulates the activity of the target cell.

B. Chemical structure ◐

Neurotransmitters can be divided into several different groups, based on their chemical properties. The best known neurotransmitter is **acetylcholine** (see p. 320). A few **amino acids** also act as neurotransmitters. **Biogenic amines** arise from the decarboxylation of amino acids (see p. 154). **Purine derivatives** with neurotransmitter function are all derived from adenosine. The largest group of neurosecreted molecules are **peptides** and **proteins**. Several of these peptides carry a cyclic glutamate residue (pyroglutamic acid or 5-oxoproline; single letter code: <G) at their *N*-terminus. The *C*-terminal carboxylate group is often blocked by an amido group (symbol: -NH$_2$). Both modifications result in better protection of the neuropeptides from non-specific degradation by peptidases. Be-

sides neuropeptides, there are also large neuroproteins.

Mode of action. Neurotransmitters and neuromodulators bind to receptors on the membranes of neighboring postsynaptic cells. Many of these molecules interact with several types of receptors that trigger different signalling pathways. In some cases, the receptors are *ligand-gated ion channels*, e.g., the "nicotinic" receptors for acetylcholine, GABA and glycine. In most cases, however, neurotransmitter receptors regulate ion channels indirectly by binding to *7-helix receptors* that transmit the signal via *G proteins* (see p. 350).

Most neurotransmitters and neuromodulators stimulate the *opening* of ion channels, whereas only a few stimulate closure. The type of ion channel involved determines the direction of the change in the membrane potential of the postsynaptic cell. A decrease in the membrane potential, e.g., due to opening of Na$^+$ channels, stimulates postsynaptic action potentials, whereas an increase, e.g., due to opening of Cl$^-$ channels, inhibits their formation.

C. Biosynthesis of the catecholamines ◑

The catecholamines are *biogenic amines*. Their common feature is the 3,4-dihydroxyphenyl ("catechol") group. Their biosynthesis begins with the amino acid **tyrosine**. Hydroxylation [1] gives rise to **dopa** (3,4-dihydroxyphenylalanine). Subsequent decarboxylation [2] yields **dopamine**. This is then converted to **norepinephrine** (noradrenalin) by a further hydroxylation step [3]. The hydrogen donor for this reaction is *ascorbic acid* (see p. 334). Finally, **epinephrine** (adrenalin) is produced by methylation of norepinephrine [4]. Dopamine, norepinephrine, and epinephrine are important neurotransmitters. The latter two also act as hormones.

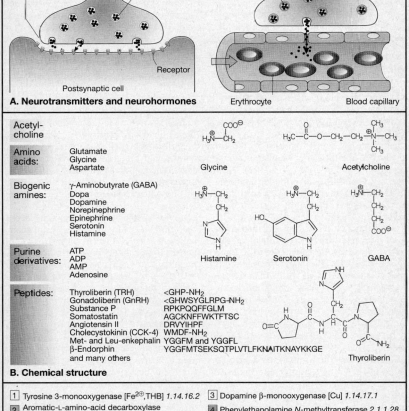

A. Neurotransmitters and neurohormones

Acetyl-choline			
Amino acids:	Glutamate Glycine Aspartate		
Biogenic amines:	γ-Aminobutyrate (GABA) Dopa Dopamine Norepinephrine Epinephrine Serotonin Histamine		
Purine derivatives:	ATP ADP AMP Adenosine		
Peptides:	Thyroliberin (TRH) Gonadoliberin (GnRH) Substance P Somatostatin Angiotensin II Cholecystokinin (CCK-4) Met- and Leu-enkephalin β-Endorphin and many others	<GHP-NH₂ <GHWSYGLRPG-NH₂ RPKPQQFFGLM AGCKNFFWKTFTSC DRVYIHPF WMDF-NH₂ YGGFM and YGGFL YGGFMTSEKSQTPLVTLFKNAITKNAYKKGE	

Glycine   Acetylcholine   Histamine   Serotonin   GABA   Thyroliberin

B. Chemical structure

1. Tyrosine 3-monooxygenase [Fe²⁺,THB] *1.14.16.2*
2. Aromatic-L-amino-acid decarboxylase (Dopa decarboxylase) [PLP] *4.1.1.28*
3. Dopamine β-monooxygenase [Cu] *1.14.17.1*
4. Phenylethanolamine *N*-methyltransferase *2.1.1.28*

Tyrosine   Dopa   Dopamine   Norepinephrine   Epinephrine

C. Biosynthesis of the catecholamines

Sight

The retina of man contains two types of photoreceptor cells, *rods* and *cones*. The rods are sensitive to low light levels, while the cones mediate color vision at high light intensities.

A. Photoreceptor ◑

The illustration shows the complex structure of one of the two types of photoreceptors, a rod. The membrane disks constituting the outer part of the rod contain **rhodopsin**, a membrane-spanning protein with 7 α helices — a feature characteristic of one particular group of signal-transducing receptor molecules (see p. 322). Rhodopsin is a light-sensitive chromoprotein (see p. 322). **Opsin**, the proteinaceous part of rhodopsin, carries **11-cis-retinal** bound to the ε-amino group of a lysine residue. Rhodopsin absorbs visible light, with maximum absorption at 500 nm.

Absorption of the energy of a photon triggers isomerization of 11-*cis*-retinal to yield all-*trans*-retinal. As a result of this *photochemical process*, the shape of the retinal molecule changes, and, within about 10 ms, this induces the allosteric conversion of inactive rhodopsin to its active form. This form, **rhodopsin***, interacts with G proteins to trigger a *signal tranduction cascade*, which culminates in decreased release of neurotransmitters at the synapses of the rod cells. As a result of this, bipolar neurons associated with the rod cells send an *altered signal* to the brain, which is perceived as the presence of light.

B. Signal cascade ○

The G protein of rod cells is known as **transducin**. Binding of transducin to light-activated rhodopsin* stimulates the exchange of bound GDP for GTP. This gives rise to an activated form called **transducin***. This active form dissociates into its α* and βγ subunits (see p. 350). The α **subunit*** mediates further transfer of the signal by binding to an inhibitory protein, thus indirectly activating a *cGMP-phosphodiesterase* [1].

In rod cells not exposed to light, the concentration of the cyclic nucleotide **cGMP** is relatively high (70 μM). This *second messenger* (see p. 352) undergoes constant synthesis and degradation catalyzed by *guanylate cyclase* and the aforementioned phosphodiesterase, respectively. Light-mediated activation of phosphodiesterase causes a rapid decline in the level of cGMP (within milliseconds).

After a short period of time, the α subunit of transducin causes its own inactivation by catalyzing the enzymatic hydrolysis of the bound GTP to GDP. This is followed by the re-association of the α subunit with the βγ subunit, and a return to the original conformation. Rhodopsin* dissociates to form opsin and all-*trans*-retinal. The latter then undergoes isomerization to 11-*cis*-retinal, and the two components once again associate with one another to form rhodopsin. At this point, the molecule has returned to its original conformation.

In the **dark** (below left), the cGMP level in the rods is high, due to high levels of guanylate cyclase activity. As a result of this, cGMP-gated cation channel proteins in the plasma membrane remain open, allowing sodium and calcium ions to enter the cell. Under these conditions, there is a constant release of amino acid-type neurotransmitters from the synapses of the rods.

In the **light** (below right), there is a decline in cGMP levels as a result of *phosphodiesterase** activity. This leads to closing of the ion channels. Sodium and calcium ions are constantly being pumped out of the cell, and therefore their concentrations decline rapidly. This leads to a *hyperpolarization* of the cell membrane, and interrupts the release of neurotransmitters. The lowered calcium concentration causes an activation of guanylate cyclase, and as a result there is a rapid rise in the cGMP level to the point at which the ion channels open again.

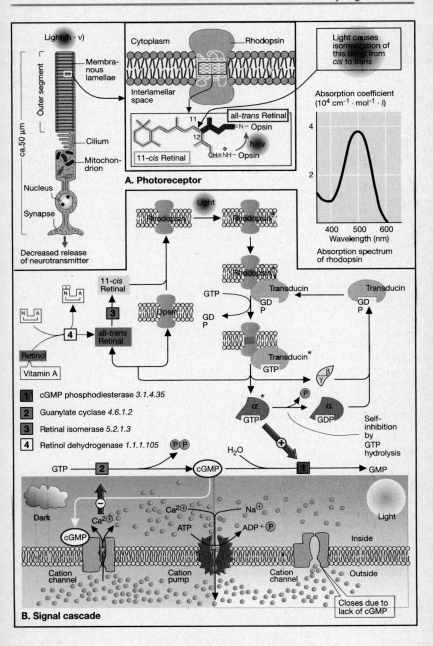

Light (h · ν)

Outer segment

Membranous lamellae

ca. 50 µm

Cilium

Mitochondrion

Nucleus

Synapse

Decreased release of neurotransmitter

Cytoplasm

Rhodopsin

Interlamellar space

all-*trans* Retinal

=N− Opsin

11-*cis* Retinal

CH=NH− Opsin

Light causes isomerization of this bond from *cis* to *trans*

A. Photoreceptor

Absorption coefficient ($10^4 \text{ cm}^{-1} \cdot \text{mol}^{-1} \cdot \text{l}$)

Wavelength (nm)

Absorption spectrum of rhodopsin

Light

Rhodopsin → Rhodopsin

Rhodopsin

Rhodopsin

11-*cis* Retinal

3

Opsin

all-*trans* Retinal

Retinol

Vitamin A

4

GTP

GDP

Transducin

GDP

Transducin

Transducin*

GTP

β
γ

α*
GTP

P

α
GDP

Self-inhibition by GTP hydrolysis

1 cGMP phosphodiesterase 3.1.4.35

2 Guanylate cyclase 4.6.1.2

3 Retinal isomerase 5.2.1.3

4 Retinol dehydrogenase 1.1.1.105

GTP → **2** → cGMP → **1** → GMP

P P

H_2O

(+)

Dark

cGMP

Ca^{2+}

Ca^{2+}

Na^+

ATP

ADP + P

Light

Inside

Cation channel

Cation pump

Cation channel

Outside

Closes due to lack of cGMP

B. Signal cascade

Organic Nutrients

A balanced diet for humans has to contain a large number of different nutrients. These include **proteins, carbohydrates, fats, minerals** (including water), and **vitamins**. The absolute and relative amounts of these substances may vary greatly depending on the diet. Some nutrients are *essential for life*, and must therefore be provided in the diet on a regular basis. The WHO (World Health Organization) and a number of national expert committees have proposed recommendations for daily minumum requirements of the various different nutrients.

A. Energy requirement ●

The amount of energy required by a human is dependent on his or her age, gender, weight, health, and especially on the extent of physical activity. It is recommended that about half of the energy intake should be in the form of carbohydrates, a third at most in the form of fat, and the rest as protein. It is often overlooked that *alcoholic beverages* can make a major contribution to daily energy intake. Ethanol has a "caloric value" of about 30 kJ · g^{-1} (see p. 294).

B. Nutrients ●

Proteins are the form in which the body obtains amino acids for the synthesis of its own proteins. Excess amino acids are degraded to provide energy. *Glucogenic amino acids* can be used for the synthesis of carbohydrates, and *ketogenic amino acids* for the synthesis of ketone bodies and isoprenoids (see p. 154).

The *minimum daily protein requirement* for a man is 37 g, and for a woman 29 g. However, the recommended amount is twice these values. The minimum requirement for pregnant and lactating women is higher still. Not only the quantity, but also the quality, of the protein is important. Proteins lacking or containing only small quantities of some of the essential amino acids are said to be of low biological value. Thus, they are needed in greater quantities. For example, legume proteins contain only small amounts of methionine, whereas wheat and corn proteins are poor in lysine. In contrast, animal proteins are considered to be high value (exceptions: collagen, gelatine).

Proteins are indispensable nutrients, because of the **essential amino acids** they provide. Essential amino acids are those that cannot be synthesized by humans (see table and p. 172). Certain amino acids, including cysteine and histidine, are not considered essential, but nevertheless promote growth. Some amino acids can substitute for others in the diet. For example, humans can synthesize the essential amino acid tyrosine by hydroxylation of phenylalanine, and cysteine from methionine.

Carbohydrates serve as a general energy source. In the diet, they are present as *monosaccharides* in honey and fruit, or as *disaccharides* in milk and in all foods sweetened with sugar (sucrose). *Polysaccharides* are found in plant (starch) and animal (glycogen) products. Although carbohydrates provide a large proportion of the energy derived from the diet, they are not essential.

Fats are the most important *energy-yielding fuels* in the diet. Their relative energy content is about twice as high as that of proteins and carbohydrates. In addition, fats are essential as carriers for vitamins, and as sources of the *polyunsaturated fatty acids* required for the synthesis of eicosanoids (see p. 354).

Minerals are a very heterogeneous group of essential nutrients. They can be divided into **macroelements** and **microelements** (trace elements, see p. 328).

Vitamins are also essential components of the diet. Animals require them, in small quantities only, for the synthesis of coenzymes (see pp. 330–334).

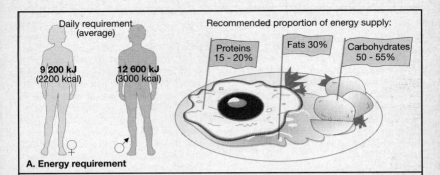

A. Energy requirement

	Quantity in body (kg)	Energy content kJ·g⁻¹ (kcal·g⁻¹)	Daily requirement (g) a b c	General function in metabolism	Essential constituents
Proteins	10	17 (4.1)	♂ 37 55 92 ♀ 29 45 75	Supplier of amino acids Energy source Daily requirement in mg per kg body weight	Essential amino acids: Val (14) Leu (16) Ile (12) Lys (12) Phe (16) Trp (3) Met (10) Cys and His stimulate growth Thr (8)
Carbohydrates	1	17 (4.1)	0 390 240-310	General source of energy (glucose) Energy reserve (glycogen) Roughage (cellulose) Supporting substances (bones, cartilage, mucus)	Non-essential nutritional constituent
Fats	10-15	39 (9.3)	10 80 130	General energy source Most important energy reserve Solvent for vitamins Supplier of essential fatty acids	Polyunsaturated fatty acids: Linoleic acid Linolenic acid Arachidonic acid (together 10g day)
Water	35-40	0	2500 - -	Solvent Cellular building block Dielectric Reaction partner Temperature regulator	
Minerals	3	0		Building blocks Electrolytes Cofactors of enzymes	Macrominerals Microminerals (trace elements)
Vitamins	-	-		Often precursors of coenzymes	Lipid-soluble vitamins Water-soluble vitamins

a: Minimum daily requirement b: Recommended daily intake c: Actual daily intake in industrialized nations

B. Nutrients

Minerals and Trace Elements

A. Minerals ◑

In quantitative terms, **water** is the most important inorganic nutrient in the diet. For an adult, the average daily requirement for water is 2.4 l. This is obtained from water in beverages and solid foods, and also from the *water of oxidation* produced by the respiratory chain (see p. 130). The central importance of water for life is discussed in more detail elsewhere (see p. 20).

The other mineral nutrients are usually divided into **macroelements** (daily requirement > 100 mg) and **microelements** (daily requirement < 100 mg). The macroelements include sodium (Na), potassium (K), calcium (Ca), magnesium (Mg), chlorine (Cl), phosphorus (P), and sulfur (S). The essential microelements, which are required in trace amounts, include iron (Fe), zinc (Zn), manganese (Mn), copper (Cu), cobalt (Co), chromium (Cr), molybdenum (Mo), selenium (Se), and iodine (I). Fluoride (F) is not essential for life, but a daily supply promotes the health of bones and teeth.

The second column in the table summarizes the average amounts of each of the elements present in the body of a 65 kg adult. The fourth column contains a list of the respective **daily requirements**. However, it should be noted that these are only *average* values. Many minerals can be readily stored in the body. Thus, fluctuations in the daily intake average out over time. Minerals stored in the body include *water*, which is distributed throughout the body, *calcium*, stored in the form of apatite in the bones, *iodine*, incorporated in thyroglobulin in the thyroid, and *iron*, stored in the form of ferritin and hemosiderin in bone marrow, spleen, and liver. Many trace elements are also stored in the liver.

The metabolism of many minerals is regulated by *hormones*, e.g., the uptake of H_2O, Ca^{2+} and phosphate, the storage of Fe^{2+} and I^-, and the excretion of H_2O, Na^+, K^+, Ca^{2+}, and phosphate. Children, pregnant and lactating females, or people in poor health, have a higher mineral requirement than normal.

The amounts of mineral nutrients that are absorbed from the diet usually depend on the requirements of the body for a particular substance. However, the overall composition of the diet can also play a role. An example of dietary influence involves calcium. The absorption of calcium is stimulated by lactate and citrate, whereas dietary phosphate, oxalic acid, and phytol inhibit Ca^{2+} absorption due to complex formation and the production of insoluble Ca^{2+} salts.

Mineral deficiencies are not uncommon, and can have various causes, e.g., poorly balanced nutrition, disturbances in mineral absorption, and hereditary diseases. *Calcium deficiency* can arise as a result of pregnancy, rickets or osteoporosis. *Chloride deficiency* is observed as a result of major losses of Cl^- during vomiting. *Iodine deficiency*, which is widespread even in highly industrialized nations, can lead to goiter. *Magnesium deficiency* can be caused by digestive disorders or a poorly balanced diet, e.g., in the case of alcoholism. Mineral deficiencies often become apparent as disturbances in blood cell formation, i.e., they lead to anemias.

The last column lists some of the functions of the minerals. Clearly, almost all of the **macroelements** function either as building blocks or electrolytes. Iodine (as a result of its incorporation into iodothyronins) and calcium act as *signaling substances*. Most **trace elements** are essential *cofactors for proteins*, especially for enzymes. On a quantitative basis, the most important metal-dependent proteins are the *iron proteins* hemoglobin, myoglobin, and the cytochromes. In addition, there are more than 300 different *zinc-dependent proteins*. At least 100 important enzymes require K^+ or Mg^{2+} for activity.

There is still no consensus as to whether vanadium, nickel, tin, boron, and silicon should or should not be considered essential trace elements.

Mineral	Content* (g)	Major source	Daily requirement (g)	Functions/Occurrence
Water	35 000-40 000	Drinks Water in solid foods From metabolism 300g	1200 900	Solvent, cellular building block, dielectric, coolant, medium for transport, reaction partner
Macroelements (daily requirement >100 mg)				
Na	100	Table salt	1.1-3.3	Osmoregulation, membrane potential, mineral metabolism
K	150	Vegetables, fruit, cereals	1.9-5.6	Membrane potential, mineral metabolism
Ca	1 300	Milk, milk products	0.8	Bone formation, blood clotting, signal molecule
Mg	20	Green vegetables	0.35	Bone formation, cofactor for enzymes
Cl	100	Table salt	1.7-5.1	Mineral metabolism
P	650	Meat, milk, cereals, vegetables	0.8	Bone formation, energy metabolism, nucleic acid metabolism
S	200	S-containing amino acids (Cys and Met)	0.2	Lipid and carbohydrate metabolism,
Microelements (trace elements)			(mg)	
Fe	4-5	Meat, liver, eggs, vegetables, potatoes, cereals	10	Hemoglobin, myoglobin, cytochromes, Fe/S clusters
Zn	2-3	Meat, liver, cereals	15	Zinc enzymes
Mn	0.02	Found in many foodstuffs	2-5	Enzymes
Cu	0.1-0.2	Meat, vegetables, fruit, fish	2-3	Oxidases
Co	<0.01	Meat	Traces	Vitamin B_{12}
Cr	<0.01		0.05-0.2	Not clear
Mo	0.02	Cereals, nuts, legumes	0.15-0.5	Redox enzymes
Se		Vegetables, meat	0.05-0.2	Selenium enzymes
I	0.03	Seafood, iodized salt, drinking water	0.15	Thyroxin
Requirement not known				▓ Metals ▓ Non-metals
F		Drinking water (fluoridated), tea, milk	0.0015-0.004	Bones, dental enamel

A. Minerals ★ Content in the body of a 65 kg adult

Lipid-Soluble Vitamins

Vitamins are essential organic compounds, which are required in very small quantities. They have to be provided with the diet, because humans and other animals have lost the capability to synthesize them for themselves. Most vitamins are precursors of **coenzymes**, and some are signaling substances. The average daily requirement differs from one vitamin to another, and it is influenced by age, gender, pregnancy, lactation, physical stress and nutritional state.

A. Vitamin supply ●

A healthy diet usually satisfies the average daily requirement for vitamins. However, vitamin deficiences may arise due to malnutrition or disturbances in the absorption of vitamins. Frequent causes of vitamin malnutrition are a shortage of food or an unbalanced diet, such as may be experienced by elderly people, alcoholics, or people consuming mainly processed foods. A vitamin deficiency can lead to **hypovitaminosis**, and in extreme cases avitaminosis. As intestinal bacteria are essential for the synthesis of particular vitamins (K, B_{12}, H), medical treatments that severely harm the intestinal flora can also lead to vitamin deficiencies.

Only some of the vitamins can be stored in the body (A, D, E, B_{12}). Thus, most vitamin deficiencies rapidly lead to **deficiency diseases**. These often affect the skin, blood cells, and nervous system.

Vitamin deficiencies can be treated by improving nutrition, or by supplying vitamins in tablet form. **Hypervitaminoses**, with symptoms of vitamin poisoning, are only observed after overdoses of vitamins A and D. Excesses of most other vitamins are rapidly excreted in the urine.

B. Lipid-soluble vitamins ◗

The vitamins can be classified either as lipid-soluble or water-soluble. The lipid-soluble vitamins include vitamins A, D, E, and K. Chemically, they belong to the **isoprenoids** (see p. 50).

Vitamin A is **retinol**, the parent substance of the "retinoids" retinal and retinoic acid. They can be synthesized by cleavage of the provitamin β-*carotene*. The retinoids are found in meat-containing diets, whereas β-carotene occurs in many fruits and vegetables (carrots). **Retinal** is the substance responsible for the color of the visual pigment *rhodopsin* (see p. 324). **Retinoic acid** is an important growth factor. Symptoms associated with vitamin A deficiency are *nightblindness, eye damage* and *disturbances in growth.*

Vitamin D is **calciol**. It is hydroxylated in the liver and kidneys to yield the hormone calcitriol ("1α,25-dihydroxy vitamin D"). Together with two other hormones (parathyrin, calcitonin), calcitriol is involved in the regulation of calcium metabolism (see p. 302). Calciol can be synthesized from 7-dehydro-cholesterol by a photochemical reaction in the skin. When the skin receives too little ultraviolet light, and when vitamin D is lacking in the diet, deficiency symptoms, such as *rickets* in children and *osteomalacia* in adults, may arise. Both diseases are characterized by disturbances in the mineralization of the bones.

Vitamin E is a collective term for **tocopherol** and related compounds, all of which contain a chromane ring. These compounds are only found in plants. They are especially abundant in germinating wheat, and provide effective protection of unsaturated lipids against oxidation.

Vitamin K is a collective term for **phylloquinone** and related substances with modified side chains. Vitamin K is synthesized by bacteria of the intestinal flora, and deficiency of this vitamin is therefore very rare. Vitamin K is involved in the carboxylation of glutamate residues of plasma proteins, which are important in blood clotting (see p. 262). Blood clotting can be inhibited by administering vitamin K antagonists (e.g., coumarin derivatives). This is one of the strategies used in the treatment of thrombosis.

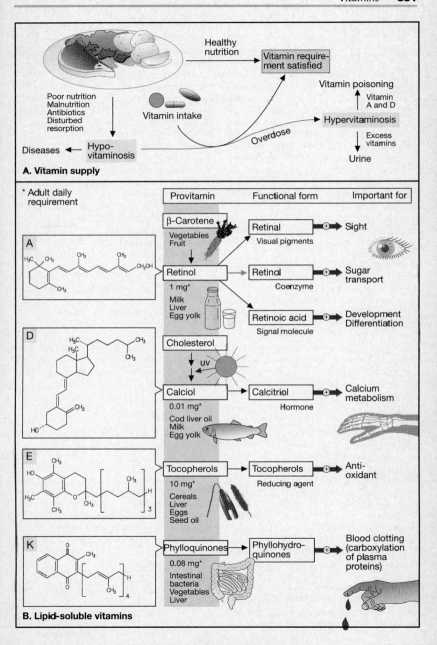

A. Vitamin supply

Healthy nutrition → Vitamin requirement satisfied

Poor nutrition
Malnutrition
Antibiotics
Disturbed resorption

Vitamin intake

Vitamin poisoning
Vitamin A and D

Overdose → Hypervitaminosis

Excess vitamins

Diseases ← Hypovitaminosis

Urine

B. Lipid-soluble vitamins

* Adult daily requirement

	Provitamin	Functional form	Important for
A	β-Carotene — Vegetables, Fruit	Retinal — Visual pigments	Sight
	Retinol — 1 mg* Milk, Liver, Egg yolk	Retinal — Coenzyme	Sugar transport
		Retinoic acid — Signal molecule	Development Differentiation
D	Cholesterol → uv → Calciol — 0.01 mg* Cod liver oil, Milk, Egg yolk	Calcitriol — Hormone	Calcium metabolism
E	Tocopherols — 10 mg* Cereals, Liver, Eggs, Seed oil	Tocopherols — Reducing agent	Antioxidant
K	Phylloquinones — 0.08 mg* Intestinal bacteria, Vegetables, Liver	Phyllohydroquinones	Blood clotting (carboxylation of plasma proteins)

Water-soluble Vitamins

A. Water-soluble vitamins I ●

Vitamin B₁ or **thiamine** consists of two ring systems, a *pyridine ring* (a 6-membered aromatic ring with two nitrogens) and a *thiazol ring* (a 5-membered aromatic ring with nitrogen and sulfur) joined by a methylene group. The active form of vitamin B₁ is **thiamine diphosphate** (TPP), which acts as a coenzyme in the transfer of hydroxyalkyl residues ("active aldehydes"), e.g., in the oxidative decarboxylation of 2-oxoacids (see p. 124) and the transketolase reaction in the hexose monophosphate pathway (see p. 142). Vitamin B₁ deficiency leads to *beriberi*. Symptoms of this disorder are neurological disturbances, impaired cardiac function, and muscle atrophy.

Vitamin B₂ is a collective term for several different vitamins — riboflavin, folate, nicotinate and pantothenate.

In metabolism, **riboflavin** is a precursor for the *prosthetic groups* flavin mononucleotide (**FMN**) and flavin adenine dinucleotide (**FAD**). FMN and FAD are tightly bound to various oxidoreductases and flavoproteins, where they participate in the transfer of hydrogen atoms. There is no disease known that can be specifically attributed to riboflavin deficiency.

Folate, the anion of folic acid, is made up of three different moieties, a *pteridine derivative*, a *4-aminobenzoate* ring structure, and one or more *glutamate* residues. Following reduction to tetrahydrofolate (THF), it serves as a coenzyme in C₁ metabolism (see p. 375). Folate deficiency is relatively common. Mild forms primarily affect the blood. Erythropoesis is impaired due to disturbed nucleic acid biosynthesis, and abnormal erythrocyte precursors appear (*megaloblastic anemia*). More severe folate deficiency leads to generalized tissue damage as a result of impairments in phospholipid synthesis and amino acid metabolism.

Unlike animals, bacteria have the ability to synthesize folate *de novo*. Due to this fact, the growth of micro-organsims can be inhibited by *sulfonamides*, synthetic antibiotics that competitively inhibit the incorporation of 4-aminobenzoate into folate (see p. 232). Since animals are unable to synthesize folate, sulfonamides have no effect on their metabolism.

Nicotinate and **nicotinamide**, together referred to as "niacin," are required for the biosynthesis of the coenzymes nicotinamide adenine dinucleotide (**NAD⁺**) and nicotinamide adenine dinucleotide phosphate (**NADP⁺**). The central importance of these molecules in the transfer of hydride ions (reducing equivalents) is discussed in the context of metabolism (see p. 100). Animals can convert tryptophan into nicotinamide, but the yields are very low. Thus, niacin deficiency is only observed when all three nutrients — nicotinate, nicotinamide, and tryptophan — are absent from the diet. Niacin deficiency manifests itself as skin damage (*pellagra*), digestive disturbances, and depression.

Pantothenate is an amide made up of β-*alanine* and *2,4-dihydroxy-3,3-dimethylbutyrate* (pantoinate). The vitamin is a building block of **coenzyme A**, which in turn is central to the metabolism of carboxylic acids (see p. 102). The prosthetic group of acyl carrier protein (ACP) also contains pantothenate (see p. 166). Pantothenate deficiency is very rare, because of its widespread availability in the diet.

Further information

It is still unclear why man and many other animals require vitamins. Presumably, in animals mutations have resulted in the loss of some of the steps of coenzyme biosynthesis, whereas this has not occurred in micro-organisms and plants. The fact that the animal diet contains sufficient quantities of intermediates of coenzyme biosynthesis (i.e., the vitamins) has prevented detrimental effects of these mutations on higher organisms.

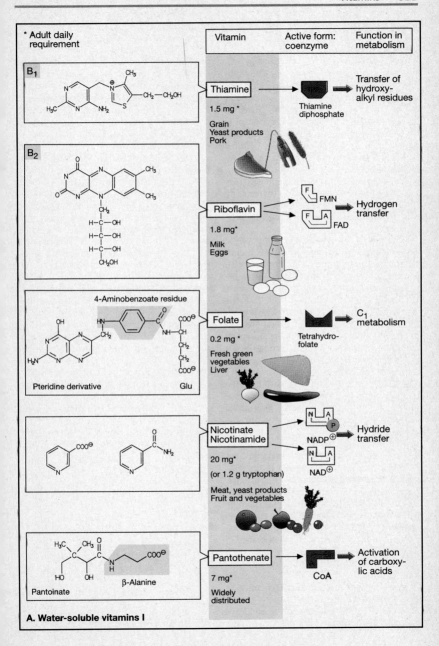

A. Water-soluble vitamins I

Water-soluble Vitamins

A. Water-soluble vitamins II ◐

Vitamin B$_6$ is a collective name for three substituted pyridines — **pyridoxal, pyridoxol** and **pyridoxamine**. The illustration shows the structure of pyridoxal, which carries an aldehyde group (-CHO) at C-4. Pyridoxol is the corresponding alcohol (-CH$_2$OH), and pyridoxamine the homologous amine (-CH$_2$NH$_2$). The active form of vitamin B$_6$, **pyridoxal phosphate**, is the most important coenzyme in amino acid metabolism (see p. 152). *Glycogen phosphorylase*, the enzyme catalyzing glycogen degradation, also contains pyridoxal phosphate. Vitamin B$_6$ deficiency is very rare.

Vitamin B$_{12}$ or **cobalamine** is one of the most complex metabolites in nature. It contains a complicated ring system (*corrin*), with cobalt as the central atom. Only micro-organisms can synthesize this vitamin *de novo*. It is particularly abundant in liver, and it is also found in meat, eggs and milk, but it is absent from plant products (vegetarians beware!). Vitamin B$_{12}$ can only be absorbed when the gastric mucosa is secreting the so-called *intrinsic factor*. This is a glycoprotein that binds cobalamine (the "extrinsic factor"), and thus protects it from degradation. In the blood, vitamin B$_{12}$ is bound to a special protein called *transcobalamine*. Substantial amounts of vitamin B$_{12}$ can be stored in the liver.

In metabolism, derivatives of cobalamine are involved in rearrangement reactions. For example, they act as coenzymes in the conversion of methylmalonyl-CoA to succinyl-CoA (see p. 148), and in the formation of methionine from homocysteine. Cobalamine derivatives are also involved in the reduction of ribonucleotides to deoxyribonucleotides.

Vitamin B$_{12}$ deficiency or a disturbance in vitamin B$_{12}$ absorption is usually due to a lack of intrinsic factor. This is the cause of *pernicious anemia*.

Vitamin C is **L-ascorbic acid** (chemically: 2-oxogulonolactone). As the hydroxyl groups of ascorbic acid are quite acidic, the acid loses a proton to become the anion, **ascorbate**.

Humans, apes, and guinea pigs have a requirement for vitamin C. This is because they all lack the enzyme L-*gulonolactone oxidase* (*1.1.3.8*), which catalyzes the last step in the conversion of glucose to ascorbate.

Vitamin C is particularly abundant in fresh fruit and vegetables. Many beverages and foodstuffs also contain vitamin C as an additive. Vitamin C is slowly destroyed by boiling. In the body, it serves as a reducing agent. The most common reactions in which it is involved are hydroxylation reactions. Among the processes affected are *collagen synthesis*, *tyrosine degradation, catecholamine synthesis*, and *bile acid synthesis*. The average daily requirement for ascorbic acid is about 60 mg, which is relatively high for a vitamin. Vitamin C deficiency, which causes a disease known as *scurvy*, is now relatively uncommon. Its symptoms, which include damage to connective tissue, bleeding, and the loss of teeth, become apparent after a few months.

Vitamin H or **biotin** is present in liver, egg yolk, and other foods, and it is also synthesized by the intestinal flora. In the body, biotin is covalently attached, via a lysine side chain, to enzymes catalyzing carboxylation reactions, e.g., *pyruvate carboxylase* or *acetyl-CoA carboxylase*. During the reaction, one of the two nitrogens of the biotin moiety binds a CO_2 molecule in an ATP-dependent reaction, and transfers it on to the acceptor molecule.

Biotin binds with high affinity and specificity to *avidin* (K$_d$ = 10^{-15} M), a protein found in egg white. Since boiling denatures avidin, eggs do not cause vitamin H deficiency unless eaten raw.

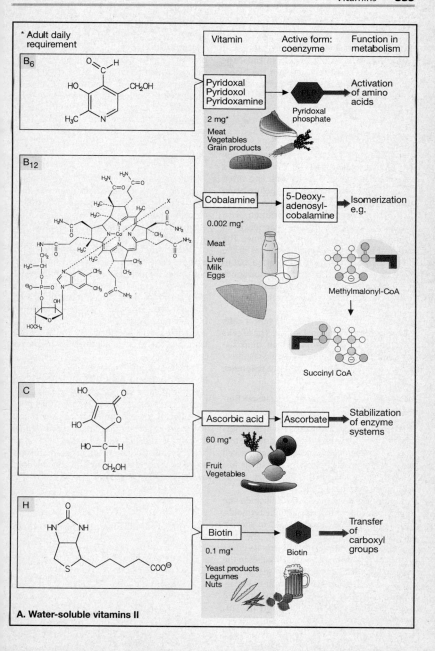

* Adult daily requirement

| | Vitamin | Active form: coenzyme | Function in metabolism |

B6

Pyridoxal
Pyridoxol
Pyridoxamine

2 mg*

Meat
Vegetables
Grain products

Pyridoxal phosphate → Activation of amino acids

B12

Cobalamine

0.002 mg*

Meat

Liver
Milk
Eggs

5-Deoxy-adenosyl-cobalamine → Isomerization e.g.

Methylmalonyl-CoA

Succinyl CoA

C

Ascorbic acid → Ascorbate → Stabilization of enzyme systems

60 mg*

Fruit
Vegetables

H

Biotin

0.1 mg*

Yeast products
Legumes
Nuts

Biotin → Transfer of carboxyl groups

A. Water-soluble vitamins II

Principles

Hormones are *chemical signaling substances* synthesized by specialized cells in *endocrine glands* before being released into the blood. The blood transports them to their target organs, where they exert specific physiological and biochemical effects.

A. Hormonal regulation system ●

Each hormone is the center of a complex system of hormonal regulation. The hormones are synthesized from precursors, and often stored, in specialized **gland cells** before being released into the blood as required. Most of them are transported in association with plasma proteins (**hormone carriers**), to which they bind in a reversible manner. The chemical interconversion of hormones in the liver usually leads to their inactivation, but in some cases it can cause activation. Hormones and their degradation products are ultimately disposed of via the excretory system, usually the kidneys.

In the target organs, there are **target cells** that receive the hormonal signal. These cells have **hormone receptors**, which bind the hormone. Binding of a hormone passes information on to the cell and triggers a **response**.

B. Principles of hormonal signal transduction in target cells ◑

There are two different ways in which a message can be transmitted from a hormone to a target cell. *Lipophilic hormones* enter the cell and exert their effect in most cases in the nucleus, whereas *hydrophilic hormones* act at the cell membrane.

Lipophilic hormones, which include the steroid hormones, thyroxin, and the retinoids, cross the cell membrane and bind to specific **receptors** inside their target cell. The hormone-receptor complex then exerts its effect on the **transcription** of particular genes in the nucleus (*transcriptional control,* see p. 108). The increased (or decreased) synthesis of particular mRNA molecules leads to altered amounts of the corresponding proteins. This then triggers a cellular response (see p. 224).

The **hydrophilic hormones** involve amino acid–derived hormones, peptide hormones, and proteohormones. Hydrophilic hormones bind to specific receptors on the outside of the cell membrane. This then triggers the synthesis of so-called **second messengers** on the inside of the cell membrane. It is these second messengers, which bring about the response of the target cell to the hormone (see p. 352).

Further information ◑

The distinction between *hormones* and signal substances, such as *mediators, neurotransmitters,* and *growth factors* is not particularly rigid. In many cases, these substances share common modes of synthesis, degradation, and action.

Besides the classical "long-distance" hormones, there are also **tissue hormones** (parahormones), which only act in the immediate vicinity of the gland cells that produce them. They arrive at their target cell by diffusion through the extracellular space, rather than by transport in the blood. Tissue hormones are particularly abundant in the digestive tract, where they regulate digestive processes (see p. 246).

Mediators are signal substances produced by many different types of cells, rather than being formed in specialized gland cells. Following their release, they can have hormone-like effects on their immediate surroundings. The most important mediators are *histamine* and the *prostaglandins* (see p. 354).

Neurohormones and **neurotransmitters** are signal substances produced and released by nerve cells (see p. 322).

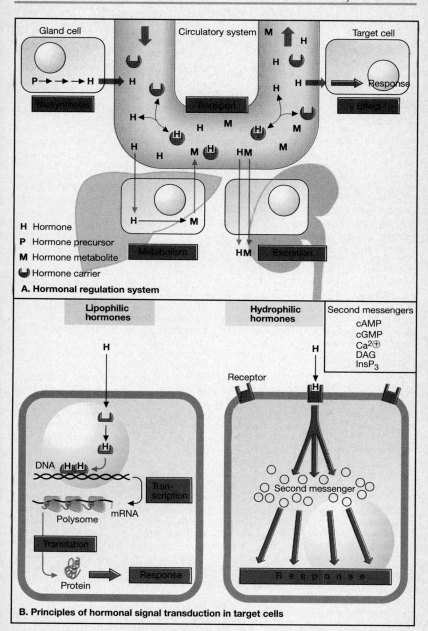

A. Hormonal regulation system

H   Hormone
P   Hormone precursor
M   Hormone metabolite
◖   Hormone carrier

Gland cell · Circulatory system · Target cell
P → → → H · Biosynthesis · Transport · Response · Effect
H → M · Metabolism
HM · Excretion

B. Principles of hormonal signal transduction in target cells

Lipophilic hormones · Hydrophilic hormones · Second messengers
cAMP
cGMP
$Ca^{2\oplus}$
DAG
$InsP_3$

DNA · Transcription · mRNA · Polysome · Translation · Protein · Response
Receptor · Second messenger · Response

Plasma Levels and Hormone Hierarchy

A. Endocrine, paracrine, and autocrine hormone effects ◑

Hormones transfer signals by moving from their site of synthesis to their site of action. They are usually transported in the blood. In this case, they are said to have an **endocrine effect** (**1**, example: insulin). In contrast, *tissue hormones*, whose target cells are immediately adjacent to the gland cells that produced them, are said to have a **paracrine effect** (**2**, example: hormones of the gastrointestinal tract). When signal molecules exert their effects on the *same* cells that synthesized them, they are said to have an **autocrine effect** (**3**, example: prostaglandins). Insulin, which is formed in the B cells of the pancreas, has both paracrine and endocrine effects. Many hormones show dual effects of this type. In the regulation of glucose and fat metabolism, insulin acts as an endocrine hormone (see p. 164), whereas it employs a paracrine mechanism to inhibit the formation and release of glucagon from the neighboring A cells of the pancreatic islets.

B. Plasma level dynamics ◑

The blood concentrations of hormones that act as signal substances are very low (between 10^{-7} and 10^{-12} M). However, these concentrations can show marked variations. Hormone levels can vary **periodically** in cycles or rhythms, which may be dependent on the time of day, month, or year, or on the sexual cycle. The illustration shows the *circadian rhythm of cortisol* as an example. Other hormones are released into the blood in an irregular manner. This leads to **episodic** or **pulse-like** changes in the concentration of the hormone in the blood. The plasma concentration of still other hormones is **event-regulated**. These hormones are released as part of the response of the organism to altered conditions either within or outside the body.

Hormone concentrations in the blood are precisely regulated. This is made possible by controlled synthesis and release. The rates of these processes are determined either by simple feedback loops or by a complex hierarchy of effects.

C. Regulatory circuit ◑

The synthesis and release of insulin by the B cells of the pancreas is stimulated by high blood glucose levels (> 5 mM). The release of insulin stimulates increased uptake and utilization of glucose by the cells of the muscle and adipose tissues. As a result, the blood glucose level falls again to its normal value (5 mM), and insulin release ceases (see p. 164).

D. Hormone hierarchy ◑

It is quite common for different hormone systems to be linked to one another. In some cases, they form a hierarchy. The most important example is the so-called *pituitary-hypothalamus axis*, which is controlled by the central nervous system (CNS). Nerve cells in the hypothalamus react to either positive or negative stimuli from the nervous system by releasing either activators or inhibitors, which are referred to as either **liberins** (releasing hormones) or **statins**, respectively. The neurohormones travel via short blood vessels to the adenohypophysis, where they either stimulate (liberins) or inhibit (statins) the biosynthesis and release of tropins. The **tropins,** or glandotropic hormones, then stimulate peripheral glands to synthesize glandular hormones. Finally, the **glandular hormones** exert their effects on the target cell. The glandular hormones usually inhibit synthesis or secretion of other hormones in the regulatory cascade. This, mostly negative, feedback affects the concentrations of the hormones higher up in the hierarchy.

Many important hormones are members of such **hormone axes,** e.g., thyroxin, cortisol, estradiol, progesterone, and testosterone.

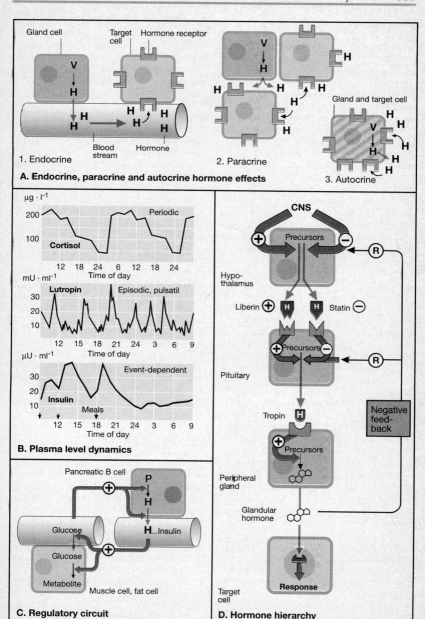

A. Endocrine, paracrine and autocrine hormone effects

Gland cell Target cell Hormone receptor

1. Endocrine

Blood stream Hormone

2. Paracrine

Gland and target cell

3. Autocrine

B. Plasma level dynamics

μg · l⁻¹

Periodic

Cortisol

12 18 24 6 12 18 24
Time of day

mU · ml⁻¹

Lutropin Episodic, pulsatil

12 15 18 21 24 3 6 9
Time of day

μU · ml⁻¹

Event-dependent

Insulin Meals

12 15 18 21 24 3 6 9
Time of day

C. Regulatory circuit

Pancreatic B cell

P
H

Glucose

H Insulin

Glucose

Metabolite

Muscle cell, fat cell

D. Hormone hierarchy

CNS

Precursors

Hypo-thalamus

Liberin Statin

Precursors

Pituitary

Tropin

Negative feed-back

Precursors

Peripheral gland

Glandular hormone

Target cell

Response

Lipophilic Hormones

There are many hormones and hormone-like substances — in man alone more than 100 have been found. From the point of view of biochemistry, a classification of the hormones into *lipophilic* and *hydrophilic* makes sense, because it also reflects the differences in their mode of action (see p. 336).

Lipophilic hormones are relatively small molecules (mass 300–800 Da). They are poorly soluble in aqueous media, and are not stored in gland cells. Instead, they are released directly after their synthesis (an exception is thyroxin). During their transport in the blood, they are bound to specific plasma proteins (hormone carriers). All lipophilic hormones share a common mode of action, i.e., they bind to intracellular receptors and thereby affect transcription (see p. 346).

A. Steroid hormones ●

The important steroid hormones in vertebrate animals are *progesterone*, *cortisol*, *aldosterone*, *testosterone*, and *estradiol* (formulae on p. 55). Nowadays, *calcitriol* (cholecalciferol, "vitamin D hormone," see p. 330) is also included in this group, even though it has a modified steroid structure. The most important steroid hormone of the invertebrates is *ecdysone* (see p. 55).

Progesterone is a female sexual hormone belonging to the *progestin* (*gestagen*) family. It is synthesized in the *corpus luteum* of the ovaries. Its concentration in the blood varies with the female cycle. Progesterone prepares the uterus for a possible pregnancy. Following fertilization, the placenta also starts to synthesize progesterone in order to maintain the pregnant state. In addition, progesterone stimulates the development of the mammary glands.

Estradiol is the most important representative of the *estrogens*. Like progesterone, it is synthesized by the ovaries and, during pregnancy, by the placenta as well. Estradiol controls the menstrual cycle. It promotes the proliferation of the mucosa of the uterus, and is also responsible for the development of the female sexual characteristics (breast, fat distribution, etc.).

Testosterone is the most important of the *androgens*, a group of male sexual steroids. It is synthesized in the Leydig cells of the testes, and controls the development and function of the male gonads, as well as the secondary sexual characteristics of men (muscles, hair, etc).

Cortisol, the most significant *glucocorticoid,* is synthesized by the adrenal cortex. It is involved in the regulation of protein and carbohydrate metabolism by promoting protein degradation and the conversion of amino acids to glucose. As a result, the blood glucose level rises (see p. 162). Synthetic glucocorticoids are used in medicine, because they inhibit inflammation and also act as immunosuppressants.

Aldosterone, a *mineralocorticoid*, is synthesised in the adrenal cortex. It has its effect on the kidneys, where it promotes Na^+ reabsorption mainly by inducing Na^+/K^+-ATPase. At the same time, it leads to elevated K^+ excretion. In this way, aldosterone indirectly increases blood pressure.

Calcitriol is a derivative of vitamin D. In the presence of ultraviolet light, an important precursor can be synthesized in the skin. The hormone itself is produced in the kidneys (see p. 304). Calcitriol promotes the absorption of calcium in the intestine and the incorporation of calcium into the bones (anti-rickets effect).

B. Iodothyronines ●

Thyroxine (tetraiodothyronine, T_4) and its active form *triiodothyronine* (T_3) are the only lipophilic signal substances derived from an amino acid. They are synthesized in the thyroid gland by post-translational iodination of protein-bound *tyrosine residues* and condensation of two such residues via an ether bond. In thyroxine, both rings are substituted with iodine atoms at positions 3 and 5 and 3' and 5', respectively.

The iodothyronines increase the basal metabolic rate and promote embryonic development.

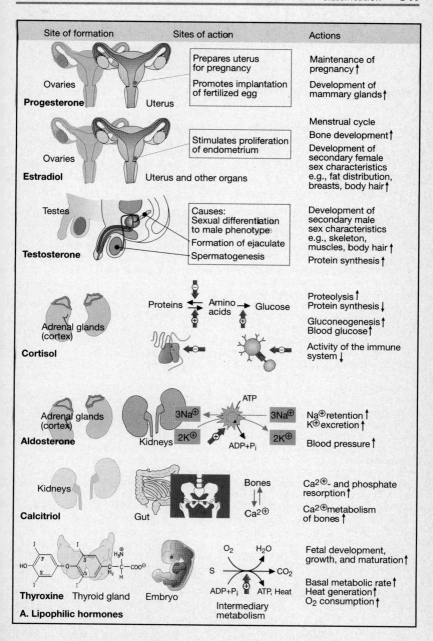

Site of formation	Sites of action	Actions
Progesterone — Ovaries, Uterus	Prepares uterus for pregnancy; Promotes implantation of fertilized egg	Maintenance of pregnancy ↑; Development of mammary glands ↑
Estradiol — Ovaries, Uterus and other organs	Stimulates proliferation of endometrium	Menstrual cycle; Bone development ↑; Development of secondary female sex characteristics e.g., fat distribution, breasts, body hair ↑
Testosterone — Testes	Causes: Sexual differentiation to male phenotype; Formation of ejaculate; Spermatogenesis	Development of secondary male sex characteristics e.g., skeleton, muscles, body hair ↑; Protein synthesis ↑
Cortisol — Adrenal glands (cortex)	Proteins ⇌ Amino acids → Glucose	Proteolysis ↑; Protein synthesis ↓; Gluconeogenesis ↑; Blood glucose ↑; Activity of the immune system ↓
Aldosterone — Adrenal glands (cortex)	Kidneys — ATP, 3Na⊕, 2K⊕, ADP+Pi	Na⊕ retention ↑; K⊕ excretion ↑; Blood pressure ↑
Calcitriol — Kidneys, Gut	Bones ⇌ Ca²⊕	Ca²⊕- and phosphate resorption ↑; Ca²⊕ metabolism of bones ↑
Thyroxine — Thyroid gland, Embryo	O_2 → H_2O; S → CO_2; $ADP+P_i$ → ATP, Heat; Intermediary metabolism	Fetal development, growth, and maturation ↑; Basal metabolic rate ↑; Heat generation ↑; O_2 consumption ↑

A. Lipophilic hormones

Hydrophilic Hormones

The hydrophilic hormones and hormone-like substances are derived from amino acids, or they are peptides and proteins composed of amino acids. They are usually stored in significant quantities in the gland cells and released in a controlled manner. Most of them do not require carriers in the blood (exceptions: ocytocin, vasopressin, somatomedins, calcitonin). Hydrophilic hormones exert their effects on their target cells by binding to receptors on the cell membrane (see p. 350).

A. Amino acid-derived signal substances ◑

The *biogenic amines* histamine, serotonin, melatonin, and the *catecholamines* dopa, dopamine, norepinephrine, and epinephrine are signal substances produced by decarboxylation of amino acids (see p. 322).

Histamine, an important *mediator* (local hormone) and *neurotransmitter*, is stored mainly in tissue mast cells and basophilic leukocytes of the blood. Histamine is involved in inflammatory and allergic reactions. Its release is stimulated by "histamine liberators," such as tissue hormones, allergens, and medicinal drugs. Histamine acts via various different receptors. Binding to H_1 receptors promotes contraction of the smooth muscle tissues of bronchia, as well as the dilation and permeability of blood capillaries. Binding to H_2 receptors slows down the heart rate, and promotes hydrochloric acid formation in the stomach. In the brain, histamine functions as a neurotransmitter via H_3 receptors.

Epinephrine is a hormone synthesized from tyrosine in the medulla of the adrenal glands (see p. 322). Its release is subject to neuronal control. This "fight or flight" hormone has its principal effects on the blood vessels, the heart, and intermediary metabolism. It constricts the blood vessels and thereby increases the blood pressure (via α_1 and α_2 receptors), it promotes the degradation of glycogen to glucose in the liver and muscles (via β_2 receptors, see p. 110) and it dilates the bronchia (also via β_2 receptors).

B. Peptides and proteohormones: examples ◑

The members of this, the numerically most important group of signal substances, are synthesized by translation and protein processing (see p. 348). The smallest peptide hormone, thyroliberin, is a tripeptide (mass 362 Da). The largest achieve molecular masses of more than 20 kDa, e.g., thyrotropin (about 28.3 kDa). Similarities in their *primary structures* show that several peptide hormones are related to one another. They can be classified into families that share *sequence homologies*. Homologous hormones have arisen from common ancestors during evolution.

Thyroliberin (thyrotropin-releasing hormone, TRH) is one of the neurohormones of the hypothalamus (see p. 32). It stimulates the secretion of thyrotropin (TSH) by cells of the pituitary gland. TRH is made up of three amino acids, which are modified in characteristic ways, i.e., the *N*-terminal glutamate residue is converted to pyroglutamate, a cyclic compound, while the *C*-terminal proline is modified to the amide. These alterations, which are also found in other small peptide hormones, provide protection against attack by exopeptidases.

Thyrotropin (thyroid-stimulating hormone, TSH) and the related proteohormones *lutropin* (luteinizing hormone, LH) and *follitropin* (follicle-stimulating hormone, FSH) are shown as examples of hormones of the adenohypophysis. They are composed of two polypeptide subunits and carry carbohydrate chains ("glycoprotein hormones") that are important for the rapid removal of the hormones from the circulation. Thyrotropin stimulates the synthesis and secretion of thyroxin by the thyroid gland.

Insulin (structure, see p. 70) is produced and released by the B cells of the pancreas. Its metabolic functions are discussed on p. 164.

Glucagon, a peptide of 29 amino acids, is a product of the A cells of the pancreas. It is secreted when the blood concentration of glucose declines. Its main effect is to increase the blood glucose level, mainly by stimulating the release of glucose from the liver (see p. 162). In this respect, it is an antagonist of insulin.

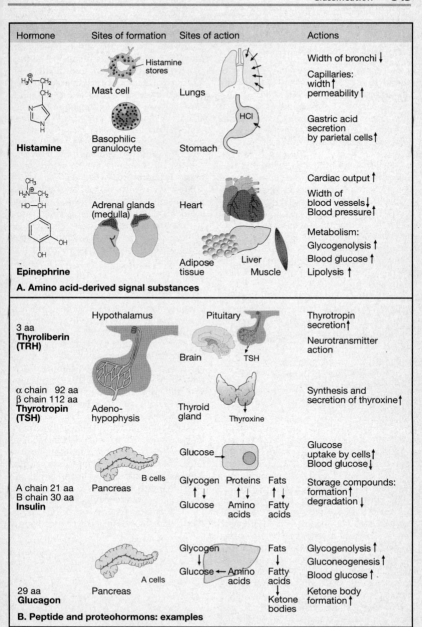

Hormone	Sites of formation	Sites of action	Actions
Histamine	Mast cell / Histamine stores / Basophilic granulocyte	Lungs / Stomach	Width of bronchi ↓ / Capillaries: width↑ permeability↑ / Gastric acid secretion by parietal cells↑
Epinephrine	Adrenal glands (medulla)	Heart / Adipose tissue / Liver / Muscle	Cardiac output↑ / Width of blood vessels↓ Blood pressure↑ / Metabolism: Glycogenolysis↑ Blood glucose↑ Lipolysis↑

A. Amino acid-derived signal substances

3 aa **Thyroliberin (TRH)**	Hypothalamus	Pituitary / Brain / TSH	Thyrotropin secretion↑ / Neurotransmitter action
α chain 92 aa β chain 112 aa **Thyrotropin (TSH)**	Adeno-hypophysis	Thyroid gland / Thyroxine	Synthesis and secretion of thyroxine↑
A chain 21 aa B chain 30 aa **Insulin**	Pancreas / B cells	Glucose / Glycogen ↑↓ Glucose / Proteins ↑↓ Amino acids / Fats ↑↓ Fatty acids	Glucose uptake by cells↑ Blood glucose↓ / Storage compounds: formation↑ degradation↓
29 aa **Glucagon**	Pancreas / A cells	Glycogen ↓ Glucose ← Amino acids / Fats ↓ Fatty acids ↓ Ketone bodies	Glycogenolysis↑ Gluconeogenesis↑ Blood glucose↑ Ketone body formation↑

B. Peptide and proteohormons: examples

Metabolism

A. Biosynthesis of steroid hormones ❶

In vertebrate animals, the biosynthesis of steroid hormones (cf. p. 369) begins with **cholesterol**. The carbon skeleton of cholesterol consists of 27 carbon atoms forming 4 rings, one of which carries a long side chain. There are standard rules for the naming of the rings and the numbering of the carbons (see p. 52).

The cholesterol required for steroid hormone biosynthesis is obtained from different sources. It can be taken up by the hormone-synthesizing gland cells in the form of lipoproteins of the LDL type (see p. 254), or it can be synthesized by these same cells from acetyl-CoA (see p. 170). Excess cholesterol is stored as fatty acid esters in lipid droplets. Rapid mobilization of cholesterol from this reserve is possible by hydrolysis of the fatty acid esters.

Partial reactions. Highly specific enzymes are involved in the biosynthesis of the steroid hormones (for details see p. 369). The reactions they catalyze can be classified as follows:

- *hydroxylations* **a, f, g, h, i, k, l, p**
- *dehydrogenations*: **b, d, m**
- *isomerization*: **c**
- *hydrogenation*: **o**
- *cleavages*: **a, e, n**
- *aromatization*: **q**

In order to illustrate the sites of action of each type of enzyme, the illustration shows the three steroids cholesterol (**1**), progesterone (**2**), and androstendione (an intermediate in testosterone biosynthesis, (**3**).

Biosynthetic pathways. Multiple sequential reactions are involved in the synthesis of the individual hormones from cholesterol. This is shown in a simplified scheme using **progesterone biosyntheis** as an example (**A**). The biosynthesis of progesterone begins with the cleavage of the side chain of cholesterol between C-20 and C-22 (**a**). The resulting steroid with the shortened side chain is called **pregnenolone**. Subsequent dehydrogenation of the hydroxyl group at C-3 (**b**) and shifting

of the double bond from C-5 to C-4 (**c**) results in the production of progesterone.

In Fig. **A**, the steroid hormones are grouped acording to the number of carbons they contain. **Cholesterol** and **calcitriol** are *27-C steroids* (C_{27}). In contrast, **progesterone, cortisol** and **aldosterone** are *21-C steroids* (C_{21}), because of the shortening of their side chains by 6 carbons. During its synthesis, **testosterone** loses its side chain completely, and is therefore classified as a *19-C steroid* (C_{19}). The aromatization of estradiol involves the loss of an angular methyl group, which makes **estradiol** an *18-C steroid* (C_{18}).

During its synthesis, **calcitriol** undergoes photohemical cleavage of the B ring. As a result, it no longer has the typical structure of a steroid, and is therefore referred to as a *secosteroid*. However, its biochemical properties are still those of a typical steroid hormone.

B. Inactivation of steroid hormones ❶

The enzymatic inactivation of the steroid hormones mainly occurs in the liver, where the steroids are either reduced or hydroxylated, and then converted to conjugates (see p. 290). The **reduction reactions** affect oxo groups and the double bond in ring A. **Conjugate formation** in the liver involves esterification with *sulfuric acid* or glycosylation with *glucuronic acid*, leading to products that are highly water-soluble (see p. 290).

The various inactivation reactions mentioned result in a range of steroid metabolites with greatly reduced hormonal activity. It is interesting to note that mammals are unable to degrade the steroid backbone. Steroids are therefore ultimately excreted with the *urine* and to some extent with the *bile*. Methods for the detection of steroids in the urine can be used to investigate the hormone metabolism.

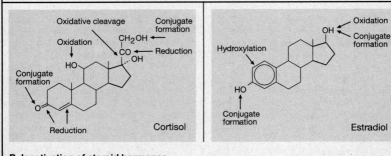

A. Biosynthesis of steroid hormones

1 Cholesterol

2 Progesterone

3 Androstenedione

C_{27} Cholesterol — d e f g → Calcitriol

C_{21} Pregnenolone — h i k b c → Cortisol

Progesterone — i k l m → Aldosterone

C_{19} Testosterone

C_{18} Estradiol

B. Inactivation of steroid hormones

Oxidative cleavage
Conjugate formation
Oxidation
Reduction
Conjugate formation
Reduction
Cortisol

Oxidation
Conjugate formation
Hydroxylation
Conjugate formation
Estradiol

Mechanism of Action

A. Mechanism of action of lipophilic hormones ◑

The most important apolar signal substances are the steroid hormones, thyroxin, and retinoic acid (see p. 340). Their major site of action is in the *nucleus* of the target cell.

In the blood, lipophilic hormones are usually bound to *carrier proteins*. However, only free hormone molecules can cross the cell membrane. This may occur either by direct permeation, or by facilitated diffusion. It is still not known exactly how the steroids hormones reach the nucleus, the site where most of them first encounter their receptor.

The target cells of steroid hormones contain a small number of **hormone receptors** (10^3–10^4 molecules per cell), which show a high affinity (K_d = 10^{-8}-10^{-10} M) as well as a high degree of *specificity* for their hormone ligands. The binding of the hormone is thought to lead to a conformational change of the receptor protein, which then leads to the following responses: A *heat shock protein* (**hsp-90**) dissociates from the receptor, allowing it to dimerize, which in turn increases its affinity for its cognate DNA sequence.

The key event that triggers the response of the cell to the hormone is the *binding of the hormone receptor dimer* to the DNA double strand. This complex binds to short stretches of nucleotides, referred to as **hormone response elements** (HRE). These are palindromic DNA sequences (see p. 236) that act as enhancer elements in the regulation of transcription (see p. 224). The illustration shows the HRE for glucocorticoids ("n" stands for any nucleotide). Sequence differences between different HREs provide for specificity in the hormone receptor-HRE interaction, i.e., only one HRE is recognized by any one hormone receptor. Nevertheless, the same HRE can control different genes, depending on the presence of other transcription factors. This explains why the same hormone can stimulate different responses in different tissues.

The binding of a hormone receptor dimer to an enhancer sequence results in increased transcription of the gene to which it belongs.

The activation of transcription may occur as a result of an alteration in the nucleosome structure, or via a direct interaction of the hormone receptor dimer with the *transcription complex*, i.e., RNA polymerase and various protein factors (see p. 224). The ultimate effect of the hormone on the cell is to alter the amount of specific mRNA species encoding key proteins that affect the functions of the cell.

B. Receptors of lipophilic hormones ○

There is a high degree of similarity between the receptors for the various lipophilic signalling substances. They all belong to a single *protein superfamily*. The receptors are made up of different **domains** with different sizes and functions. Each receptor has a *regulatory domain*, a *DNA-binding domain*, a short *nuclear-targeting domain*, and a *hormone-binding domain*. The greatest degree of homology between the different receptors is seen in the DNA-binding domain. In this region, the hormone receptors have repeated clusters of cysteine residues. The cysteines coordinate zinc ions, and therefore these sequence motifs are known as *"zinc fingers."*.

Zinc finger proteins form a group of transcription factors that not only includes the receptors for steroid hormones, thyroxin, and retinoic acid, but also a receptor that binds the environmental toxin *dioxin*, the protein product of the oncogene *erb*-A (see p. 358), and a whole list of other factors, the ligands of which are not yet known.

It is possible to synthesize compounds, which are not identical to the hormone of interest, but bind to its receptor. If the binding of this ligand triggers the same effect as the natural hormone, then it is said to be a *hormone agonist*. An example of this are oral contraceptives, which often contain estrogen and progesterone agonists. In contrast, *hormone antagonists* are compounds that bind to the receptor, but do not trigger a hormonal effect, i.e., they block the effect of the endogenous hormone.

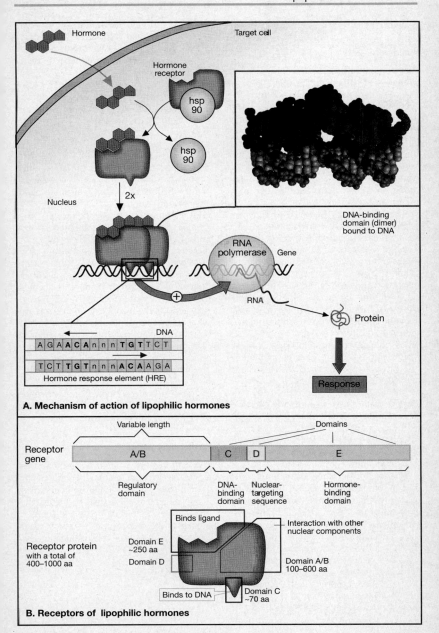

A. Mechanism of action of lipophilic hormones

DNA-binding domain (dimer) bound to DNA

A	G	A	**A**	**C**	A	n	n	n	**T**	**G**	**T**	**T**	**C**	T
T	C	T	**T**	**G**	**T**	n	n	n	**A**	**C**	A	A	G	A

DNA

Hormone response element (HRE)

B. Receptors of lipophilic hormones

Receptor gene

Variable length — A/B — Domains — C — D — E

Regulatory domain · DNA-binding domain · Nuclear-targeting sequence · Hormone-binding domain

Receptor protein with a total of 400–1000 aa

Binds ligand

Interaction with other nuclear components

Domain E ~250 aa
Domain D
Domain A/B 100–600 aa

Binds to DNA · Domain C ~70 aa

Metabolism

A. Biosynthesis ◗

Peptide hormones and proteohormones are primary gene products. The information for their synthesis is encoded by nuclear **DNA**. This information is transcribed into hnRNA, which then undergoes splicing to remove introns (**1**, see p. 224). The **mRNA** thus formed specifies a peptide that is usually much larger than the mature hormone. It codes for the sequence of a **signal peptide**, followed by a **propeptide,** followed by the mature **hormone precursor**. Translation of the mRNA occurs on ribosomes, exactly as is the case with other proteins (see p. 228). The signal peptide is synthesized first. Its function is to ensure that the ribosomes attach themselves to the rough endoplasmic retiulum (rER) and to direct the growing peptide into the lumen of the rER (**3**, cf. p. 210). Translation of the mRNA results in the synthesis of a large precursor of the hormone, the **prohormone**. Maturation of the prohormone is brought about by limited proteolysis and *protein modification*, e.g., glycosylation, formation of disulfide bonds, and phosphorylation (**4**). The mature hormone is stored in vesicles until it is released by *exocytosis* in response to the appropriate stimuli.

The biosynthesis and secretion of peptides and proteohormones is under the control of hierarchical regulatory systems (see p. 338). In these systems, calcium acts as a *second messenger*. An increase in the concentration of calcium stimulates synthesis and secretion of the hormones.

Analyses of hormone genes have shown that, in some cases, multiple, completely unrelated peptides and proteins are encoded by the same gene. A good example of this is the **pro-opiomelano-cortin (POMC)** gene. Besides encoding the hormone *corticotropin* (ACTH = adrenocorticotropic hormone) this gene also codes for several smaller peptide hormones, namely α, β, and γ *melanotropin* (MSH), β and γ *lipotropin* (LPH), β *endorphin* and *Met-enkephalin* (see p. 322). The last of these can also arise from β endorphin.

The prohormone derived from the POMC gene is a so-called *polyprotein*. The decision as to which peptide should be produced and secreted is not made before the synthesis of the large pre-propeptide is complete. The most important secretory product of the POMC gene in the pituitary is the hormone *corticotropin* (ACTH). The functions of the other peptides are not yet fully understood.

B. Inactivation and degradation ◗

The degradation of peptide hormones often starts in the blood or at the walls of the blood vessels. Peptides containing disulfide bonds (e.g., insulin) can be inactivated by reductive cleavage of such cystine residues (**1**). Other peptides and proteins are digested by *peptidases* (see p. 150). Peptidases that start digestion at one end of the peptide are called *exopeptidases* (**2**), and those that attack within the peptide chain *endopeptidases* (**3**). Proteolysis gives rise to a range of degradation products, some of which are still biologically active. Other peptide hormones and proteohormones are removed from the circulation when the hormone binds to a membrane receptor (see p. 350) and becomes endocytosed as a hormone-receptor complex (**4**). The subsequent degradation then takes place in lysosomes. The end products of the degradation of peptide hormones and proteohormones are amino acids, which once again become available to metabolism.

Further information

The *half-lives* of lipophilic hormones are much longer (hours to days) than those of hydrophilic hormones (minutes to hours). They mainly depend on the activity of the degradation systems. Medicinal drugs and injury to tissues can affect the level of a particular hormone by *induction* of the degradation systems.

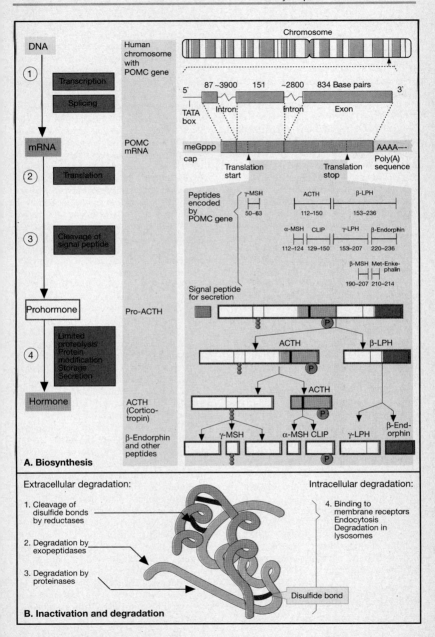

DNA

① Transcription
Splicing

mRNA

② Translation

③ Cleavage of signal peptide

Prohormone

④ Limited proteolysis Protein modification Storage Secretion

Hormone

Human chromosome with POMC gene

Chromosome

5' 87 ~3900 151 ~2800 834 Base pairs 3'

TATA box Intron Intron Exon

POMC mRNA

meGppp cap

Translation start

Translation stop

Poly(A) sequence AAAA—-

Peptides encoded by POMC gene

γ-MSH 50–63

ACTH 112–150

β-LPH 153–236

α-MSH 112–124

CLIP 129–150

γ-LPH 153–207

β-Endorphin 220–236

β-MSH 190–207

Met-Enke-phalin 210–214

Signal peptide for secretion

Pro-ACTH

ACTH

β-LPH

ACTH (Cortico-tropin)

β-Endorphin and other peptides

γ-MSH

α-MSH CLIP

γ-LPH

β-End-orphin

A. Biosynthesis

Extracellular degradation:

1. Cleavage of disulfide bonds by reductases

2. Degradation by exopeptidases

3. Degradation by proteinases

Intracellular degradation:

4. Binding to membrane receptors Endocytosis Degradation in lysosomes

Disulfide bond

B. Inactivation and degradation

Mechanisms of Action

A. Mechanism of action of hydrophilic hormones ◗

Most hydrophilic signal substances (see p. 342) are unable to cross the cell membrane. Instead, signal transmission to the inside of the cell occurs via *membrane-localized receptors* (**"signal transduction"**). The receptors are integral membrane proteins that bind the signal substance on the outside of the membrane and then undergo an alteration in their structure, which triggers the release of a second signal on the inside of the membrane. These receptors can be classified into three different types.

(1) **Type I receptors** have enzyme activity. In many cases, they contain intracellular domains with *tyrosine kinase activity*. These domains are activated by binding of the hormone to the extracellular part of the receptor, and then phosphorylate tyrosine residues in other proteins. In addition, the receptor usually phosphorylates itself. Further proteins bind to the phosphorylated tyrosine residues, become activated too, and transmit the signal to other parts of the cell. Examples of type I receptors with tyrosine kinase activity are the receptors for *insulin* and various *growth factors*.

(2) **Type II receptors** are *ion channels*. Binding of signal substance causes rapid opening of the channel, allowing specific ions, e.g., Na+, K+ or Cl-, to pass through. The cell responds to the resultant change in intracellular ion concentration in a specific way. This is the mechanism used by *neurotransmitters*, such as *acetylcholine* (via nicotinic receptors = Na+ and K+ channels) and *GABA* (via A receptors = Cl- channels).

(3) **Type III receptors** are 7-helix transmembrane proteins that transfer their signal via a family of guanine nucleotide-binding proteins, the so-called *G proteins* (see **B**). Many hydrophilic hormones employ this signal transduction pathway.

B. Signal transduction by G proteins ◗

G proteins are heterotrimers consisting of three different types of subunits, α, β and γ. The α subunit can bind the nucleotides GTP or GDP. In the resting state, GDP is bound to the G protein. When a signal substance interacts with the membrane receptor, the latter undergoes a conformational change that allows a G protein to associate with the receptor at the inner membrane surface. This interaction causes an exchange of bound GDP for GTP. The receptor then releases the activated G protein, which subsequently dissociates into the α subunit and the $\beta\gamma$ dimer. After some time, the released α subunit hydrolyzes the bound GTP to GDP, and returns to its original conformation. Before this happens, however, the activated GTP complex triggers the formation of a *second messenger*. There are four alternative ways in which this may happen, depending on the type of G protein.

(1) The α subunit activates a membrane-localized *adenylate cyclase*, which converts ATP into the *second messenger* 3',5'-cyclic AMP (**cAMP**). As a result, the hormone increases the intracellular level of cAMP (see p. 109). Some G proteins do not activate, but inhibit adenylate cyclase.

(2) The α subunit stimulates a *cGMP-specific phosphodiesterase*. This enzyme increases the rate of hydrolysis of cGMP, leading to a decrease in the concentration of this cyclic nucleotide (see p. 324).

(3) The α subunit binds to an *ion channel*, e.g., a K+ or Ca2+ channel, resulting in the opening of the channel.

(4) The α subunit activates a *phospholipase*, which subsequently hydrolyses membrane lipids. The most important of these enzymes is *phospholipase C*. Its substrate, **PInsP₂** (phosphatidylinositol bisphosphate, see p. 168), is hydrolyzed to **InsP₃** (inositol 1,4,5-trisphosphate) and **DAG** (diacylglycerol). Both of these products can act as *second messengers*. The hydrophilic InsP₃ moves to the endoplasmatic reticulum (ER), where it stimulates the release of calcium from storage. The lipophilic DAG, on the other hand, remains in the membrane and activates *protein kinase C*, which, in the presence of Ca2+ ions, phosphorylates serine and threonine residues of various proteins, thereby altering their activities.

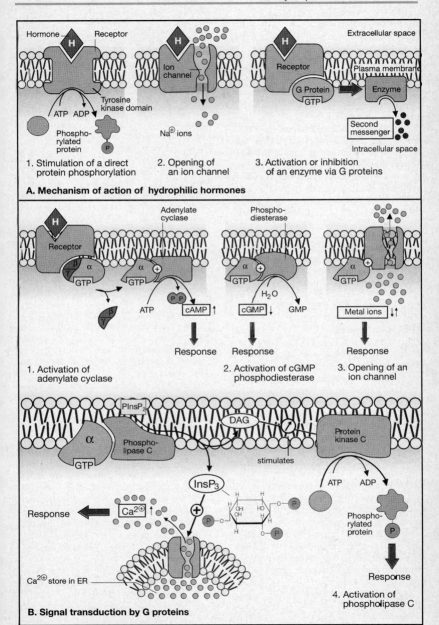

A. Mechanism of action of hydrophilic hormones

1. Stimulation of a direct protein phosphorylation

2. Opening of an ion channel

3. Activation or inhibition of an enzyme via G proteins

1. Activation of adenylate cyclase

2. Activation of cGMP phosphodiesterase

3. Opening of an ion channel

4. Activation of phospholipase C

B. Signal transduction by G proteins

Second Messengers

A. Metabolism and function of cAMP ◑

The cyclic nucleotide **cAMP** ($3',5'$-cyclic adenosine monophosphate) is an important *second messenger*.

Metabolism. cAMP is synthesized by *adenylate cyclase* [1], which catalyzes the cyclization of ATP to yield cAMP and diphosphate. Subsequent hydrolysis of diphosphate shifts the equilibrium of the adenylate cyclase reaction to the right, thus making it essentially irreversible. The degradation of cAMP to AMP is catalyzed by a *phosphodiesterase* [2], which is inhibited by high concentrations of methylated xanthines, such as *caffeine*.

The activity of adenylate cyclase is regulated by *G proteins* (see p. 350). Most G proteins stimulate adenylate cyclase, and thereby raise the cAMP level (type G_s). However, some are inhibitory (type G_i). The illustration provides a list of some of the hormones, which act through cAMP.

Mechanism of action. cAMP is an allosteric effector of *protein kinase A* [3]. The inactive form of this enzyme is a tetramer. Two catalytic subunits (C) are blocked by two regulatory subunits (R). When cAMP binds to the regulatory subunits, the tetramer dissociates, and the catalytic subunits become active. The activated enzyme phosphorylates serine residues in various other proteins. Phosphorylation of the target proteins results in either activation (e.g., in the case of phosphorylase kinase, see p. 110) or inhibition (e.g., with glycogen synthase).

There are several levels of control involved in the *switching off* of second messenger activity. The hormone dissociates from its receptor, the G protein returns to its resting state as a result of hydrolysis of GTP to GDP, and phosphodiesterase degrades cAMP to AMP. The resulting decline in the cAMP level causes a rapid return of protein kinase A to its inactive, tetrameric state.

B. Role of Ca²⁺ ◑

In the cytoplasm, the concentration of Ca^{2+} ions is usually very low (about 0.1 μM). This low level is maintained by ATP-dependent calcium "pumps." Signaling molecules can trigger a rapid rise in the calcium level in the cytoplasm as a result of the opening of calcium channels in the plasma membrane, or in the membranes of calcium storage organelles within the cell. For these hormones, Ca^{2+} is the *second messenger*.

Many of the effects of calcium are mediated by **calmodulin**. Calmodulin is a small protein (mass 17 kDa) found in most animal cells. Its three-dimensional structure (right) is characterized by two compact domains connected by an α helix. Calmodulin has 148 amino acids, which are shown as spheres in the illustration. The four sites highlighted in blue represent the Ca^{2+} binding sites. Once Ca^{2+} has bound, calmodulin becomes active and enters into regulatory interactions with other proteins, especially enzymes. This is the mechanism whereby Ca^{2+} and calmodulin regulate the activity of protein kinases, ion pumps, glycogen degradation, muscle contraction, etc.

The following molecules are currently recognized as **second messengers**: *cAMP, cGMP, DAG* (diacylglycerol), *InsP₃* (inositol triphosphate), *Ca²⁺*, and *arachidonate*. They have several properties in common:

- Second messengers arise via *cascade reactions*.
- Their intracellular concentrations are strictly regulated by extracellular signals, i.e., by *hormones, neurotransmitters, mediators, growth factors, odors,* or *light*.
- The formation of second messengers allows *signal amplification*, e.g., the binding of one hormone molecule to a single receptor can activate more than 10 G protein molecules, which, in turn, can give rise to a 10 to 100-fold higher amount of second messenger.
- The transduction of multiple signals via the same second messenger allows the *integration* of signal effects.

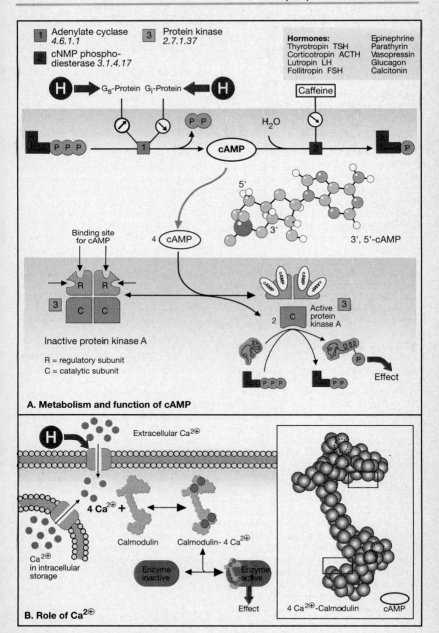

1 Adenylate cyclase
4.6.1.1

2 cNMP phospho-
diesterase 3.1.4.17

3 Protein kinase
2.7.1.37

Hormones:
Thyrotropin TSH   Epinephrine
Corticotropin ACTH   Parathyrin
Lutropin LH   Vasopressin
Follitropin FSH   Glucagon
  Calcitonin

G_s-Protein G_i-Protein

Caffeine

H_2O

cAMP

5'
3'

3', 5'-cAMP

Binding site
for cAMP

4 cAMP

R R

C C

cAMP cAMP

Active
protein
kinase A

Inactive protein kinase A

R = regulatory subunit
C = catalytic subunit

Effect

A. Metabolism and function of cAMP

Extracellular $Ca^{2\oplus}$

$4\ Ca^{2\oplus}$ +

Calmodulin   Calmodulin- $4\ Ca^{2\oplus}$

$Ca^{2\oplus}$
in intracellular
storage

Enzyme
inactive

Enzyme
active

Effect

$4\ Ca^{2\oplus}$-Calmodulin   cAMP

B. Role of $Ca^{2\oplus}$

Eicosanoids

Mediators are ubiquitous substances, which, unlike hormones, are not exclusively synthesized in gland cells, but by a whole range of cell types. Some of the most important mediators are **histamine** and the **eicosanoids**. Here, we use the eicosanoids to illustrate the basic features of mediators.

A. Eicosanoids ○

The eicosanoids constitute a large group of mediators, with a broad range of effects. They are all derived from the polyunsaturated fatty acid **arachidonic acid** (20:4) (see p. 45). In animals, arachidonic acid is synthesized from essential polyunsaturated fatty acids and incorporated into phospholipids of the plasma membrane.

The synthesis of eicosanoids begins in the plasma membrane, where *phospholipase A$_2$* [1] catalyzes the hydrolytic release of arachidonate from phospholipids. The activity of this enzyme is very tightly regulated. It is activated by hormones and other stimuli, which act via *G proteins*. The arachidonate released during eicosanoid synthesis is itself a signal substance. However, the products of its metabolism, the prostaglandins, prostacyclines, thromboxanes, and leukotrienes, are of far greater importance.

There are two different pathways for the synthesis of eicosanoids. One is initiated by *lipoxygenase* [2], and the other by *prostaglandin synthase* [3]. The reactions catalyzed by lipoxygenases produce **hydroperoxy** and **hydroxy fatty acids**, which then undergo dehydration and various transfer reactions to yield **leukotrienes**.

Prostaglandin synthase [3], a heme protein, catalyzes both a cyclooxygenase and an endoperoxidase reaction. Subsequent steps lead to the synthesis of **prostaglandins, prostacyclins**, and **thromboxans**. The ring closure, which is made possible by the four *cis* double bonds, occurs by oxygen attacking the arachidonate molecule, and involves an endoperoxide intermediate. The illustration shows the structural formulae of just a few selected eicosanoids.

Eicosanoids act as "local hormones" by binding to membrane-bound receptors in the immediate vicinity of their site of synthesis. In some cases, the target cell and the cell synthesizing the eicosanoid are one and the same (*autocrine effect*). In other cases, they act on neighboring cells (*paracrine effect*). The eicosanoids are rapidly degraded (within seconds to minutes) by the reduction of double bonds and the oxidation of hydroxyl groups. The rapidity of this process explains their limited range of action.

The **biological effects** of the eicosanoids are quite varied. They can serve as *second messengers* for hydrophilic hormones (e.g., thyrotropin, corticotropin), they can control the contraction of smooth muscle tissue (e.g., of blood vessels, bronchi, uterus), and they can influence bone metabolism, the autonomic nervous system, the immune system, cell movement and cell aggregation (e.g., of leukocytes and thrombocytes). They are also very effective pain signals for nociceptors. In some cases, their effects are mediated by cAMP and cGMP.

Further information

Under physiological conditions, **corticosteroids** are very important regulators of eicosanoid metabolism. They indirectly inhibit phospholipase A$_2$. This, in turn, results in inhibition of the synthesis of eicosanoids — a mechanism that contributes to the anti-inflammatory effects of corticosteroids.

Acetylsalicyclate (aspirin) and related antipyretics are specific *inhibitors* of prostaglandin synthase. They inactivate the enzyme by covalent acetylation of a serine residue in the active center. This explains the pain-relieving and anti-rheumatic effects of these compounds. In the stomach, they inhibit the synthesis of prostaglandins, which protect the gastric mucosa from self-digestion by stimulating the release of mucus. As a result of this, prolonged use of acetylsalicylic acid can cause stomach disorders.

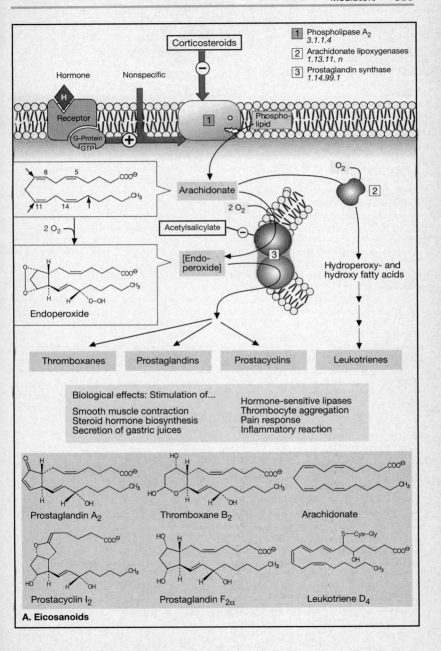

A. Eicosanoids

Cell Cycle

A. Cell cycle ●

Proliferating cells are characterized by their ability to divide. In mammals, one cycle of cell proliferation (the cell cycle) takes about 10 to 24 hours (24 h in the example shown here). The cell cycle is typically divided into four different phases, which occur sequentially. The actual division phase (the **M phase**) is readily observable under the light microscope, and has been particularly well studied (for more details, see biology textbooks). DNA replication occurs during the **S phase** (DNA synthesis, see p. 220). The M and the S phases are separated by two phases known as G_1 and G_2 (G for "gap"). During G_1, the duration of which can vary, the cell grows by synthesizing RNA, proteins, and other cellular constituents. Cells that have ceased to proliferate, because they have become fully differentiated leave the G_1 phase and enter the so-called G_0 **phase**. Here they remain, unless stimulated by mitogenic signals (e.g., growth factors, tumor viruses), in which case they return to G_1. Once they have passed a **restriction point** ("point of no return") they too can enter a new S phase.

The G_1, G_0, S, and G_2 phases are collectively referred to as **interphase**. In the cell cycle, interphase alternates with the much shorter M phase.

B. Control of the cell cycle ○

The cell cycle is regulated by the *interconversion* of regulatory proteins (see p. 106) between a phosphorylated and a dephosphorylated form. The key enzyme regulating the transition from the G_2 phase into the M phase is a specific *serine/threonine protein kinase* called **MPF** (maturation promoting factor). Various proteins involved in mitosis become phosphorylated by this enzyme, e.g., *histone H1* (a constituent of chromatin), *lamin* (a constituent of the cytoskeleton found in the nuclear membrane), *transcription factors*, mitotic *spindle proteins*, and various *enzymes*. The phosphorylation of these proteins triggers the onset of **mitosis**. At the end of mito-

sis, cyclin, the regulatory subunit of MPF (see below), is degraded, and MPF ceases to phosphorylate its protein targets. Protein phosphatases then gain the upper hand, and dephosphorylate the mitosis-related proteins. The cell returns to *interphase*.

MPF is a heterodimeric protein consisting of a catalytic subunit (**CDK** = cyclin-dependent kinase, or p34^{cdc2}; mass 34 kDa) and a regulatory subunit (**cyclin**). The cyclin subunit has to associate with the CDK subunit to form active MPF. In addition, the activity of this protein kinase is controlled by phosphorylation/dephosphorylation (a highly simplified version of this process is shown in the illustration).

The cells of vertebrate animals contain a number of different cyclins and cyclin-dependent kinases (CDK). Different combinations of the two different types of subunit regulate the point of entry into the M phase, the start of transcription in the G_1 phase, the successful passage past the restriction point in the G_1 phase, and the start of DNA replication in the S phase, as well as many other critical transitions (not shown).

In frog oocytes, the entry of the cell into the M phase (the so-called G_2/M transition) is regulated by regular oscillations in the concentration of a cyclin (upper part of Fig. **B**). There is continued synthesis of cyclin B during interphase until a maximum concentration is reached in the M phase, which triggers the MPF-catalyzed phosphorylation cascade. At the end of the M phase, the cyclin is rapidly degraded by proteolytic enzymes, which are activated by MPF itself. In other cell systems, it is the extent of phosphorylation of MPF that regulates its activity.

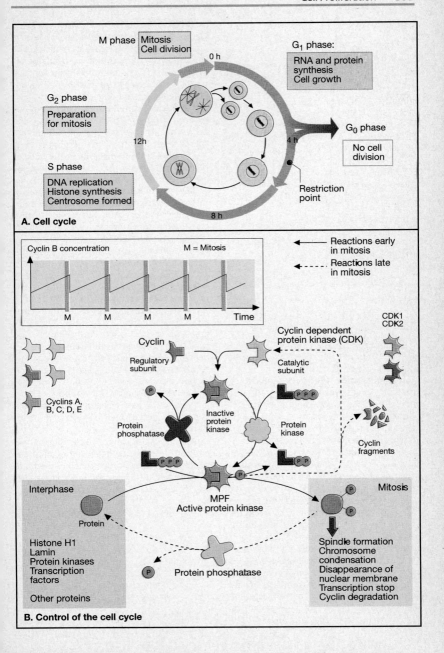

A. Cell cycle

M phase | Mitosis, Cell division

0 h

G₁ phase: RNA and protein synthesis, Cell growth

G₂ phase | Preparation for mitosis

12h

4 h

G₀ phase | No cell division

S phase | DNA replication, Histone synthesis, Centrosome formed

Restriction point

8 h

B. Control of the cell cycle

Cyclin B concentration   M = Mitosis

M   M   M   M   Time

Reactions early in mitosis

Reactions late in mitosis

CDK1
CDK2

Cyclin

Cyclin dependent protein kinase (CDK)

Regulatory subunit

Catalytic subunit

Cyclins A, B, C, D, E

Protein phosphatase

Inactive protein kinase

Protein kinase

Cyclin fragments

Interphase

Protein

MPF Active protein kinase

Mitosis

Histone H1
Lamin
Protein kinases
Transcription factors

Other proteins

Protein phosphatase

Spindle formation
Chromosome condensation
Disappearance of nuclear membrane
Transcription stop
Cyclin degradation

Oncogenes

A. Proto-oncogenes: biological role ◑

Oncogenes are genetic elements that can trigger the formation of tumors. They were first discovered in tumor-causing viruses (*viral oncogenes*). Cellular oncogenes, the so-called **proto-oncogenes**, are the cellular homologs of viral oncogenes. Their gene products, the **oncoproteins**, are involved in the regulation of normal growth and differentiation processes, especially of *cell proliferation* (see p. 356). They themselves are regulated by **tumor suppressor genes** (*anti-oncogenes*).

Proto-oncogenes can become oncogenes, i.e., trigger the formation of tumors, as a result of mutations, deletions, or overexpression. This is especially frequent when there is a simultaneous disturbance in their regulation by tumor suppressor genes.

B. Oncogene products: biochemical functions ◑

All oncogenes are alike in that they encode proteins involved in *signal transduction*

(1) Ligands encoded by proto-oncogenes are found outside the cells. They show homology to *growth factors*.

(2) Membrane receptors encoded by proto-oncogenes are similar to type I receptors (see p. 350), which bind growth factors and hormones. These receptors contain a protein kinase or guanylate cyclase domain on the inner surface of the cell membrane.

(3) GTP-binding proteins are found both in the membrane and in the intracellular space of normal cells. The *membrane-localized G proteins* mediate the effects of type III membrane receptors (receptors featuring 7 transmembrane helices, see p. 350) and thus transmit extracellular signals to effector systems localized within the cell. *Intracellular G proteins* are involved in the control of protein synthesis and protein transport (see p. 214). G proteins bind GTP, and slowly hydrolyze it to GDP, which converts them into an inactive state. The protein products of the proto-oncogene *ras* and several other oncogenes are closely related to these G proteins.

(4) Nuclear hormone receptors mediate the effects of lipophilic signal molecules (steroid hormones, thyroxin, retinoate) by regulating the transcription of specific genes (see p. 346). Several proto-oncogenes encode proteins that belong to this family of ligand-controlled transcription factors.

(5) Nuclear tumor suppressors block the reentry of fully differentiated cells into the cell cycle. The genes encoding them are therefore referred to as *anti-oncogenes*.

(6) DNA-binding proteins in the nucleus have various functions. Some oncoproteins are related to these *transcription factors*.

(7) Protein kinases play a central role in intracellular signal transduction. They catalyze the phosphorylation of specific proteins, thus modulating biological activity. The effects of protein kinases are reversed by specific *protein phosphatases*. The interplay between phosphorylation by protein kinases and dephosphorylation by protein phosphatases (*interconversion*, see p. 106) is important in the regulation of many cellular processes. A large number of proto-oncogen products are protein kinases. The protein kinases encoded by proto-oncogenes can be subdivided into various different groups, depending on their mode of activation or type of substrate. Enzymes of the *protein kinase A* group are activated by cAMP, those of the *protein kinase G* group by cGMP, the *type C protein kinases* by diacylglycerol (DAG), and the *Ca^{2+}/calmodulin-dependent kinases* by Ca^{2+} (see p. 352).

Another classification is based on the type of amino acid residue that becomes phosphorylated. According to this scheme, there are *tyrosine-specific kinases* and *serine/threonine-specific kinases*.

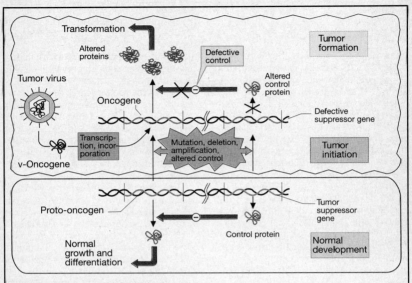

Transformation

Altered proteins

Defective control

Altered control protein

Tumor formation

Tumor virus

Oncogene

Defective suppressor gene

v-Oncogene

Transcription, incorporation

Mutation, deletion, amplification, altered control

Tumor initiation

Proto-oncogen

Tumor suppressor gene

Normal growth and differentiation

Control protein

Normal development

A. Proto-oncogenes: biological role

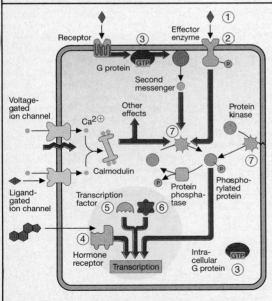

Receptor

Effector enzyme

G protein

GTP

Second messenger

Voltage-gated ion channel

Ca²⁺

Other effects

Protein kinase

Calmodulin

Ligand-gated ion channel

Transcription factor

Protein phosphatase

Phosphorylated protein

Hormone receptor

Transcription

Intracellular G protein

GTP

Oncogene products (examples)

Gene

① Ligands
sis, hst, int-2, wnt-1

② Receptors
*fms, trk, trk*B, *ros, kit, mas, neu*

③ GTP-binding proteins
Ha-*ras*, Ki-*ras*, N-*ras*

④ Nuclear hormone receptors
*erb*A, NGF1-B

⑤ Nuclear tumor suppressors
Rb, p53, *wt1*, DCC, APC

⑥ DNA-binding proteins
jun, fos, myc, N-*myc, myb, fra1, egr-1, rel*

⑦ Protein kinases
src, yes, fps, abl, met mos, raf

B. Oncogene products: biochemical functions

Tumors

A. Division behavior of cells ●

The cells of the body are normally subject to strict "social" control, i.e., they continue to divide until they make contact with a neighboring cell, and then stop dividing. This effect is known as *contact inhibition*. Exceptions include embryonic cells, cells of the intestinal epithelium (which has to be continually replaced), cells of the bone marrow (formation of blood cells), and **tumor cells**. *Uncontrolled cell proliferation* is an important indicator of the presence of a tumor.

The illustration shows the dividing behavior of cells in culture. Normal cells continue to divide *in vitro* only until they make contact with a neighboring cell (about 20 to 60 divisions in total), whereas tumor cells continue to divide, and are insensitive to contact inhibition.

B. Transformation ◑

The conversion of a normal cell into a tumor cell is referred to as **transformation**.

In medical science, a distinction is made between *benign* and *malignant tumors*. Benign tumors grow relatively slowly, and are made up of differentiated cells, whereas malignant tumors tend to grow more rapidly and form *metastases* (daughter tumors). Approximately 100 different types of tumors have been classified according to their tissue of origin. Together, they are responsible for more than 20% of all deaths in Europe and North America.

Normal cells show all of the characteristics of fully differentiated cells specialized for a particular function. They are non-dividing, and usually exist in the G_0 phase of the cell cycle (see p. 356). They are polymorphic, and their shape is determined by a rigid cytoskeleton.

In contrast, *tumor cells* have often undergone a process of de-differentiation, i.e., they resemble embryonic cells, and continue to divide unabated. The surface of these cells has undergone changes that make them insensitive to contact inhibition. The cytoskeleton has also been altered, giving the cells a more or less rounded shape. The nuclei of tumor cells can be unusual in shape, number, and size.

Tumor markers are important for the clinical detection of different types of tumors. These are proteins that are formed either by tumor cells themselves (group 1), or by other cells as a result of interactions with tumor cells (group 2). Group 1 tumor markers include *tumor-associated antigens*, secreted hormones, and enzymes. The table shows just a few examples.

The transition from the normal to the transformed state is a multi-step process.

(1) Tumor initiation. Almost every tumor begins with damage to the DNA of an individual cell. Genetic defects, such as this can be caused by *carcinogens*, i.e., tumor-inducing chemicals (e.g., constituents of tar from tobacco), physical mutagens (e.g., ultraviolet light, X-ray radiation, see p. 234) or tumor viruses (see p. 358). There are probably quite a few of the 10^{14} cells in the human body that suffer damage to their DNA during the average life-span of an individual. However, only defects in *proto-oncogenes* (see p. 358) are relevant in terms of tumor initiation. Such defects are the most important factor determining whether *transformation* will occur. However, a defect in an *anti-oncogene* (tumor suppressor gene, see p. 358) can also lead to tumor initiation.

(2) Tumor promotion is the preferential proliferation of a cell damaged by tumor-initiating factors. This process may take years to set in. *Phorbol esters* (synthetic activators of protein kinase C, see p. 347) are currently in use as model substances mimicking tumor promoters.

(3) Tumor progression is the process whereby proliferation, invasion, and metastasis production leads to the formation of a malignant tumor.

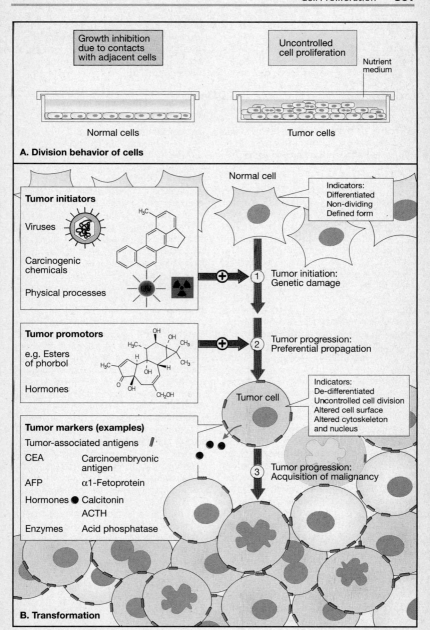

A. Division behavior of cells

Growth inhibition due to contacts with adjacent cells

Uncontrolled cell proliferation

Nutrient medium

Normal cells

Tumor cells

B. Transformation

Normal cell

Indicators:
Differentiated
Non-dividing
Defined form

Tumor initiators

Viruses

Carcinogenic chemicals

Physical processes

① Tumor initiation:
Genetic damage

Tumor promotors

e.g. Esters of phorbol

Hormones

② Tumor progression:
Preferential propagation

Tumor cell

Indicators:
De-differentiated
Uncontrolled cell division
Altered cell surface
Altered cytoskeleton and nucleus

Tumor markers (examples)

Tumor-associated antigens

CEA   Carcinoembryonic antigen

AFP α1-Fetoprotein

Hormones ● Calcitonin
           ACTH

Enzymes   Acid phosphatase

③ Tumor progression:
Acquisition of malignancy

Viruses

Viruses are *parasitic nucleoprotein complexes*. The simplest viruses consist of a single nucleic acid molecule (RNA or DNA, never both) and a protein coat. Viruses have no *metabolism of their own,* and are therefore unable to replicate themselves without the help of a living cell. Consequently, they are not considered to be living organisms. Viruses that cause damage as a result of their replication within a host cell are classified as *pathogenic.*

A. Viruses: Examples ○

Here we show only a few of the large number of viruses known. They are all represented at the same magnification. Viruses that only attack bacteria are called *bacteriophages* ("phages"). One simple example is the phage **M13** (**1a**). It consists of a single-stranded, circular DNA molecule (ssDNA) of about 7000 nucleotides encapsulated in a cylindrical coat made up of 2700 helically-arranged protein monomers. The coat of a virus is referred to as a *capsid*, and the complete structure, including the nucleic acids, as a *nucleocapsid*. M13 is an useful *vector* for foreign DNA in genetic engineering (see p. 238).

The **phage T4** (**1**), one of the largest viruses known, has a much more complex structure. Contained within its "head", there are about 170,000 bp of double-stranded DNA (dsDNA).

The plant-pathogenic **tobacco mosaic virus** (**2**) has a structure similar to that of M13, except that it contains single-stranded RNA (ssRNA) instead of DNA. The **poliovirus** (**2**), the causal agent of *poliomyelitis,* also contains ssRNA. In the **influenza virus** (**2**), the nucleocapsid is surrounded by an additional *coat* which is derived from the plasma membrane of the host cell (**C**). The coat carries membrane proteins involved in the infection process.

B. Capsid of the rhinovirus ○

Rhinoviruses are the causal agent of the common cold in humans. The capsid of the rhinovirus (**a**) is shaped like an **eicosahe-**dron, i.e., a geometrical shape made up of 20 equilateral triangles. The eicosahedron consists of three different proteins, which associate with one another to form pentamers and hexamers (**b**).

C. Life cycle of the human immunodeficiency virus (HIV) ◑

The human immunodeficiency virus (**HIV**) causes the immunodeficiency disease known as **AIDS** (acquired immunodeficiency syndrome). The structure of this virus is similar to that of the influenza virus (**A**).

The genome of HIV consists of ssRNA (2 identical strands each 9. 2 kb long), which is surrounded by a double-layered capsid and a protein-containing envelope. The failure of the immune system following HIV infection is due to the fact that this virus tends to target and incapacitate T helper cells (see p. 264)

During infection (**1**), the membrane of the virus fuses with that of the target cell, and the nucleic acid core of the nucleocapsid enters the cytoplasm of the host cell (**2**). Once there, the viral RNA is first transcribed into an RNA/DNA hybrid (**3**) and then into dsDNA (**4**). *Reverse transcriptase,* the enzyme catalyzing these reactions, is of viral (not host) origin. The dsDNA subsequently becomes integrated into the genome of the host cell (**5**), where it can remain in an inactive state for long periods of time. When it becomes active, the DNA fragment corresponding to the viral genome is transcribed by the enzymes of the host cell (**6**). Not only does the host cell replicate the viral ssRNA, but it also transcribes the mRNA molecules that encode the precursors of the viral proteins (**7**). These precursors are integrated into the plasma membrane of the cell (**8, 9**) before undergoing proteolytic modification (**10**). The cycle comes to completion with the release of new virus particles (**11**).

The group of RNA viruses that includes HIV is referred to as **retroviruses**. This is, because their life cycle starts with the synthesis of DNA from RNA, which is the reverse of the usual direction of transcription (DNA → RNA).

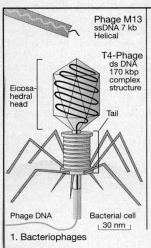

Phage M13
ssDNA 7 kb
Helical

T4-Phage
ds DNA
170 kbp
complex
structure

Eicosa-
hedral
head

Tail

Phage DNA   Bacterial cell
                   30 nm

1. Bacteriophages

Coat

Influenza virus
ssRNA (8 molecules) 3.6 kb
Nucleocapsid with coat

Tobacco mosaic virus
ssRNA 6.4 kb
Helical

Poliovirus
ssRNA 7 kb
Eicosahedral capsid

2. Plant and animal
   pathogenic viruses

A. Viruses: examples

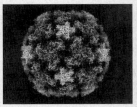

1. Structure

Pentamer

Hexamer   Capsid

2. Diagram   Eicosahedron of
                  180 monomers

B. Capsid of the rhinovirus

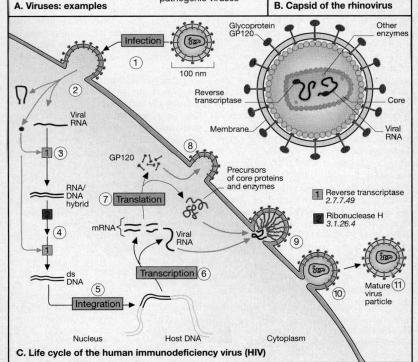

Infection
①

②

Viral
RNA

1 ③

RNA/
DNA
hybrid

2

④

1

ds
DNA

Integration ⑤

Nucleus   Host DNA

100 nm

GP120

⑦ Translation

mRNA

Viral
RNA

Transcription ⑥

⑧

Glycoprotein
GP120   Other
                         enzymes

Reverse
transcriptase   Core

Membrane   Viral
                            RNA

Precursors
of core proteins
and enzymes

1 Reverse transcriptase
  2.7.7.49

2 Ribonuclease H
  3.1.26.4

⑨

⑩   Mature ⑪
           virus
           particle

Cytoplasm

C. Life cycle of the human immunodeficiency virus (HIV)

Development and Morphogenesis

The fruit fly (*Drosophila melanogaster*) is one of the most important model organisms used in developmental biology.

A. Development of a fruit fly ○

The life-cycle of a fruit fly takes about 10 days. It begins at day 0 with the laying of the fertilized egg. *Embryonic development* takes about one day (**1**), after which a small larva hatches out of the egg. The fruit fly then goes through three larval stages, characterized by rapid growth in response to the uptake of a large quantity of nutrients (**2, 3**). The larval stages are followed by the pupal stage in which *metamorphosis* occurs, giving rise to the adult fly (imago, **5**).

The first stage in **embryonic development** (**1**, top left) is a series of nuclear divisions without concomitant cell division (*syncytial blastoderm*). The nuclei move to the borders of the egg cell (**1a**) before becoming separated from one another by the formation of cell walls (**1b**; *cellular blastoderm*). The next stage is the formation of several distinct cell layers (*gastrulation*), which will later give rise to the various different tissues of the organism (**1c**).

Imaginal development (**4/5**) occurs during the pupal stage. In the pupae, the larval tissues are degraded, and the tissues of the adult fly, e.g., the antennae, the eyes, the wings, and the genitals, begin to develop from the *imaginal disks* (highlighted in various colors). The imaginal disks are small groups of embryonic cells occurring in pairs, which are inactive in the larvae, but become activated by 20-hydroxyecdysone (see p. 54) in the pupal phase. Cell division and differentiation result in the formation of the imaginal tissue from the imaginal disks.

B. Roles played by developmental genes ◑

How does a highly differentiated organism arise from a single cell? In the future, we may become able to answer this question with the help of knowledge gained from studies of the fruit fly.

The embryonic development of the fruit fly is regulated by a hierarchical system of three classes of genes, the maternal *egg polarity genes*, the *segmentation genes*, and the *homeotic genes*. The coordinated expression of these genes produces a temporal and spatial **pattern** that dictates the position and type of each cell, as well as the stage at which it becomes active.

The **maternal genes** are already active in the mother. The products of these genes are released into the egg, where they determine, for example, the spatial axes of the embryo, i.e., the anterior/posterior and dorsal/ventral axis. The mRNA encoded by the maternal gene *bicoid* is shown here as an example. This mRNA is located at the anterior pole of the egg cell. At the beginning of embryogenesis, it is translated to give a protein that diffuses away from its site of synthesis, thus producing a gradient. The gene products of some of the other maternal genes also show a specific distribution. Their products include DNA-binding proteins, which function as *transcription factors* to either promote or inhibit the expression of other genes regulating development.

The **segmentation genes** are involved in interpreting the positional information encoded by the maternal genes. The products of these genes regulate the formation of the segments making up the insect. The segmentation genes can be subdivided into several different groups (*gap genes, pair-rule genes, segment polarity genes*). They form a coordinated system that divides the embryo into progressively smaller segments. The illustration shows the distribution of the protein encoded by the pair-rule gene *even-skipped*.

Homeotic genes are regulated by the segmentation genes. They interact with one another to determine the identity of the different segments. The products of the homeotic genes determine, which genes in each of the cells will become activated and thereby allow the cells to differentiate to become a particular segment, e.g., head, thorax, or abdomen.

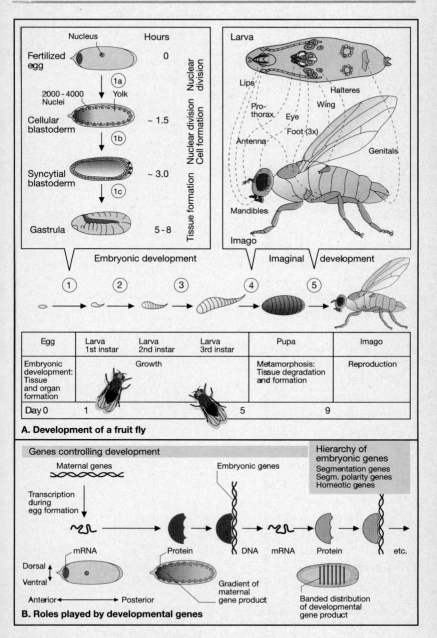

A. Development of a fruit fly

B. Roles played by developmental genes

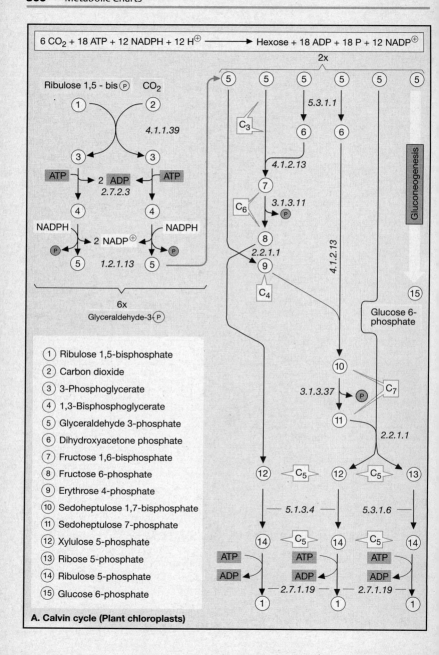

$$6\ CO_2 + 18\ ATP + 12\ NADPH + 12\ H^{\oplus} \longrightarrow Hexose + 18\ ADP + 18\ P + 12\ NADP^{\oplus}$$

2x

Ribulose 1,5 - bis Ⓟ   CO₂

① ②

4.1.1.39

③ ③

ATP → 2 ADP ← ATP
2.7.2.3

④ ④

NADPH → 2 NADP⁺ ← NADPH
1.2.1.13

⑤ Ⓟ Ⓟ ⑤

6x
Glyceraldehyde-3-Ⓟ

⑤ ⑤ ⑤ ⑤ ⑤ ⑤

5.3.1.1

C₃

⑥ ⑥

4.1.2.13

⑦

C₆   3.1.3.11   Ⓟ

⑧

2.2.1.1

⑨

C₄

Gluconeogenesis

4.1.2.13

⑮

Glucose 6-
phosphate

⑩

3.1.3.37   Ⓟ   C₇

⑪

2.2.1.1

⑫   C₅   ⑫   C₅   ⑬

5.1.3.4   5.3.1.6

⑭   C₅   ⑭   C₅   ⑭

ATP   ATP   ATP

ADP   ADP   ADP

①   2.7.1.19   ①   2.7.1.19   ①

① Ribulose 1,5-bisphosphate
② Carbon dioxide
③ 3-Phosphoglycerate
④ 1,3-Bisphosphoglycerate
⑤ Glyceraldehyde 3-phosphate
⑥ Dihydroxyacetone phosphate
⑦ Fructose 1,6-bisphosphate
⑧ Fructose 6-phosphate
⑨ Erythrose 4-phosphate
⑩ Sedoheptulose 1,7-bisphosphate
⑪ Sedoheptulose 7-phosphate
⑫ Xylulose 5-phosphate
⑬ Ribose 5-phosphate
⑭ Ribulose 5-phosphate
⑮ Glucose 6-phosphate

A. Calvin cycle (Plant chloroplasts)

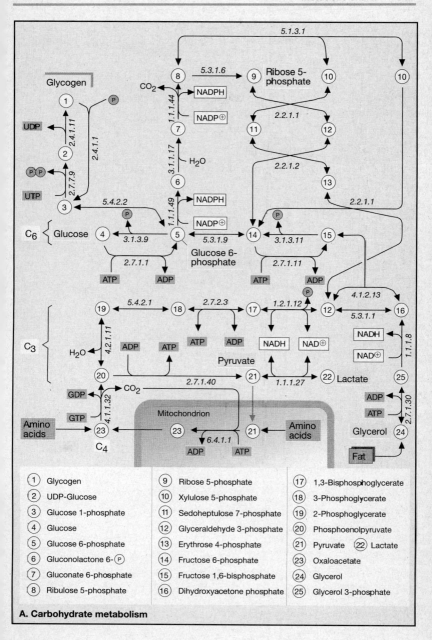

A. Carbohydrate metabolism

① Glycogen	⑨ Ribose 5-phosphate	⑰ 1,3-Bisphosphoglycerate
② UDP-Glucose	⑩ Xylulose 5-phosphate	⑱ 3-Phosphoglycerate
③ Glucose 1-phosphate	⑪ Sedoheptulose 7-phosphate	⑲ 2-Phosphoglycerate
④ Glucose	⑫ Glyceraldehyde 3-phosphate	⑳ Phosphoenolpyruvate
⑤ Glucose 6-phosphate	⑬ Erythrose 4-phosphate	㉑ Pyruvate ㉒ Lactate
⑥ Gluconolactone 6-Ⓟ	⑭ Fructose 6-phosphate	㉓ Oxaloacetate
⑦ Gluconate 6-phosphate	⑮ Fructose 1,6-bisphosphate	㉔ Glycerol
⑧ Ribulose 5-phosphate	⑯ Dihydroxyacetone phosphate	㉕ Glycerol 3-phosphate

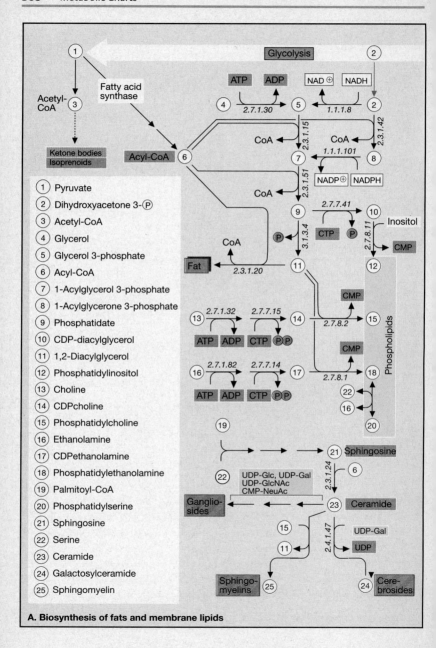

A. Biosynthesis of fats and membrane lipids

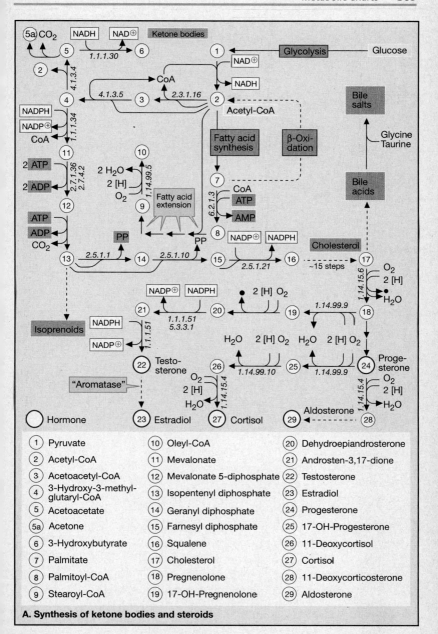

A. Synthesis of ketone bodies and steroids

① Pyruvate	⑩ Oleyl-CoA	⑳ Dehydroepiandrosterone
② Acetyl-CoA	⑪ Mevalonate	㉑ Androsten-3,17-dione
③ Acetoacetyl-CoA	⑫ Mevalonate 5-diphosphate	㉒ Testosterone
④ 3-Hydroxy-3-methyl-glutaryl-CoA	⑬ Isopentenyl diphosphate	㉓ Estradiol
⑤ Acetoacetate	⑭ Geranyl diphosphate	㉔ Progesterone
⑤ₐ Acetone	⑮ Farnesyl diphosphate	㉕ 17-OH-Progesterone
⑥ 3-Hydroxybutyrate	⑯ Squalene	㉖ 11-Deoxycortisol
⑦ Palmitate	⑰ Cholesterol	㉗ Cortisol
⑧ Palmitoyl-CoA	⑱ Pregnenolone	㉘ 11-Deoxycorticosterone
⑨ Stearoyl-CoA	⑲ 17-OH-Pregnenolone	㉙ Aldosterone

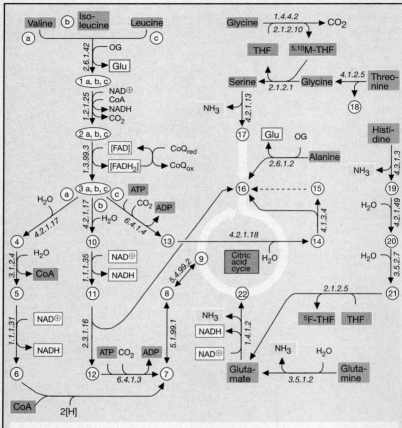

① a 2-Oxoisovalerate
① b 2-Oxo-3-methylvalerate
① c 2-Oxoisocaproate
② a Isobutyryl-CoA
② b 2-Methylbutyryl-CoA
② c Isovaleryl-CoA
③ a Methylacrylyl-CoA
③ b Tiglyl-CoA
③ c 3-Methylcrotonyl-CoA
④ 3-Hydroxyisobutyryl-CoA

⑤ 3-Hydroxyisobutyrate
⑥ Methylmalonyl-semialdehyde
⑦ (S)-Methylmalonyl-CoA
⑧ (R)-Methylmalonyl-CoA
⑨ Succinyl-CoA
⑩ 2-Methyl-3-hydroxybutyryl-CoA
⑪ 2-Methylacetoacetyl-CoA
⑫ Propionyl-CoA
⑬ 3-Methylglutaconyl-CoA
⑭ 3-Hydroxy-3-methylglutaryl-CoA

⑮ Acetoacetate
⑯ Acetyl-CoA
⑰ Pyruvate
⑱ Acetaldehyde
⑲ Urocanate
⑳ Imidazolone-5-propionate
㉑ N-Formimino-glutamate
㉒ 2-Oxoglutarate

A. Amino acid degradation I

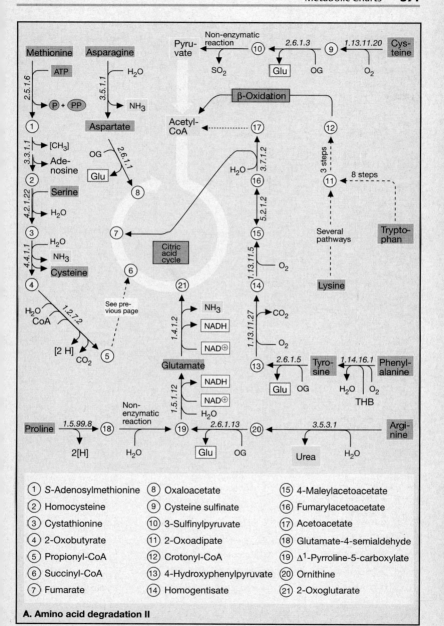

1. *S*-Adenosylmethionine
2. Homocysteine
3. Cystathionine
4. 2-Oxobutyrate
5. Propionyl-CoA
6. Succinyl-CoA
7. Fumarate
8. Oxaloacetate
9. Cysteine sulfinate
10. 3-Sulfinylpyruvate
11. 2-Oxoadipate
12. Crotonyl-CoA
13. 4-Hydroxyphenylpyruvate
14. Homogentisate
15. 4-Maleylacetoacetate
16. Fumarylacetoacetate
17. Acetoacetate
18. Glutamate-4-semialdehyde
19. Δ^1-Pyrroline-5-carboxylate
20. Ornithine
21. 2-Oxoglutarate

A. Amino acid degradation II

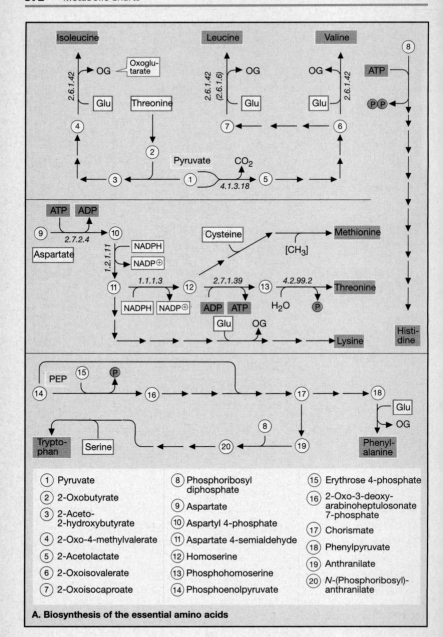

1. Pyruvate
2. 2-Oxobutyrate
3. 2-Aceto-2-hydroxybutyrate
4. 2-Oxo-4-methylvalerate
5. 2-Acetolactate
6. 2-Oxoisovalerate
7. 2-Oxoisocaproate
8. Phosphoribosyl diphosphate
9. Aspartate
10. Aspartyl 4-phosphate
11. Aspartate 4-semialdehyde
12. Homoserine
13. Phosphohomoserine
14. Phosphoenolpyruvate
15. Erythrose 4-phosphate
16. 2-Oxo-3-deoxy-arabinoheptulosonate 7-phosphate
17. Chorismate
18. Phenylpyruvate
19. Anthranilate
20. N-(Phosphoribosyl)-anthranilate

A. Biosynthesis of the essential amino acids

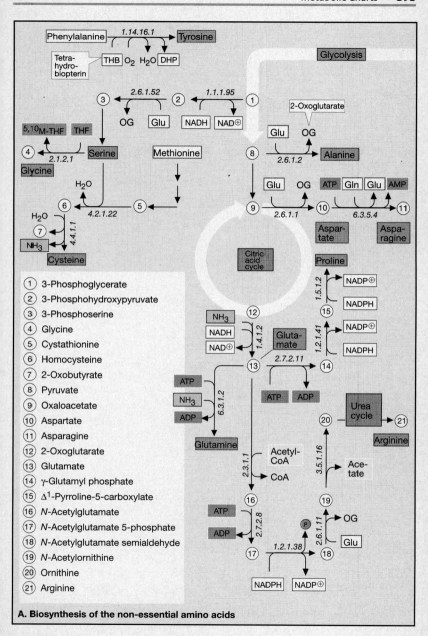

A. Biosynthesis of the non-essential amino acids

Enzyme list:

1. 3-Phosphoglycerate
2. 3-Phosphohydroxypyruvate
3. 3-Phosphoserine
4. Glycine
5. Cystathionine
6. Homocysteine
7. 2-Oxobutyrate
8. Pyruvate
9. Oxaloacetate
10. Aspartate
11. Asparagine
12. 2-Oxoglutarate
13. Glutamate
14. γ-Glutamyl phosphate
15. Δ1-Pyrroline-5-carboxylate
16. N-Acetylglutamate
17. N-Acetylglutamate 5-phosphate
18. N-Acetylglutamate semialdehyde
19. N-Acetylornithine
20. Ornithine
21. Arginine

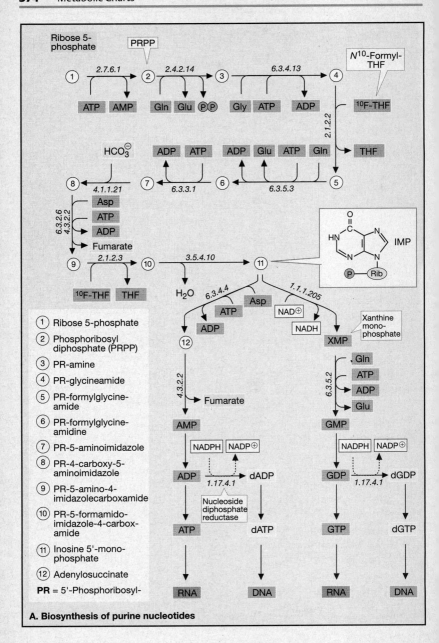

A. Biosynthesis of purine nucleotides

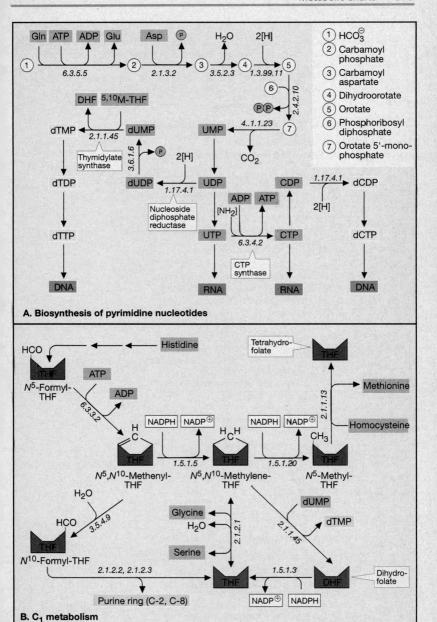

A. Biosynthesis of pyrimidine nucleotides

Legend:
1. $HCO_3^\ominus$
2. Carbamoyl phosphate
3. Carbamoyl aspartate
4. Dihydroorotate
5. Orotate
6. Phosphoribosyl diphosphate
7. Orotate 5'-monophosphate

B. C_1 metabolism

Annotated Enzyme List

Of the more than 2000 enzymes known, only the few mentioned in this book are listed here. The naming of the enzymes is based on the *Enzyme Nomenclature 1992*, published by the IUBMB. The additions in round brackets belong to the enzyme name, while prosthetic groups are enclosed in square brackets. Common names of enzyme groups are given in italics, while trivial names or abbreviations of individual enzymes are shown in quotation marks.

Class 1: Oxidoreductases (catalyze reduction-oxidation reactions)

Subclass 1.n: What is the electron *donor*?
Sub-subclass 1.n.n: What is the electron *acceptor*?

1.1	**A -CH-OH group is the donor**

1.1.1	**NAD(P)+ is the acceptor** (*dehydrogenases, reductases*)
1.1.1.1	Alcohol dehydrogenase [Zn^{2+}]
1.1.1.3	Homoserine dehydrogenase
1.1.1.8	Glycerol-3-phosphate dehydrogenase (NAD+)
1.1.1.21	Aldehyde reductase
1.1.1.27	Lactate dehydrogenase
1.1.1.30	3-Hydroxybutyrate dehydrogenase
1.1.1.31	3-Hydroxyisobutyrate dehydrogenase
1.1.1.34	Hydroxymethylglutaryl-CoA reductase (NADPH)
1.1.1.35	3-Hydroxyacyl-CoA dehydrogenase
1.1.1.37	Malate dehydrogenase
1.1.1.40	Malate dehydrogenase (oxalacetate-decarboxylating) (NADP+), "malic enzyme"
1.1.1.41	Isocitrate dehydrogenase (NAD+)
1.1.1.42	Isocitrate dehydrogenase (NADP+)
1.1.1.44	Phosphogluconate dehydrogenase (decarboxylating)
1.1.1.49	Glucose-6-phosphate 1-dehydrogenase
1.1.1.51	3(or 17)β-Hydroxysteroid dehydrogenase
1.1.1.95	Phosphoglycerate dehydrogenase
1.1.1.100	3-Oxoacyl-[ACP] reductase
1.1.1.101	Acylglycerone-phosphate reductase
1.1.1.105	Retinol dehydrogenase
1.1.1.145	3β-Hydroxy-Δ^5-steroid dehydrogenase
1.1.1.205	IMP dehydrogenase

1.1.3	**Molecular oxygen is the acceptor** (*oxidases*)
1.1.3.4	Glucose oxidase [FAD]
1.1.3.8	L-Gulonolactone oxidase
1.1.3.22	Xanthine oxidase [Fe, Mo, FAD]

1.2	**An aldehyde or keto group is the donor**

1.2.1	**NAD(P)+ is the acceptor** (*dehydrogenases*)
1.2.1.3	Aldehyde dehydrogenase (NAD+)
1.2.1.11	Aspartate-semialdehyde dehydrogenase
1.2.1.12	Glyceraldehyde-3-phosphate dehydrogenase
1.2.1.13	Glyceraldehyde-3-phosphate dehydrogenase (NADP+) (phosphorylating)
1.2.1.24	Succinate-semialdehyde dehydrogenase

1.2.1.25	2-Oxoisovalerate dehydrogenase (acylating)
1.2.1.38	N-Acetyl-γ-glutamyl-phosphate reductase
1.2.1.41	Glutamylphosphate reductase

1.2.4 **A disulfide is the acceptor**
1.2.4.1	Pyruvate dehydrogenase (lipoamide) [TPP]
1.2.4.2	Oxoglutarate dehydrogenase (lipoamide) [TPP]

1.2.7 **An Fe/S protein is the acceptor**
1.2.7.2	2-Oxobutyrate synthase

1.3 **A -CH-CH- group is the donor**

1.3.1.10	Enoyl-[ACP] reductase (NADPH)
1.3.1.24	Biliverdin reductase
1.3.1.34	2,4-Dienoyl-CoA reductase
1.3.5.1	Succinate dehydrogenase (ubiquinone) [FAD, Fe$_2$S$_2$, Fe$_4$S$_4$], "complex II"
1.3.99.3	Acyl-CoA dehydrogenase [FAD]
1.3.99.11	Dihydroorotate dehydrogenase [Fe]

1.4 **A -CH-NH$_2$ group is the donor**

1.4.1.2	Glutamate dehydrogenase
1.4.3.4	Amine oxidase [FAD], "monoamine oxidase, MAO"
1.4.3.13	Protein-lysine 6-oxidase [Cu]
1.4.4.2	Glycine dehydrogenase (decarboxylating) [PLP]

1.5 **A -CH-NH- group is the donor**

1.5.1.2	Pyrroline-5-carboxylate reductase
1.5.1.3	Dihydrofolate reductase
1.5.1.5	Methylenetetrahydrofolate dehydrogenase (NADP$^+$)
1.5.1.12	1-Pyrroline-5-carboxylate dehydrogenase
1.5.1.20	Methylenetetrahydrofolate reductase (NADPH) [FAD]
1.5.5.1	Electron-transferring-flavoprotein (ETF) dehydrogenase
1.5.99.8	Proline dehydrogenase [FAD]

1.6 **NAD(P)H is the donor**

1.6.4.2	Glutathione reductase (NADPH) [FAD]
1.6.4.5	Thioredoxin reductase (NADPH) [FAD]
1.6.5.3	NADH dehydrogenase (ubiquinone) [FAD, Fe$_2$S$_2$, Fe$_4$S$_4$], "complex I"

1.8 **A sulfur group is the donor**

1.8.1.4	Dihydrolipoamide dehydrogenase [FAD]

1.9 **A heme group is the donor**

1.9.3.1	Cytochrome-c oxidase [heme, Cu, Zn], "cytochrome oxidase, complex IV"

1.10 **A diphenol is the donor**

1.10.2.2	Ubiquinol-cyctochrome-c reductase [heme, Fe$_2$S$_2$], "complex III"

1.11	**A peroxide is the acceptor** (*peroxidases*)
1.11.1.6	Catalase [heme]
1.11.1.7	Peroxidase [heme]
1.11.1.9	Glutathione peroxidase [Se]
1.11.1.12	Lipid hydroperoxide-glutathione peroxidase [Se]

1.13	**Molecular oxygen is incorporated into the electron donor** (*oxygenases*)
1.13.11	**One donor, both oxygens are incorporated** (*dioxygenases*)
1.13.11.5	Homogentisate 1,2-dioxygenase [Fe]
1.13.11.20	Cysteine dioxygenase [Fe]
1.13.11.27	4-Hydroxyphenylpyruvate dioxygenase [ascorbate]
1.13.11.n	Arachidonate lipoxygenases

1.14	**Two donors, oxygen is incorporated into either one** (*monooxygenases, hydroxylases*)
1.14.11.2	Procollagen-proline 4-dioxygenase [Fe, ascorbate], "proline hydroxylase"
1.14.11.4	Procollagen-lysine 5-dioxygenase [Fe, ascorbate], "lysine hydroxylase"
1.14.13.13	Calcidiol 1-monooxygenase
1.14.15.4	Steroid 11β-monooxygenase [heme]
1.14.15.6	Cholesterol monooxygenase (side-chain-cleaving) [heme]
1.14.16.1	Phenylalanine 4-monooxygenase [Fe, THB]
1.14.16.2	Tyrosine 3-monooxygenase [Fe, THB]
1.14.17.1	Dopamine β-monooxygenase [Cu]
1.14.99.1	Prostanglandin synthase [heme]
1.14.99.3	Heme oxygenase (decyclyzing) [heme]
1.14.99.5	Stearoyl-CoA desaturase [heme]
1.14.99.9	Steroid 17α-monooxygenase [heme]
1.14.99.10	Steroid 21-monooxygenase [heme]

1.17	**A -CH_2- group is the donor**
1.17.4.1	Ribonucleoside-diphosphate reductase [Fe], "ribonucleotide reductase"

1.18	**Reduced ferredoxin is the donor**
1.18.1.2	Ferredoxin-NADP$^+$ reductase [FAD]
1.18.6.1	Nitrogenase [Fe, Mo, Fe_4S_4]

Class 2: Transferases (catalyze the transfer of groups from one molecule to another)

Subclass 2.n: *Which group* **is being transferred?**

2.1	**A one-carbon group is transferred**
2.1.1	**A methyl group**
2.1.1.13	5-Methyltetrahydrofolate-homocysteine S-methyltransferase
2.1.1.28	Phenylethanolamine N-methyltransferase
2.1.1.45	Thymidylate synthase
2.1.2	**A formyl group**
2.1.2.1	Glycine hydroxymethyltransferase [PLP]

2.1.2.2   Phosphoribosylglycinamide formyltransferase
2.1.2.3   Phosphoribosylaminoimidazolecarboxamide formyltransferase
2.1.2.5   Glutamate formiminotransferase [PLP]
2.1.2.10   Aminomethyltransferase

2.1.3   A carbamoyl group
2.1.3.2   Aspartate carbamoyltransferase [Zn^{2+}]
2.1.3.3   Ornithine carbamoyltransferase

2.2   An aldehyde or ketone residue is transferred

2.2.1.1   Transketolase [TPP]
2.2.1.2   Transaldolase

2.3   An acyl group is transferred

2.3.1   With acyl-CoA as donor
2.3.1.1   Amino-acid N-acetyltransferase
2.3.1.6   Choline O-acetyltransferase
2.3.1.12   Dihydrolipoamide acetyltransferase [lipoamide]
2.3.1.15   Glycerol-3-phosphate O-acyltransferase
2.3.1.16   Acetyl-CoA acyltransferase
2.3.1.20   Diacylglycerol O-acyltransferase
2.3.1.21   Carnitine O-palmitoyltransferase
2.3.1.22   Acylglycerol O-palmitoyltransferase
2.3.1.24   Sphingosine N-acyltransferase
2.3.1.37   5-Aminolevulinate synthase [PLP]
2.3.1.38   [ACP] S-acetyltransferase
2.3.1.39   [ACP] S-malonyltransferase
2.3.1.41   3-Oxoacyl-[ACP] synthase
2.3.1.42   Glycerone-phosphate O-acyltransferase
2.3.1.43   Phosphatidylcholine-sterol acyltransferase, "LCAT"
2.3.1.51   Acylglycerol-3-phosphate O-acyltransferase
2.3.1.61   Dihydrolipoamide succinyltransferase
2.3.1.85   Fatty-acid synthase

2.3.2   An aminoacyl group is transferred
2.3.2.12   Peptidyltransferase
2.3.2.13   Protein-glutamine-γ-glutamyltransferase, "fibrin-stabilizing factor"

2.4   A glycosyl group is transferred

2.4.1   A hexose residue
2.4.1.1   Phosphorylase [PLP], "glycogen (starch) phosphorylase"
2.4.1.11   Glycogen (starch) synthase
2.4.1.17   Glucuronosyltransferase
2.4.1.18   1,4-α-Glucan branching enzyme
2.4.1.25   4α-Glucanotransferase
2.4.1.47   N-Acylsphingosine galactosyltransferase
2.4.1.119   Protein glycotransferase

2.4.2   A pentose residue
2.4.2.7   Adenine phosphoribosyltransferase
2.4.2.8   Hypoxanthine phosphoribosyltransferase

2.4.2.10	Orotate phosphoribosyltransferase
2.4.2.14	Amidophosphoribosyltransferase

2.5 An alkyl or aryl group is transferred

2.5.1.1	Dimethylallyltransferase
2.5.1.6	Methionine adenosyltransferase
2.5.1.10	Geranyltransferase
2.5.1.21	Farnesyl-diphosphate farnesyltransferase

2.6 A nitrogen-containing group is transferred

2.6.1 An amino group (*transaminases*)

2.6.1.1	Aspartate transaminase [PLP]
2.6.1.2	Alanine transaminase [PLP]
2.6.1.3	Cysteine transaminase [PLP]
2.6.1.5	Tyrosine transaminase [PLP]
2.6.1.6	Leucine transaminase [PLP]
2.6.1.11	Acetylornithine transaminase [PLP]
2.6.1.13	Ornithine transaminase [PLP]
2.6.1.19	4-Aminobutyrate transaminase [PLP]
2.6.1.42	Branched-chain-amino-acid transaminase [PLP]
2.6.1.52	Phosphoserine transaminase [PLP]

2.7 A phosphorus-containing group is transferred (*kinases*)

2.7.1 With -CH-OH as acceptor

2.7.1.1	Hexokinase
2.7.1.3	Ketohexokinase
2.7.1.6	Galactokinase
2.7.1.11	6-Phosphofructokinase
2.7.1.19	Phosphoribulokinase
2.7.1.28	Triokinase
2.7.1.30	Glycerol kinase
2.7.1.32	Choline kinase
2.7.1.36	Mevalonate kinase
2.7.1.37	Protein kinase
2.7.1.38	Phosphorylase kinase
2.7.1.39	Homoserine kinase
2.7.1.40	Pyruvate kinase
2.7.1.67	Phosphatidylinositol kinase
2.7.1.68	1-Phosphatidylinositol 4-phosphate kinase
2.7.1.82	Ethanolamine kinase
2.7.1.99	[Pyruvate dehydrogenase] kinase
2.7.1.105	6-Phosphofructo-2 -kinase
2.7.1.112	Protein-tyrosine kinase

2.7.2 With -CO-OH as acceptor

2.7.2.3	Phosphoglycerate kinase
2.7.2.4	Aspartate kinase
2.7.2.8	Acetylglutamate kinase
2.7.2.11	Glutamate 5-kinase

2.7.3   **With an N-containing group as acceptor**
2.7.3.2   Creatine kinase

2.7.4   **With a phosphate group as acceptor**
2.7.4.2   Phosphomevalonate kinase
2.7.4.3   Adenylate kinase

2.7.6   **A diphosphate group is transferred**
2.7.6.1   Ribose-phosphate pyrophosphokinase

2.7.7   **A nucleotide is transferred**
2.7.7.6   DNA-directed RNA polymerase, "RNA polymerase"
2.7.7.7   DNA-directed DNA polymerase, "DNA polymerase"
2.7.7.9   UTP-glucose-1-phosphate uridyltransferase
2.7.7.12   Hexose-1-phosphate uridyltransferase
2.7.7.14   Ethanolamine-phosphate cytidyltransferase
2.7.7.15   Choline-phosphate cytidyltransferase
2.7.7.41   Phosphatidate cytidyltransferase
2.7.7.49   RNA-directed DNA polymerase, "reverse transcriptase"

2.7.8   **Another substituted phosphate is transferred**
2.7.8.1   Ethanolaminephosphotransferase
2.7.8.2   Diacylglycerol cholinephosphotransferase
2.7.8.11   CDPdiacylglycerol-inositol 3-phosphatidyltransferase
2.7.8.16   1-Alkyl-2-acetylglycerol cholinephosphotransferase
2.7.8.17   N-Acetylglucosaminephosphotransferase

Class 3: Hydrolases (catalyze bond cleavage by hydrolysis)

Sub-class 3.n: *What kind of bond* is **hydrolyzed?**

3.1   **An ester bond is hydrolyzed** (*esterases*)

3.1.1   **In carboxylic acid esters**
3.1.1.3   Triacylglycerol lipase
3.1.1.4   Phospholipase A_2
3.1.1.7   Acetylcholinesterase
3.1.1.13   Cholesterol esterase
3.1.1.17   Gluconolactonase
3.1.1.34   Lipoprotein lipase

3.1.2   **In thioesters**
3.1.2.4   3-Hydroxyisobutyryl-CoA hydrolase
3.1.2.14   Acyl-[ACP] hydrolase

3.1.3   **In phosphoric acid monoesters** (*phosphatases*)
3.1.3.1   Alkaline phosphatase [Zn^{2+}]
3.1.3.4   Phosphatidate phosphatase
3.1.3.9   Glucose-6-phosphatase
3.1.3.11   Fructose-bisphosphatase
3.1.3.13   Bisphosphoglycerate phosphatase
3.1.3.16   Phosphoprotein phosphatase
3.1.3.37   Sedoheptulose-bisphosphatase

3.1.3.43	[Pyruvate dehydrogenase]-phosphatase
3.1.3.46	Fructose-2,6-bisphosphate 2-phosphatase
3.1.3.n	Polynucleotidases

3.1.4   **In phosphoric acid diesters** (*phosphodiesterases*)

3.1.4.1	Phosphodiesterase
3.1.4.3	Phospholipase C
3.1.4.17	3',5'-Cyclic-nucleotide phosphodiesterase
3.1.4.35	3',5'-Cyclic-GMP phosphodiesterase
3.1.4.45	N-Acetylglucosaminyl phosphodiesterase

3.1.21   **In DNA**

3.1.21.1	Deoxyribonuclease I
3.1.21.4	Site-specific deoxyribonuclease (type II), "restriction endonuclease"

3.10.26–7   **In RNA**

3.1.26.4	Ribonuclease H
3.1.27.5	Pancreatic ribonuclease

3.2   **A glycosidic bond is hydrolyzed** (*glycosidases*)

3.2.1   **In O-Glycosides**

3.2.1.1	α-Amylase
3.2.1.10	Oligo-1,6-Glucosidase
3.2.1.17	Lysozyme
3.2.1.18	Neuraminidase
3.2.1.20	α-Glucosidase
3.2.1.23	β-Galactosidase
3.2.1.24	α-Mannosidase
3.2.1.28	α,α-Trehalase
3.2.1.30	N-Acetyl-β-glucosaminidase
3.2.1.33	Amylo-1,6-glucosidase
3.2.1.48	Sucrose α-glucosidase
3.2.1.52	β-N-Acetylhexosaminidase
3.2.2.n	Nucleosidases

3.3   **An ether bond is hydrolyzed**

3.3.1.1	Adenosylhomocysteinase

3.4   **A peptide bond is hydrolyzed** (*peptidases*)

3.4.11   **Aminopeptidases** (*N*-terminal exopeptidases)

3.4.11.n	Various aminopeptidases [Zn^{2+}]

3.4.13   **Dipeptidases** (act on dipeptides only)

3.4.13.n	Various dipeptidases [Zn^{2+}]

3.4.15   **Peptidyl-dipeptidases** (*C*-terminal exopeptidases, releasing dipeptides)

3.4.15.1	Peptidyl-dipeptidase A [Zn^{2+}]

3.4.17   **Carboxypeptidases** (*C*-terminal exopeptidases)

3.4.17.1	Carboxypeptidase A [Zn^{2+}]
3.4.17.2	Carboxypeptidase B [Zn^{2+}]
3.4.17.8	Muramoylpentapeptide carboxypeptidase

3.4.21	**Serine proteinases** (endopeptidases)
3.4.21.1	Chymotrypsin
3.4.21.4	Trypsin
3.4.21.5	Thrombin
3.4.21.6	Coagulation factor Xa, "thrombokinase"
3.4.21.7	Plasmin
3.4.21.9	Enteropeptidase, "enterokinase"
3.4.21.21	Coagulation factor VIIa, "proconvertin"
3.4.21.22	Coagulation factor IXa, "Christmas factor"
3.4.21.27	Coagulation factor XIa, "plasma thromboplastin antecedent"
3.4.21.34	Plasma kallikrein
3.4.21.35	Tissue kallikrein
3.4.21.36	Elastase
3.4.21.38	Coagulation factor XIIa, "Hageman factor"
3.4.21.68	Tissue plasminogen activator
3.4.21.73	Urinary plasminogen activator, "urokinase"
3.4.22	**Cysteine proteinases** (endopeptidases)
3.4.22.2	Papain
3.4.23	**Aspartate proteinases (endopeptidases)**
3.4.23.1	Pepsin A
3.4.23.2	Pepsin B
3.4.23.3	Gastricsin
3.4.23.4	Chymosin
3.4.23.15	Renin
3.4.24	**Metalloproteinases** (endopeptidases)
3.4.24.7	Collagenase
3.4.99	**Other peptidases**
3.4.99.36	Signal peptidase
3.5	**Another amide bond is hydrolyzed** (*amidases*)
3.5.1.1	Asparaginase
3.5.1.2	Glutaminase
3.5.1.16	Acetylornithine deacetylase [Zn^{2+}]
3.5.2.3	Dihydroorotase
3.5.2.7	Imidazolonepropionase
3.5.3.1	Arginase
3.5.4.6	AMP deaminase
3.5.4.9	Methylentetrahydrofolate cyclohydrolase
3.5.4.10	IMP cyclohydrolase
3.6	**An anhydride bond is hydrolyzed**
3.6.1.6	Nucleoside-diphosphatase
3.6.1.32	Myosin ATPase
3.6.1.34	H^+-transporting ATP synthase, "ATP synthase"
3.6.1.35	H^+-transporting ATPase
3.6.1.36	H^+/K^+-exchanging ATPase
3.6.1.37	Na^+/K^+-exchanging ATPase, "Na^+/K^+-ATPase"
3.6.1.38	Ca^{2+}-transporting ATPase

3.7 **A C-C bond is hydrolyzed**
3.7.1.2 Fumarylacetacetase

Class 4: Lyases (cleave or form bonds by other means than oxidation or hydrolysis)
Sub-class 4.n: *What kind of bond* **is formed or cleaved?**

4.1 **A C-C bond is formed or cleaved**

4.1.1 **Carboxy-lyases** (*carboxylases, decarboxylases*)
4.1.1.1 Pyruvate decarboxylase [TPP]
4.1.1.15 Glutamate decarboxylase [PLP]
4.1.1.21 Phosphoribosylaminoimidazole carboxylase
4.1.1.23 Orotidine-5'-phosphate decarboxylase
4.1.1.28 Aromatic-L-amino-acid decarboxylase [PLP]
4.1.1.32 Phosphoenolpyruvate carboxykinase (GTP)
4.1.1.39 Ribulose-bisphosphate carboxylase [Cu], "rubisco"

4.1.2 **Acting on aldehydes or ketones**
4.1.2.5 Threonine aldolase [PLP]
4.1.2.13 Fructose-bisphosphate aldolase, "aldolase"
4.1.3.4 Hydroxymethylglutaryl-CoA lyase
4.1.3.5 Hydroxymethylglutaryl-CoA synthase
4.1.3.7 Citrate synthase
4.1.3.8 ATP-citrate lyase
4.1.3.18 Acetolactate synthase [TPP, Flavin]

4.1.99 **Other C-C lyases**
4.1.99.3 Deoxyribodipyrimidine photo-lyase [FAD], "photolyase"

4.2 **A C-O bond is formed or cleaved**

4.2.1 **Hydro-lyases** (*hydratases, dehydratases*)
4.2.1.1 Carbonate dehydratase [Zn^{2+}], "carbonic anhydrase"
4.2.1.2 Fumarate hydratase, "fumarase"
4.2.1.3 Aconitate hydratase [Fe_4S_4], "aconitase"
4.2.1.11 Phosphopyruvate hydratase, "enolase"
4.2.1.13 Serine dehydratase
4.2.1.17 Enoyl-CoA hydratase
4.2.1.18 Methylglutaconyl-CoA hydratase
4.2.1.22 Cystathionine β-synthase [PLP]
4.2.1.24 Porphobilinogen synthase
4.2.1.49 Urocanate hydratase
4.2.1.61 3-Hydroxypalmitoyl-[ACP] dehydratase
4.2.1.75 Uroporphyrinogen III synthase

4.2.99 **Other C-O lyases**
4.2.99.2 Threonine synthase [PLP]

4.3 **A C-N bond is formed or cleaved**

4.3.1 **Ammonia lyases**
4.3.1.3 Histidine ammonia lyase
4.3.1.8 Hydroxymethylbilane synthase

4.3.2   **Amidine lyases**
4.3.2.1   Argininosuccinate lyase
4.3.2.2   Adenylosuccinate lyase

4.4   **A C-S bond is formed or cleaved**
4.4.1.1   Cystathionine γ-lyase [PLP]

4.6   **A P-O bond is formed or cleaved**
4.6.1.1   Adenylate cyclase
4.6.1.2   Guanylate cyclase

Class 5: Isomerases (catalyze changes within one molecule)

Sub-class 5.n: *What kind of isomerization is taking place?*

5.1   **A racemization or epimerization** (*epimerases*)

5.1.3.1   Ribulose-phosphate 3-epimerase
5.1.3.2   UDPglucose 4-epimerase
5.1.3.4   L-Ribulosephosphate 4-epimerase
5.1.99.1   Methylmalonyl-CoA epimerase

5.2   **A *cis-trans* isomerization**

5.2.1.2   Maleylacetoacetate isomerase
5.2.1.3   Retinal isomerase

5.3   **An intramolecular electron transfer**

5.3.1.1   Triose-phosphate isomerase
5.3.1.6   Ribose-5-phosphate isomerase
5.3.1.9   Glucose-6-phosphate isomerase
5.3.3.1   Steroid Δ-isomerase
5.3.3.8   Enoyl-CoA isomerase

5.4   **An intramolecular group transfer** (*mutases*)

5.4.2.1   Phosphoglycerate mutase
5.4.2.2   Phosphoglucomutase
5.4.2.4   Bisphosphoglycerate mutase
5.4.99.2   Methylmalonyl-CoA mutase [cobamide]

5.99   **Another kind of isomerization**

5.99.1.2   DNA topoisomerase
5.99.1.3   DNA topoisomerase (ATP-hydrolyzing), "DNA gyrase"

Class 6: Ligases (join two molecules with hydrolysis of an "energy-rich" bond)

Sub-class 6.n: *What kind of bond is formed?*

6.1   **A C-O bond is formed**

6.1.1.n   (Amino acid)-tRNA ligases (*aminoacyl-tRNA synthetases*)

6.2   **A C-S bond is formed**
6.2.1.1   Acetate-CoA ligase
6.2.1.3   Fatty-acid-CoA ligase
6.2.1.4   Succinate-CoA ligase (GDP-forming), "thiokinase"

6.3   **A C-N bond is formed**

6.3.1.2   Glutamate-NH_3 ligase, "glutamine synthetase"
6.3.2.6   Phosphoribosylaminoimidazolesuccinocarboxamide synthase (*sorry about that!*)
6.3.3.1   Phosphoribosylformylglycinamidine cyclo-ligase
6.3.3.2   5-Formyltetrahydrofolate cyclo-ligase
6.3.4.2   CTP synthase
6.3.4.4   Adenylosuccinate synthase
6.3.4.5   Argininosuccinate synthase
6.3.4.13   Phosphoribosylamine-glycine ligase
6.3.4.16   Carbamoyl-phosphate synthase (ammonia)
6.3.5.2   GMP synthase (glutamine-hydrolyzing)
6.3.5.3   Phosphoribosylformylglycinamidine synthase
6.3.5.4   Asparagine synthase (glutamine-hydrolyzing)
6.3.5.5   Carbamoylphosphate synthase (glutamine-hydrolyzing)

6.4   **A C-C bond is formed**

6.4.1.1   Pyruvate carboxylase [biotin, Mn]
6.4.1.2   Acetyl-CoA carboxylase [biotin]
6.4.1.3   Propionyl-CoA carboxylase [biotin]
6.4.1.4   Methylcrotonyl-CoA carboxylase [biotin]

6.5   **A P-O bond is formed**

6.5.1.1   DNA ligase (ATP)

Abbreviations

For	monosaccharides, see p. 35
abbreviations	amino acids, see p. 59
of	bases, nucleotides, see p. 78

aa	Amino acid residues
ACE	Angiotensin-converting enzyme (peptidyl-dipeptidase A)
ACP	Acyl carrier protein
ACTH	Adrenocorticotropic hormone (corticotropin)
ADH	Antidiuretic hormone (adiuretin, vasopressin)
ADP	Adenosine 5'-diphosphate
AIDS	Acquired immunodeficiency syndrome
ALA	δ-Aminolevulinic acid
AMP	Adenosine 5'-monophosphate
ANF	Atrial natriuretic factor
ATP	Adenosine 5'-triphosphate
b	Base (of RNA)
bp	Base pair (of DNA)
BPG	2,3-Bisphosphoglycerate
cAMP	3',5'-Cyclic AMP
CAP	Catabolite activator protein
CDK	Cyclin-dependent protein kinase (in cell cycle)
cDNA	Copy DNA
CDP	Cytidine 5'-diphosphate
cGMP	3',5'-Cyclic GMP
CIA	Chemoluminescence immunoassay
CMP	Cytidine 5'-monophosphate
CoA	Coenzyme A
CoQ	Coenzyme Q (ubiquinone)
CTP	Cytidine 5'-triphosphate
d	Deoxy-
dd	Dideoxy-
Da	Dalton (atomic mass unit)
DAG	Diacylglycerol
DH	Dehydrogenase
DNA	Deoxyribonucleic acid
EA	Ethanolamine
EIA	Enzyme-linked immunoassay
ER	Endoplasmatic reticulum
ETF	Electron-transferring flavoprotein
FA	Fatty acid
FAD	Flavin adenine dinucleotide
Fd	Ferredoxin
FFA	Free fatty acid
fMet	Formylmethionine
FMN	Flavin mononucleotide
Fp	Flavoprotein (containing FMN or FAD)
GABA	γ-Aminobutyrate (γ-aminobutyric acid)
GDP	Guanosine 5'-diphosphate
GMP	Guanosine 5'-monophosphate

GSH	Reduced glutathione
GSSG	Oxidized glutathione
GTP	Guanosine 5'-triphosphate
h	hour
HAT medium	Medium containing hypoxanthine, aminopterine, and thymidine
Hb	Hemoglobin
HDL	High-density lipoprotein
HIV	Human immunodeficiency virus
HLA	Human leukocyte-associated antigen
HMG-CoA	3-Hydroxy-3-methylglutaryl-CoA
HMP	Hexose monophosphate pathway
hnRNA	Heterogeneous nuclear ribonucleic acid
HPLC	High-performance liquid chromatography
IDL	Intermediate density lipoprotein
IF	Intermediary filament
Ig	Immunoglobulin
IL	Interleukin
$InsP_3$	Inositol 1,4,5-trisphosphate
IPTG	Isopropylthiogalactoside
kDa	Kilodalton (10^3 atomic mass units)
K_m	Michaelis constant (of an enzyme)
LDH	Lactate dehydrogenase
M	Molarity (mol · l^{-1})
Mab	Monoclonal antibody
MAP kinase	Mitogen-activated protein kinase
MHC	Major histocompatability complex
MPF	Maturation-promoting factor
mRNA	Messenger ribonucleic acid
N	Nucleotide with any base
NAD^+	Oxidized nicotinamide adenine dinucleotide
NADH	Reduced nicotinamide adenine dinucleotide
$NADP^+$	Oxidized nicotinamide adenine dinucleotide phosphate
NADPH	Reduced nicotinamide adenine dinucleotide phosphate
ODH	2-Oxoglutarate dehydrogenase
nm	Nanometer (10^{-9} m)
PAGE	Polyacrylamide gel electrophoresis
Pan	Pantetheine
PC	Phosphatidylcholine
PCR	Polymerase chain reaction
PDH	Pyruvate dehydrogenase complex
PE	Phosphatidylethanolamine
PEG	Polyethylene glycol
PEP	Phosphoenolpyruvate
P_i	Inorganic phosphate
PI	Phosphatidylinositol
pH	pH value
PLP	Pyridoxal phosphate
PQ	Plastoquinone
PRPP	5-Phosphoribosyl-1-diphosphate
PS	Photosystem
Q	Oxidized coenzyme Q (ubiquinone)

QH_2	Reduced coenzyme Q (ubiquinol)
R	Gas constant
rER	Rough endoplasmatic reticulum
RES	Reticuloendothelial system
RFLP	Restriction fragment length polymorphism
RIA	Radioimmunoassay
RNA	Ribonucleic acid
rRNA	Ribosomal ribonucleic acid
S	Svedberg (unit of sedimentation coefficient)
SDS	Sodium dodecylsulfate
sER	Smooth endoplasmic reticulum
sn	Stereospecific numbering
snRNA	Small nuclear ribonucleic acid
SR	Sarcoplasmic reticulum
THB	Tetrahydrobiopterin
THF	Tetrahydrofolate
TLC	Thin-layer chromatography
TPP	Thiamine diphosphate
tRNA	Transfer ribonucleic acid
TRH	Thyroliberin (thyrotropin-releasing hormone)
TSH	Thyrotropin (thyroid-stimulating hormone)
UDP	Uridine 5'-diphosphate
UMP	Uridine 5'-monophosphate
UTP	Uridine 5'-triphosphate
UV	Ultraviolet radiation
V	Maximal velocity (of an enzyme)
VLDL	Very low density lipoprotein

Quantities and units

1. SI base units

Quantity	SI unit	Symbol	Other units
Length	Meter	m	1 yard (yd) = 0.9144 m, 1 inch (in) = 0.0254 m
Mass	Kilogram	kg	1 pound (lb) = 0.4536 kg
Time	Second	s	
Current strength	Ampere	A	
Temperature	Kelvin	K	°C (degree Celsius), °F (degree Fahrenheit)
Light intensity	Candela	cd	
Amount of substance	Mol	mol	

2. Some derived units

Quantity	unit	Symbol	Derivation	Other units
Frequency	Hertz	Hz	s^{-1}	
Volume	Liter	l	10^{-3} m^3	1 Gallon (gal) = 3.785 l
Concentration	Molarity	M	$mol \cdot l^{-1}$	
Molecular mass	Dalton	Da	$1.6605 \cdot 10^{-24}$ g	
Molar mass			$g \cdot mol^{-1}$	
Relative molar mass		M_r	–	
Force	Newton	N	$kg \cdot m \cdot s^{-2}$	
Pressure	Pascal	Pa	$N \cdot m^{-2}$	1 Bar = 10^5 Pa, 1 mm Hg = 133.3 Pa
Energy, work, heat	Joule	J	$N \cdot m$	1 Calorie (cal) = 4.187 J
Power, wattage	Watt	W	$J \cdot s^{-1}$	
Electric charge	Coulomb	C	$A \cdot s$	
Voltage	Volt	V	$W \cdot A^{-1}$	
Enzyme activity	Katal	kat	$mol \cdot s^{-1}$	1 Unit (U) = $1.67 \cdot 10^{-8}$ kat
Specific activity			$kat \cdot (kg\ enzyme)^{-1}$	$U \cdot (mg\ enzyme)^{-1}$
Sedimentation coefficient	Svedberg	S	10^{-13} s	
Radioactivity	Becquerel	Bq	$decays \cdot s^{-1}$	1 Curie (Ci) = $3.7 \cdot 10^{10}$ Bq

3. Multiples and fractions

Factor	Prefix	Symbol	Example
10^{+9}	Giga-	G	1 GHz = 10^9 Hertz
10^{+6}	Mega-	M	1 MPa = 10^6 Pascal
10^{+3}	Kilo-	k	1 km = 10^3 Meter
10^{-3}	Milli-	m	1 mM = 10^{-3} Mol $\cdot$ Liter^{-1}
10^{-6}	Micro-	m	1 mV = 10^{-6} Volt
10^{-9}	Nano-	n	1 nkat = 10^{-9} Katal

Suggested Reading

Selected Textbooks

Alberts, B., Bray, D., Lewis, J., Raff, M., Roberts, K., and Watson, J.D. *The Molecular Biology of the Cell*, 3rd ed., Garland Publishing Inc., New York and London, 1994

Darnell, J., Lodish, H., and Baltimore, D. *Molecular Cell Biology*, 2nd ed., Scientific American Books, New York, 1990

Devlin, T.M. (ed.) *Textbook of Biochemistry*, 3rd ed., John Wiley & Sons, New York, 1992

Mathews, C.K. and van Holde, K.E. *Biochemistry*, 2nd ed., Benjamin/Cummings Publishing Comp., Menlo Park CA, 1996

Moran, L.A., Scrimgeour, K.G., Horton, R.H., Ochs, R.S., and Rawn, J.D. *Biochemistry*, 2nd ed., Neil Patterson Publ. Prentice-Hall, Englewood Cliffs, NJ, 1994

Murray, R.K., Granner, D.K., Mayes, P.A., and Rodwell, V.W. *Harper's Biochemistry*, 23rd ed., Prentice-Hall, Englewood Cliffs, NJ, 1993

Stryer, L. *Biochemistry*, 4th ed., W.H. Freeman and Company, New York, 1995

Voet, D. and Voet, J.G. *Biochemistry*, 2nd ed., John Wiley & Sons, New York, 1995

Selected Reference Works

Branden, C. and Tooze, J. *Introduction to Protein Structure*, Garland Publishing Inc., New York and London, 1991

Lenter, C. (ed.) *Geigy Scientific Tables*, Vols. 1–5, CIBA-GEIGY Ltd., Basel, 1981

Scott, T. and Eagleson, M. *Concise Encyclopedia of Biochemistry*, 2nd ed., Walter de Gruyter, Berlin and New York, 1988

Webb, E.C. (ed.) *Enzyme Nomenclature 1992*, published for IUBMB, Academic Press, San Diego, 1992

Selected Periodicals

(Journals and yearbooks)

Annual Review of Biochemistry, Annual Reviews Inc., Palo Alto USA
Most important collection of biochemical reviews

Current Biology, Current Opinion in Cell Biology, Current Opinion in Structural Biology and related journals in this series, Current Biology Ltd. London
Short articles of topical interest in the cellular and molecular life sciences

Trends in Biochemical Sciences, Elsevier Trends Journals, Cambridge, Great Britain
The "newspaper" for biochemists, official publication of the International Union of Biochemistry and Molecular Biology (IUBMB)

Source Credits

Certain graphic elements in this book are based on the following sources (with permission):

Page	Figure	Source
133	B	Schäfer, G., Biospektrum **1**, 1995, 17–21, Figs. 1 and 3
187	A	Beckman Instruments, Munich, Germany, Bulletin no. DS-555 A, page 6, Fig. 6
189	B	Goodsell, D.S., Trends Biochem. Sci. **16**, 1991, 203–206, Fig. 1a and b
191	A	Stryer, L., *Biochemistry*, Freeman and Co., New York, 1988, p. 927, Fig. 36–13
193	C	Stryer, L., *Biochemistry*, Freeman and Co., New York, 1988, p. 945, Fig. 36–47 and
		Alberts, B. et al., *The Molecular Biology of the Cell*, Garland Publ. New York and London, 1989, p. 663, Fig. 11–73B and p. 634, Fig. 11–36
201	A	Darnell, J. et al., *Molecular Cell Biology*, 2nd ed., Freeman and Co., New York, 1990, p. 689, Fig. 18–7
241	B	*The NEB Transcript*, New England Biolabs, Beverly MA, USA
255	A, B	Voet, D. and Voet, J.G., *Biochemistry*, John Wiley and Sons, New York, 1990, p. 305, Fig. 11–45 and p. 306, Fig. 11–47
265	A	Voet, D. and Voet, J.G., *Biochemistry*, John Wiley and Sons, New York, 1990, p. 1097, Fig. 34–13
273	A	Voet, D. and Voet, J.G., *Biochemistry*, John Wiley and Sons, New York, 1990, p. 1112, Fig. 34–33
307	B	Voet, D. and Voet, J.G., *Biochemistry*, John Wiley and Sons, New York, 1990, p. 1126, Fig. 34–55
315	B	Darnell, J. et al., *Molecular Cell Biology*, 2nd ed., Freeman and Co., New York, 1990, p. 923, Fig. 23–26
359	A	Darnell, J. et al., *Molecular Cell Biology*, 2nd ed., Freeman and Co., New York, 1990, p. 418, Fig. 11–24b
363	B	Rossmann, M.G. *et al.*, Curr. Biol., **2**, 1992, 86–87, Fig. 8

Index

Page numbers in **bold** type indicate a molecular structure or the definition of a particular concept. An asterisk (*) indicates a metabolic chart.

A
A band, 306
Abscisic acid, 50
Absorbance, 99
  of NAD⁺/NADH, 99
Absorption
  coefficient, 99
  photometry, 98
  of light-absorbing pigments, 325
Absorptive state, 279
Acceptor site, 228
A cells of pancreas, 280
Acetaldehyde, 294, 370*
Acetate (acetic acid), **45**, 129, 276, 295
Acetate-CoA ligase (*6.2.1.1*), 295, 321
Acetoacetate (acetoacetic acid), 154, 256, **285**, 369*-371*
Acetoacetyl-CoA, **285**, 369*
2-Aceto-2-hydroxybutyrate, 372*
2-Acetolactate, 372*
Acetone, **285**, 369*
Acetylcholine, 308, **321**, 322, 324
  mechanism of action, 350
  metabolism, 321
  receptor for, 308, 320
  synthesis, 320
Acetylcholinesterase (*3.1.1.7*), 214, 321
Acetyl-CoA, 8, **50**, 102, 125, 127, 129, 145, 154, 166, 170, **285**, 295, 368*-370*
  from amino acids, 155
  as building block for isoprenoids, 51
  from citrate, 129
  from ethanol, 285
  from fatty acids, 145, 147
  from pyruvate, 125
  as substrate for cholesterol biosynthesis, 171
  as substrate for citric acid cycle, 127, 155
  as substrate for fatty acid biosynthesis, 145, 167
  as substrate for ketone body synthesis, 285
Acetyl-CoA C-acyltransferase, (*2.3.1.16*), 147, 285, 369*, 370*
Acetyl-CoA carboxylase (*6.4.1.2*), 145, 151, **167**, 278
Acetyl-D-galactosamine, **35**

Acetyl-D-glucosamine, **35**
N-Acetyl-L-glutamate, 286, 373*
N-Acetyl-L-glutamate 5-phosphate, 373*
N-Acetyl-L-glutamate semialdehyde, 373*
N-Acteylhexosaminidase (*3.2.1.52*), 185
N-Acetylmuramic acid, 36
N-Acetylneuraminic acid, **35**, 48
N-Acetylornithine, 373*
Acetylsalicylate (aspirin), 355
Acid, **24**
  C₈, 200
  balance, 296
  constant (Kₐ), 24
  production in the stomach, 246
Acid-amide bond, 8, 64
Acid anhydride, 8
  mixed, 140, 226
Acid-base
  concept, **24**
  balance, 250, 257, 296
  catalysis, 88, 96
  reaction, 24
Acid phosphatase (*3.1.3.2*), 212
Acidic glycoprotein, 253
Acidosis, 256, 300
ACP (acyl carrier protein), **167**, 332
Aconitase (aconitate hydratase), (*4.2.1.3*), 127
Aconitate, 126
Aconitate hydratase (*4.2.1.3*), 127
Acquired immune deficiency syndrome (AIDS), 362
Acrylamide, 252
Actin, 63, 151, **191**, **306**, 308
  associated proteins, 190, 192
  myosin cycle, **306**, 308
Actinin, 190, 306
Actinomycin, 230, 233
Action potential, 308, **319**, 320, 322
Activated groups
  amino acids (aminoacyladenylate), 226
  acetic acid (acetyl-CoA), 50, **147**, 285
  glucuronate (UDP-glucuronate), 180, **291**
  sulfate (PAPS), **291**
Activation energy, **16**, 18, 88
Active center, **88**, 96

Active transport, 206
Acylcarnitine, **199**
Acyl carrier protein (ACP), **166**, 332
Acyl-CoA, 131, 145, **147**, 197, 198, 368*
Acyl-CoA dehydrogenase (*1.3.99.3*), 147
Acylglycerol, 249
Acylglycerol phosphate, 368*
Acylglycerone phosphate, 368*
Acyl residue, **47**, 102
  transport into mitochondria, 199
Adenine, **79**, 157
  as neurotransmitter, 322
Adenine phosphoribosyltransferase (*2.4.2.7*), 157
Adenohypophysis, 343
Adenosine, 78, 323
Adenosine diphosphate (ADP), 78, 105, 119, 323
Adenosine 3',5'-monophosphate (cyclic AMP), **109**, 110, 163, 337, 350, **353**
Adenosine 5'-monophosphate (AMP), **79**
  biosynthesis of, 176, 374*
  degradation of, 156
  in muscle metabolism, 310
Adenosine triphosphate, *see* ATP
S-Adenosylhomocysteine, 323
S-Adenosylmethionine, 323, 371*
Adenylate cyclase (*4.6.1.1*), 111, 350, 353
Adenylate kinase (myokinase, *2.7.4.3*), 310
Adhesion molecules, 214
Adenylosuccinate, 374*
Adipocytes, 46, 145
Adipose tissue, 279, 281
Adiuretin, 304
A-DNA, 84
ADP (adenosine diphosphate), 78, 105, 119, 323
ADP/ATP translocator, 198
Adrenal glands, 341
  cortex, 304, 341
  medulla, 343
Adrenaline, *see* Epinephrine
Adrenocorticotropin (ACTH, corticotropin), 348
Aerobic oxidation of glucose, 137
Affinity, 90
Agarose, 37
Agar agar, 36
Agonist, hormone, 320, 346
AIDS (acquired immune deficiency syndrome), 362

Alanine, **59**, 68, 153, 172, 310
  biosynthesis of, 173, 373*
  cycle, 283, 310
  degradation of, 155, 370*
  in muscle metabolism, 310
β-Alanine, 102, 155, 157, 332
Alanine transaminase (*2.6.1.2*), 151, 153, 370*, 373*
Albumin, 62, 144, **252**, 253
  complex with fatty acids, 145
  transport function of, 145, 180
  in serum, 63
Alcohol, 8, 294
  metabolism of, 294
  primary, **9**
  secondary, **9**
Alcoholism, 294, 328
Alcohol dehydrogenase (*1.1.1.1*), 139, **295**
Alcohol oxidase, 295
Aldehyde, **9**
Aldehyde dehydrogenase (*1.2.1.3*), 295
Aldehyde reductase (*1.1.1.21*), 283
Aldimine, 5, 102, 122, **153**
Alditol, 34
Aldolase (*4.1.2.13*), 141, 151
Aldose, **30**, 34
Aldosterone, **55**, 302, 304, 341, **345**, 369*
Alkali metals, 2
Alkaline earth metals, 2
Alkaline phosphatase (*3.1.3.1*), 214
Alkalosis, 256, 300
Alkane, 42
Alkanol, 42
Alkene, 5
Alkylating compounds, 234
Allantoin, **157**
Allantoic acid, 157
Allelic variability, 270
Allolactose, 109
Allopurinol, **157**
Allosteric enzyme, 107, 112
Allosteric protein, 260
Allosteric regulation, 107, **112**, 260
Allotypic variation, 271
Alpha oxidation (α-oxidation), 148
Alternative splicing, 270, 314
α-Amanitin, 222
Amide, 8

Amines, 8
  biogenic, 154, 342
  primary, **9**
  secondary, **9**
  tertiary, **9**
Amine oxidase (*1.4.3.4*), 154, 155, 296
Amino acids, **56**, 152–155, 189, 278, 279, 286
  abbreviations of, 58
  activation of, 218, **226**
  analysis of, **60**
  aromatic family of, 173
  biosynthesis of, **172**, 372*, 373*
  degradation of, **154**, 370*, 371*
  derived hormones, 342
  deamination of, 152
  in digestion, 242
  essential, 58, **172**, 326, 372*
  excretion of, 298
  glucogenic, **154**, 158, 326
  ketogenic, **154**, 326
  metabolism of, 150–154, 172, 174, 370*–373*
    biosynthesis of, **172**, 372*, 373*
    in brain of, 316
    degradation of, **154**, 370*, 371*
    in kidneys, 296, 302
    in liver, 274, 276, 278
    in muscle, 310
  as neurotransmitters, 322
  plasma concentration of, 251
  phenylthiocarbamoyl (PTC), 60
  polarity of, 58
  properties of, 56
  proteinogenic, 56, 58, **59**, 226
  resorption of, 242, 296
  sulfur-containing, 256
  stereochemistry of, 57
  structural classes, 58
  transamination of, 152
  transport systems for, 242
  in urine, 298
Amino-acid decarboxylases (*4.1.1.n*), 154, 323
Amino-acid oxidases (*1.4.3.n*), 200, 296
Amino acid tRNA ligases (*6.1.1.n*), 227
Aminoacyladenylate (aminoacyl-AMP), **227**
Aminoacyl-tRNA, **227**
Aminoacyl-tRNA synthetases (*6.1.1.n*), 188
4-Aminobenzoate, 332
4-Aminobutyrate (GABA), 154, 155, 316, **317**, **322**
4-Aminobutyrate transaminase (*2.6.1.19*), 317

Amino group, **9**
β-Aminoisobutyrate, 156, 157
5-Aminolevulinate (ALA), 178, **179**
5-Aminolevulinate synthase (*2.3.1.37*), 148, 179
Aminopeptidases (*3.4.11.n*), 150, 151, 245
Aminopropanol, 154, 155
Aminopterin, 274
Amino sugars, 34, 40, 314
Aminotransferases, 152
Ammonia, **9**, 25, 27, 115, 153, 154, 172, 287, 301
  excretion of, 298, 300
  in kidneys, 26
  metabolism of, 153
  plasma concentration of, 251
  synthesis from N_2, 173
Ammonium ions, **5**, 25, 27, 301
  excretion of, 299, 300
Ammonotelic organisms, 286
AMP (adenosine-5'-monophosphate), 78, **79**
  in muscle, 310
AMP deaminase (*3.5.4.6*), 311
Amphibolic reactions, **129**
Amphipathic helix, 74, 217
Amphipathic molecules, **22**, 202, 288
Amphiphilic molecules, 254
Ampicillin, **233**
Amplification
  of DNA, 240
  in hormone action, 352
α-Amylase (*3.2.1.1*), 138, 139, 244, 245
Amylopectin, 37, **39**
Amyloplast, 37, 39
Amylose, 37, **39**
Amylo-1,6-glucosidase (*3.2.1.33*), 161
Anabolic steroids, 311
Anabolic reactions, **104**, 129
Anaerobic glycolysis, 282
Anaplerotic reactions, **128**, 129
Anchor, lipophilic, 214
Anchor, proteins, 314
Androstendione, 344, 345, 369*
Anemia, 328, 304
  megaloblastic, 332
  pernicious, 334
Angiotensin, 304, 305, **323**
Angiotensin-converting enzyme (*3.4.15.1*), 304
Angiotensinases, 304
Angiotensinogen, 304, **305**

Angular methyl groups, 52, 340
Angular velocity, 186, 187
Anhydride
    mixed, 8, **9**
    phosphoric acid, **9**
Anomers, **32**
Antagonist, 320
    hormone, 346
    antibiotics, **232**
Anthranilate, 372*
Antibiotics, **232**
Antibodies (immunoglobulins), 244, 245,
    250, 252, 264, 268, 275, **267**, 320
    biosynthesis of, 270
    diversity of, 266, 270
    monoclonal, **274**
    serum concentration of, 267
    structures of, 267, 269
    variants, 270
Anticodon, **82**, 84, 226
Antidiuretic hormone (ADH), 302
Antigen, 264, 266, 270, 272
    tumor-associated, 360
Antigen-antibody complex, 274
Antigenic determinants, 274
Antihemophilic factor, 263
Anti-oncogene, 358, 360
α_2-Antiplasmin, 262, 263
Antiport, **199**, 207
Antiserum, 264
Antitrypsin, 253
Apatite, 328
Apolipoproteins, 254, 284
Arabinan, 37
Arabinogalactan, 38
Arabinose, 34, 35
Arachidonate (arachidonic acid), **45**, 327, **341**,
    352, **355**
    precursor of the eicosanoids, 340
Arachidonate lipoxygenases (*1.3.11.n*), 355
Archaea, 182
Arginase (*3.5.3.1*), 287, 371*
Arginine, **59**, 96, 172, 287
    biosynthesis of, 173, 286, 373*
    degradation of, 155, 286, 371*
Argininosuccinate, 286, 287
Argininosuccinate lyase (*4.3.2.1*), 287
Argininosuccinate synthase (*6.3.3.5*), 287
Aromatic family of amino acids, 173
Aromatic compounds, **4**, 94
Aromatization, 344, 369*

Arteriosclerosis, 54, 198
Ascites fluid, 274
Ascorbic acid (ascorbate, vitamin C), 312, 322,
    335
    function, 312, 322, 334
Asparagine, **59**, 172
    biosynthesis of, 173, 373*
    degradation of, 155, 371*
Aspartate (aspartic acid), 27, **59**, 60, 113, 153,
    159, 172, 174, **287**, 316, 323
    biosynthesis of, 173, 373*
    degradation of, 155, 371*, 372*
    family, 173
    as neurotransmitter, 316, 323
Aspartate carbamoyltransferase (*2.1.3.2*), **113**,
    375*
Aspartate transaminase (*2.6.1.1*), 153, 199,
    287, 317, 371*, 373*
Aspartate proteinases (*3.4.23.n*), 151, 304
Aspartate 4-semialdehyde, 372*
Aspartyl 4-phosphate, 372*
Aspirin (acetylsalicylate), 354
ATP (adenosine triphosphate), 78, 105, 113,
    115, 119, 163, 197, 310, **323**
    daily production of, 134
    energy conservation in, 116
    formation of, 116, 132, 140, 146
    free energy of hydrolysis, 115
    in muscle, 310
    in Na⁺-K⁺ transport
    level, 310
    requirement, 310
    structure of, **115**
    transport of, 199
    yield from glucose, 137
    yield from palmitate, 147
ATP/ADP translocase, 199
ATPases, 206, 208
    Ca^{2+}-transporting (*3.6.1.38*), 303, 308
    H^+/K^+ exchanging (*3.6.1.36*). 247
    H^+-transporting (*3.6.1.35*), 301
    myosin (*3.6.1.32*), 306, 311
    Na^+/K^+ exchanging (*3.6.1.37*), 185, **208**, 260,
    301, 303, 317, 318
ATP citrate lyase (*4.1.3.8*), 129
ATP synthase (*3.6.1.34*), 117, 119, 130, **133**,
    197
Atrial natriuretic peptide (ANP), 302
Atropin, 320
Autoimmune disorder, 272, 320
Autocrine hormone action, 338, 354

Autotroph, 104
Avidin, 334
Avitaminosis, 330
Axial position, **30**, 52
Axon, 316, 322

B
Bacteria, 182, 188, 264
  cell, 139, 189
  cell wall, 92, 208, 233
Bacteriophages, 238, **362**
  M13, 238, 362
  T4, 362
Bacteriorhodopsin, 117, **122**, 123
Ball-and-stick model, 6, 7
Bases, **24**
  abbreviations, 78
  nucleoside, nucleotide, 78
Base pairing, **80**, 82, 220, 226, 228, 230
B cells
  in the immune system, 264, 266
  maturation of, 270
  in pancreas, 164, 280
B-DNA, 80, 84, 85
Beer, 138, 294
Beet sugar, 36
Behenic acid, **45**
Benzene, **5**, 42
Benzo(a)pyrene, 234
  mutagenic derivative of, **235**
Beriberi, 332
Beta oxidation (ß-oxidation), 126, 144, 146, **147**, 148, 196, 280, 281
Beta-pleated sheet (ß-pleated sheet), 68, **69**
Beta turn (ß-turn), 66
Bicarbonate (hydrogen carbonate), 300
Bilayer, *see* Lipid bilayer
Bile, 54, 170, 180, 242, **244**, 248, 276, 288
  steroid metabolites in, 344
Bile acids, 46, 52, **55**, 170, 284, **289**
  biosynthesis of, coenzyme, 334
  metabolism of, **288**
  primary and secondary, 54, 288
Bile pigments, **181**, 245, 288, 291, 298
Bile salts, 242, 245, 248, **289**
Bilirubin, **181**, 290
  anemia, 180
  direct and indirect, 180
  conjugated, 180
Biliverdin, 180

Biliverdin reductase (*1.3.1.24*), 180
Biogenic amines, 154, 342
  as neurotransmitters, 322
Biological membrane, *see* Membrane
Biotin, **103**, 148, **335**
Biotransformation, 276, **290**
1,3-Bisphosphoglycerate, 120, **141**, **159**, 260, 366*, 367*
2,3-Bisphosphoglycerate, 140, **261**
Bisphosphoglycerate mutase (*5.4.2.4*), 261
Bisphosphoglycerate phosphatase (*3.1.3.13*), 261
Blood, 250–263
  buffer systems, 26
  cellular constituents of, 251
  clotting of, 250, 262
  constituents of, 251
  functions of, 251
  gas transport by, 258
  glucose, 34, **251**, 280, 282, 338
  lipids in, 250, 278
  lipoproteins in, 254
  pH of, 250, 256
  plasma, 252
  plasma pH value, 27, 250, 252, **256**
  plasma proteins, 252
Blood-brain barrier, 180, 316
Blood cells, 250, 330
Blood platelets, 250
Blood pressure
  and eicosanoids, 354
  increase of, 304
Blood sugar, 34, **251**, 280, 282, 338
  determination of, 98
  role of the liver, 276
  control of level, 158, 160
Blood vessels, 262, 312
  angiotensin action on, 304
Boat conformation, 52
Bohr effect, 260
Bond, **5**
  acid-amide, 64
  angle, 64
  chemical, 4
  dipole moment, 6
  double, **4**, 5
  single, **4**, 5
  "energy-rich", 9
  glycosidic, 36
    N-glycosidic, 40, 78
    O-glycosidic, 40

length , 6
polarity, **6**
Bone marrow, 264, 360
  heme biosynthesis in, 178
  stem cells, 304
Bones, 312, 314, 328
  mineralization, 330
Boron, 328
2,3-BPG, *see* 2,3-Bisphosphoglycerate
Brain, 280, **316**
  angiotensin action on, 304
  glucose requirement of, 280
  metabolism of, 316
  nutrient supply, 280
Branched-chain dehydrogenase complex,
  124, 370*
Branching enzyme, 161
Brassinosteroids, 50
Brunner glands, 244
Buffers, **26**
  capacity, 26
  physiological, 26
  plasma, 256, 300
  systems, **26**, 256
  in urine, 300
Buoyant density, 186
Butyric acid, **45**

C
C_1 metabolism, 375*
  coenzymes, **103**, 333
Caffeine, 352
Calcidiol, 305
Calcidiol-1-monooxygenase (*1.14.13.13*),
  305
Calciol (cholecalciferol, vitamin D), **331**
Calcitonin, 302, 330
Calcitriol, **55**, 302, **305**, 330, 340, **345**
Calcium, 328
  absorption of, 302
  channels, 308, 320
  daily requirement, 329
  excretion of, 299, 302
  homeostasis, 304
  metabolism and calcitriol, 331
  pumps, 308, 352
  salts, 288
  storage of, 196, 199, 351, 353
  active transport of, 199, 208
Calcium ion, 41, 135, 197, 263, 302, 308, 310,
  350

concentration of, 198, 308, 352
  role in exocytosis, 321
  role in hormone secretion, 348
  mechanism of action, 352
  as second messenger, 337, **352**
  role in sight, 324
Calcium oxalate, 298
Calcium phosphate, 298
Calmodulin, 151, 192, **353**, 358
Calorie (cal), 10, 11
Calorimetry, 14
Calsequestrin, 308
Calvin cycle, 121, 366*
N-CAM (cell adhesion molecule), 214
cAMP (cyclic adenosine 3',5'-mono-
  phosphate), **109**, 110, 163, 337, 350, **353**
  mechanism of action, 352
  phosphodiesterase (*3.1.4.17*), 111, 353
Camphor, **50**
Cancer cells, 360
Cane sugar, 36
Cap, 222, **224**
CAP (catabolite activator protein), 63, **109**
Capric acid, **45**
Caproic acid, **45**
Caprylic acid, **45**
Capsid, 364
N-Carbamoylaspartate, **113**, 174, 375*
N-Carbamoyl ß-isobutyrate, 157
Carbamoylphosphate, **113**, 174, **287**, 375*
Carbamoyl phosphate synthase (*6.3.4.16*),
  287
Carbohydrates, **30**
  digestion of, 242
  dietary, 326
  in membranes, 202
  metabolism of, 140, 142, 158, 160, 278, 280,
  367*
    in liver, 276, 278
    regulation of, 162
Carbon, 3
  electronic configuration, 2
  fixation, 120
Carbon dioxide (CO_2), 105, 121, 127, 129, 143,
  250, 301, 366*
  transport of, 258
Carbon dioxide-bicarbonate buffer, 256
Carbonic acid, 246, 256, 286, 301
  pK_a value, 257
Carbonic anhydrase (carbonate dehydratase)
  (*4.2.1.1*), 246, 256, 301

Carbon monoxide, 180
Carbonyl group, **9**
γ-Carboxyglutamate, 262
Carboxyl group, **9**
Carboxylation, 262
  coenzyme of, 334
  of glutamyl residues, 330
Carboxylic acids, 8, **45**
Carboxylic acid amides, 9
Carboxylic acid esters, **9**, 19
Carboxypeptidases (*3.4.17.n*), 150, 245
Carcinoembryonic antigen, 361
Carcinogens, 234, 360
  metabolism of, 290
Cardiac glycosides, 52
Cardiolipin, 48, 196, 205
Carnitine, 144, **198**
  shuttle, 144
Carnitine *O*-palmitoyltransferase, (*2.3.1.21*), 144
β-Carotene, 330
Carotenoids, 42
Carrageenan, 37
Carrier, 135, 196, **198**, 286, 342
  of hormones, 252, 336, 342, 346
  proteins in plasma, 252
Cartilage, 314
Cascades
  in blood clotting, 262
  in glycogen metabolism, 110
  in immune response, 264
  in signal transduction, 350–355
Catabolic, 104, 129
Catabolism
  of carbohydrates, 140–143
  of lipids, 144–149
  of nucleic acids, 156–157
  of proteins, 150–155
Catabolite activator protein (CAP), 63, 109
Catabolite repression, 106
Catalase (*1.11.1.6*), 180, 185, **201**, 294
Catalysis
  enzymatic, 88
  covalent, 88
  acid-base, 88
Catalyst, **18**
Cataract, 164, 282
Catecholamine, **322**, 342
  biosynthesis, 322
  excretion, 298
cDNA (copy DNA), 238

CDPcholine, 168, 368*
CDPdiacylglycerol, 168, 368*
CDPethanolamine, **168**, 368*
Cells, **182**, 188
  aggregation of, 354
  animal, 183
  bacterial, 189
  composition of, **189**
  cycle, **356**, 360
  disruption of, 184
  division, 192, 194, 357
    control of, 360
  fractionation, 185
  fusion, 274
  growth, 357
  hybrids, 274
  movement, 192
  multiplication, 360
  proliferation, 358, 360, 361
  structure of, 182
  volume, 188
Cell adhesion molecules, 272
Cell membrane, 170, 202–209, 216, 350
  lipids, 42, 204
Cell surface, 266
  receptors, 314
Cellular organelles, 184
Cellulose, 34, 37, 38, 242
Cell wall, 119, 182, 209
  of plants, 38
Center of chirality, 30, **56**
Centrifugation, 186
Centrifugal force, 184, **186**
Centrosome, 192
  formation in the cell cycle, 357
Cephalin (phosphatidylethanolamine), **48**
Ceramides, 48, 368*
Cerebrosides, 42, 368*
Ceruloplasmin, 253
Cesium chloride, 186
cGMP (cyclic guanosine 3',5'-mono-phosphate), 324, 337, 350, 352
cGMP phosphodiesterase (*3.1.4.35*), 324
Chain termination method, 238
Chair conformation, 30, 52
Channel, *see* Ion channel
Chaperones, 214
Chemical bonds, 4
Chenodeoxycholate (chenodeoxycholic acid), **55**, **289**
Chief cells, 246

Chirality, **56**
Chitin, 37
Chloramphenicol, 230
Chlorine, 3, 328
  electronic configuration, 2
Chloride ions, 246, 302
  channel, 350
  concentrations of, 319
  daily requirement of, 329
  deficiency of, 328
  excretion of, 299
  loss of, 328
  permeability coefficient of, 319
  transport of, 312
Chlorophyll, 50, **119**, 120, 122
Chloroplasts, 39, 119, 182, 366*
Cholate (cholic acid), **55**, **289**
Cholecalciferol (calciol, vitamin D), **331**
Cholecystokinin, **323**
Cholestane, **52**
Cholesterol, 42, 50, **53**, **55**, 63,171, 202, 205,
  245, 254, 276, **285**, 288, **345**, 369*
  acyl esters, 254, 284, 344
  biosynthesis of, **170**, 276
  metabolism of, 254, 276, **284**, 288
  plasma concentration, 251
  as precursor of bile acids, 288
  as precursor of steroid hormones, 340,
  344
  stereochemistry of, 53
  transport of, 254, 276
Cholesterol esterase (*3.1.1.13*), 245
Choline, **48**, 131, 204, 320, 368*
Cholinergic synapse, 320
Choline-*O*-acetyltransferase (*2.3.1.6*), 321
Choline esterase (*3.1.1.8*), 253
Cholic acid (cholate), **55**, **289**
Chondroitin 6-sulfate, **315**
Chorion gonadotropin (hCG), 298
Chorismate, 372*
Christmas factor (*3.4.21.22*), 263
Chromium, 328
Chromatin, 194, **218**
  and cell cycle, 356
Chromatography, 52
  ion exchange, 60
  partition, 60
  reversed phase, 60
  thin-layer, 53
Chromogen, 274
Chromosomes, 194, 218

Chylomicron, 242, **254**, 279, 284
  remnants, 254, 284
Chymosin (*3.4.23.4*), 245
Chymotrypsin (*3.4.21.1*), 245
Cirrhotic liver, 295
cis-active element, 224
cis-configuration, 52
cis-trans isomerization, **4**, 44, 122
Citrate (citric acid), 94, **127**, **129**, 135, 144, 163,
  256, 294, 316
Citrate lyase (*4.1.3.7*), 127, 128, 135
Citrate synthase (*4.1.3.7*), 127, 128, 135
Citric acid cycle, **126**, 144, 154, 196, 286
  and amino acid biosynthesis
  anaplerotic pathways, 128
  and ethanol metabolism, 294
  metabolic functions, 128
  reactions, **127**
  regulation of, 134
  role in calcium uptake, 328
Citronellol, **50**
Citrulline, **287**
Clathrate structure, 22
Clathrin, 212, 214, 254
Clearance, renal, 296
Clonal selection, 264
Clone, 236, 270
Cloning, 236, 274
  of cells, 274
  of DNA, **236**
CMP (cytidine monophosphate), 157
CO_2 (carbon dioxide), 166
  fixation, 120
CoA, *see* Coenzyme A
Coagulation factors, 253, 263
Coagulation pathway, 263
Coated pits, 254
Cobalamin (coenzyme B_{12}), 102, **335**
Cobalt, 328, 334
  daily requirement of, 329
Code, genetic, **250**
Coding strand of DNA, 82, 222, 238
Codon, **82**, 226
  start, 228
  stop, 228
  synonymous, 226
Coenzyme
  A, 78, **103**, 124, 126
    biosynthesis of, 332
  B_{12}, 148
  Q (ubiquinone), 50, **101**, 130, 131

Coenzymes
  group-transferring, **103**
  redox, **101**
  vitamins as precursors of, 330
Colchicine, 191
Colipase, 46, 242, 245, 249
Collagen, 62, 68, 263, 312–315, **313**, 326
  helix, 67, 68
  structure of, 69, **313**
  synthesis of, 313
  triple helix, 66, 68
Colloidal osmotic pressure, 252
Colon, 288
Compartmentation, 182, 184, 188
Competitive inhibition, 92
Complementary DNA, **80**, 220
Complement system, 266
Cones (eye), 324
Configuration, 30, 34, **56**
Conformation, **6**, **70**, 52
  R and T state, 113
Conjugates
  formation of, **290**, 345
  of bile acids, 288
  of bilirubin, 180
  in urine, 298
Connective tissue, 312, 314
Constant region of immunoglobulins, 266, 272
Contact inhibition, 360
Contraction, muscular, 306, 308
Control
  of enzymatic activity, 106
  of gene expression, 108
Control element, 108, 224
Cooperative interaction, 260
Copper, 132, 328
Coproporphyrinogen, 179
Copy DNA (cDNA), 238
Cori cycle, **283**, 310
Corrin, 334
Corticosteroids, 55, 345
  and eicosanoid metabolism, 354
Corticotropin, 304, 348
Cortisol, **55**, 150, 159, 163, 302, 311, 340, **345**, 354
  action of, 162
  biosynthesis of, 344, 369*
  hormonal axis of, 338
  metabolism of, 345
  plasma level of, 339

Cosubstrate, 100
Cotransport, 242
Cotton, 38
Coupling
  electromechanical, 308
  energetic, **10**, 11, 102, 104, 114, 135
Covalent radius, 6, 7
C peptide (in proinsulin), 70, 164, 210
C-reactive protein, 253
Creatine, **117**, **299**, **311**
  excretion of, 299
Creatine kinase (*2.7.3.2*), 117, 197, 311
Creatine phosphate, 117, **311**
Creatinin, 298, **311**
  excretion of, 296, 299
  plasma concentration of, 251
Cristae of mitochondria, 196
Crotonyl-CoA, 371*
C-terminus, 64, 70
CTP (cytidine triphosphate), 113
CTP synthase (*6.3.4.2*), 177, 375*
Curare, 356, 357
Cyclic AMP, *see* cAMP
Cyclic GMP, *see* cGMP
Cyclins, 357
Cyclin-dependent kinases, 356
Cyclic phosphorylation, 120
Cyclic 3',5'-AMP, *see* cAMP
Cyclic 3',5'-GMP, *see* cGMP
Cyclooxygenase (*1.14.99.1*), 354
Cystathionine, 371*, 373*
Cysteamine, 102, 155
Cysteine, **59**, 166, 172, 327
  biosynthesis of, 173, 373*
  degradation of, 155, 371*
Cysteine proteinases (*3.4.22.n*) 151
Cysteinesulfinate, 371*
Cystine, 58, 298
Cystinuria, 296
Cytochalasine, 191
Cytochromes, 100, 132, 180,
  a, 131
  b/f complex, 119
  c, 100, 117, 130, 132, 178
  P450 enzymes, 170, 290, **292**
Cytochome-c oxidase (*1.9.3.1*), 116, 131, 151, 185
Cytokeratin, 190
Cytokinin, 50
Cytoplasm, 184, **188**
  biochemical functions of, 188

in gluconeogenesis, 158
pH value of, 26, 27
Cytoplasmic pathway of protein sorting, 216
Cytosine, **79**
5-methyl derivative, 78
Cytoskeleton, 182, 184, **190**-193, 306
anchoring of, 204
building blocks of, 190
functions of, 192
structure of, 192
in tumor cells, 360
Cytosol, 185, 188
Cytostatic agents, 176
Cytotoxic T-cells, 264

D
Daunomycin, 230
Dealkylation, 290
Deamination, **152**, 154, 290, 300
eliminating, 152, **154**
hydrolytic, 152, 154
oxidative, 152
Debranching enzyme (4α-glucanotransferase, 2.4.1.25), 161
Decarboxylases (4.1.1.n), 155
Decarboxylation
of amino acids, 102
in citric acid cycle, 27
in fatty acid synthesis, 145
in gluconeogenesis, 159
in hexose monophosphate pathway, 143
role of pyridoxal phosphate, 102
oxidative, 124
role of thiamin (vitamin B_1), 332
Dedifferentiation, 360
Defense systems, 250, 264
7-Dehydrocholesterol, 330
Dehydroepiandosterone, 369*
Dehydrogenases, 87, 100
Dehydroxylation, 288
Deletion, 358
Denaturation, 72
Dendrites, 316
Density, 187, 254
Density-gradient centrifugation, 186
5'-Deoxyadenosylcobalamine, 148, 335
Deoxyaldose, 34
Deoxycholic acid, **55**, 289
Deoxycorticosterone, 369*
Deoxycortisol, 369*

Deoxyribonucleoside, 78
biosynthesis, 176, 374*, 375*
Deoxyribonuclease (3.1.21.1), 245
Deoxyribonucleic acid (see DNA), **78**, 80
Deoxyribonucleoside triphosphates, 220
Deoxyribose, **35**, 78
Deoxythymidine triphosphate (dTTP), 176, 375*
Dephosphorylation, 356
Depolarization, 308, 318
Dermatan sulfate, **315**
Desmin, 190, 306
Desulfatation, 290
Detergent effect, 288
Detoxification, 200, 276, **290**
Development, 364
Developmental genes, 357
Dextran, 37
D hormone (calciol, see also Vitamin D), 304, 330, 340
Diabetes mellitus, 76, 144, **164**, 256, 298, 300
insulin-dependent, 164
lipid metabolism in, 284
non-insulin-dependent, 164
Diacylglycerol (DAG), **46**, 48, 168, 248, 350, 352, 368*
Diacylglycerol 3-phosphate (phosphatidate), 49
Dicoumarol, 264
Dideoxynucleoside triphosphates, 238
Dielectric constant, 20
Dietary
deficiencies, 280
fats, 46, 168, 244, 254
Differential centrifugation, 184, 186
Differentiation, 270, 358, 364
Diffusion, 206
Digestion, **242–249**
enzymes, 242, 244
secretions, 242, **244**
of lipids, 288
process, **246, 248**
Dihydrofolate (DHF), 375*
Dihydrofolate reductase (1.5.1.3), 177, 274, 375*
Dihydrogen phosphate, **9**, 27
Dihydrolipoamide, 131
Dihydrolipoamide acetyltransferase (2.3.1.12), 124
Dihydrolipoamide dehydrogenase (1.8.1.4), 125

Dihydroorotate, 131, 174, 375*
Dihydrothymine, 157
Dihydrouracil, 157
Dihydrouridine, 226
Dihydroxyacetone phosphate (glycerone 3-phosphate), 141, 366*-368*
1α,25-Dihydroxycholecalciferol (calcitriol), **305**, 330
3,4-Dihydroxyphenylalanine (Dopa), 154, 322
1α,25-Dihydroxy-vitamin D (calcitriol), **305**, 330
Dimethylallyl diphosphate, 171
Dinucleotide, **78**
Dioxin, 346
Dioxygenases, 87, 312, 355
Dipeptide, **64**
Dipeptidases (*3.4.13.n*), 151, 245
Diphosphatidylglycerol (cardiolipin), 48, 196, 205
Dipole moment, 7
Disaccharides, **36**
    dietary, 326
Dissociation curve, 56
Disulfide, 9
Disulfide bonds, 58, 68, 70, 72, **75**, 100, 176, 210, 260, 266
    in antibodies, 266
    in collagen, 312
    in insulin, 74, 164, 348
DNA, 78, **81**, 83, 185, 189, 194, 218, **221**
    A-conformation, 84
    base pairing in, 80, 81
    B-conformation, 80, **85**
    binding proteins, 220–224, 346, 358
    cDNA (copy DNA), 238
    circular, 182
    cloning of, **236**
    coding strand, 82, 219
    damage, 360
    denaturation of, 81
    double helix, 80
    functioning of, **82**
    germline, 271
    grooves in, 80, 84
    molecular models of, **85**
    mtDNA (mitochondrial DNA), **201**
    packaging, 216
    polymerases, 194, 222, 346
    renaturation of, 81
    repair of, 230
    replication of, 220
    sequencing of, **238**
    somatic, 271
    strands, 80
    structure of, 81, 85
    transcription of, 222
    van der Waals model of, **85**
    Z-conformation, 84
DNA-binding domains, 346
DNA ligase (*6.5.1.1*), 221, 234, 236
DNA polymerase, **221**, 234, 240
    DNA-dependent (*2.7.7.7*), 220
    forms of, 221
    mechanism of, 220
    RNA-dependent (*2.7.7.49*), 362
DNA topoisomerases (*5.99.1.2/3*), 221
Dolichol, **50**, 170, 210
Domains, 112, 346
    in proteins, 166, 216, 266, 346
Dopa (3,4-dihydroxyphenylalanine), **7**, 154, **323**, 342
Dopamine, 155, **323**, 342
Dopamine-β-monooxygenase (*1.14.17.1*), 323
Double bond, 44
    conjugated, 4, **44**
Double helix, 80
Double membrane, 23
Drosophila melanogaster, 364
Duodenum, 244
Dyes, 252, 291
Dynein, 192
Dysproteinemia, 252

E
Eadie-Hofstee plot, 91
Ecdysone, 52, **55**, 340
EC number, **86**
*Eco*RI endonuclease, **237**
Educt, 11
Effector, 112
EF-G (Elongation factor G), 230
EF-Ts (Elongation factor Ts), 230
EF-Tu (Elongation factor Tu), 230
Egg cell, 364
EIA (enzyme immunoassay), 274
Eicosanoids, 42, 44, 326, 340, **354**
    metabolism of, 354
Elastase (*3.4.21.36*), 244, 245
Elastin, 314
Electrochemical gradient, **116**, 196, 206
Electrochemical potential, 318
Electrode, 28

Electrolyte, 250
  balance, 296
  excretion, **302**, 303
  shifts, 284
  plasma concentration of, 251
Electronic configuration, 3
Electron transport chains, 116, 118, 130, 132
Electronegativity, **6**
Electron-transferring flavoprotein, 146
Electrophoresis, 238, 240, **252**
Elementary fibrils, 38
Elements, **3**
Elongation
  of proteins, 230
  of RNA, 222
Elongation factors, 230
Embryonic development, 314, **364**
Emulsification, 242, 244
Enantiomer, 30, **56**
Encephalins, **323**, 348
Ends
  of nucleic acids (3' and 5'), 78
  of peptides and proteins (N and C), 64
Endergonic, **10**, 104
Endocytosis, 212
Endoglycosidase, 244
Endoplasmic reticulum (ER), 164, 183, **210**, 290
  as calcium store, 351
  in gluconeogenesis, 158
  in protein synthesis, 211
β-Endorphine, **323**, 348
Endosome, 182, 183, 185, 212
Endosymbiotic theory, 200, 204
Endothermic, 14
Endolysosomes, 212
End-product inhibition, 162,
Energetics, 16
Energetic coupling, **10**, 11, **114**
Energy, **10**, 11
  chemical, 10
  content of foodstuffs, 326
  conservation at membranes, 116
  diagram, 18
  electrical, 10, 28
  free (G), **12**, 90
    change of (ΔG), 12, 14, 114
  mechanical, 10
  metabolism, 114–139
    in muscle, 311
    in nerve cells, 316

  regulation of, 134
  potential, 11
  requirement, 134, 278, 280, 326, **327**
  reserves, 42, 146, 278, 280, 310
  storage, 280
  supply, 46, 326
    during starvation, 280
Enhancer element, 346
Enolase (*4.2.1.11*), 140
Enterohepatic circulation, 180, 288
Enthalpy (H), 14
Enthalpy change (ΔH), 14
Entropy (S), 14
Entropy change (ΔS), 14
Envelope conformation, 52
Enzyme catalogue, 86
Enzyme immunoassay, 274
Enzymes, 62, **86–103**, 104
  active site of, 88
  activity of, **86**
    unit, **86**
  activity determination of, 98
  allosteric, 112
  analysis, 98
  assay methods, **98**
  catalysis, **88**
  classification of, **86**
  functional specificity, **86**
  inhibitors of, **92**, 106, 246
  interconverting, 106
  isosteric, 112
  kinetics of, **90**
  list, 378
  precursors of, 246
  regulation
    allosteric, 106
    isosteric, 106
  substrate specificity of, **86**
  suicide inhibition, 92
Enzyme-substrate complex, 90
Epimers, 32
Epimerization of galactose, 282
Epinephrine (adrenaline), 111, 144, 154, 159, 163, 311, **323**, **343**
  biosynthesis of, 323
  effects of, 162
Episodic hormone release, 338
Epithelial cells, 192
Epithelium, intestinal, 360
Epitope, 274
Equatorial position, **30**, 52

Equilibrium, chemical, 12
Equilibrium constant, **12**, 24
*erb*B oncogene, 359
Ergosterol, **55**
Erucic acid, 45
Erythrocytes, 180, **251**, 256, 260, 305
    gas transport by, 258
    glucose requirement of, 260, 280
    glycolysis in, 260
    hexose monophosphate pathway in, 260
    membrane, 204
    metabolism of, 260
    precursor cells of, 304
Erythropoietin, 305
Erythrose 4-phosphate, 173, 366*, 367*, 372*
Escherichia coli, 124, 188, 236
    composition of, 182
    replication in, 220
    translation in, 228
ES (enzyme substrate) complex, 90
Essential nutrients, 326–335
    amino acids, 326
    fatty acids, 44
    vitamins, 330–335
Ester, **9**, 19, 42
Esterification, 46
Estradiol, **55**, 340, 344
    biosynthesis of, 345, 369*
    hormonal axis of, 338
ETF (electron-transferring flavoprotein), 146
Ethanal (acetaldehyde), **139**, 200, 276, **295**, 326
Ethanol (ethyl alcohol), **139**, 200, 276, **295**, 326
    biosynthesis in yeast, 138
    blood level of, 294
    content, 294
    degradation of, 282, 294
    metabolism of, 294
Ethanolamine, **48**, 155, 204, 368*
Ethyl alcohol (*see* ethanol), 139
Ether, **9**
Eubacteria, 182
Euchromatin, 194, 218
Eukaryotes, 182
Evolution, 250
    of hormones, 342
Excision repair, 234
Excitation, transfer of, 316, 320–321
Excretion, 276
    of ammonia, 300

    of electrolytes, 302
    of protons, 300
    of urine, 298
Exergonic process, **10**, 12
Exocytosis, 210, 216, 320
    of peptide hormones, 348
Exon, 218, **222**, 224
Exopeptidases, **150**, 244, 342
Exosomes, 182
Exothermic, 14
Expression, 82, 218, 240
Expression plasmid, 240
Extensin, 38
Extinction, 98
Extracellular space, 250
Extracellular matrix, 314
Extrinsic factor, 244, 334

F
F_{ab} fragment, 266
F-actin, 306
FAD (flavin adenine dinucleotide), 78, **101**, 125, 132, 146, 332
FAD/$FADH_2$ reduction potential, 101
Faraday constant, **29**, 319
Farnesol, **50**, 214
    as lipid anchor of proteins, 215
Farnesylation of proteins, 214
Farnesyl diphosphate, 171, 369*
Fasting, 158, 278, **280**
    lipid metabolism during, 284
Fat (triacylglycerol), 8, **43**, **47**, 48, 50, 144, 278, 284
    absorption of, 248
    biological functions of, 47
    biosynthesis of, **169**, 249, 276, 368*
    body fat, 47
    dietary, 326
    digestion of, 248, 288
    as energy reserve, 280
    hydrolysis of, 46
    metabolism of, 144–171, **145**, 254, 276, 278, 280, 284
        in liver, 278, **284**
    neutral, **46**, 168, 284
    nutritional fat, 47
    from plants, 46
    saponification of, 18, 46
    structure, 47, 49
    soluble vitamins, 330
    transport of, 254

Fatty liver, 295
Fatty acids, 42–45, **45**, 46, 129, 135, 144, 204, 214, 248, 276, 279, 281, 284
　amphipathic properties, 23
　ATP yield from, 147
　biosynthesis of, 128, **145**, **167**, 279
　branched, 44
　complexed with albumin, 145
　degradation of, 144–149, **145**, 310
　　energy balance of, 146
　　minor pathways of, 148, 200
　　regulation of, 146
　essential, 42, 326
　free (FFA), 44, 145
　list of, 45
　long-chain, 148, 200
　metabolism of, 144–149, **145**
　mobilization, 280
　in muscle, 310
　nomenclature of, 45
　odd-numbered, 148
　β-oxidation of, 144, 146
　polyunsaturated, 46, 326, 354
　release of, 144, 248
　short-chain, 249
　solubility of, 23
　structure of, **45**
　transport into mitochondria, 199
　unsaturated, 44, 148, 326
Fatty-acid-CoA ligase (*6.2.1.3*), 145, 249
Fatty-acid synthase complex (*2.3.1.85*), 145, **166**
Fatty acyl carnitin, 199
Fatty acylation of proteins, 214
Fatty stool, 244
F$_c$ fragment, 266
Fed state, **278**
Feedback
　inhibition, 107, 178
　negative, 338
Fe-protoporphyrin, 178
Fermentation, 104, **139**
　alcoholic, 138
　lactic acid, 138
　propionic acid, 138
Ferredoxin, 119, 173
Ferredoxin-NADP$^+$ reductase (*1.18.1.2*), 119
Ferritin, 328
Ferrochelatase (*4.99.1.1*), 178
Fe-S centers, 130, **133**, 172
α-Fetoprotein, 361

Fibrin, 262, 263
Fibrinogen, **63**, 253, 262
Fibrinolysis, 250, **263**
Fibroblasts, 314
Fibroin, 69
Fibronectins, 315
Filaments, 190
　model, 306
　thick and thin, 306
Filtration, 184
Fimbrin, 192
Fingerprinting, 240
Fischer projection, **30**, 56
Flagellae, 192
Flavin, 100
Flavin adenine dinucleotide (FAD), 78, **101**, 125, 132, 146, 332
Flavin mononucleotide (FMN), **101**, 132, 332
Flavodoxin, **73**
Flavoprotein, 146
Flip/flop of lipids, 202
Fluidity, 202
Fluid mosaic model, 201
Fluoride, 328
5-Fluorouracil, **177**
fMet (formylmethionine), 228
FMN (flavin mononucleotide), **101**, 132, 332
Fodrin, 192
Folate (folic acid), 102, **333**
Folding of proteins, **73**, 215
Follitropin (FSH), 342
Formaldehyde, 200
Formate (formic acid), **5, 45**, 200
N-Formiminoglutamate, 370*
N-Formylmethionine, 228
N^5-Formyl-THF (N^5-formyltetrahydrofolate), 375*
N^{10}-Formyl-THF (N^{10}-formyltetrahydrofolate), **103**, 174, 375*
Fos oncogen, 357
Fragin, 190
Free energy (G), 10, 12
　change (ΔG), 12, 14
　concentration dependence of, 12
Frame-shift mutation, 234
Frictional coefficient, 186
Fruit fly, 364
Fructose, 32–35, **35**, 249, 283, 302
　intolerance, 282
　metabolism of, 282

transport of, 249
Fructose 1,6-bisphosphate, 94, **141**, **159**, 163, 366*
Fructose-1,6-bisphosphatase (*3.1.3.11*), **159**, 163, 367*
Fructose 2,6-bisphosphate, 163
Fructose-bisphosphate aldolase (*4.1.2.13*), 141, 283
Fructose-2,6-bisphosphatase (*3.1.3.46*), 163
Fructose 1-phosphate, 282
Fructose 6-phosphate, **141**, 366*, 367*
    in gluconeogenesis, 159
    in glycolysis, 141,
    in hexose monophosphate pathway, 143
    in regulation of carbohydrate metabolism, 163
Fucose, **35**, 40
Fuel metabolism in starvation, 280
Fumarase (*4.2.1.2*), 127
Fumarate, **127**
    in amino acid metabolism, 154, 371*
    in citric acid cycle, 127, 129, 131
    in purine synthesis, 174
    in urea cycle, 287
Fumarate hydratase (*4.2.1.2*), 127, 287
Fumarylacetate, 371*
Functional group, 8
Fungi, 182
Furanose, 34, 78
Furan ring, 30
Fusion, 274

G
GABA (4-aminobutyrate), 155, 316, **318, 322**
    mechanism of action, 350
    shunt, 317
G-actin, 306
Galactokinase (*2.7.1.6*), 283
Galactosamine, *N*-acetyl-, 35
Galactose, **35**, 40, 249, 283
    metabolism of, 282
    resorption of, 242
    transport of, 249
Galactosemia, 36, 298
Galactose 1-phosphate, 282
β-Galactosidase (*3.2.1.23*), 185
Galactosyl ceramide, 48, 368*
Galacturonic acid, 34, 38
Gall bladder, 244, 288
Gall stones, 54, 180, 288, 298
Gangliosides, 42, 48, 368*

Gap gene, 364
Gap (G) phase, 357
Gas transport, 250, **259**
Gastric juice, 245
Gastrointestinal tract, 242–249
    enzymes of, 150, 244
    hormones of, 336
Gastrula, 364
GDP (guanosine diphosphate), 176, 350, 374*
Gelatine, 326
Gel electrophoresis
    of DNA, 239, 241
    of proteins, 252
Gelsolin, 190
Genes, **82**, 218, 223, 365
    in development, 365
    duplication of, 266
    embryonic, 364
    eukaryotic, 222
    expression of, 219, 364
    germline, 270
    β-globin, 222
    homeotic, 364
    housekeeping, 224
    in regulation of development, 364
    library, 238
    maternal, 364
    multiple, 270
    organization of, 223, 271
    probe, 238, 240
    segment, 270
    segmentation of, 364
    somatic, 270
    tandem structure of, 270
    technology, 236–241
    transcription of, 222
    tumor suppressor, 358
Genetic code, 200, **227**
Genetic engineering, 236–241
    applications of, 240
Genetic fingerprint, 240
Genome, **218**
Genomic library, 238
Geraniol, **50**
Geranyl diphosphate, 171, 369*
Geranylgeranol, 215
Germline, 270
Gestagens, 340
Gibbs free energy, 10, 12, 14
Gibbs-Helmholtz equation, 14
Glands, endocrine, 336

Glia protein, 190
Globin, 180
β-Globin gene, 222
Globular proteins, 71
Globulins, 252
Glomerulus, 296
Glucagon, 111, 144, 159, 163, 170, 278, 281, 338, 343
  effects, 145, 162
4α-Glucanotransferase (*2.4.1.25*), 161
Glucan-branching enzyme (*2.4.1.18*), 161
Glucocorticoids, 54, 340
  biosynthesis, 344, 369*
  mechanism of action, 346
Glucogenic amino acids, **154**, 326
Glucokinase (hexokinase, 2.7.1.1), 137, 141, 163
Gluconate (gluconic acid), 32
Gluconate 6-phosphate, 367*
Gluconeogenesis, 128, 140, **159**, 276–283, 303, 367*
  role in amino acid degradation, 154
  compartmentation of, 189
  control of, 162
  in kidneys, 302
  in liver, 276–283
  pathway, 158
Gluconolactonase (*3.1.1.17*), 143
Gluconolactone, 33, 99
Gluconolactone 6-phosphate, 367*
Gluconolactone oxidase (*1.1.3.8*), 334
α-D-Glucopyranose, 31
Glucosamine, *N*-acetyl-, 35, 40
Glucose, **31**, 35, 63, 99, 111, 139, 141, 159, 161, 163, 279, 283, 367*
  aerobic oxidation of, 137
  anaerobic oxidation of, 137
  ATP yield from, 137
  ball-and-stick model, 31
  biosynthesis of, 128, 158
  conformation of, 31
  determination, 99
  formation from glycogen, 160, 310
  formation from pyruvate/lactate, 158, 282
  excretion, 299
  Haworth projection, 31
  metabolism of, 136–143
    in liver, 278, 282
    in muscle, 310
    in nerve cells, 316
  plasma concentration of, **251**, 280, 282, 338

reabsorption of, 296, 303
transport of, 207, **209**, 242, 249
uptake of, 206, 242
van der Waals model, 31
Glucose-6-phosphatase (*3.1.3.9*), 158, 160, 163, 185, 367*
Glucose 1-phosphate, 111, 161, 163, 367*
Glucose 6-phosphate, 33, 111, 121, 141, 143, 159, 161, 163, 366*, 367*
Glucose-6-phosphate dehydrogenase (*1.1.1.49*), 158, 160, 367*
Glucose-6-phosphate isomerase (*5.3.1.9*), 141, 367
Glucose oxidase (*1.1.3.4*), 99
1,6-Glucosidase (*3.2.1.10*), 160, 245
Glucosuria, 164
Glucosyl ceramide, 48
Glucuronate (glucuronic acid), 32–35, **34**, 40, 180, 291
  conjugates with, 291, 298, 344
Glucuronosyltransferase (*2.4.1.17*), 180
GLUT (glucose transporter), 208
Glutamate (glutamic acid), 27, **59**, **115**, 153, 172, 199, 286, 287, 301, 316, 323
  biosynthesis, 173, 373*
  degradation, 155, 370*
  family, 173
  as neurotransmitter, 316, 322
Glutamate decarboxylase (*4.1.1.15*), 317
Glutamate dehydrogenase (*1.4.1.2*), 153, 154, 287, 301, 317, 370*, 373*
Glutamate 4-semialdehyde, 371*
Glutamic acid γ-phosphate, 115
Glutaminase (*3.5.1.2*), 153, 296, 301
Glutamine, **59**, **115**, 153, 172, 174, 290, 301, 302, 311
  biosynthesis of, 173, 373*
  degradation of, 155, 370*
  metabolism of
    in kidneys, 300
    in liver, 152
    in muscles, 310
  plasma level, 300
Glutamine synthetase (*6.3.1.2*), 63, 114
γ-Glutamyl phosphate, 373*
γ-Glutamyltransferase (*2.3.2.13*), 262
Glutathione, 100, 260, **261**
Glutathione peroxidase (*1.11.1.9*), 261
Glutathione reductase (*1.6.4.2*), 260
Glycan, 36
Glyceral (glyceraldehyde), **31**, 282

Glyceraldehyde 3-phosphate, 366*, 367*
Glyceraldehyde-3-phosphate dehydrogenase (NADP+) (*1.2.1.13*), 121, 137, 366*
Glyceraldehyde-3-phosphate dehydrogenase (NAD+) (*1.2.1.12*), 140, 151, 367*
Glycerate (glyceric acid), 282
Glycerol, 21, 42, **46**, 159, 168, 204, 248, 282, 302, 367*, 368*
  in gluconeogenesis, 158
Glycerol 3-phosphate, 168, 367*, 368*
Glycerone 3-phosphate (dihydroxyacetone 3-phosphate), **141**, **159**, 168, 282
α-Glycerophosphate, 131
Glycine, **59**, 68, 172, 174, 178, 289, 290, 316, 323
  biosynthesis of, 173, 373*, 375*
  conjugates with, 288, 298
  degradation of, 155, 370*
  in collagen, 312
Glycinuria, 296
Glycocalyx, 202
Glycocholate (glycocholic acid), **289**
Glycogen, 37, 110, **161**, 163, 242, 244, 278, 311, 367*
  balance, 161
  biosynthesis, 161
  cleavage, 244
  in diet, 326
  as energy reserve, 280
  in liver, 158, 276
  metabolism of, 110, **161**
    hormonal control of, 111
    regulation of, **111**, 160, 162, 352
    in muscle, 310
    in nervous tissue, 316
  phosphorylase, 111
  store, 280
  structure of, 160
Glycogenoses, 212
Glycogen phosphorylase (*2.4.1.1*), 111, 161, 163, 367*
  interconversion of, 110
Glycogen synthase (*2.4.1.11*), 111, 161, 163, 367*
Glycolipids, 42, **49**, 50, 202, 204
  structure, 49
Glycolysis, **141**, 162, 189, 279, 283, 310, 367*
  aerobic, 137, 140
  anaerobic, 137, 140, 310
  balance of, 140

energy profile of, 140
  in erythrocytes, 140
  reactions of, 141
Glycoproteins, **41**, 202, 252, 266
  complex type, 40, **203**
  hormones, 342
  mannose-rich type, 40, **203**
  N-linked, 40, 202
  O-linked, 40, 202
Glycosaminoglycans, 34, **41**, 314
Glycoside, 34
  N-glycoside, 32, 211
  O-glycoside, 32
Glycosidases (*3.2.n.n*), 210, 242
Glycosidic bond, 36
Glycosylation, 40, 211, 348
Glycosyl transferases (*2.4.n.n*), 210
Glyoxylate (glyoxylic acid), 129
  cycle, 128, 200
Glyoxysomes, 128
GMP (guanosine monophosphate), 157
  biosynthesis of, 176, 374*
  cyclic (cGMP), 324, 337, 350, 352
  degradation, 156
Goiter, 328
Golgi apparatus, 183, 185, **211**, 216, 312
Gonadoliberin (GnRH), **323**
Gonan, **52**
GOT (aspartate transaminase, *2.6.1.1*), 152
Gout, 156
G_0, G_1, and G_2 phase in cell cycle, 356
GPI anchor, **215**
G protein, 111, 214, 320, 322, 324, **351**, 353, 358
GPT (alanine transaminase, *2.6.1.2*), 152
Gradient, 186
  electrochemical, **117**, 118, 130, 196
  of morphogen, 364
Granulocytes, 251, 343
β-Granula in pancreas, 164
Gravitational force, 184
Grooves of DNA, 81
Group-transferring coenzymes, 102
Growth
  disturbances of, 330
  cell, 360
Growth factors, 330, 336, 340, 356, 358
  mechanism of action, 350
  second messengers of, 352
GTP (guanosine triphosphate), 126, 177, 190, 350

binding protein (G protein), 111, 214, 320, 322, 324, **351**, 353, 358
biosynthesis, 177, 374*
formation in citric acid cycle, 126
hydrolysis, 352
role in sight, 324
role in translation, 230
Guanine, 79, 157
Guanine nucleotide-binding protein (G protein), 111, 214, 320, 322, 324, **351**, 353, 358
Guanosine, 78
Guanosine diphosphate (GDP), 176, 350, 374*
Guanosine monophosphate (GMP, guanylate)
biosynthesis of, 176, 374*
cyclic (cGMP), 324, 337, 350, 352
degradation of, 156
Guanosine triphosphate (*see also* GTP), 126, 177, 190, 350
Guanylate cyclase (*4.6.1.2*), 325, 358
Gut, 341
Gyrase inhibitors, 233

H
H+, *see* Hydrogen ion; Proton
Hair, 68, 312
Hagemann factor (coagulation factor XIIa, *3.4.21.38*), 263
Half-life
of angiotensin II, 304
of hormones, 348
of proteins, **151**
Halobacterium halobium, 122
Halogen, 2
Haptoglobin, 253
Hartnup's disease, 296
HAT medium, 274
Haworth projection, 30
H chains of immunoglobulins (heavy chains), 266
HCl (hydrochloric acid), 242, 244
secretion of, 247
HCO₃⁻ ion (bicarbonate, hydrogen carbonate), 27, 245, 246, **257**, 258, 286, 301
transport, 207
H+ concentration (*see also* pH value), **24**, 256
HDL (high-density lipoprotein), **254**, 284
Heart, 279, 281
atrium, 302
attack, 198, 262, 310
muscle, 310
Heat formation, 135, 341

Heat-shock proteins, 214, 346
Heat transport, 250
Heavy chains
of immunoglobulins, 266
of myosin, 306
Helix
α helix in proteins, **67**, 74
amphipathic, 74
in collagen, 66, 312
double helix of DNA, 80
left-handed, 66, 68
pitch, 66, 80
right-handed, 66
in starch 38
triple, 68, 312
7-Helix motif, 324
Helper T cells, 264, 362
Hematocrit, 250
Heme, **101**, 130, 133, **179**, **181**
biosynthesis of, 178
in respiratory chain, 133
degradation of, 180
Heme oxygenase (*1.14.99.3*), 180
Hemicellulose, 38
Hemiacetal, **9**, 31
Hemoglobin, 63, 151, 178, 256, **261**, 276, 328
allosteric effects, 260
binding of
2,3-bisphosphoglycerate, 260
CO₂, 258–261
H+, 257, 261
O₂, 59, 261
buffering function, 257
carbamino derivative, 258
degradation of, 180, 276
deoxy-, 258, 269
oxy-, 258, 269
R-form, 260
structure of, 258, **269**
T-form, 260
Hemolysis, 298
Hemoproteins, 180
Hemosiderin, 328
Hemostasis, 250, **263**
Henderson-Hasselbalch equation, **257**
Heparin, **315**
Hepatocytes, 145, 276–295, **277**
Heterochromatin, 194, 218
Heteroglycan, 36
Heterotrophic organisms, 104
Hexokinase (*2.7.1.1*), 137, 141, 163

Hexose-1-phosphate uridyltransferase (*2.7.7.12*), 283
Hexose monophosphate pathway, **143**, 189, 279, 366*
  reactions of, 143
Hexoses, 30
High-density lipoprotein (HDL), **254**, 284
High-molecular weight nuclear RNA, 218, **222**, 270
High-performance liquid chromatography (HPLC), 60
Hill coefficient, 112
Hinge region in immunoglobins, 266
Hippurate (hippuric acid), **299**
Histamine, 155, **323**, 336, **343**, 354
Histidine, 27, **57**, **59**, 60, 172, 327
  biosynthesis of, 173, 372*
  buffering action of, 26
  degradation of, 155, 370*, 375*
  dissociation curve of, 57
  3-methyl-, 298, 310
  residues, 256
  titration curve of, 57
Histones, 62, 151, 194, **218**
  and cell cycle, 356
  octamer, 218
HIV (human immunodeficiency virus), 363
HLA (human leukocyte-associated) antigens, **273**
HMG-CoA (*see* hydoxylmethylglutaryl-CoA), 171, **285**, 369*, 370*
3-HMG-CoA reductase (*1.1.1.34*), 151, **170**
hnRNA (high-molecular weight nuclear RNA), 218, 222, 270
Homeostasis, 204, 250, 285, 296
Homeotic genes, 364
Homocysteine, 371*, 373*, 375*
Homogenization, 184
Homogentisate, 371*
Homoglycan, 36
Homoserine, 372*
Honey, 34, 36, 326
Hormone, 336–351, **337**, 355
  analysis, 274
  binding domains, 346
  biosynthesis of
    eicosanoids, 354
    peptide hormones, 349
    steroid hormones, 345
  classification of, 340
  control by, 111

degradation of, 343, 349
  glandular, 338
  half-life of, 348
  hierarchy, 339
  hydrophilic, 336, **343**
    concentration of, 338
  of kidneys, 305
  lipophilic, 336, **341**
  mechanism of action of, 336, 347, 351, 352
    lipophilic, 347
    hydrophilic, 351
  metabolism of, 344, 348
  plasma level of, 339
  receptors, 194, 214, **347**
    nuclear, 346, 358
  receptor complex, 346
  regulation of metabolism, 110
  regulatory circuit, 339
  and signal transduction, 337
  system, 337
  and second messenger, 337, 352
Hormone response element (HRE), 346
HPLC (high-performance liquid chromatography), 60
House-keeping genes, 224
Hsp (heat-shock proteins), 214, 346
H+-transporting ATP synthase (*3.6.1.34*), 116, 130
Human immunodeficiency virus (HIV), 363
Human leukocyte-associated antigens (HLA), 273
Hyaluronate (hyaluronic acid), 37, **41**, **315**
Hybridization
  of nucleic acids, 236, 238, 240
  of orbitals, 4
Hybridoma cells, 274
Hydration, 20, 88
Hydride ion (H−), **95**, 96, 100
  coenzyme for transfer of, 95, 333
Hydrocarbons, 42
Hydrochloric acid (HCl), 242, 244, 342
  formation of, 247
Hydrogen
  bonds, **7**, 20, 38, 66, 72, 80
  electrode, 28
  ion (proton, H+), 24, 256, 300
    balance, 256
    channel, 135
    concentration of, 256
    excretion of, 257, 296, **301**
    gradient, 117, 130, 134, 199

production, 256
 secretion, 247, 301
 transport of, 246
 transfer, coenzymes of, 333
Hydrogen carbonate (bicarbonate, HCO_3^-), 27, 245, 246, **257**, 258, 286, 301, 393*
Hydrogen peroxide, 156, 200
Hydrogen phosphate (HPO_4^{2-}), 27
Hydrogen sulfide, **9**
Hydrolase (*3.n.n.n*), **86**, 87, 212, 242, 244
Hydrolysis, **18**, 86
 of ATP, 12, 114
Hydronium ion (H_3O^+), 25
Hydroperoxy fatty acids, 354
Hydrophilicity, 22
Hydrophobic effect, 22, 72
Hydrophobicity, 22
Hydroxyacyl-CoA, 3, 131
Hydroxyacyl group, 102
Hydroxyalkyl thiamine diphosphate, 374
3-Hydroxybutyrate, 131, 256, **285**, 369*
3-Hydroxybutyrate dehydrogenase (*1.1.1.30*), 285, 369*
Hydroxy fatty acids, 354
20-Hydroxyecdysone, 364
Hydroxyurea, **177**
Hydroxylamine, 234
Hydroxylation, 290, 293, 312
 coenzyme of, 293, 334
5-Hydroxylysine, 312
Hydroxyl ion (OH^-), 24, 25
Hydroxylysine, 312
Hydroxyacyl-CoA dehydrogenase (*1.1.1.35*), 147
3-Hydroxyisobutyrate, 370*
3-Hydroxyisobutyryl-CoA, 370*
Hydroxymethylbilane synthase (*4.3.1.8*), 179
3-Hydroxy-3-methylglutaryl-CoA (3-HMG-CoA), 171, **285**, 369*, 370*
Hydroxymethylglutaryl-CoA lyase (*4.1.3.4*), 285, 369*, 370*
Hydroxymethylglutaryl-CoA reductase (*1.1.1.34*), 151, 170
Hydroxymethylglutaryl-CoA synthase (*4.1.3.5*), 285, 369*
4-Hydroxyphenylpyruvate, 371*
Hydroxypregnenolone, 369*
Hydroxyprogesterone, 369*
Hydroxyproline, 68, 298, 312
5-Hydroxytryptophan, 154
Hyperbilirubinemia, 180

Hyperglycemia, 164
Hyperlipidemia, 164
Hyperpolarization, 318
Hypertonic solution, 184
Hyperuricemia, 156
Hypervariable region, 270
Hyperventilation, 256
Hypervitaminosis, 330
Hypothalamus, 343
Hypoxanthine, **157**, 176, 274
Hypoxanthine phosphoribosyltransferase (*2.4.2.8*), 157
Hypoxia, 198, 304
H-zone, 306

I
I-band, 306
Ice, structure of, **21**
Idiotypic variability, 271
IDL (intermediate-density lipoprotein), **254**, 284
Iduronic acid, 35, 40
IEP (isoelectric point), 56
IgA (immunoglobulin), **267**
IgD (immunoglobulin), **267**
IgE (immunoglobulin), **267**
IgG (immunoglobulin), 40, 62, **267**, **269**
IgM (immunoglobulin), **267**, 273
Imaginal development, 364
Imidazole, 19
Imidazolone 5-propionate, 370*
Imino acid, 59
Immune cells, 265, 272, 274
Immune response, 265
Immune system, 250, 264–275
Immunity, 264
Immunoassays, 274
Immunoglobulins, 41, 63, 72, 245, 250, 253, 266–271, **267**, 269
 biosynthesis of, 271
 carbohydrates in, 41
 C_H2 domain, 73
 classes of, 257
 constant region, 266
 domain structure of, 267
 gene superfamily, 272
 generation of diversity, 270
 hypervariable region in, 268, 271
 IgA, **267**
 IgD, **267**
 IgE, **267**

IgG, 40, 62, **267, 269**
IgM, **267**, 273
monoclonal, 274
oligosaccharide in, 41
serum concentration of, 267
structure of, **267, 269**
superfamily, 273
tertiary structure of, 268
variability of, 271
variable region of, 266
IMP (inosine monophosphate), 157, **175**, 176, 310,
in muscle, 310
Indicator reaction, 274
Induction of proteins, 106, 240
Influenza virus, 362
Inhibition constant, 92
Inhibition of enzymes, 92
Inhibition, kinetics of, 93
Inhibitors, 92, 304
allosteric, 92
competitive, 92
non-competitive, 92
Initial velocity, 91
Initiation
of transcription, 222
of translation, **229**
of tumors, 361
Initiation factor, 228
Inner mitochondrial membrane, 197, 199
Inosine monophosphate (IMP, inosinic acid,
isonate), 157, **175**, 176, 310, 374*
Inorganic ions
excretion of, 298
as constituents of cells, 189
as constituents of the diet, 242, 326, 328
plasma concentration of, 251
myo-Inositol, **48**, 204, 368
Inositol-1,4,5-trisphosphate (InsP$_3$), 48, 168,
351, 352
role as second messenger, 337
Insect development, 364
Insulin, 63, 71, **75, 77**, 144, 159, 163, 170, 210,
278, 281, 316, 343
biosynthesis of, 165
deficiency of, 165
in diabetes mellitus, 164
effects of, 145, 162, 164
expression of, 76
hexamer, 70, 76
invariant residues in, 76

mechanism of action of, 350
metabolism of, 340
mode of action, 338
molecular models of, 77
primary structure of, 71
quarternary structure of, 71
receptor, 62, 63
receptor binding, 76
level of, 339
secondary structure of, 71
structure of, **71**, 75, **77**, 343
treatment with, 76
Zn^{2+} centers, 76
Insulin/glucagon ratio 278, 280
Insulator, 42, 46
Integral membrane proteins, 202
Integration of viral DNA, 363
Integrin, 314
Interactions
electrostatic, 72, 74, 96, 260
hydrophobic, 72
Intercalation in DNA, 233
Intercellular substance, 314
Interconversion, **106**, 110, 134, 162, 356
Interleukines, 264
Intermediary metabolism, **105**, 276
regulatory mechanisms of, 107
switching of, 280
Intermediate filaments, 69, **191, 193**
Intermediate-density lipoprotein (IDL), **254**,
284
Intermembrane space, 196
Interphase, 194, 218, 356
Internal milieu, 204, 250
Intestinal
bacteria, 288
epithelium, 242, 246, 248
flora, 330, 334
mucosa, 242, 254
Intestine, 242, 254, 281, 288
Intrinsic factor, 245, 334
Introns, 182, **222**, 218, 224, 270
sequences, 348
Inulin, 37, 296
Invasion, 360
Invert sugar, 36
Iodine, 328
daily requirement of, 329
deficiency of, 328
reaction with starch, 38
Iodothyronine, 252, 340

Ion, 20
  permeability of membranes for, 319
Ion channel, 63, 116, 207, **318**, 351
  ligand-gated, 318, 320, 322, 325, **350**
  for calcium ions, 308, 319,
  for chloride ions, 319, 322,
  for potassium ions, 319
  for sodium ions, 319, 322
  protein, 206, 324
  for protons, 132, 135
  voltage-gated, 318, 320
Ion exchanger, 60
Ionizing radiation, 234
Iron, 3, 121, 328
  daily requirement of, 329
  electronic configuration of, 2
  as heme constituent, 258
  metabolism of, 180
  proteins, 328
  role in collagen biosynthesis, 312
Iron-sulfur cluster or center, 130, 133, 172
Irreversible inhibitors, 92
Ischemia, 198
Islets of Langerhans, 164
Isobutyryl-CoA, 370*
Isocitrate, **127**, 129, 131, 135
Isocitrate dehydrogenase (*1.1.1.41*), 127, 137
Isoelectric point (IEP), 56
Isoenzymes, 94
Isoleucine, **59**, 172, 311, 327
  biosynthesis of, 173, 372*
  degradation of, 155, 370*
Isomerase (*5.n.n.n*), 86, 87
Isomers
  cis-trans, 4, 44
    of fatty acids, 44
    of retinal, 116, 122, 324
  mirror-image, 56
N^6-Isopentenyl-AMP, 50
Isopentenyl diphosphate, **50**, 171 369*
Isopeptide bonds, 262
Isoprene, **51**
Isoprenoids, **51**, 214
  biosynthesis of, 128, 171
  structure of, **51**
  as vitamins, 330
Isopyknic centrifugation, 186
Isosteric enzyme, 107
Isotonic solution, 184

Isotype, 270, 272
Isotypic variability, 271
Isourea, 286
Isovaleryl-CoA, 370*
Isozyme, 94

J
Jaundice, 180
Jejunum, 242
Joule, 10, 11
J-protein, 266
J-segment, 270
Jun oncogene, 359
Juvenile hormone, 50
Juxtaglomerula cells, 304

K
Kallikrein (*3.4.21.34/35*), 263
Katal, **86**
KDEL sequence, 217
Keratan sulfate, 315
Keratin, 69, 190, 192
Ketoacidosis, 284
Ketogenesis, 144
  inhibited by ethanol metabolism, 294
Ketogenic amino acids, **154**, 326
α-Ketoglutarate (2-oxoglutarate), **127**, **129**, 131, 135, 153, 155, 199, 287, 301
  in amino acid metabolism, 129, 153, 155, 301
  in citric acid cycle, 127
  in malate shuttle, 199
Ketohexokinase (*2.7.1.3*), 283
Ketohexoses, 32
Ketone, 8
Ketone bodies, 144 151, 256, 276, 279, 280, **285**
  biosynthesis of, 145, 281, **285**, 369*
  degradation of, 310
  in diabetes mellitus, 165
  excretion of, 299
  in muscle, 310
  in nerve cells, 316
  in starvation, 280
Ketonemia, 284
Ketose, 34
Ketosuria, 284
β-Ketothiolase (acetyl-CoA acyltransferase, *2.3.1.16*), 146
Kidneys, 256, 286, 290, 296–305, 341
  functions of, 296

gluconeogenesis in, 158
hormones of, 305, 341
metabolism of, 296
role for plasma pH, 256
urine formation in, 297
Killer T cells, 265
Kinesin, 192
Kininogen, 263
Kinetics, 16, 90, 113
Kininogen, 263
K_m, 91
Krebs cycle, *see* Citric acid cycle

L

Lactate (lactic acid), 12, 97, **99**, 135, **139**, 141,
159, 282, 311, 367*
and calcium uptake, 328
fermentation to, 139
formation of, 134, 261
gluconeogenesis from, 158
in kidneys, 302
in muscle, 310
plasma concentration of, 251
L-Lactate dehydrogenase (*1.1.1.27*) 62, 86, **95**,
137, 139, 151, 159, 185, 367*
activity measurement, 98
assay of, 99
catalytic cycle, 97
coenzyme, 95
isoenzymes of, 94
mechanism of, **97**
structure of, 95
Lactic acid fermentation, 138
Lactobacillus, 138
Lactone, 32, 142
Lactose, 36, 109, 138
operon, 108
Lac repressor, 109
Lambert-Beer law, 99
Lamina, 194
Laminin, 314
Lamins, in cell cycle, 356
Lanosterol, 171
Laurate (lauric acid), **45**
Lariat intermediate, 225
LCAT (lecithin-cholesterol acyltransferase,
2.3.1.43), 254
LDL (low-density lipoprotein), **254**, 284
receptors, 254
Leader sequence (signal peptide), 164, 210,
214, **217**
Lecithin, **48**, 168

Lecithin-cholesterol acyltransferase (LCAT,
2.3.1.43), 254
Leghemoglobin, 172
Lesch-Nyhan syndrome, 156
Leucine, **59**, 172, 311, 327
biosynthesis of, 173, 372*
degradation of, 155, 370*
Leucine aminopeptidase (*3.4.11.1*), 93
Leu-enkephalin, **323**
Leukocytes, **251**, 354
Leukotrienes, **355**
LH (lutropin), 343
plasma level of, 339
Liberins (releasing hormones, releasing fac-
tors), 338
Library of genes, 239
Lieberkühn glands, 244
Ligase (*6.n.n.n*), 86, 87, 102
Light
absorption of, 98
effects via second messengers, 324, 352
monochromatic, 98
ultraviolet, 234
in vision, 324
Light-harvesting complex, 121
Light reactions, 118
Lignin, 242
Lignocerate (lignoceric acid), **45**
Linker DNA, 218
Linoleate (linoleic acid), **45**, 46, 148, 327
Linolenate (linolenic acid), **44**, 327
Lipase (*3.1.1.3*), 46, 145, 168, 242, 245, **248**,
250
Lipid alcohols, 42
Lipid anchor, 50, **215**, 220
Lipid bilayer, 202
Lipids, **42-55**, 189, 202, 204
biosynthesis of, 51 **169**, 368*, 369*
biological role of, 43
classification of, 43
complex, 168
degradation of, 144
digestion of, 242
hydrolyzable, 42
in membranes, 203
metabolism of, 144–149, 166–170, 254,
368*
in liver, 276, **285**
in nerve cells, 316
non-hydrolyzable, 42
plasma concentration of, 25

resorption of, 249
and steroid metabolism, 344, 369*
storage of, 144
thin-layer chromatography of, 53
transport of, 254
vesicles, 23
water-solubility of, 23, 42
Lipidoses, 212
Lipid-water interface, 204
Lipoamide, **101**, 125
Lipoate (lipoic acid), 100
Lipocortin, 256
Lipogenesis, 144,
Lipolysis, 172, 280
control by hormones, 144
Lipoxygenases (*1.13.11.n*), 354
Lipophilic compounds, 22
Lipoprotein lipase (*3.1.1.34*), 145, 254
Lipoproteins (lipoprotein complexes), 54, **255**, 276, 284
Lipotropins, 348
Lithocholate (lithocholic acid), **55**, 289
Liver, 254, 276-295
amino acid metabolism in, 286
bile formation by, 288
biotransformation in, 290
cirrhosis of, 295
damage of, 295
detoxification reactions in, 276, **290**
excretory function of, 276
extracellular matrix, 314
functions of, 276
gluconeogenesis in, 158
glycogen in, 160
heme biosynthesis in, 178
metabolism, 277, **279**, **281**, 295
of amino acids, 276, 286
of bile acids, 288
of carbohydrates, 276, 278
of ethanol, 295
of lipids, 276, 284
of NH_3 152, 286
storage
of trace elements, 328
of vitamins, 334
urea cycle in, 286
Liver bile, 288
Low-density lipoprotein (LDL), **254**, 284
Lungs, 300
alveoli, 256
role for plasma pH, 256

Lutropin (LH), 342
plasma level of, 339
Lyases (*4.n.n.n*), 86, 87
Lymph system, 242, 248, 254
Lymphocytes, **251**, 264, 274
Lynen cycle, 284
Lysine, 27, **59**, 172, 327
biosynthesis of, 173, 372*
degradation of, 155, 371*
Lysophosphatidate (Lysophosphatidic acid), 168
Lysophospholipids, **49**
Lysosomes, 183, 185, **213**, 216, 254
in degradation processes, 212
functions of, 213
pH value of, 27, 213
proteins in, 213
storage disorders, 212
Lysozyme (*3.2.1.17*), 245

M
Macroglobulin, 253
Macroelements, 2, 326, 329
Macrophages, 264
Magnesium, 119, 328
excretion of, 299
deficiency, 328
Major groove of DNA, 81
Major histocompatability complex (MHC) proteins, 273
Malate, **127**, 131, 158, 286
in citric acid cycle, 126, 128
shuttle, 198
in transport of oxaloacetate, 159, 199
in urea cycle, 287
Malate-aspartate shuttle, 198
Malate dehydrogenase (*1.1.1.37*), 127, 128, 137, 287
4-Maleylacetoacetate, 371*
Malic enzyme (*1.1.1.40*), 128, 166
Malignancy, 360
Malnutrition, 328, 330
Malonyl-ACP, 167
Malonyl-CoA, 102, 145, 166
Maltose, 139, 244
Maltotriose, 244
Mammary glands, 341
Manganese, 122, 328
daily requirement, 329
in chlorophyll, 119, 121
Mannitol, 35

Mannose, 33, 35, 40
  residues, 212
  rich glycoproteins, 202
Mannose 6-phosphate, 212
  receptors of, 212
α-Mannosidase (*3.2.1.24*), 185
Marker molecules for cell fractions, 184
Mast cells, 343
Matrix, extracellular , 312, 314
Matrix space in mitochondria, 196, 198
Maturation of RNA, 194, 218, **223**, **225**
Maximum velocity of enzyme reactions, 91
  determination of, 90
Mechanism of action
  of hydrophilic hormones, 350
  of lactate dehydrogenase, 96
  of lipophilic hormones, 346
Mediators, 336, 340, 342, 354
  second messengers of, 352
Medulla of suprarenal gland, 322
  glucose requirement of, 280
Megakaryocytes, 250
Megalocytes, 332
Melanotropins, 348
Melatonin, **343**
Membrane, 198, 202–209
  anchor, 214
  assymmetry of, 204
  carbohydrates in, **203**
  channels, 207, 319
  components of, 202
  composition of, 204
  damage of, 235
  and energy conservation, 117
  fluidity of, 54, **202**
  fluid mosaic model of, 203
  functions of, 204
  fusion, 215
  glycoproteins in, 203
  hormone receptors located in, 348, 350
  lipids, 42, **203**
  lipid bilayers in, 203
  of mitochondria, 132, 197, 199
  permeability of, 206, 318
  potential, 116, 199, 318
  proteins, 203, 209, 272, 318
    biosynthesis of, 210
    integral, 214, 216
    peripheral, 214
  pumps, 207, 209
  receptors, 254, 348, 350, 358

  structure of, 202
  synaptic, 320
  transport process across, 206, 208, 216
  viruses enveloped by, 363
Memory cells, 264
Menachinon, **50**
Menstrual cycle, 341
Menthol, **50**
6-Mercaptopurine, **177**
Messenger RNA (mRNA) , **83**, 109, 189, 194,
  218, 223, 225, 229
  maturation of, 82
  modification of, 225
Messenger, second, 322, 324, 337, 348, 350,
  353, 354
  metabolism of, 352
Metabolic acidosis, 164, 256
Metabolic pathway, **86**, **104**
  amphibolic, 128
  anabolic, 104
  anaplerotic, 128
  catabolic, 104
Metabolism, **104**
  in brain, **317**
  of erythrocytes, 260
  hormonal regulation of, 110
  intermediary, 105, 276
  in kidneys, 29
  in liver, 276
  in muscle, 310
  major pathways of, 105
  overview of, 104
  regulatory mechanisms in, **107**
Metabolite, 102, **105**
  activated, 102
  pool, 104
Metallothionein, 290
Metabolic regulation, 107
Metal proteinases (*3.4.24.n*), 151
Metamorphosis, 364
Metastasis formation, 360
Met-enkephalin, **323**
Methane, 20
  properties of, 21
  water-solubility of, 22
Methemoglobin, 258, 260
  reduction of, 260
N^5,N^{10}-Methenyl-THF (N^5,N^{10}-methenyl-
  tetrahydrofolate), 375*
Methionine, **59**, 172, 327, 375*
  biosynthesis of, 173, 372*

degradation of, 155, 371*
N-formyl, 228
Methotrexate, **177**
2-Methylacetoacetyl-CoA, 370*
Methylacrylyl-CoA, 370*
Methylation, 290
2-Methylbutyryl-CoA, 370*
3-Methylcrotonyl-CoA, 370*
N^5,N^{10}-Methylene-THF (N^5,N^{10}-methylene-
tetrahydrofolate), **103**, 174, 375*
Methylglucoside, 33
3-Methylglutaconyl-CoA, 370*
Methyl group, angular, 52
7-Methyl-GTP residue, 224
3-Methyl histidine, 298, 310
2-Methyl-3-hydroxybutyryl-CoA, 370*
Methylmalonyl-CoA, 148, **335**, 370*
Metylmalonyl-semialdehyde, 370*
Methyl nitrosamine, 234
N^5-Methyl-THF (N^5-methyltetrahydrofolate),
103, 375*
Mevalonate, **171**, 369
Mevalonate diphosphate, **171**, 369*
MHC (major histocompatibility complex)
proteins, 264, **273**
receptors of, 264
Micelles, 23, **289**
formation of, 46, 300
Michaelis constant, 91
Michaelis-Menten equation, 91
Microelements (trace elements), **2**, 326, **329**
Microfibrils, 38
Microfilaments, 191, **193**
ß₂-Microglobulin, 273
Micro-oganisms, 294, 334
Microsomal fraction, 185
Microtubules, 191, **193**
Microvilli, 192
Milk, 36, 138, 242, 266, 326
Mineralization, 330
Mineralocorticoids, 340
Minerals, 326, 329
Minor groove of DNA, 81
Mitochondria, 182, 185, 196–201, **197**, 216
composition of, 196
division of, 200
DNA in, 200
functions of, 196
genetic code of, 200, 226
and gluconeogenesis, 158
and heme biosynthesis, 178

intramembrane space of, 130
isolation of, 185
malate shuttle in, 197
matrix of, 130, 144
membranes of, 130, 132, 135, **196**, 198, 205
structure of, 197
transport systems of, 199
Mitosis (M phase), 218, **357**
Mixed acid anhydride, **9**, 114
Mixed-function oxygenase, 87, 100
Modification, 210
post-translational, 312
Mol, 11
Molecular mass
determination of, 186
of plasma proteins, 252
Molecular movement, 188
Molecular orbitals, 4
Molecular structure, **7**
Molting, 364
Molybdenum, 172, 328
Monoacylglycerol, **47**, 248
Monoamine oxidase (1.4.3.4), 154
Monoclonal antibodies, 274
Monocytes, 250
Monomer, 70
Monooxygenases (1.14.n.n), 87, 100, **293**
Monosaccharides, 30, **35**
abbreviations of, 34
dietary, 326
list of, 35
nomenclature, 34
reactions of, 32
resorption of, 242, **249**
ring forms of, 35
Morphogenesis, 364
Motor neurons, 308
Motor proteins, 192, 320
M phase (Mitosis), 218, **357**
mRNA (messenger RNA), **83**, 109, 189, 218,
223, 225, 229
maturation of, 82
modification of, 225
mtDNA (mitochondrial DNA), 201
Mucines, 245, 246, 248
Mucoid, 354
Mucopolysaccharides, 212
Mucosa, 46, 247
Multi-enzyme complex, 124, 166
Muramoylpentapeptide carboxypeptidase
(3.4.17.8), 233

Mureins, 37, 209
Muscle, 279, 281, 306–311
  amino acid metabolism in, 310
  contraction of, 307, 308, 352
    control of, 309
  energy metabolism of, 311
  fibers of, 306
  glycogen, 160
  heart, 310
  metabolism of, 310
  protein metabolism in, 310
  striated, 306
  structure of, 306
Mutagens
  chemical, 234
  physical, 234
Mutarotation, **33**
Mutations
  frame-shift, 234
  point, 234
  repair of, 235
  somatic, **234**, 270, 358
Myasthenia gravis, 320
Myelin, 316
Myeloma cells, 274
Myocardial infarction, 354
Myofibrils, 306, 308
Myoglobin, **73**, 151, 178, 180, 310, 328
  oxygen saturation curve of, 260
Myo-inositol, **49**
Myokinase (adenylate kinase, 2.7.4.3), 310
Myosin, 63, 151, **307**
  ATPase (3.6.1.32), 311
  filament, 306
  head of, 308
Myristate (myristic acid), **45**, 214
  as lipid anchor of proteins, 215

N
Na⁺ channel, 319, 322
Na⁺/Ca²⁺ exchanger, 199
Na⁺/K⁺ pump (Na⁺/K⁺-exchanging ATPase,
  3.6.1.37), 185, **209**, 249, 260, 301, 303, 316,
  318
NAD⁺ (nicotinamide adenine dinucleotide),
  78, **95**, 97, 99, **101**, 105, 130, 132, 135, 294,
  332
  biosynthesis of, 194, 332
  functions of, 100
  hydride transfer by, 95

  light absorption of, 98
  metabolism of, 146
  structure of, 95, 100
NADH, 132
  light absorption of, 98
NADH dehydrogenase (1.6.5.3), 131
NADP⁺ (nicotinamide adenine dinucleotide
  phosphate), 78, 95, **100**, 105, 119, 121, 332
  structure of, 100
NADPH, 105, 119, 143, 166, 170, 278
  in cholesterol biosynthesis, 170
  in deoxyribonucleotide formation, 177
  in hydroxylation reactions, 293
  in fatty acid biosynthesis, 167
  formation by hexose monophosphate path-
    way, 143
NAD(P)⁺/NAD(P)H reduction potential, 28,
  101
Na⁺/K⁺-exchanging ATPase (3.6.1.37), 185,
  209, 249, 260, 301, 303, 316, 318
Nephron, 296
Nephritis, diabetic, 164
Nernst equation, **319**
Nerve cells, 286, 316–325
  amino acid metabolism in, 316
  structure of, 316
  energy metabolism in, 316
  glucose requirement of, 280, 316
  neurohormones from, 336
  plasma membrane of, 205
  structure of, 317
Nerve toxins, 320
Nervonic acid, **45**
Nervous tissue, 279, 281, 284, 316–325
Neu oncogene, 359
Neuraminic acid, *N*-acetyl-, 34, 40
Neurofilament, 190
Neurohormones, **323**, 336
Neuromodulators, 318, 322
Neuromuscular junction, 308
Neuron, 294, **317**, 318, 320
Neuropeptides, neuroproteins, 322
Neurotransmitters, 154, 304, 306, 316, 318,
  323, 342
  second messenger, 352
  mechanism of action, 350
Neutral fats (triacylglycerols), **46**, 168, 278,
  278
NH₃ (ammonia), **9**, 25, 27, 115, 153, 154, 172,
  287, 301
  excretion of, 298, 300

in kidneys, 26
metabolism of, 153
plasma concentration of, 251
synthesis from N_2, 173
Niacin, 332
Nickel, 328
Nicotinamide, 94, **333**
Nicotinamide adenine dinucleotide, *see* NAD$^+$
Nicotinamide adenine dinucleotide
phosphate, *see* NADP$^+$
Nicotinate (nicotinic acid) 94, **333**
Nicotinate mononucleotide (NMN), 194
Nicotinic acetylcholine receptor, 320
Night blindness, 330
Nitrogen
balance **150**
fixation, 172
metabolism of, 286
Nitrogenase (*1.18.6.1*), 173
Nitrous acid, 234
Noble gases, 2
Nonessential amino acids, 173
Nonheme iron proteins, 130, 132, 172,
Nonpolar compounds, 22
Non-histone proteins, 194, 218
Norepinephrine (noradrenaline), 135, 145,
154, **323**, 342
Nuclear
envelope, 194
lamina, 194
localization signal, 195, 346
membrane, 194
pores, 194
proteins, import of, 195
receptors, 346, 358
skeleton, 218
Nucleases, 242
Nucleic acids, 80–85, 188
biosynthesis of, 82
digestion of, 242
functions, 83
molecular models of, **85**
Nucleocapsid, 362
Nucleoid, 188
Nucleolus, 194
Nucleophilic substitution, 6, 8, 18
Nucleosidases (*3.2.2.n*), 245
Nucleosides, 78
abbreviation of, 78
Nucleoside diphosphate (NDP), biosynthesis
of, 176, 374*, 375*

Nucleoside-diphosphate reductase (*1.17.4.1*),
176
Nucleoside triphosphate (NTP)
biosynthesis of, 176, 374*, 375*
role as coenzymes, 114
Nucleosomes, 218
Nucleotides, **79**, 189
biosynthesis of, 176, 374*, 375*
degradation of, 156
nomenclature of, 78
role of, 176
Nucleotide kinases (*2.7.4.n*), 197
Nucleus, 183, 185, **195**, 216, 218
interaction with cytoplasm, 195
Nutrients, 327
Nutrition, 46, 242, 326–335, **326**

O
O_2, *see* Oxygen
Octadecanoate (stearate), **45**, 46
Octadecenoate (oleate), **45**, 46
Octadecadienoate (linoleate), **45**, 46, 148, 327
Octadecatrienoate (linolenate), **45**, 327
Odd-chain fatty acids, 148
Oil, 44, 46
Oil-drop effect, 23, 77
Okazaki fragment, 220
Oleate (oleic acid), **45**, 46
structure of, 45
transport into mitochondria, 199
Oleyl-CoA, 369*
Oligomer, 70
Oligonucleotides, 78
Oligosaccharides, 30, 36, 41, 202
Oligosaccharidases, 249
Oligo-1,6-glucosidase (*3.2.1.10*), 245
Oncogenes, 214, 346, **359**, 360
Oncoproteins, 346, 358
One-carbon metabolism, 103, 375*
Operator, 109
Operon, 108
Opsin, 324
Optical activity, 56
Optical rotation, 32
Orbital, **2**, 4
hybridization, 5
Organelles, **182**, 182–217
isolation of, 185
marker molecules for, 185
Ornithine, 173, **287**, 371*, 373*
Ornithine carbamoyltransferase (*2.1.3.3*), 287

Orotate, 174, 375*
Orotate 5'-monophosphate, 375*
Osteomalacia, 330
Osteoporosis, 328
Ouabain, 209
Outer membrane
  of bacteria, 209
  of mitochondria, 197
  of nuclei, 195
Ovary, 341
Overexpression of proteins, 240, 358
Overhanging ends, 236
Oxalate (oxalic acid)
  role in calcium uptake, 328
  in gall stones, 298
Oxaloacetate, 102, **127**, **129**, 135, 154, 158,
  163, 173, 199, 287
  in amino acid metabolism, 155, 173, 371*,
  373*
  in carbohydrate metabolism, 367*
  in citric acid cycle, 127, 129, 135
  condensation of, 127
  in gluconeogenesis, 159, 163
  transport of, 159, 199
Oxidases (1.1.3.n), 87, 100
  mixed-function, 148
Oxidation, **28**
  α-oxidation of amino acids, 148
  β-oxidation of fatty acids, 126, 144, 146,
  147, 148, 196, 281
  ω-oxidation of fatty acids, 148
  protection against, 330
Oxidizing agents, 28
Oxidative phosphorylation, 104, 116, **130**, 196
  ATP generation by, 117, 131
  control of, 135
  electron carriers in, 131
  location of, 197
  overview of, 117, 131
  uncoupling of, 135
Oxidoreductases (1.n.n.n), **87**, 100
Oxoacid dehydrogenases, 102, **124**, 134
2-Oxoadipate, 371*
2-Oxobutyrate, 371*-373*
2-Oxo-3-deoxy-arabinoheptulosonate
  7-phosphate, 372*
2-Oxogluconolactone, 334
2-Oxoglutarate (α-ketoglutarate), **127**, **129**,
  131, 135, 153, 155, 199, 287, 301
  in amino acid metabolism, 129, 153, 155,
  301, 370*, 371*, 373*

in citric acid cycle, 127
in malate shuttle, 199
2-Oxoglutarate dehydrogenase complex, 124,
  127, 135, 137
2-Oxoisocapronate, 370*, 372*
2-Oxoisovalerate, 370*, 372*
2-Oxo-3-methylvalerate, 370*
2-Oxo-4-methylvalerate, 372*
5-Oxoproline (pyroglutamate), 322, 342
Oxygen (O_2), **3**, 105, 117, 119, 130, 135, 197,
  250, 258, 260
  activated, 293
  in amino acid degradation, 151
  in cholesterol biosynthesis, 169
  electronegativity of, 20
  formation from water, 119, 121
  in hydroxylation reactions, 293, 317
  partial pressure of, 305
  redox potential of, 28
  saturation with, 259, 261
  solubility of, 258
  storage of, 310
  supply of, 310
  transport of, 258
    regulation of, 260
  utilization of, 276
    by liver, 276
    by muscle, 310
    by nerve cells, 316
Oxygenation, 258
Oxyhemoglobin, 269
Oxysterols, 170

P

P450 (cytochrome P450), 170, 290, **292**
Pain, role of eicosanoids, 354
Pair-rule genes, 364
Palindrome, 236, 346
Palmitate (palmitic acid), **45**, 46, 144, 146,
  166, 215, 369*
  ATP yield from, 147
  biosynthesis of, 145, 167
  degradation of, 145, 147
  as lipid anchor of proteins, 215
  transport into mitochondria, 199
Palmitoyl-CoA, 368*, 369*
Pancreas, 164, 244, 247, 279, 281, 338, 343
  digestive enzymes from, 247
  lipases from, 288
  secretions of, 242, **245**, 247
Pantetheine, 103, **167**

Pantothenate (pantothenic acid), **333**
Pantoinate (pantoinic acid), **103**, **333**
Papain (*3.4.22.2*), 266
PAPS (phosphoadenosinephosphosulfate), 291
Parahormone, 336
Paracrine hormone activity, 339, 354
Parathyrin, 302, 305, 330
Parathyroid gland, 302
Parietal cells, 246
Parkinson's disease, 154
Partial pressure, 258
Passive transport, 207
Pattern formation, 364
PCR (polymerase chain reaction), 240
Pectins, 38
Pellagra, 332
Penicillin, 93, **233**
    excretion of, 296
Pentoses, 30
    digestion of, 242
Pentose phosphate pathway (hexose monophosphate shunt), **143**, 189, 279
Pepsins (*3.4.23.n*), 244, 246
Pepsinogens, 244
Peptidases (*3.4.n.n.*), **151**, 242, 305
    in digestive tract, 244
    lysosomal, 150
    role in degradation of neuropeptides, 322
Peptide bond, 8, **65**, 288
    biosynthesis of, 84, 230, 288
    chemical synthesis of, 65
    conformation of, 65
    dimensions, 65
    hindered rotation of, 65
    structure of, 65
Peptide chain, conformations of, 65
Peptide hormones, 343
    biosynthesis of, 349
    degradation of, 349
    inactivation of, 349
    mechanism of action, 351
    metabolism of, 349
Peptides
    as neurotransmitters, 322
    as hormones, 343, 349, 351
    secondary structure of, 67
Peptidyl-dipeptidase A (*3.4.15.1*), 201
Peptidyl site, 228, 230
Peptidyl transferase (*2.3.2.12*), 230
Perinuclear cleft, 194

Periodic table, **3**
Peripheral membrane proteins, 202
Periplasmic space, 209
Peristaltic movement, 242
Permeability coefficient, 318
Permeability of membranes, 207, 319
Permeases, 206
Pernicious anemia, 334
Peroxidases (*1.11.1.n*), 87, 99, 180, 185, 201, 275, 355
Peroxides, 260
Peroxisomes, 148, 185, **201**, 216
pH (*see also* pH value), **24**
Phages, 238, 362
    M13, 362
    T4, 362
Phagocytosis, 192, 212, 266
Phalloidin, 191
Pharmaceuticals, 276, 290
    excretion of, 296
    induction of degradation systems by, 348
    metabolism of, 290
Phase partitioning, 22
Phase I and II reactions, 290
Phenols, 200
Phenylalanine, **59**, 83, 172, 327, 373*
    biosynthesis of, 173, 372*
    degradation of, 155, 371*
    tRNA^{Phe}, 82, 83, 85
        van der Waals model of, 84
Phenylethanolamine *N*-methyltransferase (*2.1.1.28*), 323
Phenylisothiocyanate, 61
Phenylpyruvate, 372*
Pheophytin, 120, 122
Phi (φ) angle, 65
Phorbol ester, **361**
Phosphatase
    acid (*3.1.3.2*), 212, 361
    alkaline (*3.1.3.1*), 214, 245
    in control of phosphorylase, 111
Phosphate (phosphoric acid), **5**, 115, 300, 302
    in calcium uptake, 328
    daily requirement of, 328
    excretion of, 299, 302
    transport of, 199
Phosphate buffer
    pK$_a$ values of, 257
Phosphate group, transfer potential, 115
Phosphatidate (phosphatidic acid), 42, **49**, 168, 368*

Phosphatides, 42
Phosphatidylcholine, **49**, 168, 202, 205, 368*
Phosphatidylcholine-sterol acyltransferase
  (LCAT, *2.3.1.43*), 254
Phosphatidylethanolamine, **49**, 168, 205
Phosphatidylinositol, 168, 205, 214, 368*
  glycosylated, **215**
1-Phosphatidylinositol-4,5-bisphosphate
  (PInsP$_2$), 48, 168, 350
  as precursor of second messengers, 351
1-Phosphatidylinositol-4-phosphate, 168
Phosphatidylserine, **49**, 168, 205, 368*
Phosphoadenosinephosphosulfate (PAPS),
  291
Phosphoanhydride bond, 115
Phosphodiesterase (*3.1.4.n*), 111, 185, 324, 351,
  353
  cAMP specific (*3.1.4.17*), 111, 353
  cGMP specific (*3.1.4.35*), 324, 350
Phosphoenolpyruvate, 129, 135, **141**, **159**, 163,
  367*, 372*
Phosphoenolpyruvate carboxykinase (GTP)
  (*4.1.1.32*), 159, 163, 367*
6-Phosphofructokinase (*2.7.1.11*), 137, 141,
  163, 278
6-Phosphogluconate, 143
Phosphogluconate dehydrogenase (decarboxylating) (*1.1.1.44*), 143
6-Phosphogluconolactone, 142, 143
2-Phosphoglycerate, **141**, **159**, 367*, 373*
3-Phosphoglycerate, **121**, **141**, **159**, 366*, 367*
Phosphoglycerate kinase (*2.7.2.3*), 121, 137,
  141, 366*, 367*
Phosphoglycerides (phospholipids based on
  glycerol), 49
Phosphoserine, 372*
3-Phosphohydroxypyruvate, 373*
Phospholipase, 245, 350
  A$_2$ (*3.1.1.4*), 244, 247, 355
  C (*3.1.4.3*), 351
Phospholipids, 42, **49**, 50, 203, 204, 245, 248,
  254, 263, 284, 355
  biosynthesis of, 168, 368*
  in membranes, 202, 204
  model of, 203
  as precursors of second messengers, 351,
  355
  structure of, 49
4-Phosphopantethein, 166
Phosphoprotein phosphatase (*3.1.3.16*), 111
Phosphopyruvate hydratase (*4.2.1.11*), 141

Phosphoribosylamine, 374*
Phosphoribosyl-5-aminoimidazole, 374*
Phosphoribosyl-5-amino-4-imidazolecar-
  boxamide, 374*
N-(Phosphoribosyl)-anthranilate, 372*
Phosphoribosyl-4-carboxy-5-aminoimida-
  zole, 374*
5-Phosphoribosyl 1-diphosphate (PRPP), 157,
  372*, 374*, 375*
Phosphoribosyl-5-formamido-imidazole-4-
  carboxamide, 374*
Phosphoribosylformylamide, 374*
Phosphoribosylglycineamide, 374*
Phosphoribulokinase (*2.7.1.19*), 121
Phosphoric acid-anhydride bond, 114
Phosphoric acid diester, 8
  bond, 114
Phosphoric acid ester, **9**, 32
Phosphorolysis of glycogen, 111, 161
Phosphorus, dietary, 328
Phosphorylase (*2.4.1.1*), 310
  a-form, 111
  b-form, 111
  coenzyme of, 334
  interconversion of, 111
Phosphorylase kinase (*2.7.1.38*), 111, 310, 352
Phosphorylation, **102**, 110
Phosphorylation/dephosphorylation of pro-
  teins, 111, 348, 357
Phosphorylation, oxidative, 104, **130**
  ATP generation by, 117, 131
  control of, 135
  electron carriers in, 131
  location of, 197
  in muscle, 310
  overview of, 117, 131
  uncoupling of, 135
Phosphorylation, substrate-level, **116**, 140
Phosphorylations, cascade of, 111, 356
Phosphoserine, 373*
Photoautotrophic organisms, 118
Photolyase (*4.1.99.3*), 234
Photometry, 98
Photorespiration, 200
Photoreceptor, 324
Photosynthesis, 104, 118–123, **119**
  dark reactions of, 118, 120
  light reactions of, 118
  overview of, 118
  reaction center in, 123
  redox series in, 121

Photosystems, 118, 122
  PSI, 119, 121
  PSII, 119, 121
Photoreactivation, 234
Phylloquinone, **331**
Phytol, **51**, 118
  role in calcium uptake, 328
pH value, 24
  decline of, 310
  of body fluids, 26
  maintenance of constant, 256, 300
  of plasma, 256, 257
  of urine, 298
PInsP₂ (1-phosphatidylinositol 4,5-bis-
  phosphate), 48, 168, 350
Pitch of helices, 66, 80
Pituitary, 343
Pleated sheet, 66
  anti-parallel, 67
  in fibroin, 68
  parallel, 67
pKₐ value, **25**, 56
  of amino acids, 56, 59
  definition of, 25
Platelet factor, 263
Plaque, 238
Plasma (blood), 250–257
  composition of, 251
  hormones in, 338
  pH value of, 27, **257**
Plasma cells, 25, 266
Plasma proteins, 250, **253**
  half-life of, 252
  lipoproteins, 254
  list of, 253
  site of synthesis of, 252, 276
Plasma membrane, 183, 185, 202, 204, 318,
  354
Plasma thromboplastin precursor, 263
Plasmid, **237** , 240
Plasmin (*3.4.21.7*), 262, 263
Plasminogen, 253, 262
Plasminogen activator (*3.4.21.68/73*), 262
Plastocyanin, 119
Plastoquinone, 50, **117**, 119, 121
Pleated sheat, 67
Point mutation, 271
Polar group, 22
Polar head group, 45, 203
Polarimetry, 32
Polarity

  of amino acids, 59
  of membranes, 204
  of molecules, 20, 22, 206, 290
Polarity genes, 364
Poliovirus, 362
Polyacrylamide gel, 252
Polyadenylate tail, 224
Polyadenylation sequence, 222
Polyethylene glycol, 274
Polymerase, 194, **221**, 222, 346
Polymerase chain reaction (PCR), 240
Polymorphism, 272
  genetic, 252
Polynucleotide, 78
Polynucleotidases (*3.1.3.n*), 245
Polyol pathway, 282
Polypeptide chain, 66
Polyprotein, 348
Polysaccharides, 30, **37**, 38 188, 249
  dietary, 326
  in plants, 38
  list of, 37
  reserve, 36
  structure of, 36
  water-binding, 36
Polysome, 228
POMC (pro-opiomelanocortin) gene, 348
Pool of metabolites, 104, 198
Pore complex, 194
Porin, 197, 209
Porphobilinogen, **179**
Porphobilinogen synthase (*4.2.1.24*), 179
Porphyria, 178
Porphyrins, 180
  biosynthesis of, 128, **179**
  degradation of, 180
Portal vein, 242, 249, 276, 278
Postresorption state, 281
Postsynaptic membrane, 320, 322
Post-translational modifications, 56, 68, 210,
  312
Potassium channels, 318
Potassium ions, 328
  concentration of, 319
  daily requirement of, 329
  excretion of, 296, 299
  permeability coefficient of, 319
Potential, 11
  action, 308, **319**, 320, 322
  chemical, 16
  electrical, 28

electrochemical, **116**, 198, **206**, 318
  resting, 319
Potter-Elvehjem homogenizer, 184
Power stroke, 309
Pre-albumin, 252
Pregnancy, 341
  test, 298
Pregnenolone, **345**, 369*
Prekallikrein, 263
Prenylation of proteins, 50, 132, 214
Pre-propeptide, 348
  in collagen, 312
  in hormone biosynthesis, 348
  of insulin, 164
Presynaptic membrane, 320
Presynaptic neuron, 320
Preservatives, 291
Primary bile acids, 288
Primary culture, 275
Primary structure of protein, 70
Primary transcript (hnRNA), 222
Primary urine, 296, 302
Primary wall of plant cells, 39
Primase (*2.7.7.6*), 220
Primer, 220, 240
Proaccelerin, 263
Procollagen, 312
Procollagen-lysine 5-dioxygenase (*1.14.11.4*), 313
Procollagen-proline dioxygenase (*1.14.11.2*), 313
Proconvertin, 263
Product, **11**
Product inhibition, 134
Proenzymes, 244, 247
Profilin, 190
Progesterone, **55**, 341, **345**
  biosynthesis of, 369*
  hormone axis, 338
Progestins, 340
Progression of tumors, 361
Proinsulin, **71**, 164, 210
Prokaryotes, 182
Proliferation of
  cells, **357**, 358, 361
  tumors, 360
Proline, **59**, 68, 172
  biosynthesis of, 173, 373*
  in collagen, 312
  degradation of, 155, 371*
  hydroxylation, 313

Promotion, 360
Promoter, 109
Promoter region, 222, 224
Pro-opiomelanocortin (POMC) gene, 348
Propionate (propionic acid), **45**, **139**
  fermentation to, 139
Propionibacterium, 138
Propionyl-CoA 148, 370*, 371*
Prostacylins, **355**
  autocrine effects of, 338
Prostaglandins, **355**
Prostaglandin synthase (*1.14.99.1*), 355
Prosthetic group, 100, 132
Proteases (proteinases) 151
Protease inhibitors, 262
Proteins, 62–77, 189, 203
  amino acid sequence of, 70
  apolar core of, 76
  backbone of, 67
  biosynthesis of, *see* Protein biosynthesis
  buffering activity of, 26
  cleavage of, 150
  conformation of, 72
  degradation of, 150
  denaturation of, 72, 247
  dietary, 326
    value of, 172
    minimum requirement, 326
    in starvation, 280
  digestion of, 242
  DNA-binding, 358
    domains, 270
  as energy store, 280
  excretion of, 299
  folding of, 73, 214
  functions of, 62
  globular, 70
  glycosylated, **40**, 202, 252, 266
  half-life of, 150, 252
  import, 194
  localization of, 214
  lysosomal, 213
  in membranes, 202, 209
  metabolism of, 151
  modification of, 211, 216, 262, 312
  as neurotransmitter, 322
  nuclear, 195
  peptide bonds in, 64
  in plasma, 253
  overexpression of, 240
  regulatory, 109

requirement, 150
secondary structure of, 62–77
single strand-binding, 221
sorting of, 217
structural, **69**
synthesis of, *see* Protein biosynthesis
targeting, 217
translocation through membranes, 214
transport of, 210, **214**, 216
Proteinases (*3.4.21.n-3.4.99.n*), 151,
in digestive tract, 245
Protein biosynthesis, 150, 189, 211, 216, 226–
231, **228**
antibiotic inhibitors of, 233
in cell cycle, 357
initiation of, 229
elongation of, 230
essential amino acids, 326
role of hormones in, 348
in rough endoplasmic reticulum, 211
initiation of, **228**
termination of, 230
Protein kinases (*2.7.1.n*), 135, 350, 353, 356,
358
classification, 358
type A, 111, 352
type C, 351, 360
Protein-lysine 6-oxidase (*1.4.3.13*), 313
Protein modification, 211, 216, 262, 312
Protein phosphatases (*3.1.3.16*), 134, 356, 358
Protein phosphorylation, 110, 356, 358
Protein targeting, 217
Protein translocation, 215
Proteoglycans, 40, **315**
Proteohormones, 343
Proteolysis, 150
in hormone inactivation, 348
limited, 304
Prothrombin, 253, 263
Protofilaments, 190
Protons (H+), 24, 256, 300
balance, 256
channel, 135
concentration of, 256
excretion of, 257, 296, **301**
gradient, 117, 130, 134, 199
production of, 256
secretion of, 247, 301
transport, 246
Proton pumps
electron-driven, 117, 131

light-driven, 117, 119, 121
in lysosomes, 212
Proton motive force, 116
Proto-oncogenes, 359
Protoporphyrin, 179
Protoporphyrinogen, 179
Provitamin, 330
D₂, 248
Pseudouridine, 226
Psi (ψ) angle, 65
PTC (phenylisothiocyanate) amino acids,
61
Pteridine, 102, 332
Pulsed hormone release, 338
Purine bases, **79**
biosynthesis of, 156, **175**, 374*, 375*
degradation of, 156
as neurotransmitters, 322
salvage pathway for, 156
Puromycin, 230, 233
Pyranoses, 30, 34
Pyran ring, 30
Pyridine nucleotides, 101
Pyridoxal, 102, **335**
Pyridoxal phosphate, 152
Pyridoxamine, 334
Pyridoxamine phosphate, 152
Pyridoxol, 334
Pyrimidine bases, **79**
biosynthesis of, **175**, 375*
degradation of, 156
Pyroglutamate (5-oxoproline), 322, 342
Pyrrole, 178, 258
Pyrroline-5-carboxylate, 371*, 373*
Pyruvate, 12, 95, 97, 99, **125**, 131, 135, 137, 139,
141, 153, 155, 159, 163, 197, 199, 367*-370*,
372*, 373*
conversion to acetyl-CoA (oxidative decarboxylation), 125, 137
conversion to alanine (transamination),
153
conversion to ethanol (fermentation),
139
conversion to lactate (fermentation), 95,
97, 99, 137, 139
conversion to oxaloacetate (gluconeogenesis), 129, 159
family of amino acids, 155, 173
formation from glucose (glycolysis), 141
plasma concentration of, 251
transport into mitochondria, 199

Pyruvate carboxylase (*6.4.1.1*), 128, 154, 158, 163, 367*
  coenzyme 334
Pyruvate decarboxylase (*4.1.1.1*), 139
Pyruvate dehydrogenase (PDH, *1.2.4.1*), **125,** 135, 137
Pyruvate dehydrogenase complex, 125
Pyruvate dehydrogenase kinase (*2.7.1.99*), 135
Pyruvate dehydrogenase phosphatase (*3.1.3.43*), 135
Pyruvate kinase (*2.7.1.40*), 135, 137, 141, 158, 163, 278, 367*
Pyruvate/lactate reduction potential, 28

Q

Q (coenzyme Q, ubiquinone), **50**, **101**, 126, 131, 132, 170
Q_{10} rule, 16
Quaternary structure, 70

R

Racemate, 56
Radicals, 234
  formation of, 200
Radioimmunoassay, 274
Radius, atomic, 7
Ras protein, 359
Rate constant, 16, 90
Reabsorption, 296
Reaction center, 120, 122
  photosynthetic, 123
Reaction, heat of, 14
Reaction kinetics, 16, 90
Reaction order, 16
Reaction rate, 16
  of enzymes, 91
Reaction specificity, 86, 346
Receptors, 216, 358, 264
  β-adrenergic, 111
  for catecholamines, 242
  7-helix receptors, 322, 350
  for hormones, 336, **347**, **351**
  ligand interaction, 214
  for light, 324
  membrane-bound, 351, 352
  for neurotransmitters, 322
  for odor, 324
  for peptide hormones, 351
  postsynaptic, 320

  for steroid hormones, 108, 224, **347**
  for taste, 324
Receptor-mediated endocytosis, 254
Receptor tyrosine kinases, 351
Recombination, somatic, 270
Recombination repair, 234
Red algae, 37
Red cells, *see* Erythrocytes
Redox coenzymes, 101
Reduction potential, **28**, 130, 132
Redox reaction, 14, **28**, 100
Redox scale, 28
Redox series, 28
Redox system, 28, 121, 132
  of photosynthesis, 121
  of respiratory chain, 133
Reducing agent, 28, 334
Reducing equivalents, 100
Reduction, 28
Reduction assays, 32
Reductones, 32
Refsum syndrome, 148
Regulated secretion, 216
Regulation
  allosteric, 107, **112**, 260
  of carbohydrate metabolism, 163
  of citric acid cycle, 135
  of energy metabolism, 135, 137
  of gene expression, 106, 108, 346
  isosteric, 107
  metabolic, 107
Regulation system, hormonal, 336
Regulator protein, 106, 108
Regulatory enzyme, 106
Regulatory mechanisms, 107
Regulatory subunit, 112, 352
Releasing factor, 230
Renin (*3.4.23.15*), 304
Renin-angiotensin system, 305
Repair of DNA, 234
Replication, 194, 218, **221**
  direction of, 220
  fork, 220
  mechanism of, 220
  origin of, 220
Repression, 106
Repressor, 108, 109
rER (rough endoplasmic reticulum), 185, 211, 216
RES (reticuloenthothelial system), 180
Residual body, 212

Resistance against antibiotics, 236
Resonance, **4**
Resorption, 242
  of amino acids, 248
  of bile acids, 288
  of carbohydrates, 249
  of dietary constituents, 242, 249
  disturbances of, 328
  of ethanol, 294
  of lipids, 249
Respiration, 130
  rate of, 256
Respiratory acidosis, 256
Respiratory chain, 116, **131**, **133**, 134, 146, 196
  complexes of, 130
  control of, 135
  organization of, 131
  redox systems of, 133
  regulation of, 134
Resting potential, 319
Restriction endonucleases (*3.1.21.4*), 237, 238, 240
Restriction fragment length polymorphism (RFLP), 240
Restriction point, 356
Retention signal, 217
Reticuloendothelial system (RES), 180
Retina, 324
Retinal, 122, **325**, 330
Retinal isomerase (*5.2.1.3*), 325
Retinoate (retinoic acid, vitamin A acid) 330, 340
  mechanism of action of, 346
Retinoids, 330
Retinol, 325, **331**
Retinol-binding protein, 253
Retinol dehydrogenase (*1.1.1.105*), 325
Retroviruses, 362
Reverse transcriptase (*2.7.7.49*), 362
RFLP (restriction fragment length polymorphism), 240
R_f value, 53
RGD sequence, 314
Rhinovirus, 362
Rhizobium, 172
Rhodopseudomonas viridis, 122
Rhodopsin, 324, 330
Riboflavin, **333**
Ribonuclease (*3.1.27.5*), 245
Ribonuclease H (*3.1.26.4*), 362
Ribonucleic acid, *see* RNA

Ribonucleoside-diphosphate reductase (*1.17.4.1*), 177
Ribonucleotide reductase (*1.17.4.1*), 176
Ribose, **35**
Ribose 5-phosphate, 143, 366*, 367*, 374*
Ribosomal RNA (rRNA), **83**, 185, 194, 218, 228
Ribosomes, 82, 183, 184, 188, 210, 218, **229**
  free, 216
  isolation of, 185
  marker molecules of, 185
  in protein synthesis, 231
  RNAs of, 83
  in rough endoplasmic reticulum, 211
  structure of, 189, 228
  subunits of, 228
  translocation of, 230
Ribozyme, 18, 86, **224**
Ribulose, 35
Ribulose 1,5-bisphosphate, **121**, 366*
Ribulose 5-phosphate, 121, 143, 366*, 367*
Ribulose-bisphosphate carboxylase (*4.1.1.39*), 121, 366*
Rickets, 328, 330
Rifampicin, 230
Rifamycin, 233
Ring conformations, 52
RNA (ribonucleic acid), **79**, **83**, 185, 189
  biosynthesis of, 222
  in cell cycle, 357
  functions of, 82
  hnRNA (high molecular weight nuclear RNA), 218, **222**, 270
  mRNA (messenger RNA), 83, 109, 189, 194, 218, 223, 225, 229
  maturation of, 82, 194, 218, 223, **225**, 348
  rRNA (ribosomal RNA), **83**, 185, 194, 218, 228
  snRNA (small nuclear RNA), **83**, 218, 222, 224
  structure of, 226
  transcription of, 223
  tRNA (transfer RNA), 83, **85**, 189, 195, 218, **227**
    van der Waals model of, 85
  types of, 83
RNA polymerase, 109, 188, 222, 346
  DNA-directed (*2.7.7.6*), 223
    forms of, 222
Rods, 324
Root nodules, 172
Rotor, 186

Roughage, 242
Rough endoplasmatic reticulum (rER), 185, 210, 216
rRNA (ribosomal RNA), **83**, 185, 194, 218, 228
Rubber, 50
Rubisco (ribulose-bisphosphate carboxylase, *4.1.1.39*), 121, 366*

S

Saccharomyces cerevisiae, 138
Salicylate, 355
Saliva, 244, 266
Salvage pathway, 157
Saponins, 54
Sarcomere, 306
Sarcoplasmic reticulum, 308
Saturation, 144, 258
Saturation curve, 91
  sigmoidal, 106, 112, 258
Schiff's base, 102, 152
Schwann cells, 316
Scurvy, 312, 334
SDS (sodium dodecyl sulfate), 252
Second messengers, 336, 350, **352**, 354
  in vision, 324
  metabolism of, 352
Secondary-active transport, 206, 302
Secondary bile acids, 288
Secondary structure, 66, **70**
  of collagen, 313
  formation of, 210
  of insulin, 71
  of polypeptides, 67
Secosteroid, 54, 344
Secretion, 296
  of hormones, 348
Secretions, 244
  intestinal, 266
Secretory pathway, 216
Secretory vesicle, 210
Sediment, 184
Sedimentation, 186
  coefficient, 187
  constant, 187
  velocity 187
Sedoheptulose 1,7-biphosphate, 366*
Sedoheptulose 7-phosphate, 143, 366*, 367*
Segmentation genes, 364
Segment polarity genes, 364
Selection, clonal, 264
Selenium, 328

Self-digestion, 212
Self-splicing RNA, 224
Semiconservative replication, 220
Sequencing of DNA, 238
Serine, **59**, 68, 172, 368*
  biosynthesis of, 173, 373*
  degradation of, 155, 370*, 375*
  family, 17
  in phospholipids, 48, 204
  as precursor of cysteine, 373*, 375*
  as precursor of glycine, 373*
  residue in active center, 320
Serine proteinases (*3.4.21.n*), 150, 151, 262
  clotting factors, 262
  chymotrypsin, (*3.4.21.1*), 245
  thrombin, (*3.4.21.5*), 263
Serotonin, 155, **323**, 343
  excretion of, 298
Serum, 250
Serum albumin, 63, 253
Serum proteins (plasma proteins), 252
  half-life of, 252
  lipoproteins, 254
  list of, 253
  site of synthesis of, 252, 276
Sesquiterpenes, 50
Sex characteristics, 341
Sex hormone-binding globulin, 253
Sialic acid, 34, 48, 252
Sight, 324
Sigmoidal saturation curve, 106, 112, 258
Signal peptidase (*3.4.99.36*), 210, 217
Signal peptide (leader peptide), 164, 210, 214, 217
  of immunoglobulins, 270
  of peptide hormones, 348
Signal reception, 204
Signal recognition particle, 210
Signal region, 216
Signal sequence, 216
Signal substances, 320, 336
  hydrophilic, 343
    mechanism of action, 350
  lipophilic, 340
    mechanism of action, 346
  vitamins as precursors of, 330
Signal transduction, 204, 272, 320, 324, **351**, 358
Signals, structural, 216
Silica gel, 186
Silicon, 328

Silk, 68
Single-strand binding proteins, 220
β-Sitosterol, **55**
Skeletal muscle, 306
Skin, 312, 330
SKL sequence, 217
Sliding-filament model, 306
Small intestine
  pH value in, 27
  secretions of, 244
Small nuclear ribonucleoprotein particles (snRP), 224
Small nuclear RNA (snRNA), **83**, 218, 222, 224
Smell, second messengers for, 352
snRNA (small nuclear RNA), **83**, 218, 222, 224
sn-number, 46
Soap, 46
  bubbles, 23
Sodium-calcium exchanger, 199
Sodium channel, 319, 320
Sodium dodecyl sulfate (SDS), 252
Sodium ions (Na⁺), 300, 328
  concentration of, 319
  daily requirement of, 329
  excretion of, 299
  permeability coefficient of, 319
  plasma concentration of, 251
  reabsorption of, 302
  retention of, 304
  transport of, 206, 208
Sodium-potassium transport (*see also* Na⁺/K⁺-exchanging ATPase), 209
Solenoid, 218
Solubility of fatty acids, 23
Somatic mutation, 270
Somatic recombination, 270, 272
Somatostatin, **323**
Sorbitol, 33, **35**, 282
Sorting of proteins, 214, **217**
Spectrin, 190, 192
Spectroscopy, 99
S phase, 356
Sphingolipids, 42, 48
Sphingomyelins, **49**, 205, 368*
Sphingophospholipids, 48
Sphingosine, 42, 48, 368*
Spliceosome, 224
Splicing of mRNA precursors, **225**, 270, 348
  alternative, 314
Squalene, 50, **171**, 369*
SR-foot, 308

Standard conditions, 12
Standard free energy change (ΔG°), 12, 14
Standard potential, **28**
Starch, **39**
  cleavage of, 244
  digestion of, 242, 244
  dietary, 326
Start codon, 227, 229
Start of transcription, 225
Starvation, **281**, 300, 310, 316
  carbohydrate metabolism in, 159
  fuel metabolism in, 159, 280, 310, 316
  lipid metabolism in, 144, 284
Statins, 338
Stationary phase, 60
Steady-state conditions, 12
Stearate (stearic acid), **45**, 46
Stearyl-CoA, 369*
Stem cell, 305
Stercobilinogen, 180
Stereochemistry
  of carbohydrates, 31
  of steroids, 53
Sterines, 54
Steroid alkaloids, 54
Steroid hormones, **55**, 290, 336, **341**, 345
  biosynthesis of, **345**, 369*
  inactivation of, 276, **345**
   in liver, 276, 290
  mechanism of action of, 347
  metabolism of, 290, **345**
  regulation of development by, 364
  structure of, 55
Steroids, 42, 50, 52–55, **52**
  biosynthesis of, 171, 369*
  building blocks of, 53
  3D structure of, 53
  excretion of, 298
  nomenclature of, 52
  overview of, **54**
  structure of, **52**
Sterols, 42, 50, 52, **55**
  carriers of, 170
  esters of, 42
Stigmasterol, **55**
Stomach
  glands of, 246
  secretions of, 242–247
  pH value in, 27, 245
Stop codon, 227, 229

Stop-transfer signal or sequence, 214, 216
Storage diseases, 212
Streptococcus, 138
Streptokinase, 262
Streptomycin, 230
Striated muscle, 306
Stroma, 118, 119
Structural proteins, 66, **68**, 312
Structure
  primary, 71
  secondary, 71
  tertiary, 71
  quarternary, 71
  molecular, 71
Stuart-Prower factor, 263
Substance P, **323**
Substitution, nucleophilic, 6, 18
Substrate, **86**, 90
  analogue, 92, 304
  saturation curve, 90, 106, 258, 260
  specificity, 86
Substrate-level phosphorylation, **116**, 140
Subunit, 70
  catalytic, 112, 352
  regulatory, 112, 352
Succinate (succinic acid), **127**, **129**, 131
Succinate-CoA ligase (*6.2.1.4*), 127, 137
Succinate dehydrogenase (*1.3.5.1*), 127, 131, 137, 185
Succinate-semialdehyde   dehydrogenase
  (*1.2.1.24*), 317
Succinyl-CoA, **127**, 129, 135, 149, 155, 179, 370*, 371*
  in amino acid degradation, 155
  in citric acid cycle, 127, 129, 135
  in degradation of odd-numbered fatty acids, 149
  in heme synthesis, 179
Sucrose, 34, **37**
  gradient, 186
Sugars, 30–41, 189
  conformation of, 31
  derivatives of, 33
  metabolism of, 140–143, 158–165, 282
  in nutrition, 326
  sulfates, 315
  transport, role of retinol in, 331
Sugar alcohols, 32, 34
Suicide substrate, 92, 233
Sulfate, 256
  esters, biosynthesis of, 290

  excretion of, 299
  residues in sugars, 314
Sulfathiazole, **233**
Sulfatides, 48
3-Sulfinylpyruvate, 371*
Sulfolipids, 48
Sulfonamides, 333
Sulfur, 328
Sulfur-containing amino acids, 59
  role in acid-base balance, 256
Sulfuric acid, 256
  conjugate formation with, 298, 344
  formation of, 256
Super helix, 68
Supernatant, 184
Suppressor cells, 264
Svedberg unit, 186
S value (sedimentation coefficient), 186
Sweat, 266
  pH value of, 27
Symbiosis, 172, 200
Symbol, chemical, 3
Symport, 199, **206**
Synapse, 316, **321**, 324
Synaptic cleft, 308, 320, 322
Synthases, 87
Synthetases, 87

T
Target cells of hormones, 336, 346
Targeting of proteins, 217
TATA box, 222
Taurine, **289**
Taurocholate (taurocholic acid), 289
T cells, **264**, 272
  receptors of, 273
Temperature, 12
Template, 218
  strand of DNA, 220
Tendons, 312, 314
Termination of protein biosynthesis, 230
Terminus of DNA, 3'- and 5'-, 78
Terminus of protein, C- and N-, 64
Terpenes, 50
Tertiary structure, **70**
  of collagen, 313
  formation of, 210
  of insulin, 71
Testes, 341
Testosterone, **55**, **341**
  biosynthesis of, 344, 369*

hormonal axis, 338
  function of, 150, 311, 341
Tetracycline, **233**
Tetrahydrofolate (THF), **103**, 332, 375*
  metabolism of, 375*
  N^5-methyl-, **103**, 375*
  N^5-N^{10}-methylene-, **103**, 375*
  N^{10}-formyl-, **103**, 174, 375*
Tetraiodothyronine (thyroxine), 170, 252, 274, 336, **341**
Tetrose, 30
T-helper cells, 264, 362
Thermodynamics, 12–15
Thiamine, **333**
Thiamine diphosphate of, **103**, 124, 332
  hydroxyethyl derivative of, 125
Thiazole ring, 332
Thick myosin filament, 307
Thin actin filament, 307
Thin-layer chromatography, 52
Thioester, **9**, 102
Thiokinase (succinate-CoA ligase, *6.2.1.4*), 128, 294
Thioclastic cleavage, 146
Thiol, 9, 184, 260
Thioredoxin, 177
Threonine, **59**, 60, 172, 327
  biosynthesis of, 173, 372*
  degradation of, 155, 370*
Thrombin (*3.4.21.5*), 263
Thrombocytes, **251**, 354
Thromboplastin, 263
Thrombosis, 262, 330
Thromboxanes, **355**
Thrombus, 262
Thylakoid, 119
Thymidine, 78, 274
Thymidylate synthase (*2.1.1.45*), 177, 375*
Thymine, **79**, 157
Thymine dimer, **235**
Thymus, 264
Thyreoglobulin, 328
Thyroid gland, 302, 341, 343
Thyroliberin (TRH), **323**, 343
Thyrotropin (TSH), 343
Thyroxine, 170, 252, 274, 336, **341**
  hormonal axis, 338
  mechanism of action of, 346
  binding globulin, 253
Tiglyl-CoA, 370*
Tin, 328

Tissue factor, 263
Tissue hormones, 336, 338
Tissue plasminogen activator (TPA), 262
Titration curve, 25
T lymphocytes, 265, 272
Tobacco, 360
Tobacco mosaic virus, 362
Tocopherol, **331**
Toxic compounds, 276
TPA (tissue plasminogen activator), 262
Trace elements, 2, **329**
Tracer, 274
Trans-active factor, 224
Transaldolase (*2.2.1.2*), 142, 367*
Transaminases (*2.6.1.n*), 152, 286
Transamination, 102, **153**, 154, 282
Transcobalamin, 253
Transcortin, 253
Transcriptase, reverse (*2.7.7.49*), 362
Transcription, 82, 108, 218, **223**, 225
  control of, 109, 225
  during cell cycle, 356
  direction of reading, 222
Transcription factors, 108, 218, 346, 358, 364
  basal, 222, 224
Transcriptional control, 106, **109**, 222, **225**, 336, 346
Transducin, 324
Transferases (*2.n.n.n*), **87**, 290
Transferrin, 253
Transfer RNA (tRNA), 83, **85**, 189, 195, 218, **227**
Transformation, 236, 240, **361**
Transketolase (*2.2.1.1*), 102, 142, 366*
  role of vitamin B_1, 332
Transition metal, 2
Transition state, 18, 88, 92, 96
Translation, 210, 214, 218, 226–231, **229**
Translocation
  in protein synthesis, 230
  across membranes, 198, 202, 215
Translocators, 198, 202, 214
Transmembrane helices, 122, 202, 209
Transmembrane proteins, 122, 202, 209
Transmitter, 316, 320, 323
Transplantation antigens, 272
Transport, 192, 199, 204, **207**, 209, 216, 302
  active, 207, 296, 209
  across membranes, 207, 215
  process, 210
    in kidneys, 302

in mitochondria, 199
secondary active, 302
systems, 198, 296
vesicles, 210, 212
Transport ATPase (*3.6.1.n*), 207, **209**
Transport of metabolites/ions, 204, **207**, 319
of Ca²⁺, 199, 208
of cholesterol, 257
of fatty acids, 145, 199
of gases, 259
of glucose, 209, 249
of Na⁺ and K⁺, 209, 319
of proteins, 215
of sugars, 249
of triacylglycerols, 255
Transport proteins (transporters), 206
Transthyretin, 253
Trehalase, 245
Triacylglycerols (triglycerides, neutral fats),
    47, 144, 168, 248, 254
biosynthesis of, **145**, 278
digestion of, 249
metabolism of, 145
plasma concentration of, 251
transport of, 255
Triacylglycerol lipase (*3.1.1.3*), 245, 249
Triiodothyronine (T₃), 340
Trimming, 210
Triose-phosphate isomerase (*5.3.1.1*), 141,
    366*, 367*
Triple helix of collagen, 312
Triplet, 226
Triterpenes, 50
tRNA (transfer RNA), 83, **85**, 189, 195, 218, **227**
formation of, 227
structure of, 226
van der Waals model of, 85
Tropines, 338
Tropomyosin, 306, 308
Troponin, 309
Trypsin (*3.4.21.4*), 150, 245, 247
inhibitor, 246
structure of, 151
Trypsinogen, 247
Tryptophan, **59**, 327
biosynthesis of, 173, 372*
conversion to nicotinate, 332
degradation of, 155, 371*
5-hydroxy-, 154
Tubule, 296
cells, 300

Tubules, transverse, 308
Tubulins, 191
Tumors, 358, **361**
chemotherapy of, 176
initiation of, 360
markers, 361
progression of, 360
promoters, 361
promotion, 360
suppressor genes, 358, 360
therapy of, 346
viruses, 356, 360
β-Turn of polypeptides, 67
Turnover number
of enzymes, 90
of acetylcholine esterase, 320
Tyrosine, **59**, 172, 322, **323**
biosynthesis of, 173, 373*
degradation of, 155, 371*
as precursor of thyroxine, 340
Tyrosine 3-monooxygenase (*1.14.16.2*), 323
Tyrosine kinase (*2.7.1.112*), 351

U
Ubiquinol (coenzyme Q, reduced form), 50,
    101, 127, **131**, 133
Ubiquinol-cytochrome-c reductase (*1.10.2.2*),
    131
Ubiquinone (coenzyme Q, oxidized form), 51,
    101, 127, 131, 133
UDP (uridine diphosphate), 175
UDPgalactose, 283
UDPglucose, 111, **161**, 283, 367*
UDPglucose 4-epimerase (*5.1.3.2*), 283
UDPglucuronate, 181, 291
Ultrafiltration, 296, 302
Ultracentrifugation, 184, 186
Ultraviolet light
in calciol (vitamin D) synthesis, 330
as mutagenic agent, 234, 360
UMP (uridine monophosphate), 157, **175**, 177,
    375*
Uncouplers, 134
Uniport, 207
Unsaturated fatty acids, 44
degradation of, 148
Uracil, **79**, 157
Urate (uric acid), 157, 286, **299**
excretion of, 296, 299
formation of, 157
plasma concentration of, 251

Urate oxidase (*1.7.3.3*), 200
Urea, 151, 153, **287**, 299
  cycle, 197, **287**
  diffusion through membranes, 207
  excretion of, 299
  plasma concentration of, 251
  as product of amino acid degradation, 151, 153
  in urine, 299
Ureotelic animals, 286
Uric acid, **157**, 286, **299**
  excretion of, 296, 299
  formation of, 157
  plasma concentration of, 251
Uricase (*1.7.3.3*), 156
Uricemia, 157
Uricotelic animals, 286
Uridine, 78
Uridine diphosphate (UDP), 282
Uridine diphosphate-glucuronic acid (UDP glucuronate), 181, 291
Uridine monophosphate (UMP), 157, **175**, 177, 375*
Uridine triphosphate (UTP), 177, 375*
Uridylate (UMP), **175**, 177, 375*
Urine, 296–303, **299**
  in acid-base balance, 256, 301
  buffer systems of, 26, 301
  composition of, 299
  formation of, 297
  hormones in, 299
  pH value of, 26, 27, 299
  steroid metabolites in, 344
Urobilinogen, 180
Urocanate, 370*
Urochrome, 298
Urokinase (*3.4.21.73*), 262
Uroporphyrinogen, **179**
Uroporphyrinogen-III synthase (*4.2.1.75*), 179
Uterus, 341
UTP (uridine triphosphate), 175, 375*
UTP glucose-1-phosphate uridyltransferase (*2.7.7.9*), 161
UV light, 234, 330, 360

V
V of enzymes, 91
Vacuole, 39, 119, 182
Valerate (valeric acid), **45**

Valine, **59**, 173, 311, 327
  biosynthesis of, 173, 372*
  degradation of, 155, 370*
Van der Waals model, **7**
Van der Waals radius, **7**
Vanadium, 328
Variable region of immunglobulins, 266–271
Variation
  allelic, 270
  allotypic, 270
Vasoconstriction, 304
Vasopressin (adiuretin), 302, 304
Vector for cloning, 236, 238
Very low density lipoprotein (VLDL), 145, **255**, 279, 285
Vesicles, 23, 182, 184, 210, 216
  secretory, 210
  synaptic, 320
  transport, 214
Villin, 190, 192
Vimentin, 190, 306
Vinblastine, 191
Vincristine, 191
Viral oncogenes, 358
Viruses, 264, **363**
Viscosity, 186
Visual excitation, 325
Visual pigments, 325
Vitamins, 326, 330–335
  A (retinol) **331**
  A acid (retinoate), as signal substance, 330, 340
  B$_1$ (thiamine), **333**
  B$_2$ (riboflavin, folate, nicotinic acid and pantothenate), **333**
  B$_6$ (pyridoxal), **335**
  B$_{12}$ (cobalamin), 244, 329, **335**
  C (ascorbic acid), 312, 323, **335**
  D (calciol), **331**
  deficiency of, 330
  dietary, 326
  E (tocopherol), 100, **331**
  fats as carriers of, 326
  fat-soluble, 42, 244, **331**
  H (biotin), 103, **335**
  hydrophilic, 333, 335
  isoprenoid, 50
  K (phylloquinone), **50**, 100, 262, **331**
  lipophilic, 42, 244, **331**
  and liver, 276
  reabsorption of, 242

supply of, 330
  water-soluble, 333, 335
Vitreous humor, 40
VLDL (very low density lipoprotein), 145, **255**, 279, 285
Voltage, electrical, 11
Volume, specific, 186

W
Water, **20**, 105, 131, 189, 328
  acid-base properties of, 25
  boiling point of, 21
  balance, 250
  cluster, 20
  daily requirement of, 328
  dipolar properties of, 20
  dipole moment of, 21
  excretion of, 296, 303
  ion product of, 25
  physical properties of, 21
  recycling of, 303
  as solvent, 20
  structure of, 20
  transport through membranes, 207
  vaporization temperature of, 21
Water-cleaving enzyme, 121
Wavelength, 98
Waxes, **43**
Well-fed state, 279
White blood cells, 250
Wine, 138, 294
Wool, 68

Work, **10**
  chemical, 11
  electrical, 11
  mechanical, 11
Wound healing, 314

X
Xanthine, **157**
  methylated, 352
Xanthine oxidase (*1.1.3.22*), 157
Xenobiotics, 276, 290
  metabolism of, 204, 290
X-ray chrystallography, **70**, 260
X-ray radiation, 360
Xylanes, 38
Xyloglucan, 37, 38
Xylose, 35
Xylulose 5-phosphate, 143, 366*, 367*

Y
Yeast, 138

Z
Zinc finger, 346
Zinc ions, 71, 113, **329**, 346
  daily requirement, 329
Zinc proteins, 328
Z line, 307
Zonal centrifugation, 186
Zwitterion, 56
Zymogens, 247, 262